FOOD PRODUCTS

by

SAUL BLUMENTHAL

Food Technologist and Consultant,
Shirley Laboratories
Fellow, American Institute of Chemists

1947

CHEMICAL PUBLISHING CO., INC.

BROOKLYN N. Y.

LONDON
Food Trade Press Ltd.

Food Products

ISBN: 978-0-8206-0036-9

Chemical Publishing Company:
www.chemical-publishing.com
www.chemicalpublishing.net

First Edition:

© **Chemical Publishing Company, Inc.** - New York 1947

Second Impression:

Chemical Publishing Company, Inc. - 2011

Printed in the United States of America

FOREWORD

This book is written for everybody interested in the preparation of food from the manufacturing plant executive, the food chemist, and the food salesman, right down to the housewife.

The author's knowledge of the subject is based on scientific training and many years of practical, first-hand experience in the manufacture of foods. In addition to compiling the results of his own experience, he has included methods developed and used by other specialists in the field.

Whether the food manufacturer has an old, well-established business or is just starting up a small plant, it is hoped that this book will be an essential part of his equipment. To the established manufacturer, it presents the latest developments and the newest scientific methods in the field of food production. To the beginner, it offers the fundamentals on materials and methods, without which he cannot hope to be successful.

Although the production of good, nutritious food is still an art, as it has always been, in this age of rapid scientific development and ever-increasing knowledge, it is fast becoming an exact science. The food manufacturer must plan wisely and prepare his products scientifically and with understanding. This book presents the manufacture of food as a series of operations governed by well-defined physical, chemical, and bacteriological principles which must be understood and carefully followed for good results.

Here are scientific, factory-tested formulas for making hundreds of commercial food products in the canning, flavoring, beverage, confectionery, bakery, condiment, dairy, meat, fish, and allied industries. The sections on plant sanitation and food handling present the latest scientific methods of hygienic control in the food industry. The most modern developments in the fields of dehydration and freezing of fruits and vegetables are completely discussed in a comprehensive chapter on the subject.

ACKNOWLEDGMENTS

Grateful acknowledgment is made to the following individuals and firms from whose writings and publications useful information was drawn:

American Cyanamid & Chemical Corp.
American Lecithin Company
Ansul Chemical Company
Armour & Company
Central Soya Company
W. A. Cleary Corporation
L. Russell Cook
The Dow Chemical Company
Glyco Products Company
Krim-Ko Company
Dr. K. H. Lorenz
The Mathieson Alkali Works, Inc.
Musher Foundation
Pillsbury Flour Mills Company
Practo-Fax, Nulomoline Company
Procter & Gamble Bakery Service
Rohm & Haas Company
The Safety Car Heat & Lighting Co., Inc.
Stein-Hall and Company, Inc.
A. E. Staley Manufacturing Company
U. S. Office of Education
Winthrop Chemical Company, Inc.

In addition, thanks are due to all food manufacturers, research chemists, and technicians whose publications were of assistance in writing this book.

TABLE OF CONTENTS

CHAPTER I

BASIC FOOD INGREDIENTS

Ethyl or Grain Alcohol

Among the raw materials used in the food industries, ethyl alcohol stands out as one of the most important. It is also known as grain alcohol and cologne spirits. Ethyl alcohol is obtained by the fermentation of sugar solutions and saccharified mashes of starch-containing substances. There is also a substantial quantity produced synthetically.

The sugar content of molasses is a cheap source of alcohol. It accounts for almost 75% of the alcohol produced. Blackstrap molasses, a by-product of cane sugar mill operations, containing 50 to 60% sugar by weight, is principally used. The sugar in blackstrap molasses is present as sucrose, $C_{12}H_{22}O_{11}$, and glucose or grape sugar (invert sugar), $C_6H_{12}O_6$.

Alcohol is produced from molasses by a fermentation process in which the sugars are converted to alcohol by the action of enzymes. The molasses is made into a mash, so that the final concentration of sugar falls within a range 15 to 25% by weight. Cultured yeast is then added in quantities ranging from 3 to 10% by volume. This results in a vigorous fermentation.

The yeast has a double action upon the sugar. This is brought about by the presence of two enzymes in the yeast, invertase and zymase. Invertase converts the sucrose into invert sugar.

$$\underset{\text{Sucrose}}{C_{12}H_{22}O_{11}} + \underset{\text{Water}}{H_2O} \xrightarrow{\text{Invertase}} \underset{\text{Invert sugar}}{2\ C_6H_{12}O_6}$$

The zymase then converts the invert sugar into ethyl alcohol and carbon dioxide.

$$\underset{\text{Invert sugar}}{C_6H_{12}O_6} \xrightarrow{\text{Zymase}} \underset{\text{Ethyl Alcohol}}{2\ C_2H_5OH} + \underset{\text{Carbon dioxide}}{2\ CO_2}$$

Temperature and the amount of nutrients added to the mash are carefully controlled to sustain the vigorous activity of the yeast. The yeast itself does not undergo any change during the 36 to 48 hours that it takes to yield the maximum amount of alcohol, but merely acts as the agent for converting sugar to alcohol. The resulting alcohol content may amount to from 6 to 12%. The mash is then redistilled twice to remove the aldehydes, esters and fusel oils, which impart an undesirable odor to alcohol. Among the by-products recovered from the purification of alcohol are carbon dioxide, potash for fertilizers, vitamin concentrates for feedstuffs, and binders for the foundry and coal processing industries.

Starch, in the form of grain, when cooked with water and then treated with barley malt is converted by the diastatic enzymes into sugar. Yeast will then ferment this sugar into alcohol. A method, known as the amyl process, saccharifies and then ferments starches into sugars and alcohol. This process had its origin in a study made by Dr. Calmette of the so-called *Chinese yeasts*. The molds are able to saccharify starch rapidly and the combination of yeasts and mold converts the starch to sugar and the sugar to ethyl alcohol in the mash. Very little alcohol, however, is made in this country by this method. Synthetic ethyl alcohol is also produced by a number of methods.

PROPERTIES OF ETHYL ALCOHOL

Formula	C_2H_5OH
Molecular weight	46
Also known as	Grain alcohol Fermentation alcohol Cologne spirits Spirits of wine
Color	Colorless
Odor	Ethereal, vinous
Taste	Pungent
Other properties	Liquid, volatile
Specific gravity	0.7851

Boiling point	78.4° C.
Solubility	Soluble in water, ether, etc.
Grades	Absolute alcohol 99.80%
	Rectified spirit 95% or more
	Cologne spirit of pure distillate 95–96% alcohol content.

Aldehydes

These compounds are aromatic fruity liquids, utilized to reinforce flavor, especially of strawberry, raspberry, cherry, pineapple, orange, lemon, lime and many other fruit products. Aldehydes are expensive and, being very powerful, must be used with care.

Amyl Alcohol

This is a colorless oily liquid having a penetrating characteristic odor. One part is soluble in 40 parts water and it is miscible with alcohol, ether and chloroform. It is obtained in the purification of ethyl alcohol and extensively used for flavoring whisky, other liquors and fruit essences.

Vanillin

Vanillin is obtained as the methyl and ethyl derivatives, ethyl vanillin being the stronger of the two in flavor. They are both crystalline and excellent for flavoring liquids and foods. They should be used sparingly since the flavor becomes more pronounced upon standing.

Coumarin

Coumarin occurs as colorless aromatic crystals. It is used for flavoring liquids and foods, and to reinforce the vanillin flavor in food products. Generally coumarin may be substituted up to one third of the total vanillin content.

Glycerin (Glycerol)

Glycerin is a liquid obtained by the hydrolysis of vegetable or animal fats or fixed oils and then purified by distillation. If olive oil, which consists principally of glyceryl oleate, is saponified with alkali, e.g., sodium hydroxide, sodium oleate, a soap, and glycerin are the resulting products. Glycerin is also produced by the fermentation of sugars, wines, beer, bread and other similar products.

Pure glycerin is a colorless, odorless, sweet-tasting, viscous, high-boiling, substantially non-volatile liquid, containing not less than 95% glycerin. It finds wide employment in the preparation of base extracts for flavoring purposes, in food color concentrates, in the extraction of drugs and in many other procedures. Glycerin is soluble in all proportions in water and alcohol and is insoluble in most other organic solvents.

Because of its efficiency as a moistening agent, glycerin is very useful to the baker and confectioner. It retards the drying out of cakes and icings, keeping the products fresh. It has been recommended for the prevention of mold growth in bakery goods, and replaces part of the egg in cake recipes, making the product lighter. Glycerin is used in candy both as a sweetening and preservative ingredient. Many delicatessen formulas for pickles, fish and meat products also use glycerin. In another section, there will be found advanced information on the application of glycerin in many important branches of the food industry.

Propylene Glycol

Propylene glycol is a viscous hygroscopic liquid that is now extensively used in the food field as a solvent for flavors. It dissolves synthetic flavoring substances and some essential oils. These solutions can be diluted with water. Due to the penetrating properties of propylene glycol, flavoring agents dissolved in it are carried into and distributed evenly through the product to be flavored. It also serves as a good moistener in bread, cake, confectionery, chocolate, coconut and many other products. It is now approved by the Pure Food and Drug Department of the United States.

SOLUBILITIES OF FLAVORING AGENTS IN PROPYLENE GLYCOL—WATER MIXTURES AT 25° C

Propylene Glycol	100	80	60	40	20
Water	—	20	40	60	80
Flavoring Agent			Per Cent by Weight		
Amyl acetate	Miscible	Miscible	Miscible	1.48	1.34
Benzaldehyde	Miscible	18.97	4.62	1.80	0.82
Cassia oil	Miscible	3.31	0.85	0.69	0.21
Citral	Miscible	0.35	0.17	0.10	0.04
Clove oil	Miscible	1.19	0.26	0.24	0.12
Coumarin	7.70	5.05	2.30	0.50	0.32
Ethyl acetate (99%)	Miscible	Miscible	Miscible	11.65	8.09
Ethyl formate	Miscible	Miscible	Miscible	—	17.45
Ethyl vanillin	14.20	10.80	5.20	1.84	0.79
Eucalyptol	19.90	4.75	1.73	0.35	0.25
Isoamyl formate	Miscible	5.22	4.51	1.68	1.53
Lemon oil	0.81	0.52	0.32	0.13	0.03
Methyl anthranilate	Miscible	Miscible	Miscible	—	—
Methyl salicylate	17.05	—	—	—	—
Nutmeg oil	1.53	0.34	0.17	0.14	0.11
Orange oil	0.26	0.13	0.08	0.06	0.03
Phenyl ethyl alcohol	Miscible	Miscible	Miscible	18.95	3.11
Sassafras oil	2.02	1.21	0.20	0.12	0.08
Vanillin	20.20	20.10	12.60	5.85	2.09

SOLUBILITY OF ETHYL VANILLIN IN PROPYLENE GLYCOL—WATER MIXTURES

Per Cent Propylene Glycol by Volume	*Ounces Ethyl Vanillin per Gallon Solvent Mixture at °F*					
	32	*50*	*68*	*77*	*86*	*104*
0	0.20	0.27	0.37	0.45	0.55	0.84
20	0.33	—	—	1.07	—	—
40	0.61	—	—	2.54	—	—
60	2.40	—	—	7.48	—	—
80	6.41	—	—	16.55	—	—
100	8.00	—	—	23.00	—	—

SOLUBILITY OF VANILLIN IN PROPYLENE GLYCOL—WATER MIXTURES

Per Cent Propylene Glycol by Volume	*Ounces Vanillin per Gallon Solvent Mixture at °F* 32	50	68	77	86	104
0	0.47	0.68	1.07	1.40	1.80	3.10
20	0.67	1.33	1.60	2.87	4.00	8.42
40	1.66	2.66	4.00	8.41	11.68	41.70
60	5.10	8.01	13.35	19.75	33.38	88.00
80	10.57	16.69	26.70	32.03	46.73	93.50
100	10.32	16.[illegible]	28.04	32.70	46.73	80.10

SOLUBILITY OF COUMARIN IN PROPYLENE GLYCOL—WATER MIXTURES

Per Cent Propylene Glycol by Volume	*Ounces Coumarin per Gallon Solvent Mixture at °F* 32	50	68	77	86	104
0	0.12	0.17	0.22	0.27	0.34	0.52
20	—	0.18	0.33	—	0.53	0.87
40	—	0.33	0.53	—	1.20	2.40
60	—	1.60	2.68	—	3.74	6.67
80	—	3.34	5.34	—	9.35	17.4
100	—	4.94	8.80	—	14.41	26.7

Triacetin

Glycerin forms esters with acetic acid, known as mono-, di-, and triacetin, which are used as solvents.

Triacetin is a colorless liquid having a boiling point of 275 to 280° C. It is free from odor, non-toxic and a very good solvent for oils, ethers and esters. It hydrolyzes in the presence of water and should not be used with it.

Amyl and Ethyl Esters, Ethers and Aldehydes

These are all aromatic chemicals used to augment certain flavors.

The acids used in combination with flavoring substances are formic, acetic, butyric, propionic, lactic and others.

The esters result from the action of organic acids on alcohols. They differ in the number of carbon atoms and physical properties, such as odor, boiling point, etc. A pure ester is always free from water and alcohol. It should be clear when dissolved in benzene and, in some cases, in liquid paraffin.

ESTERS, ALDEHYDES, ETHERS AND KETONES

Amyl acetate	Imparts fruity, ethereal flavor
Amyl benzoate	Imparts cherry, plum, prune flavor
Amyl butyrate	Imparts fruity, ethereal flavor
Amyl formate	Imparts apricot, peach flavor
Amyl isobutyrate	Imparts fruity, ethereal flavor
Amyl propionate	Imparts apple, pineapple, strawberry flavor
Amyl valerate	Imparts peach, pineapple flavor
Butyl acetate	Imparts pear, raspberry, pineapple flavor
Butyl butyrate	Imparts banana, pear, butter flavor
Cinnamyl propionate	Imparts strawberry, raspberry, grape flavor
Ethyl acetate	Imparts fruity, ethereal flavor
Ethyl anthranilate	Imparts grape, winelike flavor
Ethyl benzoate	Imparts strawberry, cherry flavor
Ethyl butyrate	Imparts pineapple flavor
Ethyl formate	Imparts apricot, peach, pineapple flavor
Ethyl propionate	Imparts apple, pineapple flavor
Ethyl sebacate	Imparts grape, peach flavor
Ethyl valerate	Imparts apple, peach, pineapple flavor
Ethyl oxyhydrate (Rum Ether)	Imparts rum, ethereal flavor
Ethyl anthranilate	Imparts grape flavor
Isobutyl acetate	Imparts raspberry flavor
Methyl anthranilate	Imparts grape flavor
Methyl ionone	Imparts raspberry, strawberry flavor
Methyl salicylate	Imparts wintergreen, root beer flavor
Oenanthic ether	Imparts cognac flavor
Phenyl acetic aldehyde	Imparts apricot, cherry, pistachio flavor
Phenyl butyl ketone	Imparts cherry, peach, strawberry flavor

Phenyl ethyl butyrate	Imparts strawberry, raspberry, peach flavor
Phenyl ethyl cinnamate	Imparts peach, apricot, cherry flavor
Rhodinol butyrate	Imparts in traces, cherry, raspberry, strawberry flavor

HIGHER ALDEHYDES AND ALCOHOLS

Octyl alcohol C_8	Imparts rose, tuberose flavor
Nonyl alcohol C_9	Imparts lemon, lime, orange flavor
Decyl alcohol C_{10}	Imparts orange flavor
Undecyl alcohol C_{11}	Imparts lilac, lily flavor
Duodecyl alcohol C_{12}	Imparts lemon, orange flavor
Heptyl aldehyde C_7	Imparts strawberry flavor
Octyl aldehyde C_8	Imparts jasmine, rose flavor
Nonyl aldehyde C_9	Imparts orange, lemon, lime flavor
Decyl aldehyde C_{10}	Imparts citrus flavor
Undecyl aldehyde C_{11}	Imparts lily, violet flavor
Duodecyl aldehyde C_{12}	Imparts orange, lemon, lime flavor
Peach aldehyde C_{14}	Imparts cherry, peach, plum flavor
Apricot aldehyde C_{15}	Imparts apricot, peach flavor
Strawberry aldehyde C_{16}	Imparts raspberry, strawberry flavor
Cherry aldehyde C_{17}	Imparts cherry pit-like flavor
Cocoanut aldehyde C_{18}	Imparts cocoanut flavor
Pineapple aldehyde C_{19}	Imparts pineapple flavor
Raspberry aldehyde C_{20}	Imparts raspberry, cherry, currant flavor
Fruit aldehyde C_{22}	Imparts fruity flavor
Citral	Imparts lemon flavor
Diacetyl	Imparts butter flavor

VOLATILE OR ESSENTIAL OILS

For all practical purposes, volatile or essential oils may be described as being aromatic, oily, generally liquid but sometimes semi-liquid or solid. The oils volatilize or, upon heating, evaporate without decomposition.

Essential oils may be prepared as follows:

1 By pressure, as the oils of lemon, lime, orange and grape fruit.

2 By distillation with water, or passing a current of steam

through a distilling flask containing the matter to be extracted.

3 By fermentation and subsequent distillation with water, as with essential oils of mustard and bitter almonds. These seeds contain no ready essential oil so they are crushed, left in contact with water, and, due to the action of nitrogenous ferments, the essential oil is formed. This is then recovered by distillation with water.

4 By transferring scents of various plants by immersion in a fixed oil devoid of odor, such as olive oil or lard. The aromatic odor is finally collected by extraction with ethyl alcohol, is used as is or redistilled.

5 By extraction with low-boiling solvents, either by percolation or maceration, recovering the solvent by distillation, and collection of the essential oil.

The component ingredients of essential oils consist of terpenes, aldehydes, ketones, acids, phenols, alcohols, esters, ethers and sulfur compounds. Essential oils, generally, are liquids lighter than water but there are several that are heavier.

ESSENTIAL OILS

Angelica oil

Solubility in alcohol:	1 part in 12 parts of 90% strength
Flavor:	Spicy
Use:	Gin, cordials

Anise oil

Solubility in alcohol:	1 part in 3 parts of 90% strength
Flavor:	Sweet, licorice
Use:	Candy, cordials

Bergamot oil

Solubility in alcohol:	1 part in 2 parts of 80% strength
Use:	Flavoring medicines and perfume

Birch tar oil

Solubility in alcohol:	1 part in 3 parts dehydrated alcohol
Use:	Flavor in leather, soap, for insect bites

Bitter almond oil

Solubility in alcohol:	1 part in 2 parts 70% strength
Flavor:	Almond
Use:	Cherry syrups

Bitter orange oil

Solubility in alcohol:	1 part in 5 parts 95% strength
Flavor:	Orange
Use:	Flavor

Calamus oil

Solubility in alcohol:	1 part in 2 parts 70% strength
Flavor:	Spicy
Use:	Spice

Cardamom oil

Solubility in alcohol:	2 parts in 5 parts 70% strength
Flavor:	Spicy
Use:	Cordial, spice

Cassia oil

Solubility in alcohol:	1 part in 2 parts 80% strength
Flavor:	Spicy
Use:	Fountain flavors

Caraway oil

Solubility in alcohol:	1 part in 8 parts 80% strength
Flavor:	Spicy
Use:	Cordial, spice, kummel

Cedar oil

Solubility in alcohol:	1 part in 6 parts 95% strength
Flavor:	Tobacco-like
Use:	Rum

Carrot oil

Solubility in alcohol:	1 part in 5 parts 95% strength
Flavor:	Carrot-like
Use:	Soups

Celery oil

Solubility in alcohol:	1 part in 8 parts 90% strength
Use:	Fountain syrup, soup

Cinnamon oil

Solubility in alcohol:	1 part in 3 parts 70% strength
Flavor:	Spicy
Use:	Flavor

Clove oil

Solubility in alcohol:	1 part in 2 parts 70% strength
Flavor:	Spicy
Use:	Flavor

Chamomile roman oil

Solubility in alcohol:	1 part in 2 parts 80% strength
Flavor:	Sweet
Use:	Flavor, vermouth

Coriander oil

Solubility in alcohol:	1 part in 3 parts 70% strength
Flavor:	Spicy
Use:	Cordial, spice

Cognac

Solubility in alcohol:	1 part in 8 parts 80% strength
Use:	Whiskey flavors

Fennel oil

Solubility in alcohol:	1 part in 9 parts 90% strength
Flavor:	Sweet, licorice
Use:	Cordial

Galanga

Solubility in alcohol:	1 part in 7 parts 95% strength
Flavor:	Camphor-like
Use:	Ginger

Ginger oil

Solubility in alcohol:	1 part in 7 parts 95% strength

Flavor: Spicy
Use: Ginger ale

Juniper oil

Solubility in alcohol: 1 part in 4 parts 95% strength
Use: Gin flavors

Hop oil

Solubility in alcohol: 1 part in 1 part 95% strength
Use: Beer flavor, coffee, maple, celery

Lemon oil

Solubility in alcohol: Clearly soluble in 95% strength; forms cloudy liquid in weaker concentrations
Flavor: Lemon
Use: Lemon drinks, bakery goods

Lemon oil; terpeneless

Solubility in alcohol: Clearly soluble in 80% strength; maybe as much as 14 times stronger than lemon oil; will not oxidize
Flavor: Lemon
Use: Baking goods, gelatin and drink powders

Lovage oil

Solubility in alcohol: 1 part in 3 parts 80% strength
Flavor: Maple
Use: Condiments, foods

Lime oil

Solubility in alcohol: Clearly soluble in 95% strength; forms cloud in 90%
Flavor: Lime
Use: Lime drinks, ginger ale, candy

Laurel oil

Solubility in alcohol: 1 part in 2 parts 70% strength
Flavor: Camphor-like, strongly aromatic odor
Use: Liquors

Mace oil

Solubility in alcohol:	1 part in 3 parts 90% strength
Use:	Doughnut flour, liquors, foods

Marjoram oil

Solubility in alcohol:	1 part in 1 part 85% strength
Use:	Food seasoning

Mandarin oil (Tangerine oil)

Solubility in alcohol:	Clearly soluble in 95% strength
Use:	In orange flavors for bouquet

Neroli oil (Orange flower oil)

Solubility in alcohol:	1 part in 2 parts 80% strength
Flavor:	Bitter
Use:	Flavors and perfume

Neroli oil (Portugal oil)

Solubility in alcohol:	Clearly soluble in 95% strength The pressed oils are Italian, West Indian and Californian; there is also the extracted kind
Flavor:	Orange-like odor
Use:	Flavor tone in soft drinks, perfume

Nutmeg oil

Solubility in alcohol:	1 part in 3 parts 90% strength
Use:	Doughnut flour, soft drinks

Orris oil

Solubility in alcohol:	1 part in 1½ parts 80% strength
Odor:	Violet-like odor
Use:	Fruit flavors

Origanum oil

Solubility in alcohol:	1 part in 3 parts 70% strength
Use:	Liquors, sometimes used where thyme is called for

Pennyroyal

Solubility in alcohol:	1 part in 2 parts 70% strength
Use:	Stimulant, protection against insect bites

Palmarosa oil

Solubility in alcohol:	1 part in 3 parts 70% strength
Flavor:	Sweet
Use:	Honey flavor

Peppermint oil

Solubility in alcohol:	1 part in 4 parts 70% strength There are Japanese, English and American grades; Japanese quality used for medicinal purposes primarily
Flavor:	Peppermint, cooling to taste
Use:	Candy flavor, chewing gum

Parsley oil

Solubility in alcohol:	1 part in 7 parts 70% strength
Flavor:	Spicy, bitter
Use:	Spice flavor

Pepper oil

Solubility in alcohol:	1 part in 12 parts 90% strength
Flavor:	Spicy
Use:	Food flavor

Peru oil

Solubility in alcohol:	1 part in 3 parts 90% strength
Flavor:	Sweet, aromatic
Use:	Food flavors

Pettigrain oil

Solubility in alcohol:	1 part in 1½ parts 80% strength
Flavor:	Lemon-like, bitter, aromatic
Use:	Liquors, gin

Pimento oil

Solubility in alcohol:	1 part in 2 parts 70% strength
Use:	Food flavor

Rose oil

Solubility in alcohol:	Soluble in 90% strength with formation of cloud due to separation of paraffin hydrocarbons
	Prepared from red and white roses, in Germany, Bulgaria, and France
Use:	Food flavors, perfumes

Rose geranium oil

Solubility in alcohol:	1 part in 3 parts 70% strength
	Varieties are Reunion, African, French, and Spanish Geraniums, also Bourbon
Odor:	Pleasant rose-like odor
Use:	Flavor, perfume

Sassafras oil

Solubility in alcohol:	1 part in 2 parts 90% strength
Use:	Food flavor, perfume in soap making

Spearmint oil

Solubility in alcohol:	1 part in 1 part 80% strength
Use:	Food flavor, chewing gum

Star anise oil

Solubility in alcohol:	1 part in 3 parts 90% strength
Flavor:	Sweet, licorice
Use:	Cordials, vermouth

Sweet birch oil (Wintergreen)

Solubility in alcohol:	1 part in 7 parts 70% strength
Use:	Root beer, liniments

Thyme oil

Solubility in alcohol:	1 part in 2 parts 80% strength
Use:	Seasoning flavor

Wormwood oil

Solubility in alcohol:	1 part in 2 parts 80% strength
Use:	Liquors, cordials

Ylang-ylang oil

Solubility in alcohol:	1 part in 2 parts 90% strength
Flavor:	Pineapple, peach
Use:	Fruit flavors

From the following table of equivalents one can readily calculate the quantity of essential oil represented by a known weight of spice.

Quantity	*Spice*	*Essential Oil*
100 lb	Allspice	3½ lb
100 lb	Angelica root	¾ lb
100 lb	Angelica seed	1 lb
100 lb	Anise seed (Russian)	2½ lb
100 lb	Anise star (Chinese)	3 lb
100 lb	Bitter almonds	½ lb
100 lb	Black pepper (oleoresin)	6 lb
100 lb	Calamus root	2½ lb
100 lb	Capsicum (oleoresin)	8 lb
100 lb	Caraway seed	5 lb
100 lb	Cardamom seed	5 lb
100 lb	Cassia cinnamon	1 lb
100 lb	Celery seed	3 lb
100 lb	Cinnamon ceylon	1 lb
100 lb	Cloves	17 lb
100 lb	Coriander seed	½ lb
100 lb	Cumin seed	3 lb
100 lb	Dill seed	3½ lb
100 lb	Estragon	½ lb
100 lb	Fennel seed	5 lb
100 lb	Garlic (imitation flavor)	4 oz
100 lb	Ginger (imitation flavor)	10 lb
100 lb	Horse radish	1 oz
100 lb	Laurel leaves (bay)	2 lb
100 lb	Lovage root	½ lb
100 lb	Mace	12½ lb
100 lb	Mustard seed	¾ lb
100 lb	Nutmeg	12½ lb
100 lb	Onion	4 oz
100 lb	Parsley seed	3 lb
100 lb	Sage	2 lb

Quantity	*Spice*	*Essential Oil*
100 lb	Sweet basil	3/20 lb
100 lb	Sweet marjoram	½ lb
100 lb	Thyme	2 lb
100 lb	Valerian root (oleoresin)	1 lb
100 lb	Wild marjoram	2 lb

The oil can be added directly to the products and distributed during the mixing operation or, if the quantity of oil is small, it can be mixed with a quantity of a fixed oil, such as corn oil, to increase the bulk to a desired quantity. For example, a calculated amount, say 3 drams 49 minims, can be dissolved in sufficient corn oil to make the entire bulk 1 fluid ounce, which would be the total quantity used and would contain just the 3 drams 49 minims of the flavoring oil.

Other methods are also applicable. An emulsion can be made in calculated proportions for the addition of any desired quantity to a batch, or the oils can be absorbed on salt, sugar, flour, starch or any other dry material used in the process.

For pickle makers, a solution of the oil in 75% acetic acid forms a terpeneless solution of the flavor bodies that is readily distributed in the various liquors and products.

SUGGESTED USES OF SYNTHETIC AND NATURAL AROMATIC CHEMICALS IN FLAVOR FORMULAS

Acetaldehyde 50%	Use less than 10% in apple, peach and apricot formulas
Aldehyde C_{10}	Use less than ½% in orange and lemon formulas
Aldehyde C_{14}	Use less than 3% in peach and apricot formulas
Aldehyde C_{16}	Use less than 10% in strawberry and raspberry formulas
Benzyl butyrate	Use less than 3% in strawberry and raspberry formulas
Benzyl formate	Use less than 2% in raspberry and cherry formulas
Benzyl phenyl acetate	Use less than 3% in honey formulas
Benzyl propionate	Use less than 2% in raspberry and cherry formulas
Benzyl propyl acetate	Use less than 30% in banana formulas
Butyl acetate	Use less than 20% in raspberry, pineapple and grenadine formulas
Butyl anthranilate	Use less than 5% in grape formulas
Butyl butyrate	Use less than 5% in banana, butter and pear formulas
Cinnamyl anthranilate	Use less than 12% in grape formulas
Cinnamyl butyrate	Use less than 5% in grape, raspberry and strawberry formulas
Cinnamyl formate	Use less than 2% in peach, raspberry and strawberry formulas
Cinnamyl propionate	Use less than 5% in grape, raspberry and strawberry formulas
Diethyl anthranilate	Use less than 20% in grape formulas
Ethyl anthranilate	Use less than 25% in grape formulas
Ethyl cinnamate	Use less than 2% in strawberry and raspberry formulas
Ethyl lactate	Use for taste in grape and wine flavor
Ethyl methyl anthranilate	Use less than 25% in grape formulas
Ethyl pelargonate	Use as substitute for cognac oil

Geraniol (pure, absolute)	Use less than 5% for rose notes, peach, strawberry, raspberry, grenadine and apple formulas
Geranyl acetate	Use less than 3% in apple formulas
Geranyl formate	Use less than 3% in apple formulas
Isobutyl acetate	Use less than 20% in raspberry, banana and grenadine formulas
Isobutyl benzoate	Use less than ¼% in raspberry and strawberry formulas
Isoeugenol	Use in strawberry and raspberry formulas
Jasmin synthetic	Use less than 10% in strawberry, raspberry, cherry and other flavors where fruity tones are desired
Linalool	Use less than 3% in lemon, lime, pineapple, banana, orange and honey formulas
Linalyl acetate	Use less than 5% in apricot, peach, strawberry, lemon and lime formulas
Linalyl formate	Use less than 3% in peach and apricot formulas
Methyl anthranilate	Use less than 20% in grape formulas
Methyl isoeugenol	Use less than 3% in raspberry and strawberry formulas
Oenanthic ether	Use less than 15% in cognac, grape, butter and wine formulas
Phenyl ethyl acetate	Use less than 5% as sweetener in fruity formulas
Phenyl ethyl butyrate	Use less than 10% for fruity flavor in strawberry, raspberry and peach formulas
Phenyl ethyl cinnamate	Use less than 5% in peach, apricot and cherry formulas
Phenyl ethyl formate	Use less than 5% in strawberry formulas
Phenyl ethyl phenyl acetate	Use less than 3% in honey, cherry and grenadine formulas
Phenyl propyl acetate	Use less than 1% in peach, apricot and pineapple formulas

Terpeneless Essential Oils

Terpeneless oils are manufactured from the finest essential oils. In the process of their manufacture, terpenes, pectins, waxes, etc., all of which are non-essential from the standpoint of aroma and flavor, are removed. Terpeneless oils are readily soluble in dilute solutions of ethyl alcohol and have the stability, flavor and odor desired by the discriminating manufacturer.

USES OF TERPENELESS ESSENTIAL OILS

Oil bay, terpeneless	In small amounts, for ginger ales
Oil bergamot, terpeneless	In traces, for fine citrus flavors
Oil cloves, terpeneless	Used for fine flavors instead of oil of cloves; much smoother and more soluble; twice as strong
Oil coriander, terpeneless	Used instead of oil of coriander for ginger ale where solubility and a finer note are desired
Oil fennel, terpeneless	Used instead of oil of fennel because of easy solubility and fine note
Oil ginger, terpeneless	Used instead of plain oil of ginger because of its high solubility; excellent for ginger ale of the dry type
Oil grapefruit, terpeneless	For drink powder and nectar flavor; blends well with lime
Oil lavender flower, terpeneless	In small amounts, for ginger ale
Oil lemon, terpeneless	Especially recommended for flavoring gelatin, drink powders and nectars; economical for making extract due to high strength and solubility
Oil limes, distilled, terpeneless Oil limes, expressed, terpeneless	Recommended for same uses as given above for lemon

Oil mace, terpeneless Oil orange, terpeneless	Recommended for same uses as given above for lemon
Oil peppermint, terpeneless Oil pimento, terpeneless Oil spearmint, terpeneless	Used in ginger ale because of high solubility and strength

Fruit Acids and Other Acids Used in Food Industries

All fruits contain some malic, citric and tartaric acids in various proportions. Malic acid is found in apples, cherries, etc., citric acid in lemons, and tartaric acid in grapes. Lactic, phosphoric, acetic, benzoic, butyric, succinic and adipic acids do not belong to the true fruit acids. However, some of them, such as benzoic acid, are found in a number of fruits.

Acetic acid is a colorless liquid with a vinegar-like odor. Benzoic acid is a white crystalline solid somewhat aromatic but slightly soluble in water. Butyric acid is colorless, rancid-smelling and soluble in water. The solid acids, such as citric and tartaric, are soluble in water as are the liquid lactic and phosphoric acids.

Citric acid is the most widely used in the field of foods and beverages. It occurs abundantly in nature in the juice of lemons and limes. Most of the citric acid now being used commercially is prepared by fermentation of sugar.

Tartaric acid is also widely used in foods and beverages. However, it is being replaced, to a considerable extent, by citric acid. The commercial acid is the *d*-isomer and occurs in nature in the juice of the grape. During the fermentation of this juice to wine, tartaric acid is precipitated as the acid potassium salt and, to some extent as the calcium salt, in crystalline deposits termed argols. These argols, imported from the larger wine-producing countries of the world, serve as the raw material for the production of tartaric acid.

Gluconic acid is generally sold as a solution in water but can be obtained as a white crystalline solid. It is made by a fermentation oxidation of glucose. Indications are that it will be a valuable addition to the list of acids used in the food industry. It has already been suggested as a mild preservative for fruit juice.

Ascorbic acid is recognized as Vitamin C. The U.S.P. and international units of Vitamin C are defined as 0.05 milligrams of ascorbic acid. One gram of ascorbic acid is, therefore, equal to 20,000 U.S.P. or international units. In aqueous solution, it is slightly oxidized on exposure to atmospheric oxygen. The presence of even small amounts of iron or copper salts in solution accelerates this oxidation. The acid is found in citrus juice, such as that of lemon, orange and lime.

SOLUBILITY OF FRUIT ACIDS

Citric acid	100 grams soluble in 50 grams water at room temperature
	100 grams soluble in 180 grams alcohol at room temperature
Tartaric acid	100 grams soluble in 76 grams water at room temperature
	100 grams soluble in 330 grams alcohol at room temperature
Malic acid	Possesses a similar degree of solubility as above

APPROXIMATE PERCENTAGE OF ACIDS IN FRUITS

Fruit	*Per Cent Total Acid*	*Summary*
Apple	0.71	Malic
Banana	0.26	Malic
Cantaloupe	0.18	Citric
Cherry	1.04	Malic
Crabapple	1.50	Malic
Cranberry	2.73	Citric mostly, malic little
Currant	2.43	Citric mostly, malic little
Gooseberry	2.12	Malic and citric
Grape	1.00	Tartaric
Lemon	6.50	Citric
Orange	1.35	Citric
Peach	0.75	Malic
Persimmon	0.09	Malic
Pineapple	0.75	Citric

Fruit	*Per Cent Total Acid*	*Summary*
Plum	1.27	Malic
Pomegranate	4.52	Citric
Quince	1.22	Malic
Raspberry	1.51	Citric
Strawberry	1.00	Citric
Watermelon	0.20	Malic

Resins and Balsams

Resins represent a class of natural vegetable products composed of carbon, hydrogen and oxygen. They are closely allied to the essential oils, all of which, when exposed to the air, absorb oxygen and are finally converted to substances having the characteristics of resins. In most cases, resins are obtained from plants which yield them mixed with, and dissolved in, a corresponding essential oil. They are insoluble in water but dissolve in alcohol, ether and essential oils. Many of them possess acidic properties and form frothy solutions in alkalies. The resins are divided into three groups:

Hard or true resins
Oleo-resins, soft resins and balsams
Gum resins

The hard resins are solids at room temperature and contain little or no essential oil. Under this head are included:

Copal
Amber
Colophony or common resin
Dammar or kauri resin
Lac resin (stick lac, seed lac, shellac)

The above are not volatile, fuse when heated, and are soluble in carbon tetrachloride, chloroform, fixed and volatile oils.

Resins, which exude from the cell ducts or from the plant itself and are usually accompanied by the volatile oil itself, are classified as follows:

Dragon's blood
Mastic
Sandarac
Guaiac resin
Benzoin (from styrax benzoin)

The oleo resins, or soft resins, are mixtures of resins and volatile oils. The ones containing benzoic or cinnamic acids in addition are generally called balsams and have a vanilla-like aroma. They are as follows:

Peru balsam from a tree in Central America
Tolu balsam from a tree in Venezuela
Storax (styrax) from a tree in Asia Minor and Syria
Turpentines (Canada balsam, Venice turpentine)
Burgundy pitch
Copaiba balsam
Elemi

The gum resins are milky exudations from plants and contain gum, which is entirely or partly soluble in water and alcohol. Many of these resins contain essential oil. They are:

Galbanum
Asafetida
Olibanum
Myrrh
Gamboge
Scammony
Euphorbium

There are also so-called manufactured oleo resins. The seed, plant, root or bark is extracted with ethyl alcohol or another suitable solvent, and all the volatile and aromatic principles are carefully separated in the form of resins. These are then very readily soluble in alcohol.

Spice and Flavor Aids

It is the immediate contact of the substance with the tongue that informs us through the sense of taste, whether the product has a good or just ordinary flavor. The sense of smell is also important in detecting flavor. There are many kinds of flavor such as sweet, sour, bitter and salty. Each in itself may excite the appetite and in that way increase the amount of food consumed. It is important that the flavor is pleasant and agreeable and in harmony with the food. The transfer of flavor may be realized by dissolving essential oils, ethers, esters, vanillin, etc., in ethyl alcohol or by adding the flavor in emulsion form. Thorough mixing together of flavor and sugar and then adding gradually to the foodstuff is an inexpensive way of introducing flavor.

Flavoring ingredients are sometimes susceptible to oxidation, which imparts an "off" odor or a rancid taste. Anti-oxidants can be used to prevent or retard this. Fatty food preparations are the ones to be watched for rancidity or flavor change, and to be fortified with anti-oxidants.

With but few exceptions, cane sugar, dextrose, honey, molasses and corn syrup are the usual sweetening agents in foods. Sour taste is due to the presence of acids, which may be of organic or mineral origin. The so-called fruit acids commonly used are citric, tartaric and malic. A sour taste is generally refreshing and, with the addition of sugar, aids in rounding out the flavor. A salty taste is necessary for foods because it increases palatability. A bitter taste stimulates and excites the appetite and promotes digestion.

Aromatic substances are stimulating food additions. No household or food factory could possibly get along without these invaluable spices. In general, flavors, salt and spices are necessary additions, without which many foods would be uninteresting. These food adjuncts may have little nutritive value, but they render foods enjoyable to the taste and stimulate lazy appetites.

Spices are aromatic, often pungent, vegetable substances used chiefly for the seasoning of food. They are derived from different parts of their respective plants. They may be contained in the stems, leaves, buds, flowers, barks, root stock, seeds and fruits. For example, ginger is a root stock, cinnamon the inner bark, cloves the flower buds, nutmegs the seeds, and sage, thyme and other herbs the entire upper part of the plant. The aromatic qualities are contained chiefly in their essential oils.

Condiments are manufactured foods composed of spices, sweetening agents and thickeners; they may be salty, spicy or aromatic. These products are also made with vinegar, sauces and various collections of pickles. The addition of condiments to foods rounds out the taste and improves the flavor.

SUGGESTED USES FOR SOME SPICES AND FLAVOR AIDS

Benzoin	Aromatic addition to beverages
Capsicum	Spicy flavor
Cubebs	Spicy flavor
Celery	Flavor
Cherry bark (wild)	Blends well with vanillin
Coffee	Blends well with other flavors
Fenugreek	Maple flavor
Gentian	Good base for bitters
Ginger Jamaica	Ginger ale
Haw	Good in maple flavor
Hops	Good with oil of hops
Horehound	Good for cough drops
Lemon peel	True flavor
Licorice	Maple flavor conditioner
Lovage	Condiments, truer flavor than the oil
Olibanum	Flavor to beverages
Patchouly	Candy flavor
Orange peel	True flavor, produces natural cloud
Opoponax	Good addition to beverages
Orris	Spicy flavor
Paprika	Spicy flavor
Peru	Aromatic addition
Sassafras	Used in root beer, birch beer and sarsaparilla
Tonka	Fixative for vanilla
Vanilla	For making vanilla flavor

SPICES AND CONDIMENTS

For convenience, the spices and condiments may be classified with reference to the source from which they are obtained.

From Stems or Leaves

Bay leaf	Peppermint
Burnet	Parslev
Sage	Chervil

From Stems or Leaves

Spearmint
Sweet marjoram
Sweet basil
Savory
Tarragon
Wintergreen
Chives
Thyme

From Buds and Flowers

Capers
Cloves
Saffron

From Barks

Cassia
Cinnamon

From Roots and Root Stocks

Turmeric
Garlic
Ginger
Horseradish
Sassafras
Onion
Shallot

From Immature and Ripe Fruits

Allspice
Capsicum
Paprika
Juniper
Pepper
Vanilla

From Seeds

Tonka
Bitter almond
Anise
Cardamom
Caraway
Coriander
Celery
Cumin
Dill
Fennel
Grains of paradise
Mustard
Nutmeg
Mace
Foenugreek

Allspice

Allspice is the fruit of a West Indian tree. It is about the size of a small pea and is gathered when fully grown and nearly ripe. It is then sun-dried. It is called allspice because of its resemblance in flavor to a mixture of all spices, particularly cinnamon, nutmeg and cloves. It is also known as pimento because of its resemblance to the peppercorn and as Jamaica pepper because it is largely grown in Jamaica.

Angelica

This is an aromatic plant, native to the Alps, and growing freely in many other parts of Europe as far north as Iceland and Lapland. In the latter countries, the fleshy roots and stalks are used respectively for food and medicine. Commercially, the young and tender leaf stalks are candied and are used for decorating candies and cakes.

Annatto

This is a substance extracted from the reddish pulp which surrounds the seeds of the annatto tree. Found principally in South America and the West Indies, it is a yellowish-red dye used for coloring butter, cheese, ice cream and liquors. It is exported in cakes of 2 or more pounds in weight and wrapped in leaves.

Aniseed

This is a small seed of an annual plant cultivated chiefly in Spain, Egypt, Syria, Germany and Mediterranean countries. It is used as a condiment, for medicine and for flavoring candies and syrups.

Anise

This is the fruit of a Chinese tree known as star or Chinese anise, or badian. It is similar, in flavor and properties, to the common anise but being star-shaped, is very different in appearance. In China, it is often added to tea.

Basil

Basil is a widely distributed, highly aromatic and sweet herb native to India. It is used especially in turtle soups, and its flavor resembles that of the clove. Basil vinegar is made by steeping basil leaves in vinegar. Mountain mint is sometimes called basil.

Bay Leaves

These leaves are aromatic and come from the sweet bay or

laurel tree, which grows wild in Greece, Italy, Mediterranean countries and in some southern sections of the United States. The dried leaf is used for flavoring soups, pickles, fish products, etc.

Sweet bay and its leaves have no connection with bay rum. Bayberry leaves, when distilled with rum, give bay rum its characteristic flavor.

Burnet

This is a perennial garden herb whose young leaves are used for flavoring soups and in salads, where they impart a cucumber flavor. When crushed with other mild spices, like chervil or parsley, it lends a splendid flavor.

Capers

Capers are pickled flower buds of the caper bush growing in Mediterranean countries and in India. They are used as pickles and added to sauces. After the crop is gathered they are carefully dried and then stored in barrels of brine or vinegar. The vinegar is sometimes flavored with tarragon sprigs, elder flowers, cloves or peppercorns. During the following winter, the capers are grated through sieves. The smallest ones have the best quality. They are then replaced in the barrels and preserved until sold. Before shipment, they are washed in vinegar to make them firm, and the ones of finer quality are repacked in small bottles. The buds of the bean-caper and nasturtium are sometimes used for adulteration.

Caraway Seeds

These are deep brown seeds of an aromatic plant which grows wild in Holland, Germany and many other countries, including the United States, especially in California. They are utilized as a culinary flavor, in confections, liquors and rye bread. The oil distilled from the seed is used chiefly in making perfume and in compounding household flavoring extracts. The root of the caraway plant resembles parsnips and, when young and tender, serves as an excellent vegetable. The young shoots may be used for flavoring soups and stews.

Cardamom

This is a three celled, ovate, triangular, yellowish pod, enclosing a number of small, dark, wrinkled angular seeds, with an aromatic and agreeably pungent flavor. The cardamon fruit is native to India and Jamaica. The seeds are used for culinary purposes, confections, flavoring extracts, etc. Other cardamoms, similar in character, come from several East Indian plants of the Amomum genus, especially the round or cluster cardamom, the Siamese bastard cardamom, and the Java cardamom.

Cassia Buds

These are the dried flower buds of the tree which yields cassia cinnamon. The flavor is similar to that of cinnamon. In appearance, they somewhat resemble cloves.

Chervil

This is an aromatic low-growing herb cultivated in all temperate climates and very popular in the south. The young leaves are used for seasoning soups and salads. It resembles parsley.

Cassia

Cassia is obtained from China and the Dutch West Indies, and is darker in color, rougher in appearance and about 4 times as thick as true cinnamon. Cassia is a bark, the heaviest parts of which are ground, and the thinner ones used whole. Its flavor is more pronounced than that of cinnamon. The best grades are Saigon, Batavia and China.

Chili (Chilies)

This consists of the pods of several species of small, fruited, pungent capsicums. The name is an adoption from the Mexican. Chilies are prepared, both ripe and unripe, as a separate ingredient to be added to mixed pickles. They are extensively consumed in hot countries. The two Mexican dishes containing them are known as chile con carne and chicken tamale. Chili is also used as a group name for highly seasoned condiments made either with whole capsicums or pepper.

Chili Con Carne

This is a Mexican dish, very popular in the southwest, the chief ingredients of which are beef, beans, chili, garlic and spices.

Chives

This is a plant of a species allied to the onion and leek, cultivated principally for the leaves, which grow in thick tufts resembling grass but are hollow like onion leaves. Chives can be used as a substitute for onions, especially in soups, stews and salads.

Celeriac (Knob Celery, Turnip Celery or German Celery)

This plant is cultivated solely for its large turnip-like roots. It is grown principally in Europe. It is eaten as a separate vegetable or added to soups, stews and salads for flavor. When served as a salad, it is boiled, sliced, and served with oil, vinegar or other desired ingredients.

Celery

Celery is a cultivated plant, found wild in many parts of this country and Europe. The present varieties are all of modern development. They are constantly being improved to meet the demands of an ever-expanding market. At one time an expensive article, it has become, in recent years, a lower-priced product for general consumption. It is grown in California, New York, Michigan and many other states.

Celery requires constant care for cultivation; rich, moist, soft and mucky soil gives it the best qualities. Every part of the plant is used to advantage. The stalks are eaten raw and the trimmings, fresh or dried, are excellent for flavoring broths.

Celery Seeds

These seeds are used for celery salt, numerous pickles, seasonings and flavoring soups. They are imported from France.

Cinnamon

This is the inner bark of the young wood of a cinnamon tree. Ceylon cinnamon is chiefly of Ceylonese cultivation; cassia cinnamon is the principal East Indian and Chinese type. Both varieties are sold in quill form and then ground. Its fragrant aromatic flavor makes cinnamon a popular adjunct to cookery, etc. Ceylon cinnamon has a pale yellowish-brown color. Its quills are lighter, cleaner and smoother in appearance than those of cassia. One specific trade difference between ceylon cinnamon and cassia is that the latter has a single quill, whereas the former has many quills enclosed in an envelope. In the ground it is frequently impossible to tell the difference, especially if the cassia is of good quality. The cinnamon tree is also cultivated in Java, Sumatra and tropical Asia.

The preparation of cinnamon is simple. The branches, about 1 inch in thickness, are cut and trimmed to a length of 4 feet when the bark is ripe. The bark is stripped, peeled and cut down two sides with a sharp-pointed, slightly hooked knife and, after a little gentle prying, is pulled off in two long strips. The strips are piled in heaps and allowed to ferment. The outer skin is then removed, exposing the pale yellow inner bark, which is the cinnamon. The strips are permitted to dry for ½ hour and are then formed into the quills, in which form they are marketed.

Cloves

Cloves are the dried flower buds of the clove tree. When plucked, they are reddish but change to brown in the process of drying by the smoke of wood fires or in the sun. The cloves contain about 18% volatile oil. The principal sources of supply are Zanzibar and the East Indies. The flower buds, upon drying or curing, lose about 50% of their weight. The best cloves should be large, plump, purplish-brown in color and unbroken. They are widely used for flavoring confectionery and desserts.

Coriander Seeds

This is the smooth, round, yellowish or brownish seed of an annual plant, native to Germany, Hungary, Italy and Asia. It is a little smaller than the peppercorn. The best seeds have a yellow color, whereas those of poorer quality are brown or black. They have a spicy taste and odor, and are used in cakes, cordials and, in Scotland, as a popular sugar-coated candy.

Cumin Seeds

These are the seeds of an herb of the caraway type. They are of aromatic odor and flavor, popular in Europe and Asia, but seldom used here, except in curry powders. In Germany, the seeds are particularly liked for flavoring bread, in Holland for cheese, and in Norway for anchovies and as additions to soup, pastry and pickles. They are used by meat packers for sausage flavoring.

Curry Powder

This is a highly seasoned condiment extensively used in India and other eastern countries. In India and Ceylon, it commonly consists of black and cayenne peppers and a number of spices, e.g., nutmeg, cinnamon and cloves, made into a powder or paste with tumeric. Curry powders sold here usually contain additional ingredients, such as coriander seed, fenugreek seed, cumin and ginger.

Dill Seeds and Leaves

The dill is a hardy, easily grown herb cultivated chiefly for its aromatic, pungent seeds which are employed in the manufacture of pickles and sauces. The leaves, having the combined flavor of fennel and mint, are used for their flavoring quality. The plant is grown in the United States, southern Europe, Egypt and South Africa.

Dulse

This is an edible purplish-red seaweed abundant along the New England coast. It is eaten raw as a relish. It may be plain boiled (either fresh or dried) or cooked in butter or milk. It is served as a vegetable with fish or meat and added to stews.

Fennel

The fennel is cultivated chiefly for its young leaves which are excellent additions to salads and garnishes. It is known as the common garden or sweet fennel. The seeds are used for seasoning and contain from 3 to 7% of an ethereal oil called anethol. The seeds are sweet and spicy, and are used in medicine and food.

Fenugreek Seeds (Foenugreek)

These seeds are the fruit of an annual herb, from 1 to 2 inches in length, brownish-yellow, with a peculiar odor, bitter in flavor and mucilaginous. In Egypt, they are used as flour in breadmaking and, in Greece, are added to honey. The seeds are 1/8 inch long and have a strong, unpleasant, aromatic odor suggestive of the onion. They are used in the preparation of imitation maple.

Galangal

This is the aromatic rootstock of several plants of the ginger family imported chiefly from China. At one time it was used extensively as a spice and in medicine.

Garlic

The garlic is similar to the onion in odor. The bulb is white-skinned or rose-tinged and divided into ten or twelve sections known as cloves. At certain seasons, garlic abounds in many pastures, imparting a rank flavor to the milk and butter of cows which feed upon it. Its main uses are in flavoring soups, sauces, salads and pickles. It is an important ingredient in many food combinations used in southern Europe. The plant is native to Italy, France and Spain, and is also grown in the United States. The strong flavor is due to an oil which is rich in sulfur.

Ginger

Ginger is the rootstock or rhizome of a perennial reed-like plant with annual leafy stems, 3 to 4 feet high, which grows freely in most tropical places. The rootstocks are gathered when the stalks wither. There are two kinds of ginger found in commerce. The first is the coated, or black, ginger with a wrinkled surface. These roots are scalded or washed and then allowed to dry in the sun. This method is usually employed for the older, poorer roots. The second kind, uncoated or white ginger, is washed after scraping. The natural color of white scraped ginger is a pale buff, which is often whitened by bleaching or treating with lime water. This method sometimes reduces its quality.

First-grade white ginger should be large, light buff, soft and of strong characteristic flavor. The present supply comes chiefly from Jamaica, the Malabar coast of India, East Indies, Africa, China and Japan. Ginger coming from Africa is dark and coarse in texture, but has an excellent strong flavor, and is largely used in the manufacture of ginger ale extracts. Japan ginger is usually bleached or limed before shipment. It is large, smooth and of finer appearance, but is inferior in strength. Green ginger has not been scraped, washed or dried. It is used in preserving.

Ginger which has been preserved in sugar syrup, also known as Canton or sugar-candied ingber or ginger, consists of young green roots which, previous to curing with sugar, have been macerated with seawater. Preserved ginger is sold in tins and jars. The imported Chinese product is highly prized. Crystallized ginger is also made from the young root. Flaked ginger is sold as leaf ginger. Other ginger products are ginger beer, ginger tea and Jamaica ginger, which is a good stimulant frequently employed for dyspepsia and colic.

Hops

Hops were cultivated in France and Germany as early as the eighth century and are now also planted in the United States, where the producing sections are New York and the Pacific coast states. The hop is the fruit of a climbing plant also found wild in Europe and Asia. It is a multiple fruit consisting of an oblong, oval, or conical mass of dry overlapping scales, like pine cones. It imparts the distinctive flavor to beer beverages. Hops are used in compounding a large number of essential oil extracts. The essential principle is the yellowish powder, lupulin. In selecting choice hops, those, which are full of lupulin, free from mold, bright and silky in appearance, rich in odor, free from leaves, stems, scales, or sticks, and which, when rubbed between the hands, impart a yellowish glint and gummy feeling to the skin, should be selected. The hops are harvested in the latter part of August, cured, kiln-dried and then baled.

Horseradish

This plant is grown in many temperate countries, essentially for the white flesh of its very pungent root, which is ½ to 2 inches in diameter and 1 foot or more in length. It grows in the United States, especially in New Jersey, Long Island, Missouri and the northwest. It is grated, mixed with vinegar and used as a condiment with oysters, meat and fish dishes. The strong taste and odor are due to an oil rich in natural sulfur compounds. Grated horseradish, bottled in vinegar, loses its pungency in about 3 to 4 weeks, becomes dark and takes on an earthy odor. The roots are easily grown and cared for, and may be left in the ground over the winter and dug up as needed. They may then be kept fresh for some time by burying in cool sand. The root should not be exposed to the sun or air. Horseradish powder is prepared by cleaning and grating the root, then drying with gentle heat or dry hot air. Protection and care should be given the eyes when grating the root.

Hyssop

This is a small bushy herb of strong odor and pungent flavor. The green young parts are sometimes used in salads or boiled as a vegetable. The dried flowers are pulverized and used to flavor soups. The dried flower spikes, cut just before opening, are used in hyssop tea and in home-made cough mixtures.

Juniper Berries

These are dried aromatic berries, generally blue, growing in numerous varieties on this continent and throughout much of the eastern hemisphere. They are too pungent to eat raw, but they impart a very nice flavor to corned beef. The small wood of the shrub is used for smoking hams and flavoring food. The dried berries are used in large quantities to flavor gin. The oil, which is distilled from the dried berry, is a powerful diuretic and is used in medicine.

Kolanut (Soudan Coffee)

The seed, not properly the nut, of Cola acuminata, a large tree of the family Sterculiaceae, native to western tropical Africa, is cultivated in the West Indies, Brazil, and other tropical countries. The seeds average about an inch in length, are brown or reddish-gray in color and have an odor resembling that of nutmeg. In the tropics, especially in the Sudan, they are employed as a stimulant coffee. The seed, when fully matured, is about the size of a horse chestnut but grows in pods. The natives gather them in large quantities and use them as a tonic to conserve and promote energy. The seeds contain from 2 to 3 times as much caffein as the coffee bean. For this reason, they are used in flavor extracts and beverages.

Leek

The leek is a culinary herb, closely allied to the onion, and is cultivated for the branched parts of the leaves, commonly called the stems, and the bulbous roots. Both of these are used in flavoring soups and stews. In Europe, the stems are served in the same manner as asparagus. The rank odor of the raw leek disappears in cooking, leaving behind a mild, onion-like flavor.

Lovage

This is a plant, the root and seeds of which are aromatic, and are used in flavoring vinegar, sauces, etc. It is native to Southern Europe, where it is sometimes cultivated as a salad plant. A minor use of the leaves is the production of lovage oil, used as a flavoring for condiments and as imitation maple flavor.

Mace

Mace is the covering which envelops the shell surrounding the nutmeg. It is prepared for market by being flattened and dried in the sun for several days, much of it becoming reddish-yellow during the process. It is used, both in blade and ground form, as a flavoring for sauces and puddings. It closely resembles the nutmeg in taste but has an even more pleasing flavor. The mace with a deep orange color, and a clear, transparent, wax-like appearance is the choicest variety. The dull-looking mace is undesirable. Mace also furnishes a yellow volatile oil and a buttery fixed oil. The latter is known as nutmeg balsam when mixed with other substances. A product called nutmeg butter consists of the compressed oil and fatty substances derived from off-sized, spoiled, refuse nutmegs.

Marjoram

This is an aromatic sweet herb whose fresh and dried tops and leaves are used to flavor soups, dressings, etc. They are cut as soon as the plant begins to flower. Marjoram is a low, bushy, perennial plant grown in southern Europe. The German variety is of a darker color but has a better flavor than the French. The most desirable varieties are the sweet, winter sweet and pot. The common, or wild, variety of majoram resembles the cultivated but is coarser in flavor.

Marshmallow

The marshmallow is a tall erect plant, related to the hollyhock, native to both Europe and Asia. It grows freely around marshes near the sea. A water extract of the root and other parts is in a tasteless, odorless, mucilaginous gum, used in medicine as a demulcent for children and for making marshmallow candy. Today this plant is no longer used for making marshmallows, which are now being made with sugar, corn syrup, flavor and gelatin.

Mate (Paraguay Tea)

This consists of the leaves and young shoots of a species of holly, used widely as a tea or beverage brew in Brazil and other sections of South America. The gathering, roasting, pulverization, and breaking of the leaves and shoots.is an important industrial occupation in both Brazil and Paraguay. The dry mate is greenish-yellow with an aroma suggesting green tea. The finest quality is made of the newly opened leaf buds. A cheaper product is made of the whole large leaves. The title mate is derived from the vessel in which the brews were formerly made. The chemical composition of mate very closely resembles that of coffee and tea, and includes the same stimulating caffein.

Mint

Mint is a name generally applied to a large number of strongly scented perennial plants cultivated chiefly for their essential oils, the best known of which are peppermint, spearmint and pennyroyal. About 50% of the world's commercial supply of peppermint and spearmint is produced and distilled in the acres of muckland in southern Michigan and northern Indiana. Black mint with deep green leaves is the variety richest in peppermint oil and has the most delicate flavor. The plants are cut when in full bloom, allowed to cure like hay, then steamed in wooden vats. The heat breaks the oil cells and permits the oil to escape with the steam; the oil is then separated from the condensed water. The spearmint leaf, used wet or dried, is the ordinary mint for kitchen use. Dried mint is sold in packages, tins or bottles and should always be kept in a dry place. Mint grows easily under almost any conditions. If it is to be used dry, the stalks should be cut just prior to full bloom and spread out in a shady location where they can dry slowly.

Mountain Mint (Basil)

This is a general name for several species of a genus of perennial North American herb with aromatic leaves of a mint-like flavor. The true basil is native to India and southern Asia.

Mustard

This popular condiment is obtained by grinding and blending the velvety flour extracted from the interior of the crushed seeds of one or more species of the mustard plants. It consists chiefly of the small, irregular-shaped seeds, dark brown on the outside and yellow on the inside, of the black or brown mustard (Brassica nigra), together with a certain portion of the smooth yellow or pale brown seeds of the white mustard (Sinapis alba).

England, Holland, Russia and Germany supply us with the best white or yellow mustard seeds. The brown seeds come from Italy, England, Russia, Holland and India. The percentage of fixed oil in mustard seed may run as low as 18% and as high as 40%. The oil is mild, tasteless and is used to adulterate more expensive vegetable oils. The essential oil of the mustard, which gives it its pungency, is developed only when the seed comes in contact with water. The manufacture of mustard flour is somewhat similar to that of wheat flour. The black mustard flour must be blended with the yellow mustard flour to bring about the right taste and flavor. Pleasant taste and odor are sometimes more important in this condiment than pungency. A good mustard flour should contain all three qualities properly blended.

Nutmeg

The nutmeg is the kernel of the fruit of a tropical tree native to the East Indies. The fruit is round and pear-like in shape, 2 inches in diameter and has a reddish-yellow skin. It splits in two when mature, exposing a pink or red covering around the nutmeg kernel, which, when dried, is popularly known as mace. After the fruits have been gathered and the outer coverings of flesh and mace have been removed, the kernels are placed upon gratings over a slow fire and dried at about 140° F. The nutmegs are removed from the shells as soon as they are loosened by the heat treatment, which may take several weeks. After drying by exposure to the sun, they are put up in tight casks ready for shipment. The nutmeg is a native of the Malay Archipelago, but is also cultivated in Singapore, Penang, Sumatra, Java and the West Indies. In purchasing nutmegs, the round and compact varieties of oily appearance and greater weight are desirable. They are graded by the number per pound which may vary from 80 to 140. The larger ones are more showy, but those of moderate size are just as good. Light-weight, dried and dull fruits, or those which are long and oval, should be avoided. Grated nutmeg should be stored in a cool place. It does not keep as well as whole nutmeg.

Okra (Gumbo)

Okra is a large herbaceous plant cultivated in the southern United States and the West Indies. When dried, it yields a mucilaginous product used as a thickener for soups, etc. Occasionally, after proper roasting, it is used as a substitute for coffee. Iron and tin cooking utensils discolor the okra.

Olive

The olive is the fruit of an evergreen tree cultivated in all Mediterranean countries and in the United States, particularly in California. The fruit is green until it attains maturity. Then it gradually becomes yellowish and, finally, in most varieties, changes to a dark purplish-brown. There are numerous varieties which differ markedly in size and oil content. The oil content may be from 10 to 30% of the flesh of the fruit. For oil pressing, it is important to use olives rich in oil. In pickling, the more fleshy fruits are desirable.

The finest imported pickled green olives come from the south of Spain. California and other western states lead the market in the production of the pickled ripe or black olive. California is also making great progress in the pickling of green olives. The varieties chiefly cultivated on the coast are the Mission, Manzanillo, Ascolano and Sevillano. The Manzanillo and the Sevillano are Spanish varieties. The Sevillano comprises the imported Queen olives of Spain, and the Ascolano is an Italian strain. The Mission and Sevillano varieties are typically oval in shape, whereas the other two are rounded-oval.

Olives, used for pickling, are gathered just before they begin to change to their ripening color. They are then sorted according to size and quality, soaked in lye solution to reduce bitterness, then washed with sufficient water to remove the lye flavor, and finally pickled. The method of pickling varies with the place of manufacture. In some localities, brine salt solutions, or salt and vinegar mixed, are used. Some manufacturers add a little sugar to aid fermentation; fennel, thyme, coriander, laurel leaves, etc., may also be added in small amounts. A perfect pickled olive is yellowish-green, firm, with a pinkish pit and an agreeable flavor. Inferior fruit is somewhat dark, soft, mushy and woody-tasting. These defects are caused by age or imperfect curing. Pickled or salted ripe black olives are purplish-black on the inside, dark in pulp, with a bland flavor due to the mellowing of the oil in ripening. They are treated in much the same way as the green olives to remove the bitterness. Then they are kept in a strong brine solution under proper conditions for several months. However, the preferred method is to sterilize the dark olives in cans after the lye treatment; this is the procedure in general use in California. Both green and ripe olives are relishes and constitute a wholesome and nutritious food. Green olives come stuffed with peppers, celery, almonds, etc., and are sold in paste form in jars.

There are more than a dozen different sizes of olives that come from the grading machines. Most of them are listed below:

QUEENS

70– 80's	run from	44	to	50	olives to a quart
80– 90's	run from	51	to	56	olives to a quart
90–100's	run from	57	to	62	olives to a quart
100–110's	run from	63	to	69	olives to a quart
110–120's	run from	70	to	75	olives to a quart
120–130's	run from	76	to	81	olives to a quart
130–140's	run from	82	to	87	olives to a quart
140–150's	run from	88	to	94	olives to a quart
150–160's	run from	95	to	100	olives to a quart
160–180's	run from	101	to	108	olives to a quart
180–200's	run from	109	to	116	olives to a quart
240–260's	run from	151	to	162	olives to a quart

The Manzanillas are graded in sizes similar to the Queen grades. In bottling olives, both Queens and Manzanillas are used. They are either added directly to the bottle or placed in very carefully by the use of wooden placing tongs. The olives are washed with several changes of fresh water and covered with a 7 to 8% salt solution, to which has been added ½ to ¾% lactic acid. The lactic acid comes in liquid form of about 50% strength. It is an off-white liquid, practically odorless, which is now being used extensively in food manufacturing.

Onion

The onion is a common field and garden vegetable native to southern Asia. Today, it is grown in every part of the world, and in particularly large quantities in Germany, Spain and parts of Africa. In the United States, it is grown in California, Texas, New York and Ohio.

Varieties grown for the market are the white and yellow globe and the red Wethersfield types. The strong flavor and smell of the onion are due to an oil rich in sulfur compounds.

Bermuda onions are imported into this country in large quantities. They are large, mild, Spanish-type onions and are now grown successfully in the southern states.

Spanish onions are very much like the Bermuda variety and are now also grown in this country. The Egyptian onion is now being cultivated along the mucky banks of the Mississippi River.

The multiplier onion, also known as the potato onion, is good for flavor and represents the underground bulb of the seedless variety. The larger specimens do not keep very well. These onions furnish the early green onions of the market known as peelers or bunchers.

Onions should be stored in a dry, well-ventilated place; on foggy days, all windows should be closed. They should not be piled too close together as they are apt to sweat, grow and rot. A temperature of 32 to 40° F offers the best protection. Curing in the sun for several days is advisable before storing the onions in a cellar. Plants closely allied to the onion are chives, garlic, leek, shallot and Welsh onion. Onion powder is now being manufactured. It is a white product and is finding numerous uses in flavoring sauces, condiments and soups.

Orange Blossoms

These blossoms are used chiefly for the manufacture of neroli and in the preparation of many puddings, icings and other desserts. The blossoms are crushed first to bring out flavor.

Orange flower water is the fragrant liquid left after distillation of the neroli. It is used in making syrups, perfumes and, at one time, was used quite extensively in the manufacture of various desserts.

Parsley

This is a kitchen herb used for garnishing, flavoring and the preparation of parsley oil, a mild stimulant. Parsley grows in Sardinia and southern Europe. It is easily grown anywhere. A concentrated solution of the herb forms a viscous liquid, somewhat resembling pectin.

Hamburg parsley, or turnip rooted parsley, is a variety of common parsley grown for its white root which looks like a small parsnip, and is used in the same way.

Parsnip

This is a fleshy root vegetable, white in color and of a peculiar pleasing flavor. It grows wild in Europe and Asia. It is rich in sugar, starch and vitamin content, and makes a pleasant sauce with mustard, vinegar and a small amount of salt. Dug roots should be kept stored in sand or moss, in a cool moist cellar or similar place, protected from light and air, to prevent them from becoming spongy and too tough.

Petsai (Chinese Cabbage or Celery Cabbage)

This is a salad vegetable popular in China, which has come to America. It resembles lettuce in appearance. The inner leaves are crisp and delicate. Like lettuce, it is excellent as a salad and is tender when used as a boiled vegetable.

Pepper

The black and white powdered peppers used as condiments consist of the ground dried fruits of Piper nigrum. This perennial is a climbing shrub native to the forests of western and southern India. For centuries, it has also been cultivated on the Malay Peninsula, Sumatra, Java, Ceylon, Siam and, to a limited extent, in Borneo and other tropical countries. The fruit is a small round berry. The peppercorn grows in loosely packed clusters of twenty to thirty berries closely attached to a common fruit stalk. Pieces of the stalk may often be found in whole black peppercorns and, upon examination, depressions where the berries were attached can be found. The shrub or vine grows to a height of 8 to 12 feet and is supported either by poles or trees.

Black Pepper

This variety consists of the whole immature berries, gathered as soon as one or two on the clusters commence to turn red. It is then dried in the sun or near a mild fire. The outer fleshy portion of the berry shrivels in the process, turns brown or black and becomes hard, brittle and adherent to the stony inner portion, in this way forming a hardened, wrinkled cortex. The Dutch East Indies are the main source of supply, with the British Straits Settlement and India next in importance. The principal commercial classifications are Malabar, Singapore, Penang, Trang, Acheen and Lampong. Others, of less importance, include Japan and Siam. As a general rule, the heavier the peppercorns are, the better the grade. They should be firm, round, clean, uniform in appearance and very furrowed.

White Pepper

This is made by grinding peppercorns, after removal of their dark outer shells. Either ripe or immature berries are used. This variety is less pungent but has more flavor than black pepper. It is manufactured on the pepper plantation or by merchants who buy peppercorns in the open market.

Decorticated white pepper is the peppercorn from which the

seed coats have been partly or entirely removed. There are several qualities classified to the length to which the process is carried. The most expensive variety consists of smooth, hard, pearly kernels, which have been freed from the entire seed coat. The classification of white pepper is similar to that of black pepper: Singapore, Penang, Tellicherry, Siam and others. Coriander white pepper is a product of especially attractive appearance, which has been screened to uniform size and then bleached.

Long Pepper

This is the dried immature fruit spike of a plant of a different species, P. longum, native to the hottest parts of India, Ceylon and Malaysia. The spikes are about 1½ inches in length and ¼ inch in diameter and have a central stem covered with closely packed, very small, round berries. The whole plant is hardened. The product is crudely prepared and, generally, includes a large amount of dirt and other extraneous matter. It does not possess the fine flavor or strength of true pepper and is sometimes used to adulterate either whole or ground pepper. It is principally used for pickles.

Red Pepper

Ordinary red pepper is the powdered ripe pod, both flesh and seeds or the flesh alone, of any variety of the capsicum species which gives the edible fresh pepper. It bears no relation to the true pepper plant. The capsicums constitute a native American genus and were unknown to the Old World until after Columbus's voyage to America. Supplementing the use of red pepper as a condiment is its use in such products as gingersnaps, ginger ale, etc., to enhance their pungency.

Cayenne Pepper

This variety is distinguished from red pepper by the fact that it is obtained only from small fruited capsicums. They are small narrow red pods, straight or curved, and from ½ to

1½ inches in length. Many varieties are now cultivated in tropical and sub-tropical countries, including our southern states, and are known chiefly as chilies.

Cayenne pepper is made by two different methods. By one method, the pods are dried and then ground to a fine powder. By the other method, the finely ground pepper, which has been mixed with flour is leavened with yeast, shaped into small cakes and baked hard. The product is then ground and sifted. It should ordinarily have a dull red color. The best known varieties of general importance are the East African, Zanzibar, Mombassa, Sierra Leone and Jampan. The last variety is less pungent than the others. Nepal pepper, from a capsicum grown in India, is a good choice variety, yellowish-red in color, very pungent and has an excellent flavor.

Cayenne is a powerful stimulant. It produces warmth in the stomach, and, when taken in small amounts, aids digestion.

Paprika

This variety is the product of Hungarian and Spanish varieties of large red peppers, and is also known as sweet cayenne pepper, Hungarian pepper, pimiento or pimento. It is an aromatic red pepper, produced by drying and grinding special varieties of capsicums. There are two distinct types. The Hungarian paprika, or Hungarian pepper, is obtained from the medium large, thick-walled, dark to bright red, conical Hungarian capsicum. It is pungent, markedly sweet and aromatic. The Spanish paprika, or pimiento, is made from the more rounded Spanish capsicum, and is sweeter, agreeably aromatic, very bright in color, but lacking in pungency. Hungarian paprika is used for flavoring many dishes, while the Spanish variety is serviceable as a coloring agent for sauces, soups, baking-products, etc. The flavor, pungency and sweetness of the many varieties of paprika depend a great deal upon geographical derivation of the plant. The color depends upon the amount of flesh, seed and seed tissue used, and upon the grinding process. In general, mild flavor and fine reddish color are the desirable qualities looked for. "Rosen" paprika is fine Hungarian paprika,

Tabasco

Tabasco is a long-podded red pepper cultivated chiefly in southern Louisiana. In commerce it is best known in the form of tabasco pepper sauce, which is a rich red, concentrated extract, generally put up in very small bottles or sold in powdered form. Tabasco sauce is excellent for flavoring soups, salads and tomato preparations. Small quantities should be used, as it is very strong. A mixture of olive oil, salt, vinegar and a small amount of tabasco sauce will result in a product, which is soft and delicate with the characteristic tabasco flavor.

Peppers, Vegetable

These peppers are the fresh, cultivated, large-fruited varieties of capsicum, which can be consumed in the ripe, fresh, canned or dried state. They are of the Spanish paprika type and mild in flavor. All varieties are used for the commercial manufacture of condiments, sauces, pickling, stuffing olives, etc. They are added to soups and catsups, for both color and flavor.

Capsicum

There are numerous varieties of two or more species of capsicum, which is a bushy, branched plant, grown in nearly every part of the world. The small fruited varieties are usually the most pungent, and are popularly known as chilies. The variety characterized by larger fruit is commonly known as pepper. Chilies are made into cayenne pepper and used whole in pickles and vinegar, while peppers are eaten as a vegetable and ground into ordinary red pepper and paprika. The pungency of the capsicum lies in the tissues enclosing the seeds, and to a lesser extent, in other parts of the fruit.

Peppercorn

The peppercorn is the whole berry of the pepper plant.

Pimento

This is the dried aromatic berry of the pimento tree, popularly known in this country as allspice. The tree is native to the West Indies, Mexico and South America.

Pimiento (Pimienta)

This is a large-fruited Spanish capsicum, used as a vegetable.

Poppy Seeds

The seeds, containing about 50% oil, are derived from an herbaceous plant sprinkled with bristly hairs. They have a white milky juice and, when bruised, give off a disagreeable narcotic odor.

Dutch poppy seeds are imported from Holland. They are deep blue in color and are considered to be of the best quality. Other sources of supply are Russia, India and Turkey. The last variety is almost white in color. The seeds are used extensively by bakers, for sprinkling rolls and, when ground up with sugar, honey and flavoring, as a cake filling. The oil pressed from the seeds is pale yellow, almost odorless, of a pleasing flavor and used as a table dressing. Inferior grades of the oil are utilized for making soap.

Rhubarb (Pie Plant)

The rhubarb is a strong perennial plant, grown for its thick, succulent, acid leaf stalks, which serve as an excellent fruit for use in sauces and pies. The wilted stalk is revived by drying in the sun, and then soaking in water. Rhubarb leaves should never be eaten because their oxalic acid content is sufficiently high to be dangerous for those with sensitive kidneys. The plant is used on a very small scale in the manufacture of rhubarb juice, which must be sweetened with sugar. The process consists of removing the oxalates and oxalic acid, pressing, concentrating and sweetening the juice.

Roselle (Red Sorrel, Jamaica Sorrel)

This is an annual bushy plant, whose seed pods, resembling cranberries in flavor and appearance, make excellent jellies. The pods are boiled until soft and made into a jelly in the usual manner. The fleshy part surrounding the seed pods can also be made into a jam or syrup or used in pies. The plant is native to Australia and has recently been introduced into this country.

Rosemary

Rosemary is a hardy evergreen plant, native to southern Europe and Asia Minor, whose leaves are very fragrant and stimulating. Its leaves are sometimes used in seasoning and flavoring preserves, but its principal use is in perfumery.

Safflower Seed

This is a product of a thistle-like herb, with an orange-red head, grown in the United States, and used in dyeing, as a substitute for saffron.

Saffron

Spain is the most important source of supply of saffron, the highly colored, dried, orange-red stigmas and styles of a purple variety of crocus. It is frequently adulterated with tumeric or safflower seed. Saffron is used principally in perfumes, for tinting confectionery, and as a general cooking ingredient. There are restrictions on its use in foods in this country.

Sage

Sage is a shrub whose dried, whitish-green leaves are used for flavoring soups, sauces, cheeses, sausages, poultry, pork, etc., because of their aromatic, bitter and slightly astringent flavor. It is cultivated in Greece and Dalmatia, the latter variety being the better. The Greek variety has a stronger odor but is inferior in taste to the Dalmatian.

Sarsaparilla

The sarsaparilla extract is made from the long, thin, dark-brown, bitter rootstocks of several species of tropical smilax. In beverages, its natural bitterness is modified by the addition of oil of wintergreen, sugar, vanillin and other flavors.

Smilax

This is a genus containing about 225 species, consisting mostly of herbs and wood-climbing trailing plants, some of which yield sarsaparilla.

Sassafras

This is the spicy bark of the trunk and root of the sassafras tree, used to make sassafras tea, and cultivated largely along the Atlantic coast and in other parts of the United States. The twigs, leaves and leaf buds of the sassafras tree have long been used for their flavor, and are now used extensively in the manufacture of beverages and confectionery.

Savory

Savory is an annual herb with a strong, agreeable aromatic smell and flavor. Its smooth leaves, both fresh and dried, are in common use for flavoring soups, sausages, sauces and dressings. Dried savory is sold in cans and bottles. The plant is a native of the Mediterranean countries.

Sloe (Blackthorn)

This is a thorny shrub of the plum group, whose fruit is nearly round in shape, about the size of a very large pea, sour and astringent in flavor. It is used as a flavoring in various liquors, especially sloe gin.

Tamarind

Tamarind is the fruit of an East African tree which reaches a height of 80 feet. It consists of thin, brown, shell pods, filled with an acrid, sweet, brownish pulp and from one to twelve large, flat, hard, glossy seeds. It contains acetic, citric and tartaric acids as well as salts of potassium and other metals. Pulp and seeds are transferred to kegs, covered with boiling syrup. and shipped to foreign markets. Sugared tamarinds are packed in alternate layers of sugar and fruit, to retain the original color and flavor.

Tamarind pulp serves as an excellent addition to chutneys,

curries, sauces and, when considerably diluted with sugar water, makes a good beverage.

Tarragon (Estragon)

Tarragon is a perennial aromatic herb used for flavoring vinegar, mustard, pickles, sauces and salads. It is grown in Siberia and Europe, and now also in America. The leaf of the tarragon plant is sold green or dried. Half a pound of green tarragon will flavor fifty or more gallons of tarragon vinegar.

Thyme

Thyme is native to the Mediterranean region and a member of the mint family. It is cultivated for its fragrant tops and leaves which are popular for seasoning. The lemon-scented is the best variety of thyme. For drying purposes, the thyme should be cut just before it starts to blossom. Commercial thyme comes in powdered form and is used in dressings and fillings.

Tonka (Tonqua Bean, Coumara Nut)

Tonka is the dark aromatic seed of a tall South American tree, formed singly in a thick pod. The essential principle is coumarin a white crystalline powder found under the coat and between the lobes. Because of similarity in aroma, the extracted coumarin is used to bolster flavor in imitation vanilla extracts. The Dutch tonka bean, or angostura, is about 1½ inches long and is brown inside. The Para, or English, tonka bean is a small seed, with very few white courmarin crystals and is white inside.

Turmeric (Curcuma)

Turmeric is the aromatic rootstock of a southern Asiatic plant, sold both in root and powder form. The dried root varies from a dark orange to a deep reddish-brown, but in powdered form appears yellow. The central portion of the root of the curcuma or turmeric plant is tuberous, sending out nearly cylindrical branches, some of which become thickened at their ends, and contain a pure starch. The principal commercial varieties are Chinese, Bengal, Madras, Java and Cochin.

Turmeric is the principal ingredient of curry powders and is also used in mustard and other spices. Its pigment is marketed as curcuma oil and curcumin and is used as a dyestuff. Its use as a coloring material in foodstuffs is somewhat restricted by the pure food laws.

Vanilla Beans

The vanilla bean is the fruit of an orchid, a parasitic climbing plant found in most tropical countries, but indigenous to eastern Mexico. The vanilla plant has been introduced into some of the West Indian Islands, Bourbon, Madagascar, and several East Indian Islands. In its native country it is probably fertilized through the agency of insects. In other countries artificial fertilization is required. The vanilla bean is from 6 to 10 inches long and about ½ inch thick. The beans are gathered when they become slightly hard and begin to turn yellow at the lowest end. It requires long experience to determine the proper stage for harvesting the beans. If the pod is too green when picked, an inferior flavor results after the curing process. If the pod is too ripe it splits while curing, losing some of its flavor. The flavor is developed in the curing processes of which there are two, the Mexican or sun process and the Bourbon or hot water process. The production of 1 pound of the cured bean requires 4½ to 5 pounds of the green beans.

The various kinds of vanilla beans listed in order of quality are:

Mexican beans	Quality A—9 in. long; rough to the touch Quality B—7¾–8 in. long; dark brown
Porto Rican beans	
South American beans	
Bourbon beans	7 in. long; not as firm as Mexican
Madagascar beans	Almost black
Java beans	
Tahiti beans	6 in. long; reddish-brown

Vanilla beans from which the vanillin has been removed by means of a solvent are sometimes offered to the trade. The fraud can be detected by the weakness of the flavor and odor. Such beans are sometimes improved in appearance and odor by the use of benzoic acid.

The crystals, as a rule, are flattened, rhomboidal and lie upon the bean. Those of vanillin are usually needle-shaped and stand out almost at right angles from the surface of the fruit. Place some of these crystals in a test tube and heat over a Bunsen burner. If they were composed of benzoic acid the irritating odor of this acid may be detected.

PER CENT VANILLIN IN VANILLA BEANS
(Busse)

Mexican vanilla bean	1.69–1.86
Bourbon vanilla bean	1.91–2.48
Java vanilla bean	2.75
Tahiti vanilla bean	2.00

The practical value of a vanilla bean is not, by any means, to be determined by the percentage of vanillin present, since frequently the very finest beans show the smallest vanillin content.

PER CENT RESIN IN VANILLA BEANS
(Brooks)

Mexican vanilla bean (whole)	1.47
Mexican vanilla bean (cuts)	1.96
Bourbon vanilla bean	1.93
Comoros vanilla bean	1.96
Mexican and South American vanilla bean	2.56

True Vanillin and Imitation Vanillin

About 1858, it was definitely established that the fine, tiny, white crystals, found on vanilla beans and regarded by many as the criterion of quality, were, in fact, the aromatic principle of the bean itself. This crystalline substance, the chemical constitution of which was unknown at that time, was given the name vanillin. Ordinary vanillin is methyl vanillin, which is naturally present in the vanilla bean. Today we have a stronger competitive flavor in ethyl vanillin. The choice between the two varities depends on the manner of preparation and the use.

Natural vanillin is found in a wide variety of common plant substances, e.g., sugar beets, asparagus, cloves, benzoin, balsam and asafetida. However, the amount of vanillin found in these products is so small, that it does not pay to process them for it. For many years, chemists had been trying to synthesize vanillin from its component elements. The first successful synthesis of vanillin was carried out using coniferin, a resin obtained from the bark of pine trees. This method of production was used until vanillin was discovered to be methyl protocatechuio aldehyde, an aldehyde of vanillic acid. Sometime later, guaiacol, or methyl catechol, was discovered in the tar of the beechwood tree. This proved to be another step forward in indicating a production method of vanillin from guaiacol of aniline origin. This new source provided sufficient raw material to take care of almost all the major requirements for artificial vanillin. By the simple addition of oxygen to the eugenol in oil of cloves it is transformed to crude vanillin.

Today a large amount of vanillin is manufactured from clove oil. Two pounds of clove oil are required to make one pound of vanillin; about 6 pounds of the clove spice are required to make one pound of oil. In England, the manufacturers use clove oil as the starting point in vanillin manufacture and, although the majority of European plants employ the so-called guaiacol or coal-tar process, it is safe to assume that most of the vanillin is manufactured from clove oil, with only a small quantity coming from guaiacol.

Since vanillin is a definite chemical compound, it will always have the same flavor and aroma, if extreme care is taken in manufacturing and refining it. Therefore, vanillins produced from eugenol or guaiacol should compare favorably.

Briefly, vanillin is produced by first distilling the oil from the clove spice. The oil passes over with the steam condensate and is separated from the water. The clove oil so obtained is largely composed of eugenol. The eugenol is then separated from the clove oil and treated with caustic soda under pressure. It is then acetylated with acetic anhydride, and the acetylation product is oxidized to crude vanillin. Numerous other methods have been advanced for the manufacture of vanillin from eugenol.

One pound of synthetic vanillin roughly approximates the flavoring value of 30 pounds of the best vanilla bean.

Ethyl Vanillin

The ethyl ester of protocatechuic aldehyde is known erroneously as ethyl vanillin and is also known commercially as Bourbonal, Ethovan, Vanillal and Vanirome. With respect to intensity of taste and odor, this compound is about 3 to 4 times as strong as vanillin. Ethyl vanillin melts at approximately 40° C (lower than regular vanillin), but the two compounds are chemically quite similar.

Ethyl vanillin consists of colorless flakes soluble in alcohol and ether and is sparingly soluble in water.

Wintergreen

Wintergreen is a small evergreen herb, used primarily for extraction of its oil. It grows wild in many parts of the United States. It is also known as oil of gaultheria and is used to flavor confectionery and soft drink beverages. The principal present source of so-called natural wintergreen oil is the sweet, or black, birch, Betula lenta, the twigs and leaves of which yield oil of betula, chemically identical with wintergreen.

ROOTS, BARKS, HERBS, LEAVES, STEMS, SEEDS, FLOWERS

Very Strong, Bitter

Common centaury (Herba Centaurii)
Taste: Very strong, bitter
Use: For bitters and liquors

Wormwood (Herba Absinthii)
Taste: Strong, bitter and aromatic
Use: For absinthe and bitters

Strong, Bitter

Holy thistle (Herba Cardui Benedicti)
  Taste: Strong, bitter and lasting bitter
  Use: For bitters

Cinochona bark (Cortex Chinae)
  Taste: Very bitter, astringent
  Use: For bitters

Gentian root (Radix Gentianae)
  Taste: First sweet, then strongly bitter
  Use: For Enzianschnaps

Quassia wood (Lignum Quassiae Surinamense)
  Taste: Not as bitter as gentian
  Use: For bitters

Aromatic, Bitter

Inule root (Radices Inulae)
  Taste: Bitter, peculiarly spicy
  Use: For bitters

Angelica seeds (Semina Angelicae)
  Taste: Sharp, bitter
  Use: For bitters

Angelica root (Radix Angelicae)
  Taste: Strongly spicy, not as bitter as the seeds
  Use: For bitters, especially chartreuse

Valerian root (Radix Valerianae)
  Taste: Sweet, spicy, somewhat bitter
  Use: For bitters

Sweet calamus (Rhizoma Calami)
  Taste: Spicy and bitter
  Use: For aromatic liquors

Lemon peel (Cortex Citri)
  Taste: Spicy, slightly bitter
  Use: Chartreuse, curacao

Coriander (Fructus Coriandri)
  Taste: Sweet, strong, spicy, then burning sensation
  Use: For liquors, spice

Cubebs (Fructus Cubebae)
  Taste: Spicy and bitter, not burning
  Use: For liquors, spice for *Ulmer Pfefferkuchen*

Fenouil batard (Fructus Anethi)
Taste: Spicy like dill and caraway, slightly bitter
Use: For hard liquors

Roman chamomile flowers (Flores Chamomillae Romanae)
Taste: Bitter, spicy
Use: In small quantities for liquors

Curled mint (Herba Menthate Crispae)
Taste: Spicy, slightly bitter, but not cooling
Use: For liquors

Caraway (Fructus Carvi)
Taste: Strongly aromatic, slightly bitter
Use: For kummel, spice

Lovage root (Radix Levistici)
Taste: Strongly spicy, first sweet, then bitter
Use: For stomach bitters and liquors

Laurel leaves (Folia Lauri)
Taste: Aromatic, spicy, bitter
Use: For liquors, spice

Marjoram (Herba Marjoranae)
Taste: Spicy, bitter
Use: For liquors, spice

Clarysage (Salvia Slarea)
Taste: Spicy, bitter
Use: For liquors, substitute for hops

Myrrh (Gummi Resina Myrrha)
Taste: Astringent, bitter
Use: For tinctures

Orange flowers (Flores Aurantii)
Taste: Pleasantly aromatic, slightly bitter
Use: For liquors

Beautressel (Herba Saturejae)
Taste: Strong, bitter
Use: Rarely for liquors
Remarks: Called *Pfefferkraut*

Peppermint (Folia Menthae Piperitae)
Taste: Burning, cool, slightly bitter, and strongly spicy
Use: For liquors, spice

Rosemary flowers (Flores Rosemarini)
Taste: Mildly spicy, very slightly bitter
Use: For liquors

Sage leaves (Folia Salviae)
Taste: Powerful, spicy and bitter
Use: For liquors, spice

Thyme (Herba Thymi)
Taste: Pleasantly spicy but bitter
Use: For liquors, spice

Juniper berries (Fructus Juniperi)
Taste: Strong, spicy and bitter
Use: Genever and gin

Levant wormseed (Flores Cinae)
Taste: Peculiar, bitter and cooling
Use: Rarely for stomach bitters

Ginger (Rhizoma Zedoariae)
Taste: Camphor-like, bitter and spicy
Use: For stomach bitters

Angostura bark (Cortex Angosturae)
Taste: Pleasantly spicy but very bitter
Use: For angostura type bitters

Curacao peel (Cortex Curacao)
Taste: Bitter but different from orange
Use: For curacao and Benedictine

Iva musk (Herba Ivae Moschatae)
Taste: Spicy but bitter and lasting
Use: Iva liquor

Orange peel (Fructus Aurantii Immaturi)
Taste: Strong, bitter
Use: Bitters

Bitter orange peel (Cortex Aurantii Fructus)
Taste: Fruity, bitter
Use: Abtei liquors

Rhubarb root (Rhizoma Rhei)
Taste: Aromatic, bitter
Use: Stomach bitters

Mild Aromatic without Bitter Taste

Arnica flowers (Flores Arnicae)
Taste: Strongly astringent and weakly bitter
Use: In small quantities for bitters and liquors
Remarks: Contains the poison arnicin

Elder flowers (Flores Sambuci)
Taste: Sweet, then scratching
Use: Liquor and tobacco aroma

Guaiacum wood (Lignum Guajaci)
Taste: Sharp and scratching, not bitter, rather sweet
Use: Strong liquors

Cocoa beans (Semen Cacao)
Taste: Pleasant, spicy, nut-like, very slightly bitter
Use: For liquor

Lavender flowers (Flores Lavandulae)
Taste: Strong, spicy, not bitter
Use: For liquor

Linden flowers (Flores Tiliae)
Taste: Pleasantly aromatic, slimy
Use: For some liquors

Beunet root (Radix Caryophyllatae)
Taste: First bitter, then sweet, later astringent
Use: Heavy liquors

Parsley (Herba Petroselini Foliorum)
Taste: Strong, spicy
Use: Liquor, spice

Celery seeds (Semen Apii Graveolentis)
Taste: Like celery
Use: Spice extracts, spice

Celery root (Rhizoma Apii Graveolentis)
Taste: Strong, spicy, slightly burning
Use: Liquors, spice

Orris root (Rhizoma Iridis)
Taste: Pleasantly aromatic, but slightly irritating
Use: Liquors

Hyssop (Herba Hyssopi)
Taste: Pleasant, spicy, slightly camphor-like, slightly bitter
Use: Liquors

Aromatic with Burning Taste

Pyrethrum root (Radix Pyrethri)
Taste: Burning, sharp, astringent
Use: Strong-tasting liquors

Galanga root (Rhizoma Galangae)
Taste: Pleasantly spicy but sharp
Use: Strong bitters

Aromatic with Sweet Taste

Sweet orange peel (Cortex Aurantii Dulcis)
- Taste: Pleasantly sweet, refreshing
- Use: Liquors

John's bread (Fructus Ceratoniae or Siliqua Dulcis)
- Taste: Very slimy but very sweet
- Use: For rum, cognac

Manne (Manna)
- Taste: Very sweet, afterwards irritating
- Use: Liquors

Liquorice root (Radix Liquiritiae)
- Taste: Peculiarly sweet but slightly bitter
- Use: Liquors

Aromatic with Vanilla Taste

Melilot (Herba Meliloti)
- Taste: Slimy, sharp and bitter
- Use: Tobacco essence

Tonka beans (Fabae Tonco)
- Taste: Strong, slimy, sharp and bitter
- Use: Waldmeister, tobacco, bitters, liquors

Vanilla (Fructus Vanillae)
- Taste: Pleasantly aromatic, bitter-sweet
- Use: Liquor, essences

Wood-ruff (*Waldmeister*) (Herba Matrisilvae or Asperulae Odorata)
- Taste: Bitter like coumarin
- Use: Waldmeister essence, tobacco flavor

Aromatic with Tannin

Limousin bark (Cortex Quercus Ilex)
- Taste: Strongly bitter, astringent
- Use: Cognac, for aging whiskey, etc.

TRUE SPICES

Allspice	Cloves
Cardamom	Mace
Cinnamon	Nutmeg
Cinnamon flowers	Vanilla bean

AROMATICS WITH SWEET TASTE

Anise	Fennel
Star anise	Liquorice root
Carob	Manna
Coriander	Sweet orange peel

AROMATICS WITH SPECIFIC TASTE

Apricot kernels	Bitter taste
Bitter almond (poisonous)	Bitter taste
Common laurel leaves	Bitter taste
Oak bark	Tannic acid taste
Tannic bark	Tannic acid taste
Tonka bean	Coumarin taste
Vanilla beans	Aromatic taste
Woodruff	Coumarin taste

COLORING MATTERS

Alkanet root	Red	Mallow flowers	Red
Buckthornberry	Yellow	Pimento	Red
Catechu	Brown	Raspberry (black)	Red, purple
Cochineal	Red	Raspberry (Colombia)	Purple
Cudbear	Yellow	Saffron	Yellow
Curcuma root	Yellow	Sandalwood	Red
Huckleberry (black)	Red	Turmeric	Yellow
Kino	Red	Yellow wood	Yellow

Gums

The natural gums may be subdivided into two classes: 1 gums which are *soluble* in water, forming a thick, viscous, sticky solution, which can be filtered, and 2 gums which do not dissolve, but swell with water to form heavy viscous liquids, which cannot be filtered. The natural gums are vegetable products which occur as amorphous, translucent or transparent exudations from plants. The gums may be dissolved out with the aid of alkalies, but upon addition of alcohol all gums are

precipitated or thrown out of solution. Gums differ from starch or cellulose by being soluble or swelling in water. They differ from sugars by being incapable of alcoholic fermentation with yeast. Three important ingredients found in gums are:
Arabin or arabic acid (the soluble form as found in gum arabic)
Bassorin or insoluble gum (found in gum tragacanth)
Cerasin (insoluble, found in cherry gum)

Gum Acacia or Arabic

This valuable gum consists mainly of arabic acid, or arabin, combined with lime, potassium or magnesium. It is bleached by exposure to the sun and comes both in powdered and lump form. The gum is an exudation from the acacia and mimosa species found as low shrubs in Africa and the Orient. The gum is used extensively for the suspension of insoluble substances in water, especially flavoring oils. The granulated gum dissolves more readily in water than the finely powdered product. Its easy solubility and the absence of the tendency to form lumps make the coarse powder preferable for solutions, emulsions, etc. Acacia is also used as a binding agent in tablets and pills.

Carragheen or Irish Moss

The mucilaginous constituent of Irish moss is carragheen. It is not precipitated out with alcohol and in this respect, differs from gum or starch. Irish moss, also called chondrus, is the dried plant of an alga that grows in the Atlantic Ocean. It is used as an emulsifying agent for cod liver oil and also to make jellies for sick patients, 1 part being sufficient to combine with 60 parts of slightly sweetened water. Carragheen can now be obtained in a highly purified state, so that it is capable of being combined with even more water, and is free from the peculiar taste which is so characteristic of Irish moss.

Algin (Alginates)

Algin is nontoxic and is extensively used in the food industry. It is used primarily as a stabilizer in ice cream where it serves to prevent the formation of large ice crystals without masking the flavor and produces an ice cream which has a smooth velvety texture of creamy consistency. It also facilitates production since ice cream mixes made with algin whip fast and require no aging before freezing. In the same way, in water ices and other frozen desserts, it stabilizes the product against the formation of large ice crystals.

In chocolate milk, it serves as a suspending agent for the cocoa fibers and, in milk puddings, it acts as a gelling agent. It is used in place of starch and various gums in salad dressing where it acts as an emulsifying and stabilizing agent and imparts the proper body. Its colloidal properties also make it a valuable addition to other foodstuffs such as doughnuts, cake, icings, buttermilk, cream and confectionery.

In the pharmaceutical field, the pronounced colloidal properties of algin render it useful as an emulsifying, bodying and suspending agent.

Carragar

Carragar is a new proprietary sea-plant derivative prepared from carragheen or Irish moss. The active principle is gelose, a gelatinous carbohydrate capable of forming rigid water gels without the aid of sugars or other dehydrating agents. The gel forms best at pH 4.0 to 7.0. It is subject to breakdown if heated in the presence of acids. This hydrolysis can be avoided by adding the acid as the last step in the process.

Carragar, if mixed with sugar before cooking, goes into solution more quickly. It can be used as a stabilizer and emulsifier in edible products such as fillings, icings, meringues, glaze mixtures and marshmallow.

Krim-Ko Gel

Krim-Ko Gel is a purified defibrinated sea-plant extractive recommended as a general-purpose colloid. It is a derivative of Irish moss. The active principle of this plant is the gel-forming carbohydrate known as gelose. This colloid thickens, gels, stabilizes, emulsifies and suspends solid particles in liquid solutions. It performs satisfactorily in the range of pH 4.0 to 7.0.

It is sensitive to sustained high-temperature processing and to acidity from pH 4.0 down to 1. The best way to use it is to put it into complete solution by sprinkling it on the surface of cold water with vigorous agitation. Heating this suspension to 160°F and holding it at this temperature for at least 15 minutes insure complete solution of the particles with no lumps. Actually, the colloid may be added at any temperature below 130°F and the subsequent heat-processing of the batch may be relied on to complete the solution. Solutions of this colloid are practically neutral in reaction and will withstand repeated heatings when the processing batch is substantially neutral. Where the user's product requires adjustment to a higher or lower pH level, it is desirable that the adjustment be made after heat-processing when the batch has been cooled down below 160°F. This colloid is now being used in food and beverage manufacturing.

Bassorin

Bassorin is the insoluble gum portion found in tragacanth, and is precipitated out of solution by means of alcohol, as in the separation of arabic acid from gum acacia.

Gum Karaya or Indian Gum

This is a very useful gum, its color varying from a light pink to grayish-white, very cheap in price, and is shipped from Bombay.

Gum Tragacanth

There are many species of low shrubs growing in the country between eastern Persia and Greece which yield tragacanth. It consists of about 33% bassorin (insoluble gum), 53% soluble gum, 11% water and 3% impurities. In water, tragacanth absorbs a certain amount, then swells very much to form a soft adhesive paste, but does not dissolve. If additional water is added, the paste forms a uniform solution, but in the course of 1 or 2 days, the greater part of the gum separates out, leaving a small amount dissolved in the supernatant liquid. There are numerous starch grains present in gum tragacanth. Tragacanth of varying colors is sold in leaf form. Common tragacanth consists of pear-shaped pieces of a brownish color.

British Gum or Artificial Gum

Starch, when acted upon by dilute acids, diastase or heat, is converted to dextrin, a substance which resembles gum in appearance and properties. Dextrin is soluble in hot or cold water and forms a mucilaginous solution from which it is precipitated by alcohol. The product, found on the market, is sold as British or artificial gum.

British gum is sold in the form of a brilliant white powder, or in small masses or fragments resembling natural gum. It may be distinguished from gum arabic by the taste and the peculiar harsh sensation when rubbed in the hand.

Solubility of Gums and Thickening Agents in Water

Gum arabic is very soluble in hot or cold water. It should be added to the water and stirred at the same time. In making concentrated solutions, the best method is to pack a linen bag with the gum and immerse it in the water.

Agar-agar, a purified seaweed, dissolves very slowly in cold water but rapidly in hot. On cooling, it forms a gel or stiff jelly. If the agar-agar is heated with acids, it breaks down forming a correspondingly weaker gel on cooling.

British gum, dextrin, is easily soluble in water and does not form a gel on cooling.

Gum karaya, Indian gum, is sparingly soluble in either hot or cold water. The particles swell, forming a heavy, stringy, sticky liquid. It is used in emulsifying oils and in suspending solids in solution.

Irish moss, carragheen, is a seaweed, sparingly soluble in water, which forms a translucent, viscous, yellowish solution. It has a great affinity for water, emulsifies oils and suspends solids in solution.

Locust bean kernel gum powder is derived from St. John's bread. It is sparingly soluble in water, swells considerably, forming an opaque colloidal solution from which a precipitate settles out on standing. It has great affinity for water, and is used in any number of ways to stabilize fruit juices from running out of pies and bakery goods.

Pectin, a carbohydrate found in fruits such as the apple, lemon and orange, is available in powder form and is used extensively in making jellies. It is easily dispersible in hot or cold water and will form a jelly only when the sugar, acid and pectin are present in correct ratio to each other. Pectin is used for thickening chocolate syrups and many other products.

Quince seed, derived by cooking the fruit in water, forms an opalescent solution, stringy, sticky and very useful for thickening water or sugar solutions.

Sodium alginate and ammonium alginate, products manufactured from sea kelp or alginic acid, are readily soluble in cold or hot water when properly prepared. A good way to prepare a solution is to add the product slowly and mix it in at the same time. The solution is a pale straw color, translucent, viscous and readily precipitated by acid or mineral salts, forming alginic acid or metal alginates. The salts are used extensively as conditioners in ice cream mixes, and for suspending solids in liquids.

Defatted soybean flour has a great affinity for water and is, therefore, a good addition to products treated with water which may have a tendency to weep out.

Gomagel, a new proprietary gum, is a white powder which forms viscous water gels and jellies. This is used as a substitute for agar, gelatin and tragacanth in certain cases.

Pectinol Enzyme for Cloudy Fruit Juices

Most fruit juices, as they come from the press, are cloudy. In almost every case they remain so. This cloudiness is due to colloidal matter which cannot be removed by ordinary methods of filtration. Later coagulation of some of the colloidal material causes the formation of a muddy sediment in the container. The resulting juice and wine have an unattractive appearance and may even give the impression that they were carelessly prepared.

Old fashioned muddy fruit juices are fast becoming obsolete. Clear, sparkling juices are growing in demand. Clarification with Pectinols is the most satisfactory way to produce this clear juice.

The Pectinols act on the pectic substances to convert them into soluble substances and into small amounts of easily filterable solids. Their action is positive and permanent. This method differs markedly from mechanical methods of clarification which usually remove color and flavor as well as suspended solids. Mechanical clarification methods consist of adding more solids to remove those already present, in this way increasing the residues remaining on filter presses. Pectinols are completely soluble and bring about solubility of the pectic substance in the juice.

Enzyme actions are not instantaneous like the neutralization of an acid with an alkali. Enzymes require time to do their work. In determining how much enzyme is required to do a certain amount of clarification, the time available for the reaction must be considered. Temperature, also, is a vital factor in the consideration. The Pectinols, like most enzymes, are active between 32 and 140°F. At the lower temperatures, their action is slow, but the rate of action in general increases as the temperature is raised, and reaches its optimum at about 100°F. Higher temperatures quickly destroy the enzyme, and 10 minutes at 140°F inactivate the Pectinols completely.

Time and temperature are, therefore, necessary factors in the determination of the amount of enzyme to use. For example, at 100°F, 10 ounces of Pectinol will do a certain amount of work in 4 hours, 5 ounces in 8 hours, or 20 ounces in 2 hours; but, if the temperature is only 70°F, about twice as much enzyme or twice as much time will be required as at 100° F.

Individual juices vary considerably in their reactivity toward the Pectinols, so the preferred amount of Pectinol and time for clarification must be determined experimentally.

Juices treated with sodium benzoate before clarification will still react to the Pectinols, although the reaction is somewhat retarded if the amount of sodium benzoate greatly exceeds 0.1%. Freshly pressed juices usually react much more quickly than those which have stood for a long time. The juice of ripe fruit reacts much better than that of unripe fruit. The juice from cold storage apples is generally more easily clarified than that from fresh fruit.

The Pectinol Group

All enzymes are specific in their nature. For example, pepsin, an enzyme found in the stomach, acts only on protein substances and will not act upon starch and fat, while diastase acts only on starch. Pectinols, as their name indicates, have been developed to work specifically on the pectic colloids of fruit juices. Pectinol enzyme preparations have been developed for the clarification of apple cider and apple wine; Pectinol W for the clarification of grape juice, grape wine and for apple wines; and Pectinol M for the clarification of miscellaneous juices and wines such as blackberry, strawberry, raspberry, citrus, peach, etc.

Antioxidants

Fats and fatty foods are attacked by oxygen and an oxidative change or deterioration known as rancidity results. This change is hastened by light, iron and copper salts. The oxygen is absorbed by the unsaturated bonds of the fatty acids with formation of peroxides and, finally, certain aldehydes and ketones are produced. The following list of antioxidants may be satisfactorily used as a protection against oxygen for fatty foods.

Antioxidant	*Per Cent*
Pyrogallol	0.01
Wheat germ oil	1.00
Tocopherol alpha	0.05
Lecithin	0.10
Guaiac gum	0.05
Guaiac gum	0.05 plus 0.002 phosphoric acid
Avene x	0.20

Lecithin

Lecithin is found widely distributed in the animal and vegetable kingdoms. Commercial lecithin is obtained chiefly from soybeans.

DIRECTIONS FOR USING LECITHIN AND EFFECTS

	ITEM	*Quantity to use*	*Time of addition*	*Dough, batter, batch condition*	*Machinability or handling quality*
1	Biscuits, hard cookies, ice box cookies	2% of fat	In mixer with fat or after flour cleared in	Drier, more pliable	Less sticky, clean mold or deposit
2	Bread, rolls and sweet goods	2-3 oz per 100 lb flour	With shortening	Drier dough, more resilient	Less sticky
3	Brittles	1% of fat	Direct to batch following fat		
4	Cakes (emulsified shortening)	1% of shortening	After obtaining full vol. in batter	More fluid	Cleaner deposit, lies flat in pans
5	Cakes (regular fats)	1-1¾% of shortening	After obtaining full vol. in batter	More fluid	Cleaner deposit, lies flat in pans
6	Caramels, nougats, toffees	1% fat	Direct to batch following fat		Less sticky
7	Chocolate	0.25-0.3%	Prior to tempering		
8	Chocolate flavor for ice cream	0.25%	Prior to molding	Softer	Easier molding
9	Cookies (molasses)	4% of fat	In mixer with fat or after flour cleared in		Less sticky, clear cut
10	Cookies (soft)	2% of fat	In mixer with fat or after flour cleared in	More pliable	Clear cut, less sticky
11	Doughnuts (cake type)	2% of fat	With fats		Clean deposit
12	Doughnuts (raised)	2% of fat	With fats		Clean deposit
13	Dry mixes (various, prepared)	*(See individual products)*	With fats		
14	Fat-bearing chocolate syrups and toppings	1% of fat	Premix in fat-containing constituents		
15	Flavors (emulsions)	1-10% of oils	In oils		Rapid dispersion in product reduces flash evap. loss
16	Flavors (oils)	1%	Unimportant		*(See above)*
17	Gravies and sauces	1-2% of fats	Variable		
18	Hot chocolate and other powdered drinks	0.5-1.0% total dry weight	Before dehydration or before pulverizing or sifting		
19	Ice creams (low solids) and sherbets	0.15%	Premixed with portion of sugar	Body less watery	
20	Ice cream coatings	0.5%	Just prior to tempering		
21	Icings and fillings of fat and suspended solids	1% of fat	Unimportant	Softer	More pliable
22	Icings of fat, moisture and solids	1% of fat	After obtaining full vol.	Softer	More pliable
23	Lard	0.1-0.25%	When under 150° F and before Rotating	Slightly softer	
24	Meats (canned with gravy)	1-2% of fat	Variable		
25	Oleomargarine	0.1-0.75%	In fats		Reduces spattering
26	Pan greases and slab grease (straight)	1% of fat	Before use		
27	Pan grease with suspended solids	1% of fat	When mixing	More fluid	Easier application
28	Peanut butter for commercial use	0.5%	After grinding	Softer	Easier deposit, bkrg. reduced
29	Peanut butter for retail sale	0.5%	After grinding	Softer	Improved
30	Pie doughs	2% of fat	With shortening		Mixing time must be reduced
31	Shortening	0.25%	When under 150° F and before Rotating	Slightly softer	
32	Soda crackers	2-3 oz per 100 lb flour	With shortening	Drier dough	Less sticky

ADDENDA: (1) Lecithin does not save fat except in items 7, 20, 21, and 27.
(2) It prolongs shelf life by inhibiting oxidation and by moisture retention.
(3) It increases fermentation tolerance of 2 and 12.

ON PRODUCT CHARACTERISTICS AND QUALITY

Viscosity	*Gloss*	*Texture*	*Grain*	*Color*	*Separation*	*Dispersion*
	Slight	More tender		More uniform		Improved; checking reduced
		More silky crumb; more tender crust	More uniform	Of crust, more uniform		
		More tender			Retards separation	
		More silky	More uniform	Of crust, more uniform		
		More silky	More uniform	Of crust, more uniform		
		More tender, less sticky			Retards fat separation	
vered	Improved					
vered						More rapid & complete in mix
	Slight	More tender	More uniform	More uniform		
		More silky	More uniform	More uniform		
		Softer	More uniform	More uniform		
		Softer		More uniform		
						Improved
iable	Improved	Smoother			Retards fat separation	
ed		Smoother			Reduced	More rapid in product; on consumer's taste buds
						(See above)
iable		Smoother			Reduced	
						More rapid & complete in reconstitution
		Smoother	Less grit		Of air, reduced	
ered; ntained	Improved					
ered if no omitted	Improved	Smoother				
able	Improved	Smoother				
					Slight improvement	More rapid and uniform
		Gravy smoother			Of fat, reduced	
				Slight tint with high %'s	Reduced weeping	More rapid and more uniform
change						More efficient coverage
ered	Unimportant	Smoother			Slightly reduced	More efficient coverage
ered	Unimportant	Smoother			Unimportant	Improved
ered	Slight	Smoother			Greater; correctable with solids	
		More flaky		More uniform		Fat in thinner layers
						More rapid and uniform
		More tender	More uniform	More uniform		

Emulsifiers in Food Products

Emulsifiers are ingredients used to disperse solids or solutions of solids within a liquid, or to disperse mutually insoluble liquids within another liquid. If the correct emulsifier is chosen, nearly all food materials can be emulsified. Common emulsifiers are lecithin, egg yolk, Irish moss, gum karaya, tragacanth, gelatin, sodium alginate, locust bean gum and many others.

Edible Fats and Oils

Fats and oils have complex structures and are compounds of glycerin and various fatty acids. The oils, when subjected for a long time to the action of light and air, develop a distinctly rancid odor and taste.

Oils and fats that are liquid at ordinary temperatures consist mostly of oleic acid; olive oil is an example. Stearic and palmitic acids are predominant in those fats that are solid at room temperature; beef and mutton tallow are good examples.

TABLE OF EDIBLE FATS AND OILS

Almond oil	Palm kernel oil
Cocoa butter	Peanut oil
Cochin oil	Poppyseed oil
Ceylon oil	Rapeseed oil
Corn oil	Soybean oil
Cocoanut oil	Sesame oil
Cottonseed oil	Sunflower oil
Olive oil	Teaseed oil

PROPERTIES OF OILS

Name of Oil	*Iodine Value*	*Specific Gravity at 15° C*	*Acidity* (Based on Oleic Acid)
Almond	93–104	0.9175–0.9195	5.8
Apricot kernel	101–108.7	0.9172–0.9200	—
Cocoanut	6.2– 10	0.9260	—
Corn (Maize)	113–129	0.9213–0.9268	1.4
Cottonseed	104–115	0.9150–0.9300	0.0–3
Lard	65– 80	0.9130–0.9190	—

Name of Oil	*Iodine Value*	*Specific Gravity at 15° C*	*Acidity* (Based on Oleic Acid)
Linseed	171–201	0.9250–0.9380	—
Olive	77.5– 91	0.9140–0.9180	0.3–5
Peach kernel	93–109	0.9180–0.9215	—
Peanut	85–103	0.9110–0.9220	0.8–4
Rape	102.1–106.8	0.9132–0.9180	0.5-7
Sesame	103–114	0.9210–0.9244	0.3-3
Soybean	121–139	0.9194–0.9290	—

PROPERTIES OF FATS

Name of Fat	*Solidifying Point at ° C*	*Name of Fat*	*Solidifying Point at ° C*
Beef marrow	29–31	Goose fat (wild)	18–20
Beef tallow	27–38	Hare fat	17–23
Beeswax	60.5–62.8	Horse fat	20–45
Bone fat	15–17	Human fat	15
Butter fat	20–23	Lard	27.1–29.9
Chicken fat	21–27	Mutton tallow	32–41
Cocoa butter	21.5–23	Rabbit fat (tame)	22–24
Cottonseed stearin	16–22	Rabbit fat (wild)	17–22
Dog fat	20–26	Vegetable tallow (Chinese)	27–35
Egg fat	—	Wool fat (purified)	38–40
Goose fat (domestic)	18–20		

Olive Oil

Olive oil has long been a very important fat in the diet of people of the Mediterranean countries. Large quantities of olive oil are produced in Spain, France, Italy, and, of late, there has been considerable production in America.

Olive oil is pressed from warm ripe olives, which have first been ground between stones. The paste thus formed is put into closely woven baskets, which are then piled one on top of the other in a hydraulic press. Great pressure is then brought to bear upon this ground mass. The first oil that is extracted in this manner is known as virgin or extra virgin olive oil. After the oil has been allowed to run for some time, the baskets are removed, and the pomace softened up considerably with hot water and replaced in the hydraulic press and squeezed again. This second pressing, which yields a slightly inferior grade of oil, is collected. Each subsequent pressing produces oil of lower grade than that obtained from the previous pressing.

Olive oil has a pale yellow or greenish-yellow color and has a specific gravity of 0.914 to 0.918; it solidifies at 2 to 5° C. Oils containing an excessive amount of solid fats solidify at higher temperatures. Olive oil is soluble in ether, chloroform and carbon disulfide. It becomes rancid on long exposure to light. One ton of olives will yield approximately 35 to 45 gallons of oil. The residue or pomace that remains behind after extraction of the oil is utilized for further recovery with various fat solvents such as ether, benzene, chloroform, carbon disulfide and gasoline. In European countries where olive oil is produced, the pomace is dried and then extracted with a volatile solvent. The oil so produced is refined and used for industrial purposes, e.g., soap manufacture.

Cocoanut Oil

This oil resembles butter fat in properties and is used for making artificial butter. It is a white, semi-solid, lard-like fat which becomes liquid at 20 to 25° C. The oil is soluble in the usual solvents such as ether, chloroform, carbon disulfide, etc.

Refined cocoanut oil is a white neutral product with an agreeable flavor and taste and should be stored in a cool moisture-free compartment.

Corn Oil

Corn oil is obtained from the germ of the Indian corn kernel. The grain is softened by allowing it to soak in warm water, and the loosened germs are then crushed to a ground meal. The oil thus obtained is refined, producing edible corn oil. The product has a clear yellow color and has a specific gravity of 0.9213 to 0.9268 at 15° C. Corn oil is extensively used in the manufacture of mayonnaise and oleomargarine.

Cottonseed Oil

Cottonseed yields from 30 to 35% oil. The meat of the seed is pressed between heavy rollers and cooked in steam-jacketed kettles. It is then extracted in a hydraulic press using a pressure of 3000 to 4000 pounds per square inch. About 85% of the oil is extracted and the balance remains behind in the oil cake. This cake is exceptionally rich in fats and proteins and is very valuable as a cattle food. The cottonseed oil, in alkaline solution, has to be refined further with the aid of heat and agitation in large tanks. In this manner, the free fatty acids are removed along with the resinous coloring materials. After the impurities settle out, a clear yellow oil is drawn off, filtered through Fullers' earth and deodorized. The refined oil is known as summer yellow oil. Winter pressed oil is made by chilling the summer yellow and removing the stearin and palmitin which crystallize out. The fat is then sold as cottonseed stearin.

Cottonseed oil is soluble in ether, chloroform and carbon disulfide. It is practically odorless and has a specific gravity of 0.915 to 0.930 at 15° C. It is extensively used in the manufacture of mayonnaise and lard substitutes. The yield from 1 ton of cottonseed is approximately 50 pounds of oil.

Peanut Oil

The oil that is first pressed from the peanut is nearly colorless. That of the second pressing is usually pale greenish-yellow. Peanut oil is used as a salad oil and in manufacturing shortening, oleomargarine and similar compounds. The specific gravity is 0.911 to 0.922 at 15° C.

Sesame Oil

The pressed oil of the sesame is yellow and has a pleasing taste. It can be used as a salad oil and for making butter substitutes.

Soybean Oil

Pressed soybean oil is yellow and has a pleasing taste and odor. It is used as a salad oil and the meal cake, left after expression of the oil, is a valuable cattle food. The soybean is high in oil content and contains very little starch.

Hydrogenation of Oils

Hydrogenation converts liquid oils into solid fats. If olein, e.g., in cottonseed oil, reacts with hydrogen, it is converted to stearin, which is a solid. Hydrogenation is conducted under pressure, at 150 to 250° C, with hydrogen gas and a finely divided nickel or platinum catalyst. The hydrogen used must be exceptionally pure and absolutely free of hydrogen sulfide, arsenic, sulfur, etc. Oils which have peculiar odors lose them, to a considerable degree, during hydrogenation. The melting point of the oil varies with the degree of hydrogenation.

Cereals and Cereal Flours

Wheat

Wheat flour, the most important cereal food, represents the vital part of bread and bakery products. It is exceptionally rich in gluten, the essential ingredient in fermented bread products.

Selective cultivation and cross-breeding have brought about the development of the excellent modern varieties of wheat. Climate, soil, type of seed, and time of planting determine the character of the wheat. Wheat flour and other wheat products supply 20% of the total food used in the United States.

The United States government classification of wheat is as follows: hard red spring, durum, hard red winter, soft red winter, white mixed wheat. Durum wheat, which possesses a high gluten content, yields a yellowish flour that is not suitable for bread making; it is used mainly for macaroni.

Winter and Spring Wheat

Winter wheat is sown during the fall, in localities having a winter climate mild enough to permit the plant to survive.

Spring wheat is sown in the spring in those localities whose climate is too rigorous for winter wheat.

Winter wheat is harvested in the latter part of June and July; spring wheat in late July, August and September. Spring wheat, grown east of the Rocky Mountains, usually represents hard wheat. Winter wheat is classified as soft winter or hard winter.

Hard and Soft Wheat

There are many varieties of both the hard and soft, spring and winter wheat. Hard wheat is relatively hard to the touch, has a high gluten content, is strong in elasticity and quality, and is used in bread making. Soft wheat, on the other hand, is relatively soft to the touch, contains less gluten, and may be very strong or weak in quality. Its greatest use is in the baking of cakes, pastries, doughnuts, etc.

Wheat Flour

Milling of wheat produces various types of flours. Before the actual milling takes place, pre-treatment of the grain is necessary, e.g., screening, scouring, washing and drying. The thoroughly clean wheat is then moistened under properly controlled temperature and moisture conditions.

Straight flour represents all the flour produced from the wheat in the milling process after the removal of the bran and feeds. Patent flour is highly refined flour, is very light in color, and comes from the central part of the wheat grain. Clear flour is the residue left after the patent portion has been removed. The patent and clear portions together represent the straight flour. Clear flour may also be sub-divided into first and second clears, the former being more highly refined and having better color and less ash than the latter.

Red dog shorts, the residue of the kernel, averaging from 3 to 5% of the entire grain, are used as animal feeds. Graham wheat is the entire crushed wheat berry. Whole wheat flour may, or may not, have had part of the bran removed. Blended flours are made by mixing together various grades of wheat. Farina, for example, represents the hard, crushed particles of wheat which are known as middlings; on the other hand, gluten flour is the product resulting from washing out part or all of the starch from wheat flour. The yield of flour secured from any common wheat usually runs 70 to 76% middlings, which are subsequently ground into flours of various grades, and 24 to 30% feed.

STANDARDS FOR ENRICHED FLOUR

Ingredients	*Milligrams per Pound of Flour*	
Required	Minimum	Maximum
Thiamine	2.0	2.5
Niacin	16.0	20.0
Riboflavin	1.2	1.5
Iron	13.0	16.5
Optional		
Calcium	500	625
Vitamin D (U.S.P. units)	250	1000

Rye Flour

There are three grades of rye flour, light, middle and dark. Rye is a species of grain resembling wheat but, generally, having much taller stalks. It is a very hardy grain and will grow in cold regions and in soil that is too poor for wheat. The chief producer is Russia. However, it thrives best under conditions that are also favorable to wheat. Rye bread is made by mixing rye flour with wheat flour, known as Bohemian rye. Rye is also used in very large quantities for the production of whisky.

Corn or Maize

The term *corn* is used generally to designate all the principal grains. In this country, however, it is specifically applied to Indian corn, or maize, the seed of the most beautiful, luxuriant and abundant of all grain grasses. It is native to tropical America and was used as a food by the Indians, centuries before the arrival of Columbus. The corn belt in America includes Iowa, Illinois, Nebraska, Indiana, Missouri, Kansas and Ohio. The other important producing countries are Hungary, Roumania, Italy, Russia, Mexico, Argentina, South Africa and Egypt.

Corn has a lower protein content than wheat and oats but is fully equal in all other respects to the other grains, and, in many cases, the fat or oil content is greater. It does not produce as good a loaf of bread as wheat grain because of its smaller gluten content.

COMPARISON OF THIAMINE, NIACIN, RIBOFLAVIN AND IRON CONTENTS OF VARIOUS BAKERY PRODUCTS

(Figures in Milligrams per Pound)

	Bread			*Sugared Doughnuts (Baking Powder)*		*Coffee Cakes* (Average of various types)*	
	Whole Wheat	Unenriched White	Enriched White (Minimum)	Unenriched Flour Used	Equivalent of Enriched Flour Used	Unenriched Flour Used	Equivalent of Enriched Flour Used
Thiamine	1.35	0.25	1.10	0.45	0.90	0.50	1.00
Niacin	12.00	3.50	10.00	3.00	7.50	5.50	10.00
Riboflavin	0.55	0.35	0.70	0.40	0.70	0.85	1.20
Iron	10.00	3.50	8.00	6.00	11.00	6.00	11.00

* Values for coffee cake are quite variable due to different types of cakes and to different kinds and amounts of fillings.

Corn grain consists of the germ (the oily part), the endosperm (the body of the kernel, mostly starch and some gluten), and the hull, or bran. The germ is used in the manufacture of corn oil and the starch provides other edible manufactured products. There are many varieties of corn, e.g., flint, dent, soft and sweet. The whole kernel minus the germ and hull constitutes the edible part of the corn and is called samp. Coarsely ground corn is called hominy. The coarsely ground product of the whole kernels, bolted to remove the bran, is known as corn meal.

Rice

The chief rice producing countries are India, China, Japan, Spain, Italy, Egypt, Honduras and America.

Type	Characteristics
Japanese	Round and short
Honduras	Long and slender
Blue rose	Medium
Brown rice	Unpolished, without hollows
Rice flour	Ground rice
Rice polish	Consists of the ground rice coating and the leavings of the polished rice; a valuable cattle food

Oats

Oats are harvested and stored in a manner similar to that used for wheat grain. Large acreage is employed for the cultivation of oats in America and Europe where they are popular as a cereal in the form of oatmeal (the ground kernel). Scotch style oatmeal consists of coarsely ground oat grains.

Barley

Barley is used in the manufacture of beer. The kernels are moistened, spread out on the floor, and allowed to sprout under controlled conditions of moisture, temperature and sprout growth. The grain is raised in America, Russia, Austria, Hungary, Chile, Germany and Scandinavia. Polished and

Scotch style barley are called pearl barley and coarsely ground barley, respectively.

Buckwheat

Buckwheat is grown in America, France, Russia and Germany. The flour is used for making buckwheat cakes and puddings.

Starches

Starch, a natural food stuff, is found in cereal grains, roots and tubers, in most vegetables and, to some extent, in fruits. Modern processed starch has a bland or neutral taste, creamy smoothness and translucence. Starches are finding wide-spread use in the manufacture of pies, pastries, sauces and confectionery.

When starch is cooked in water, the small opaque granules swell, becoming almost clear or translucent; prolonged cooking bursts these small granules or cells. The degree of cohesiveness, or shortness, of a hot or cold starch jelly will vary, depending on the kind of raw starch selected. A starch jelly, which maintains the same creaminess or efficiency in the hot and cold states, is said to be stable.

In the preparation of corn starch puddings, the starch acts as the thickening agent. Two ounces of starch are usually sufficient for 1 quart of milk. If a large egg is used in this mixture, about ½ ounce of starch may be omitted. However, if butter or cream is added, the starch content should be increased by at least 10%. On the other hand, when starch is cooked in water, 2 ounces to a quart will form a tender cream jelly, whereas 4 ounces will form a very firm jelly.

In fruit fillings of good quality, containing 50 to 60% solid fruit by weight, the starch content may vary from 2 to 6% of the batch weight, depending on the firmness desired. When the fruit content is lower, there must be a corresponding increase of starch to make up for body, handling quality and shelf life. Lemon pie fillings, which are mixtures of water, acid, starch and eggs, require less starch as the egg content is increased. The consistency of starch jellies depends, more or less, upon other ingredients such as sugar, egg, fat, dry milk, corn syrup, salt, sodium benzoate, etc. For example, if 1 pound of starch is cooked with an average amount of juice or water in the presence of 7 pounds of sugar, the starch does not gel properly, regardless of the time or temperature employed. However, if the same amount of starch and liquid is processed with 3 or 4 times its weight of sugar, and the balance of the required sugar is introduced after the starch is swelled completely, the resulting jelly will be clearer, glossier and will have a satisfactory heavy body. For the same reason as explained above, too much egg, fat or dry milk solids present at the time that the starch is being cooked with water or milk, may inhibit the proper gelling development of the starch granules. When starch is heated for a long period in contact with fruit acids, there is considerable weakening of the gelled product. What happens is that the acid converts some of the starch into sugars. It is advisable that acids be added after the starch has become clear and translucent or, better still, after the batch has been completely cooked.

Starch from Corn Syrup

Starch-making is essentially a process of mechanically separating starch from the other ingredients of the corn. The shelled corn is softened by steaming and is coarsely ground to loosen the germs which are removed by flotation. The mass is then more finely ground. The hulls and fiber are screened out, and the resulting liquor, consisting of gluten and starch suspended in water, is run along narrow tables, where the starch settles to the bottom, and the gluten and water run off. The substance left on the tables is starch which is further purified for marketing as commercial corn starch, or converted into corn syrup, corn sugar or dextrose. The starch tables are long narrow troughs, about 2 feet wide and more than 100 feet long, with sides varying from 12 inches high at the head of the table to 6 and 7 inches high at the foot. They are slightly inclined so that the starch liquor, entering on the higher end, moves very slightly to the other end where it overflows. The starch, being heavier than the gluten, settles to the bottom of the table. For further purification, it is flushed off the table with clean water, refiltered and dried in kilns through which a current of hot air circulates for about 24 hours. Air-dried starch contains from 13 to 18% water, whereas kiln-dried starch is practically moisture-free. However, it readily absorbs from 7 to 10% moisture from the atmosphere.

The form of the starch depends on the time of the drying; if it is completed within a few days, the product is a mixture of starch powder and larger particles containing about 13% moisture. Commercially, this form is known as pearl starch and is the basic starch of the corn products industry. When the drying process is prolonged, sometimes for weeks, the wet mass contracts irregularly and breaks up into distorted prisms or crystals; this product is crystal starch. Powdered starch is prepared by milling pearl starch and running it through a silk screen to remove the over-dried particles. Lump starch, used for laundry purposes, is made from powdered starch by treating it with steam, subjecting it to high pressure, crushing it and screening to size. All these starches are thick boilers that must be boiled to a thick paste. Thin-boiling starches, which are used in the textile industry, are made from ordinary, thick-boiling starch by heating with an acid to reduce the viscosity. The degree of viscosity produced depends upon the strength of the acid, heating temperature, and the duration of treatment.

In the manufacture of corn syrups and sugars, corn starch is suspended in water and heated with dilute hydrochloric acid under pressure. By this treatment, the starch is converted to dextrose, maltose, dextrin and other carbohydrates. The dextrose solution is neutralized with sodium carbonate to remove excess acid, and passed through filter presses and bone-char filters to remove all extraneous substances. Concentration of the liquid after several filtrations produces corn syrup, glucose, which is prepared for the market and sold to candy makers, bakers, etc. Corn sugar is made by allowing a concentrated sugar liquor to crystallize out on large tables. There are several grades of sugar, of different degrees of purification, on the market. When a pure dextrose is desired, a special tank crystallizer is employed; the process requires several days and produces a dextrose from 99½ to 100% pure. Dextrins are produced by roasting dry starch in cookers, usually with the addition of mineral acid. Sometimes no acid is used, however, and the product of this conversion is called British gum. These dextrins are used in the manufacture of adhesives.

CHEMICAL COMPOSITION AND PROPERTIES OF SWEETOSE SYRUP AND CORN SYRUP

	Sweetose Syrup	*Ordinary Corn Syrup*
Moisture	18.50%	19.70%
Sugars	55.73	27.70
Dextrin	25.94	52.35
Ash	0.28	0.25
Total Solids	81.95	80.30
Gravity at 100°F	43°Bé	43°Bé
Dextrose Equivalent	65	42
Sweetness	¾ that of an 82% sucrose solution	⅜ that of 82% sucrose solution

Arrowroot (Maranta)

The true arrowroot is a starch obtained from the rootstock or rhizome of a West Indian plant, Bermuda arrowroot, and two or three related species, largely cultivated in tropical countries. The plant has been introduced into the Gulf states and thrives freely there. The name has been attributed to the Indian practice of using the fresh roots to cure wounds made by poisoned arrows.

When the plants are about a year old, they are dug out of the ground. In Bermuda and Jamaica, they are first washed, the paper-like scale cleaned off, washed again, drained and finally beaten to a pulp in mortars or other machinery. The milky liquid thus obtained is passed through coarse cloths or sieves, washed many times with water, allowed to settle down, and finally dried, leaving the arrowroot of commerce.

High grade arrowroot is a light white powder, odorless when

dry but giving off a very slight odor when mixed with boiling water, and swelling into a smooth jelly when cooked. It is used in biscuits, puddings, jellies and cakes, also with beer, tea, milk, or boiled with a little flavoring added. It is an easily digested food for children and invalids.

Tapioca (Cassava, Manioc)

Tapioca flour or starch comes mainly from Java, although some of it is grown in Santa Domingo, South America and Cuba. The dry sliced roots of the cassava plant are known as gaplek. The ground roots are called gaplek meal. These products are used for food by the natives of Java and may be used for the manufacture of starch. The cassava products are tapioca flour, pearls, seeds, flakes and siftings. The flour is the form generally used for most industrial purposes. The pearls and seeds are in globular form and are used entirely for food. The larger globes are pearls and the smaller ones, seeds. Flakes and siftings are also used chiefly for food.

The starch is extracted from the root in much the same way as it is from potatoes. The starch cell is broken up by mechanical grating and the pulp conveyed, by flowing water, to a sieve through which the liberated starch grains pass, along with other impurities. As soon as the starch settles to the bottom of the settling tank, the liquid at the top is drawn off, and the residue cut into blocks which are put into wooden tubs and purified. The purification process consists of vigorous stirring of the starch, replacing, and allowing it to settle with clean water, which is changed every 24 hours. This continues for about a week, then the starch is removed to drying shelves and, at the right time, broken by hand into small pieces. Tapioca flour is simply made by drying, pulverizing and screening. The flakes and siftings are made by cooking the starch in shallow pans for 2 or 3 minutes to produce a partial dextrinization; the starch is then sieved, spread over a surface and heated to 110 to 115° F. After drying for 24 hours, the product is ready for packing. To make tapioca pearls and seeds, the starch, after being broken up in the drying shed, is sieved and transformed to hammock-like contrivances suspended from the roof. A rocking motion, imparted by hand to these hammocks, causes the starch grains to adhere to one another. The particles are then graded as to size by screening, cooked in the same way as for flakes and siftings, again screened, and finally dried.

Sago Starch or Flour

Sago is a starchy food prepared from the pith of the trunks of several tropical palms, particularly the sago palm of the East Indies. The tree has a hard, outer rind or shell about 1 inch in thickness. The whole inner portion is filled with a soft white substance that has the consistency of cheese, with numerous coarse, rather brittle fibers running through it.

In order to obtain the flour or starch, the palms are cut down and split lengthwise. The pith is removed, powdered and then washed with water to obtain the starch free from fiber and other impurities. This sago meal is then made into a paste and rubbed through sieves of various sizes, producing the small pearl sago, the larger bullet sago, etc. The smaller particles cook more readily.

Sago is valuable as a food, since it is cheap and nutritious; it is prepared in the same manner as rice puddings. The impure sago flour imported to this country is called crude flour, contains woody fibers and cooks up into a dark paste. However, some manufacturers are purifying the product, since it has been recognized for many years that sago has certain properties lacking in the other starches, which make it valuable in industry.

Potato Starch

The tubers or potatoes are ground to a pulp and freed of fiber by water washings. The plant consists chiefly of starch and a very small amount of protein. Large quantities are used, mixed with rye or wheat flour, in the making of bread. It is used for thickening gravies and in the preparation of sausage fillers.

Amioca Starch

This is a new domestic starch obtained from waxy maize. Through laboratory tests and commercial trials carried on by American Maize Products Company, National Starch Products, Inc., and Stein Hall & Company, Inc., it is now confirmed that Amioca is a satisfactory substitute for tapioca flour and other tuber starches in those cases where the properties of tuber starches make them superior to others. This product was developed as a result of the difficulty in obtaining tapioca starch during the war.

Ordinary corn starch gelatinizes at 64°C and continues over a range of 30°C or more, whereas Amioca gelatinizes sharply starting at 70°C and proceeding over a range of only 8°C. This is sufficient under conditions where a minimum of cooking time is desirable. Amioca has a higher viscosity than potato starch when equal parts of water and starch are used. Unlike corn starch pastes, it does not gel on cooling. Tapioca, on cooling to room temperature, becomes very much more viscous than Amioca, which bodies up only slightly. This property of not gelling or not setting may be quite important in some food recipes. Amioca starch contains no amylose, whereas tapioca has 20%, potato 25% and corn starch 27%. Amioca lends itself satisfactorily to the making of the pearls and beads used in the tapioca puddings.

Sweet Potato Starch

The process of manufacturing sweet potato starch is similar to that used for other group starches. The sweet potatoes are washed, ground and screened. The pulp is passed through a de-washing process and is sold for stock food. The starch milk from the screens is passed over tables to recover the starch. After the last traces of color have been removed by bleaching, the starch is dried and pulverized. Sweet potato starch can be prepared in technical form, although at present only the better starch is being marketed.

APPROXIMATE STARCH CONTENT OF COMMERCIAL INGREDIENTS BEFORE REFINEMENT

Ingredient	*Per Cent Starch*	*Ingredient*	*Per Cent Starch*
Arrowroot	23.0	Rye flour (dark)	70.0
Barley flour	77.0	(medium)	75.8
Barley (whole)	73.0	(light)	78.5
Corn flour	77.0	(whole)	75.2
Oatmeal (dry, uncooked)	68.0	Sago flour	84.0
Potatoes (fresh)	19.0	Tapioca (dry)	86.4
Rice (brown)	78.0	Wheat (whole)	72.4
(white)	79.5	(white)	77.2
		(cake)	78.5

Macaroni, Spaghetti, Vermicelli and Noodles

Although macaroni, spaghetti and similar pastes at one time were considered a typical Italian food they have become an important part of the American diet.

The essential point in the manufacture of these pastes is that the meal or semolina, from which they are made, comes from hard, very glutinous wheat, e.g., durum or macaroni wheat. The semolina is moistened with a minimum quantity of boiling water, mixed by machinery until smooth and tough, and worked up in a powerful machine kneader. The dough then goes into the cylinder of a press, under tremendous pressure, and it is slowly pressed out at the bottom of the cylinder through the small holes of a perforated plate. The plate shapes the product, which may have a pin hole in the center, may be hollow or tubular, or may be a solid sphaghetti, etc.

Pure durum wheat, semolina, etc., raw pastes are slightly golden in color and translucent. To determine purity a piece is broken; the fracture should be horny, not starchy. A more conclusive test is made by cooking the substance in water; if the product becomes mushy or sticky, or does not retain its shape, it is probably a flour product or made from a wrong variety of wheat. Flour macaroni and flour spaghetti are inferior products made from wheat flour or a mixture of semolina and wheat flour.

Noodles resemble the other forms of Italian paste described above. However, noodles are made from ordinary flour instead of semolina. If noodles are sold as *egg noodles,* they must contain not less than 5½% egg solids by weight. Plain or *water noodles* are made without the addition of eggs and must be labeled *water noodles.* Any artificial coloring is prohibited by law, and the moisture content must not exceed 13%.

Sugar Products

Sugar, sucrose, is derived from the sugar cane and sugar beet. The refined products are identical in chemical composition, appearance and flavor. The sugar cane yields about 16 to 21% sucrose, whereas the sugar beet yields about 9%. The sugar cane is generally ripe for harvest after 12 to 16 months growth; it is then cut close to the ground just before its flowering time, as the juice content is heaviest at this time. The sugar cane is cut, pressed, and the sucrose dissolved out with water. The sugar beet is treated in approximately the same manner. An average analysis of high grade sugar cane stalks shows about 72% water, 18% sugar and 10% woody and vegetable matter. One hundred tons of cane stalks will yield from 7 to 14 tons of sugar.

TYPES OF COMMERCIAL SUGARS

Cube sugar	Crystallized sugar cubes
Loaf sugar	Crystallized in loaf shapes
Powdered sugar	Coarsely powdered sugars
XXXX Powdered sugar	Very finely pulverized sugar
Confectioners' sugar	Very finely pulverized sugar with 3% added starch
Brown sugar	Cane sugar with small percentage of molasses
Yellow sugar	Cane sugar with traces of invert sugar and coloring matter Nos. 1–15
Other sugars	
Maple	Derived from maple sap
Corn	Derived from corn
Levulose	Derived from fruits

Maltose	Derived from malted barley
Lactose	Derived from milk
Corn syrup	Derived from corn starch
Malt syrup	Derived from malted barley
Invert sugar syrup	Derived from cane sugar by the action of acid
Sugar cane syrup	Derived by evaporating the juice of the sugar cane
Sugar syrup	Made by dissolving cane sugar in water
Rock candy syrup	The uncrystallized syrup in the manufacture of rock candy

Apple Syrup

This is a sugar syrup, produced from apples, that is very sweet and bland with no distinctive color or flavor. Apple juice contains about 11% sugar, a little more than half of which is levulose. Dextrose and sucrose, in about equal amounts, make up the balance of the sugar. The syrup can be produced to contain 75% solids and refined to a very light color with elimination of apple flavor. It is then somewhat similar to commercial invert sugar syrup. The syrup is made from clean sound apples and ground to a juice which is then extracted in a hydraulic press. The pomace can be pressed a second time and the extraction of the second pressing is added to the first. The combined juice is then brought to pH 7.8 to 8.2, and to this, a lime slurry containing 1 pound of hydrated lime per gallon of water, is added. The juice must be kept above 78°F to avoid possible precipitation of a lime-sucrose compound. Stir vigorously during the addition of the lime to prevent localized action of the lime on the invert sugar.

The juice is now heated to 175°F to accelerate the action of the lime on the pectin, and the suspended matter that is formed is removed in a filter press. Roughly, 4 pounds of diatomaceous filter earth per 100 gallons of juice are used. The clear filtered juice is now acidified with 25% sulfuric acid to pH 5 to 5.5 and can be further clarified by filtering again with activated carbon, if desired. The juice is now evaporated in a vacuum pan to about 75% solids, and foaming can be avoided by adding about ⅔ ounce of oil per 100 gallons of juice. A vacuum of 25 inches or better is required, at least during the finishing stages, to avoid darkening and giving the syrup a poor flavor.

Raisin Syrup

The raisins are cut or punctured, and then extracted with hot water at 145 to 150°F several times. The resulting extract is used to extract several successive lots of raisins. Each lot is extracted 4 to 5 times, and the final extraction, in each case, is carried out with water. The extract so obtained may be as high as 40°Brix.

The raisin syrup is now ready for filtering with a filter aid such as Dicalite or Hy Flo. The acidity is then adjusted with calcium carbonate to bring it down to 1/10 of 1%. The liquid and carbonate are then heated to 140 to 150°F, and agitated at this temperature until the reaction is complete. Allow the syrup to cool and settle overnight. Remove the clear liquid and filter using ½ to 1% by weight of activated vegetable charcoal heated, with stirring, to about 140°F. The syrup is now filtered hot with filter aid, and the clear product is concentrated in a stainless steel vacuum pan to 68 to 70°Brix. If calcium tartrate separates out at this concentration, it may be desirable to filter the syrup when it reaches 50 to 55°Brix and then resume concentration in vacuum.

Grape Syrup

In the manufacture of this syrup, crush and press white grapes; then heat the juice with filter aid to 170°F and filter. The acidity of the juice is then brought down to 1/10 of 1% by using a slurry of calcium carbonate. It is now heated to 140°F, cooled quickly, and the sediment allowed to settle overnight. The clear liquid is then removed from the sediment and filtered using a filter aid. Now add to this clear liquid ½ to 1% of a good decolorizing carbon. Heat to 150 to 160°F, with stirring, until sufficient decolorization takes place. Now add filter aid and then filter the product once more. The clear liquid is now concentrated in a stainless vacuum pan to 69 to 70° Brix. It is good practice to allow the concentrated syrup to settle for at least a month to deposit tartrates. Shallow containers are preferable to deep ones. Store in a cool storage to prevent excess darkening.

The use of sulfur dioxide retards darkening in the manufacture of syrups, but the use of an excess may bring about corrosion of tin containers. However, the use of approximately 500 parts per million will have little effect on tin corrosion.

Kernel Sugar Pastes

The bitter principle in apricot and bitter almond kernels is amygdalin, a hydrocyanic glucoside. It can be removed by a number of methods. One method of manufacturing the macaroon paste is to blanch apricot kernels so as to loosen the skins and then rub them off mechanically. The peeled kernels are first ground and then heated with water to 50°C. This results in the breaking down of amygdalin to benzaldehyde and hydrocyanic acid. These compounds can be removed by steam distillation, and the residual meal can be filtered to remove excess water. Cane and corn sugars are added and steam-heated to remove excess water. The hot paste is packed in cans.

Another method for breaking down amygdalin calls for removal of the skins, heating the kernels in water, drying the treated kernels to expel the benzaldehyde and hydrocyanic acid and then grinding kernels and sugar together into a fine paste.

The removal of bitterness can be hastened by heating the kernel to 140° F in running water for about 24 hours. The kernels are then dried at 140 to 145°F. Dilute alkali may also be used up to 1.5%.

Fruit Essences

Fruit essences are synthetic mixtures made to simulate, as closely as possible, the aroma of pure fruits and their juices. They are composed chiefly of ethers, esters, volatile oils and odorous substances. The fruit ethers are an intermediate product in reinforcing the final flavor with alcohol, glycerin, sugar and other flavor carriers. The following formulas are to be considered as fruit flavor essences and will be referred to as such in all formulas calling for these ingredients.

Apricot Flavor Essence

Formula No. 1

True Jasmine Oil	1/6	oz
Ethyl Butyrate	10	oz
Ethyl Valerianate	5	oz
Ethyl Salicylate	2	oz
Amyl Butyrate	1	oz
Pettigrain Oil	1½	oz
Amyl Alcohol	½	oz
Peach Flavor Essence	2	oz
Alcohol	16	oz

Formula No. 2

Anisic Aldehyde	1/3	oz
Aldehyde C_{14}	3	oz
Amyl Acetate	5	oz
Iris Concrete (10% Solution)	1¾	oz
Bitter Almond Essence	1¾	oz
Ionone (10% Solution)	1/3	oz
Jasmine Absolute (Solution)	1/30	oz
Vanillin	1¾	oz
Ethyl Caproate	1¾	oz
Ethyl Acetate	3	oz
Ethyl Caprylate	2	oz
Amyl Butyrate	1¾	oz
Amyl Valerate	1¾	oz
Eugenol	1/6	oz
Alcohol	16	oz

Apple Flavor Essence

Formula No. 1

Ethyl Nitrite	1	oz
Acetic Aldehyde	¼	oz
Ethyl Acetate	1	oz
Amyl Valerianate	10	oz
Cognac Oil	¼	oz
Ethyl Butyrate	12	oz
Amyl Acetate	12	oz
Alcohol	16	oz

Formula No. 2

Apple Flavor Essence	20	oz
Isobutyl Valerianate	3	oz
Ethyl Butyrate	2½	oz
Methylethylphenyl Glycidate	1/3	oz
Triacetin	20	oz

Formula No. 3

Cyclohexanyl Acetate	1½	oz
Cyclohexanyl Valerianate	1/3	oz
Amyl Valerianate	2/3	oz
Amyl Formate	1/10	oz
Geranyl Acetate	2/3	oz
Geraniol	½	oz
Alcohol	16	oz

Almond Flavor Essence (Bitter)

Benzaldehyde	5	oz
Benzyl Alcohol	30	oz
Synthetic Neroli Oil	1/30	oz
Bitter Almond Oil	1/3	oz
Vanillin	1/6	oz
Alcohol	16	oz

Butter Flavor Essence

Formula No. 1

Diacetyl	6	oz
Butyl Acetate	½	oz

Ethyl Butyrate	⅓	oz
Butyl Butyrate	⅙	oz
Triacetin	1½	oz

Formula No. 2

Ethyl Pelargonate	⅕	oz
Ethyl Butyrate	¼	oz
Butyl Butyrate	⅕	oz
Butyl Acetate	½	oz
Diacetyl	5	oz
Triacetin	2	oz

Banana Flavor Essence

Ethyl Butyrate	10	oz
Amyl Acetate	10	oz
Pettigrain Oil	1	oz
Benzyl Benzoate	2	oz
Orange Oil	1	oz
Amyl Butyrate	10	oz
Alcohol	16	oz

Butterscotch Flavor Essence

Ethyl Oenanthate	6	oz
Ethyl Pelargonate	12	oz
Butyl Butyrate	16	oz
Amyl Acetate	16	oz
Ethyl Butyrate	16	oz
Butyric Acid	16	oz
Orange Oil	½	oz
Alcohol	16	oz

Cherry Flavor Essence

Formula No. 1

Ethyl Benzoate	5	oz
Ethyl Acetate	10	oz
Benzyl Benzoate	1	oz
Cognac Oil	½	oz
Vanillin	½	oz
Amyl Alcohol	1	oz
Alcohol	16	oz
Bitter Almond Oil	2	oz

Formula No. 2

Synthetic Neroli Oil	⅛	oz
Vanillin	1	oz
Clove Oil	6	oz
Cinnamon Oil	2	oz
Cognac Oil	1½	oz
Amyl Alcohol	1½	oz
Benzaldehyde (Bitter Almond Oil)	4½	oz
Ethyl Pelargonate	2	oz
Benzyl Benzoate	8	oz
Ethyl Oenanthate	20	oz
Ethyl Acetate	64	oz
Alcohol	16	oz

Cherry Flavor Essence (Sour)

Eugenol	½	oz
Cognac Oil	1	oz
Cinnamic Aldehyde	1	oz
Benzoic Aldehyde	9	oz
Coumarin	2	oz
Ethyl Caprylate	2	oz
Amyl Acetate	8	oz
Ethyl Acetate	8	oz
Bourbonal	5	oz
Black Current Flavor Essence	½	oz
Rose Essence	½	oz
Sweet Orange Essence	5	oz
Lemon Essence	40	oz
Iris Essence (1 to 70)	5	oz
Vanillin	3	oz
Amyl Formate	5	oz
Amyl Butyrate	10	oz
Alcohol	16	oz

Cacao Flavor Essence

Clove Oil	3	oz
Cinnamon	3	oz
Vanillin	6	oz
Benzyl Alcohol	8	oz
Ethyl Oenanthate	12	oz
Amyl Butyrate	20	oz
Amyl Acetate	32	oz
Ethyl Acetate	40	oz
Alcohol	16	oz

Coffee Flavor Essence

Ethyl Acetate	20	oz
Ethyl Formate	20	oz
Ethyl Pelargonate	10	oz
Benzyl Benzoate	7	oz
Lemon Oil	2	oz
Cognac Oil	½	oz
Synthetic Neroli Oil	1/12	oz
Alcohol	16	oz

Cognac Flavor Essence

Amyl Alcohol	5	oz
Oenanthic Ether	15	oz
Alcohol	16	oz

Currant Flavor Essence (Red)

Formula No. 1

Amyl Acetate	4	oz
Isobutyl Acetate	14	oz
Amyl Butyrate	18	oz
Lemon Oil	8	oz
Orange Oil	12	oz
Pettigrain Oil	½	oz
Benzyl Acetate	½	oz
Cinnamon Oil	¼	oz
Synthetic Neroli Oil	1/30	oz
Ethyl Acetate	4	oz
Alcohol	16	oz

Formula No. 2

Cyclohexanyl Valerianate	90	oz
Ethyl Benzoate	2	oz
Propyl Propionate	2	oz
Amyl Acetate	1	oz
Propyl Acetate	1	oz
Alcohol	16	oz

Formula No. 3

Cyclohexanyl Butyrate	50	oz
Methyl Cinnamate	40	oz
Aldehyde C_{16}	½	oz
Oleoresin Vanilla	1	oz
Oil of Portugal	3	oz
Oil of Pettigrain	1	oz
Orris	½	oz
Alcohol	16	oz

Date Flavor Essence

Honey Flavor Oil	½	oz
Eugenol	¼	oz
Peach Flavor Essence	1	oz
Cinnamon Oil	½	oz
Synthetic Neroli	¼	oz
Amyl Butyrate	5	oz
Ethyl Benzoate	10	oz
Amyl Valerate	10	oz
Amyl Acetate	18	oz
Ethyl Acetate	20	oz
Alcohol	16	oz

Fig Flavor Essence

Fig Extract (Alcoholic)	2	lb
Peach Flavor Essence	1/16	oz

Raspberry Flavor Essence	½	oz
Peppermint Oil	1	drop

Ginger Flavor Essence

Pettigrain Oil	4	oz
Cardamom Oil	4	oz
Clove Oil	2	oz
Galanga Oil	1/39	oz
Ginger Oil	12	oz
Ethyl Oenanthate	20	oz
Amyl Butyrate	30	oz
Ethyl Acetate	20	oz
Alcohol	16	oz

Gooseberry Flavor Essence

Coumarin	1	oz
Ethyl Acetate	12	oz
Amyl Acetate	8	oz
Ethyl Butyrate	8	oz
Ethyl Benzoate	1	oz
Fennel Oil	⅛	oz
Synthetic Neroli Oil	1/30	oz
Alcohol	16	oz

Grape Flavor Essence

Formula No. 1

Ethyl Formate	2	oz
Ethyl Oenanthate	10	oz
Methyl Anthranilate	10	oz
Alcohol	16	oz

Formula No. 2

Green Cognac Oil	5	oz
Methyl Anthranilate	2	oz
Ethyl Cinnamate	3	oz
Propyl Cinnamate	3	oz
Ethyl Butyrate	½	oz
Alcohol	16	oz

Grenadine Flavor Essence

Formula No. 1

Benzaldehyde	2	oz
Pettigrain Oil	½	oz
Peach Flavor Essence	6	oz
Amyl Valerate	6	oz
Amyl Acetate	3	oz
Ethyl Acetate	2	oz
Benzoyl Benzoate	4	oz
Oil of Cloves	¼	oz
Cinnamon Oil	¼	oz
Alcohol	16	oz

Formula No. 2

Peach Flavor Essence	2	oz
Strawberry Flavor Essence	6	oz
Vanillin	1/30	oz
Alcohol	16	oz

Formula No. 3

Cyclohexanyl Valerianate	3	oz
Amyl Acetate	1½	oz
Ethyl Butyrate	1	oz
Ethyl Benzoate	1	oz
Amyl Butyrate	1	oz
Ethyl Salicylate	1	oz
Vanillin	1/10	oz
Alcohol	16	oz

Honey Flavor Essence

Amyl Valerate	15	oz
Pelargonic Ether	1	oz
Ethyl Acetate	50	oz
Phenyl Ethyl Acetate	4	oz
Ethyl Butyrate	2	oz
Alcohol	16	oz

Lemon Flavor Essence

Lemon Oil	24	oz
Orange Oil	8	oz
Amyl Acetate	6	oz
Amyl Butyrate	16	oz
Ethyl Acetate	4	oz
Alcohol	16	oz

Lime Flavor Essence

Lime Oil	16	oz
Lemon Oil	16	oz
Peppermint Oil	¼	oz
Amyl Butyrate	4	oz
Amyl Acetate	4	oz
Alcohol	16	oz

Mulberry Flavor Essence (Blackberry)

Ethyl Acetate	12	oz
Ethyl Butyrate	24	oz
Ethyl Formate	3	oz
Amyl Acetate	2	oz
Benzaldehyde	1/16	oz
Ethyl Pelargonate	¼	oz
Ethyl Oenanthate	¾	oz
Benzyl Acetate	1	oz
Alcohol	16	oz

Melon Oil Flavor Essence

Ethyl Formate	1	oz
Ethyl Butyrate	3	oz
Amyl Butyrate	2	oz
Vanillin	¼	oz
Lemon Oil	1	oz
Ethyl Valerianate	4	oz
Acetaldehyde	½	oz
Alcohol	16	oz

Nut Flavor Essence

Butter Flavor Essence	1/16	oz
Mace Oil	1½	oz
Bitter Almond Oil	½	oz
Ethyl Oenanthate	3	oz
Vanillin	1/15	oz
Orange Oil	¼	oz
Clove Oil	1/30	oz
Anise Aldehyde	1/16	oz
Alcohol	16	oz

Orange Flavor Essence (Sweet)

Lemon Oil	4	oz
Amyl Acetate	4	oz
Ethyl Acetate	2	oz
Sweet Orange Oil	12	oz
Amyl Butyrate	12	oz
Alcohol	16	oz

Orange Flavor Essence (Bitter)

Bitter Orange Oil	12	oz
Amyl Butyrate	8	oz
Pimenta Oil	⅛	oz
Pettigrain Oil	2	oz
Amyl Acetate	3	oz
Ethyl Acetate	1	oz
Alcohol	16	oz

Peach Flavor Essence

Formula No. 1

Benzaldehyde	2	oz
Ethyl Valerianate	2	oz
Amyl Alcohol	¼	oz
Pettigrain Oil	¼	oz
Pimenta Oil	¼	oz
Sweet Orange Oil	1	oz

Vanillin	1/4	oz
Lemon Oil	1	oz
Amyl Butyrate	1	oz
Ethyl Butyrate	2	oz
Amyl Acetate	4	oz
Alcohol	16	oz

Formula No. 2

Cyclohexanyl Butyrate	30	oz
Ethyl Cinnamate	7	oz
Benzyl Butyrate	7	oz
Aldehyde C_{14}	1	oz
Isobutyl Salicylate	1	oz
Amyl Butyrate	3	oz
Geranyl Formate	1/2	oz
Alcohol	16	oz

Pear Flavor Essence

Formula No. 1

Amyl Acetate	32	oz
Ethyl Acetic	10	oz
Orange Oil	2	oz
Bergamot Oil	1	oz
Vanillin	1/2	oz
Pettigrain Oil	1/2	oz
Oil of Cloves	1/16	oz
Alcohol	16	oz

Formula No. 2

Geraniol	1/2	oz
Propyl Propionate	1/4	oz
Ethyl Cinnamate	5	oz
Butyl Acetate	1	oz
Geranyl Butyrate	2	oz
Amyl Formate	1/2	oz
Alcohol	16	oz

Pineapple Flavor Essence

Amyl Butyrate	8	oz
Ethyl Butyrate	5	oz
Ethyl Acetate	2	oz
Amyl Acetate	10	oz
Ethyl Pelargonate	6	oz
Ethyl Sebacate	2	oz
Butyric Acid	1/20	oz
Lemon Oil	1/20	oz
Vanillin	1/20	oz
Alcohol	16	oz

Pistachio Flavor Essence

Pimento Oil	1/2	oz
Mandarin Oil	1/10	oz
Benzyl Benzoate	1/10	oz
Butyric Acid	1/30	oz
Heliotropin	1/5	oz
Benzaldehyde	1/2	oz
Peach Flavor Essence	1/4	oz
Nutmeg Oil	1/4	oz
Orange Oil	1/4	oz
Amyl Acetate	1/2	oz
Ethyl Butyrate	1	oz
Ethyl Acetate	2	oz
Ethyl Oenanthate	1/2	oz
Alcohol	16	oz

Plum Flavor Essence

Formula No. 1

Amyl Butyrate	3	oz
Amyl Formate	3	oz
Ethyl Formate	2	oz
Ethyl Acetate	1	oz
Benzaldehyde	1	oz
Amyl Alcohol	1/2	oz
Coriander Oil	1/3	oz
Clove Oil	1/6	oz
Mandarin Oil	1/5	oz
Lemon Oil	1/20	oz
Alcohol	16	oz

Formula No. 2 (Green Gage)

Ethyl Acetate	1	oz
Amyl Butyrate	2	oz
Benzyl Acetate	1/10	oz
Ethyl Butyrate	1	oz
Ethyl Oenanthate	1	oz
Benzaldehyde	1/2	oz
Ethyl Pelargonate	1/3	oz
Orange Oil	1/5	oz
Vanillin	1/20	oz
Cognac Oil	1/10	oz
Amyl Alcohol	1/10	oz
Cinnamon Oil	1/20	oz
Clove Oil	1/40	oz
Rose Oil	Trace	
Geranium Oil	1/50	oz
Alcohol	16	oz

Prune Juice Flavor Essence

Ethyl Acetate	3	oz
Ethyl Benzoate	2	oz
Ethyl Oenanthate	1	oz
Amyl Alcohol	1/2	oz
Amyl Acetate	1/6	oz
Amyl Butyrate	1/6	oz
Almond Oil	1/3	oz
Cinnamon Oil	1/10	oz
Clove Oil	1/10	oz
Vanillin	1/40	oz
Alcohol	16	oz

Pomegranate Flavor Essence

Tincture Coccionella	1/10	oz
Orange Oil	1/10	oz
Clove Oil	1/10	oz
Vanilla Extract	1/2	oz
Tincture of Ginger	1/3	oz
Maraschino Essence	3	oz
Alcohol	16	oz

Raspberry Flavor Essence

Formula No. 1

Terpeneless Oil of Cloves	5	oz
Orris Liquid (1% Solution)	1	oz
Cinnamon Oil	5	oz
Bitter Almond Oil	2	oz
Rose Oil	2	oz
Phenyl Ethyl Alcohol	50	oz
Benzyl Acetate	40	oz
Synthetic Neroli	2	oz
Peru Balsam	10	oz
Coumarin	1/4	oz
Vanillin	1	oz
Propyl Acetate	10	oz
Amyl Acetate	2	oz
Ethyl Acetate	1/2	oz
Alcohol	500	oz

Formula No. 2

Amyl Acetate	6	oz
Amyl Butyrate	1 1/2	oz
Synthetic Rose	1/4	oz
Methyl Benzoate	1/2	oz
Nitrous Ether	2	oz
Ethyl Acetate	6	oz
Ethyl Butyrate	1	oz
Amyl Valerate	1 1/2	oz
Methyl Cinnamate	1/2	oz
Phenyl Ethyl Alcohol	2	oz
Alcohol	16	oz

Formula No. 3

Isobutyl Acetate	12	oz
Amyl Acetate	8	oz
Ethyl Acetate	5	oz
Synthetic Rose Oil	1/64	oz
Jasmine Oil (10%)	1/16	oz
Cinnamon Oil	1/16	oz

Clove Oil	1/5 oz
Pettigrain Oil	1/5 oz
Vanillin	1/4 oz
Orris Oil (10% Solution)	1/4 oz
Benzyl Benzoate	1 oz
Ethyl Formate	1 oz
Amyl Butyrate	1/6 oz
Methyl Salicylate	1/6 oz
Ethyl Butyrate	1 oz
Ethyl Acetate	1 oz
Benzyl Acetate	1 oz
Amyl Butyrate	1 oz
Alcohol	16 oz

Formula No. 3

Ethyl Cinnamate	5 oz
Cyclohexanyl Butyrate	5 oz
Isobutyl Valerianate	1/5 oz
Methyl Benzoate	1/15 oz
Oleoresin Vanilla	1/60 oz
Alcohol	16 oz

Formula No. 4

Cyclohexanyl Butyrate	9 oz
Amyl Formate	1/5 oz
Ethyl Benzoate	1/5 oz
Amyl Acetate	1/4 oz
Aldehyde C_{14}	1/10 oz
Vanillin	1/40 oz
Alcohol	16 oz

Strawberry Flavor Essence

Formula No. 1

Yara Yara (10% Solution in Benzyl Benzoate)	1 oz
Amyl Acetate	3 oz
Aldehyde C_{14}	1½ oz
Iris Concrete (10% Solution)	3/4 oz
Bitter Almond Flavor Essence	1/20 oz
Rose Essence	1/20 oz
Jasmine Essence	1/100 oz
Ionone	1/100 oz
Bourbonal	1/12 oz
Alcohol	16 oz

Formula No. 2

Amyl Formate	½ oz
Cinnamon Oil	1/30 oz
Coumarin	1/20 oz
Synthetic Neroli	1/3 oz
Ethyl Benzoate	½ oz

Tangerine Flavor Essence

Lemon Oil	1 oz
Bitter Orange Oil	1 oz
Amyl Acetate	1 oz
Ethyl Acetate	½ oz
Mandarin Oil	2 oz
Amyl Butyrate	3 oz
Alcohol	16 oz

Universal Flavor Essence

Amyl Butyrate	4 oz
Ethyl Formate	12 oz
Amyl Acetate	12 oz
Ethyl Acetate	20 oz

This formula is generally considered to constitute a neutral flavor. It is used for making up flavor compositions not previously mentioned and serves as a supplementary flavor essence.

Walnut Flavor Essence

Formula No. 1

Cardamom Oil	1/50	oz
Lemon Oil	1/10	oz
Anisaldehyde	1/10	oz
Butyric Acid	1/30	oz
Clove Oil	1/40	oz
Orange Oil	1/6	oz
Benzaldehyde	1/6	oz
Vanillin	1/30	oz
Nutmeg Oil	3/4	oz
Amyl Valerate	3	oz
Ethyl Oenanthate	3	oz
Ethyl Acetate	3	oz
Alcohol	16	oz

Formula No. 2

Clove Oil	1	oz
Cardamom Oil	1/4	oz
Lemon Oil	1/2	oz
Anisette Oil	1/2	oz
Butyric Acid	3/4	oz
Orange Oil	2	oz
Benzaldehyde	2	oz
Vanillin	1	oz
Mace Oil	4	oz
Ethyl Oenanthate	10	oz
Amyl Valerate	10	oz
Ethyl Acetate	20	oz
Alcohol	16	oz

ARTIFICIAL FRUIT, FOOD AND SOFT DRINK FLAVORS

The artificial flavors that follow are used for flavoring bakery goods, condiments, soft drinks and various other food products.. Flavors should be used sparingly, since they are very aromatic, odorous substances. The amount to be used per unit of food or liquid must be decided upon after experimentation. Oil-soluble food colors may be added to flavors, if desired.

Fruit Flavors

Apple Flavor

Apple Flavor Essence	1	oz
Orange Flavor Essence	1	oz
Alcohol	6	oz

Apricot Flavor

Apricot Flavor Essence	4	oz
Peach Flavor Essence	1/4	oz
Benzaldehyde	1/16	oz
Oenanthic Ether	1/16	oz
Aldehyde C_{14}	1/30	oz
Alcohol	6	oz

Banana Flavor

Banana Flavor Essence	1	oz
Lemon Flavor Essence	1/4	oz
Orange Flavor Essence	1/4	oz
Alcohol	6	oz

Bitter Almond Flavor

Bitter Almond Oil	1	oz
Alcohol	6	oz

Biscuit and Cake Flavor

Benzaldehyde	1	oz
Clove Oil	1/10	oz
Vanillin	1/10	oz
Alcohol	6	oz

Bitter Orange Flavor

Lemon Flavor Essence	2	oz
Bitter Orange Flavor Essence	2	oz
Orange Flavor Essence	6	oz
Alcohol	6	oz

Blackberry Flavor

Raspberry Flavor Essence	1	oz
Currant Flavor Essence	1	oz
Lemon Flavor Essence	1/3	oz
Blackberry Flavor Essence	3	oz
Alcohol	6	oz

Butter Flavor

Butyric Acid	1	oz
Ethyl Butyrate	3/4	oz
Amyl Butyrate	1	oz
Alcohol	6	oz

Butterscotch Flavor

Butterscotch Flavor Essence	4	oz
Benzaldehyde	1/5	oz
Alcohol	6	oz

Caramel Flavor

Coffee Extract	4	oz
Maple Flavor	2	oz
Vanillin	1/8	oz
Coumarin	1/30	oz
Brown Food Color	Enough	
Alcohol	6	oz

Cherry Flavor

Benzaldehyde	1	oz
Clove Oil	1/8	oz
Cinnamon Oil	1/8	oz
Cherry Flavor Essence	12	oz
Alcohol	6	oz

Cocoanut Flavor

Walnut Flavor Essence	6	oz
Cocoanut Flavor Essence	32	oz
Butyric Acid	2	oz
Orange Oil	1	oz
Benzaldehyde	1/8	oz
Mace Oil	1/4	oz
Clove Oil	1/2	oz
Alcohol	16	oz

Black Currant Flavor

Currant Flavor Essence	7	oz
Black Currant Essence	32	oz
Synthetic Neroli	1/16	oz
Vanillin	1/8	oz
Alcohol	6	oz

Cream Soda Flavor (Vanilla)

Vanillin	3	oz
Coumarin	1	oz
Alcohol	6	oz

Coffee Cake Flavor

Coffee Extract	1/2	oz
Orange Oil	1/4	oz
Lemon Oil	1/4	oz

Benzaldehyde	1/30	oz
Cassia Oil	1/16	oz
Alcohol	6	oz

Fig Flavor

Peach Flavor Essence	1/4	oz
Raspberry Flavor Essence	2	oz
Peppermint Oil	1/60	oz
Vanillin	1/60	oz
Date Flavor Essence	6	oz
Fig Flavor Essence	12	oz
Alcohol	6	oz

Fruity Cake Flavor

Pineapple Flavor Essence	1	oz
Strawberry Flavor Essence	1	oz
Cherry Flavor Essence	1	oz
Prune Flavor Essence	1	oz
Alcohol	6	oz

Grape Flavor

Currant Flavor Essence	4	oz
Pineapple Flavor Essence	5	oz
Grape Flavor Essence	6	oz
Alcohol	6	oz

Pear Flavor

Vanillin	1/3	oz
Coumarin	1/10	oz
Bergamot Oil	1/4	oz
Lemon Flavor Essence	1	oz
Currant Flavor Essence	1	oz
Pear Flavor Essence	3	oz
Alcohol	6	oz

Grenadine Flavor

Vanillin	1/10	oz
Strawberry Flavor Essence	1	oz
Grenadine Flavor Essence	2	oz
Raspberry Flavor Essence	4	oz
Peach Flavor Essence	4	oz
Alcohol	6	oz

Ginger Cake Flavor

Apricot Flavor Essence	1	oz
Cardamom Oil	2	oz
Clove Oil	1	oz
Ginger Oil	3	oz
Lemon Oil	3	oz
Alcohol	6	oz

Green Gage Flavor

Green Gage Flavor Essence	2	oz
Eugenol	1/16	oz
Coumarin	1/16	oz
Alcohol	6	oz

Honey Flavor

Honey Flavor Essence	2	oz
Raspberry Flavor Essence	1/8	oz
Rose Water	2	oz
Alcohol	6	oz

Lemon Flavor

Lemon Oil	1/4	oz
Lemon Flavor	4	oz
Alcohol	6	oz

Lime Flavor

Lime Oil	½	oz
Lime Flavor Essence	4	oz
Alcohol	6	oz

Macaroon Tart Flavor

Benzaldehyde	3	oz
Peach Flavor Essence	½	oz
Orange Oil	1/16	oz
Alcohol	6	oz

Maraschino Flavor

Maraschino Flavor Essence	3	oz
Universal Flavor Essence	⅓	oz
Alcohol	6	oz

Melon Flavor

Vanillin	1/10	oz
Orange Flavor Essence	1	oz
Melon Flavor Essence	3	oz
Alcohol	6	oz

Mint Jelly Flavor

Peppermint Oil	3	oz
Orange Oil	⅙	oz
Alcohol	6	oz

Nectarine Flavor

Peach Flavor Essence	1	oz
Plum Flavor Essence	¼	oz
Vanillin	1/10	oz
Alcohol	6	oz

Peach Flavor

Apple Flavor Essence	2	oz
Peach Flavor Essence	12	oz
Vanillin	⅛	oz
Alcohol	6	oz

Pineapple Flavor

Pineapple Flavor Essence	1	oz
Alcohol	6	oz

Pistachio Flavor

Cognac Oil	⅛	oz
Benzaldehyde	¼	oz
Vanillin	1/30	oz
Pistachio Flavor Essence	2	oz
Peach Flavor Essence	2	oz
Butyric Acid	⅛	oz
Alcohol	6	oz

Plum Flavor

Heliotropin	1/16	oz
Plum Flavor Essence	1	oz
Alcohol	6	oz

Raspberry Flavor

Raspberry Flavor Essence	12	oz
Currant Flavor Essence	6	oz
Alcohol	6	oz

Strawberry Flavor

Strawberry Flavor Essence	4	oz

Raspberry Flavor Essence	1	oz
Alcohol	6	oz

Walnut Flavor

Formula No. 1

Walnut Flavor Essence	2	oz
Coumarin	1/16	oz
Lovage Oil	1/30	oz
Alcohol	6	oz

Formula No. 2

Mace Oil	½	oz
Benzaldehyde	⅓	oz
Vanillin	⅛	oz
Lemon Oil	½	oz
Butyric Acid	¼	oz
Peach Flavor Essence	¾	oz
Nut Flavor Essence	2	oz
Alcohol	6	oz

Spicy Flavors

Spice Meat Extract

Mace Extract	¾	oz
Clove Extract	⅛	oz
Thyme Extract	¾	oz
Basil Extract	¾	oz
Shallott Extract	7	oz
Celery Extract	4	oz

Spice Fish Extract

Onion Extract	4	oz
Pepper Extract	1	oz
Pimenta Extract	1	oz
Ginger Extract	⅛	oz
Clove Extract	⅛	oz
Bay Laurel Extract	¼	oz

Spice Sausage Extract

Savory Extract	1	oz
Pimenta Extract	2	oz
Pepper Extract	5	oz
Onion Extract	6	oz
Bay Laurel Extract	⅛	oz

Spice Smoked Meat Extract

Savory Extract	1	oz
Cardamom Extract	½	oz
Basil Extract	2	oz
Pepper Extract	4	oz
Garlic Extract	8	oz

Spice Pickle Extract

Celery Extract	1½	oz
Clove Extract	¾	oz
Dill Extract	1¼	oz
Pimenta Extract	2½	oz
Capsicum Extract	3	oz
Pepper Extract	4	oz

Spice Curry Extract

Tumeric Extract	2	oz
Ginger Extract	2	oz
Pepper Extract	4	oz

Spice Tarragon Extract

Tarragon Extract	6	oz
Celery Extract	1½	oz

Spice Mixed Pickle Extract

Tarragon Extract	½ oz
Dill Extract	1 oz
Pimenta Extract	2 oz
Capsicum Extract	2 oz
Caper Extract	3 oz
Shallott Extract	3 oz

Spice Mustard Pickle Extract

Dill Extract	1 oz
Pepper Extract	¾ oz
Celery Extract	½ oz
Capsicum Extract	½ oz
Onion Extract	½ oz
Cardamom Extract	½ oz
Volatile Mustard Oil	1/30 oz

Spicy culinary extracts are generally manufactured by dissolving ½ to 1½ ounces of the essential (spice) oil in about 12 ounces of pure alcohol, and then adding 2 to 3 ounces of water. Extracts such as garlic, onion, celery, etc., are best manufactured by allowing the latter to soak in dilute alcohol, and then drawing off the clear liquid for use as a flavor.

Spice salts are made by simply rubbing the salt and the spice extracts together and then air drying.

Spice Vinegar Flavors

Tarragon Vinegar

Tarragon Oil	1 oz
Cognac Flavor	⅛ oz
Alcohol	16 oz

Fruit Vinegar

Cognac Flavor	⅛ oz
Strawberry Flavor Essence	1 oz
Raspberry Flavor Essence	3 oz
Ethyl Acetate	½ oz
Alcohol	16 oz

Mustard Vinegar

Mustard Oil Volatile	⅛ oz
Pimenta Oil	¼ oz
Peppermint Oil	1/30 oz
Tarragon Oil	1 oz
Alcohol	16 oz

Wine Vinegar

Cognac Flavor	¼ oz
Celery Oil	½ oz
Ethyl Acetate	¾ oz
Alcohol	16 oz

Malt Vinegar

Cognac Flavor	⅛ oz
Ethyl Acetate	¼ oz
Cardamom Oil	1/30 oz
Alcohol	16 oz

Universal Vinegar Essence

Tarragon Oil	1	oz
Celery Oil	½	oz
Thyme Oil	½	oz
Pimenta Oil	⅛	oz
Clove Oil	1/30	oz
Alcohol	16	oz

Celery Vinegar

Tarragon Oil	½	oz
Celery Seed Oil	1½	oz
Alcohol	16	oz

POLISHES FOR CHOCOLATE-COATED CANDIES

Where the food laws permit, lacquer coatings are used on chocolate-coated candies. The candy takes on a high sheen, and has a better appearance; the coating also retards sticking, especially during the summer months. The simplest way of using lacquer coatings is to sponge the polish on the candies while they are being rotated in a chocolate pan. The lacquer dries rather quickly, leaving a polished surface. The pan is equipped with hot and cold air controls, and the inside is always coated with a thin layer of carnauba wax.

Shellac Lacquer Coating
Formula No. 1

Edible White Shellac	1	lb
Pure Alcohol	7	pt
Ethyl Acetate	1	pt

Formula No. 2

Gum Benzoin	1	lb
Pure Alcohol	8	pt

Formula No. 3

Edible Shellac	1	lb
Gum Sandarac	1	lb
Pure Alcohol	8	pt
Soluble Brown Dye	1	oz

Formula No. 4

Gum Mastic	4	oz
Edible Shellac	1	lb
Brown Dye	¼	oz
Gum Sandarac	1	lb
Alcohol	8	pt

SALT

Salt is essentially sodium chloride; any other ingredients are incidental impurities. The four chief methods of obtaining it are: 1 by evaporation of the brine from wells sunk into salt beds, 2 by mining, 3 by evaporation of sea water, and 4 by evaporation of the brine of natural springs. The chief producing states are Michigan, New York, Ohio and Kansas.

Mined salt is taken out of the beds in a dry state and is known as rock salt. It is ground in crushers, sifted and refined into numerous grades for marketing. Salt, to which a very small percentage of potassium iodide has been added, is called iodized salt.

COMPOSITION OF TABLE SALT

Ingredient	*Per Cent*	*Ingredient*	*Per Cent*
Sodium chloride	98.25	Magnesium chloride	0.10
Insoluble residue	0.08	Sodium sulfate	0.26
Calcium sulfate	1.31		

Salt is a condiment that is necessary to the human, and the animal, diet. It gives zest to foods, and diets deficient in salt are unpalatable. It is used for brining of pickles, olives and many other foods. The instrument used for measuring salt concentrations is the salometer. While this is the method generally used, the Baumé hydrometer can also be used and the following table shows the Baumé degrees as compared with salometer degrees.

SODIUM CHLORIDE BRINE TABLE

Salometer Reading 60° F	*Baumé Reading* 60° F	*Per Cent Salt in Brine by Weight*
0	0.0	0.000
2	0.6	0.528
4	1.0	1.056
6	1.6	1.584
8	2.1	2.112
10	2.7	2.640
12	3.3	3.167
14	3.7	3.695
16	4.2	4.223
18	4.8	4.751

Salometer Reading 60° F	*Baumé Reading* 60° F	*Per Cent Salt in Brine by Weight*
20	5.3	5.279
22	5.8	5.807
24	6.4	6.335
26	6.9	6.863
28	7.4	7.391
30	7.9	7.919
32	8.5	8.446
34	9.0	8.974
36	9.5	9.502
38	10.0	10.030
40	10.5	10.558
42	11.0	11.086
44	11.5	11.614
46	12.0	12.142
48	12.5	12.670
50	12.9	13.198
52	13.4	13.725
54	13.9	14.253
56	14.4	14.781
58	14.8	15.309
60	15.3	15.837
62	15.8	16.365
64	16.2	16.893
66	16.7	17.421
68	17.2	17.949
70	17.7	18.477
72	18.1	19.004
74	18.6	19.532
76	19.1	20.060
78	19.6	20.588
80	20.0	21.116

Food Colors

Certified food colors are employed in the preparation of foods, candies, syrups, etc. There is no longer any reason for the consumer to look suspiciously at an attractively colored food preparation, since the use of harmful chemicals for coloring purposes is now definitely prohibited by law. Any tint or shade desired can now be obtained by the wise and careful selection and blending of certified food colors. The government has specified those colors that may be used in food manufacture, and each color must now pass rigid chemical tests in order to safeguard the consumer. The use of coloring matters in foods, nevertheless, still deserves close inspection and regulation to safeguard against the use of uncertified colors by unscrupulous manufacturers. Colors may be manufactured both in paste and liquid forms.

Certified Food Colors

(Selected By The United States Department Of Agriculture)

Primary Colors

Red	Erythrosine	
	Amaranth	
	Ponceau 3R	
	Ponceau SX	
Orange	Orange I	
Green	Guinea Green	
	Light Green S. F.	
	F.C.F. Green	
Blue	Sodium Indigo Disulfonate	
Yellow	Tartrazine	
	Yellow A. B.	Soluble in oil only
	Yellow O. B.	

Secondary and Tertiary Colors

These colors can be made by blending two or more primary colors.

Solubility Chart for Certified Food Colors

When certified colors are used in foods, it is important that the color be kept uniform. However, because of the varying solubility of the dyes at different temperatures, it is highly important that standard color solutions be prepared and maintained with the foregoing principle in mind.

The following chart, prepared by the laboratories of Wm. J. Strange Co., Chicago, is useful to anyone who is concerned with formulas or quality control.

SOLUBILITY CHART FOR CERTIFIED FOOD COLORS AT °F

	Freezing	40°	50°	60°	70°	80°	90°	100°	110°	120°	130°
Ponceau 3R	6.2	10.9	13.6	15.1	16.7	17.9	19.1	20.3	21.5	22.5	22.8
Orange I	1.5	1.9	2.2	2.5	2.7	3.2	4.0	5.0	8.6	13.2	17.5
Tartrazine	5.9	6.2	8.1	10.3	14.9	21.4	35.3	41.5	45.9	48.8	49.9
Amaranth	3.1	12.1	17.7	20.9	22.9	26.0	28.2	30.4	31.9	33.6	35.3
Guinea green	12.0	18.0	24.0	30.0	33.2	36.0	38.4	42.2	46.0	48.6	51.2
Fast green	12.0	18.0	24.0	30.0	33.2	36.0	38.4	42.2	46.0	48.6	51.2
Erythrosine	5.0	6.2	7.4	8.6	10.0	11.4	12.8	14.4	16.0	17.6	19.3
Indigotin	0.8	0.9	1.1	1.3	1.5	1.7	1.9	2.1	2.3	2.5	2.8
Sunset yellow	9.2	23.0	28.0	30.9	33.7	35.6	37.4	39.2	40.9	42.6	43.8
Ponceau SX	1.25	2.5	5.0	7.4	8.6	9.8	10.4	11.0	11.5	11.8	12.1
Naph. yellow	10.9	12.1	13.2	14.3	16.0	17.7	20.3	23.0	26.0	29.5	32.8
Light green	Very soluble; 25% will remain in solution even at freezing.										
Brill blue	Very soluble; 32% will remain in solution even at freezing.										

This chart indicates the number of ounces of color which will remain in solution in distilled water to make a gallon of finished liquid at a given temperature. In dissolving the above colors, the solution must be heated to at least 50° F above the temperature indicated by the chart.

Compounded Food Colors

Yellow

Tartrazine	9	oz
Amaranth	1	oz

or

Tartrazine	15¼	oz
Orange I	¾	oz

Egg Shade

Tartrazine	14½	oz
Orange I	1½	oz

Peach Color

Ponceau	1¼	lb
Tartrazine	10	oz

Golden-Yellow

Tartrazine	75	oz
Orange I	25	oz

Yellow-Orange

Orange I	9	oz
Tartrazine	1	oz

Orange (Yellowish)

Tartrazine	50	oz
Orange I	50	oz

Reddish-Orange

Orange I	9	oz
Amaranth	1	oz

Orange-Red

Ponceau	75	oz
Tartrazine	30	oz

Violet

Amaranth	88	oz
Sodium Indigo disulfonate	42	oz

Royal Purple

Amaranth	8	oz
Sodium indigo Disulfonate	2	oz

Grape

Amaranth	2	oz
Sodium Indigo Disulfonate	¼	oz

Green

Tartrazine	6 or 6½	oz
Sodium Indigo Disulfonate	4 or 3½	oz

Brilliant Green

Tartrazine	10	oz
Sodium Indigo Disulfonate	6	oz

Yellow-Green

Sodium Indigo Disulfonate	4	oz
Tartrazine	12	oz

Mint Green

Tartrazine	7	oz
Sodium Indigo Disulfonate	3	oz

Creme de Menthe Green

Light Green S.F.	48	oz
Tartrazine	52	oz

Olive Green

Sodium Indigo Disulfonate	4	oz
Orange I	1½	oz
Tartrazine	10½	oz

Leaf Green

Light Green

Deep Red

Ponceau	75	oz
Amaranth	25	oz

Ruby Red

Amaranth

Striping Red

Erythrosine	10	oz
Amaranth	30	oz
Ponceau	60	oz

Carmine Red

Erythrosine	90	oz
Ponceau	10	oz

Cherry Red

Ponceau

Sherry Red

Amaranth	12	oz
Orange	4	oz

Raspberry Red

Amaranth

Grenadine Red

Amaranth

Strawberry Red

Amaranth	8½	oz
Orange I	1½	oz

Brown

Orange I	13	oz
Sodium Indigo Disulfonate	8	oz
Amaranth	3	oz
Tartrazine	2	oz

Almond Brown
(Reddish-Brown)

Orange I	13¾	oz
Amaranth	2¼	oz

Chocolate Brown

Sodium Indigo Disulfonate	28	oz
Amaranth	20	oz
Tartrazine	40	oz
Orange I	12	oz

Liquid Colors

Liquid colors can be made by dissolving 4 ounces to 1 gallon of any of the dye formulas outlined on the previous pages. The addition of 1 pound of glycerin to each gallon of liquid color will be sufficient to retard the development of molds or fermentation. However, in order to positively guarantee no fermentation in a liquid color preparation, ¼ ounce of sodium benzoate, dissolved in a little water and ⅛ ounce of citric acid, should be added.

Paste Colors

Recently, manufacturers have put out paste color preparations, sold as a specified weight tablet or block, which, when added to a definite volume of a liquid or solid, will result in the desired shade. These paste colors are easily made by adding 1 pound of color to 3½ pints of warm glycerin. The mixture is stirred until all of it is dissolved and then 8 pounds of confectioner's sugar, free from lumps, is slowly added. The heating is continued until a smooth paste is obtained.

Vegetable Colors

Greens	1. Chlorophyll 2. By mixing Tincture of Saffron or Tincture of Turmeric with a solution of Indigo Carmine (made from the paste) in various proportions a variety of green shades can be obtained
Reds	Cochineal Syrup and Carmine
Yellows	Tincture of Saffron and Tincture of Turmeric
Browns	Caramel (Burnt Sugar)
Blues	Indigo Carmine Paste
Violet	By mixing Red + Blue
Oranges	Tincture of Annatto and by mixing Red + Yellow
Grays	By mixing Red + Green or Red + Blue + Yellow

Tincture of Chlorophyll

Extract green spinach with diluted alcohol and evaporate to paste. Use 1 part paste to 10 parts alcohol and filter, or moisten fine hacked grass with water and press. Boil the juice and decant the greenish liquid. Press the residue off from the surplus liquid and extract the residue with alcohol. Filter and evaporate. Take up the residue in hot water. The insoluble, separating, greenish, flocky mass is dissolved in hydrochloric acid and finally precipitated with water.

Tincture of Saffron

Saffron	3 oz	Alcohol	1 qt
Water	1 qt		

Macerate several days and filter.

Tincture of Turmeric

Turmeric (Bruised)	1 oz	Diluted Alcohol (50–50)	6 oz

Macerate in a closed vessel for 7 days and filter.

Cochineal Syrup

Powdered Cochineal	12 oz	Distilled Water	30 oz
Potassium Bicarbonate	4 oz	Alcohol	24 oz

Rub the potassium bicarbonate with the cochineal powder. Mix the alcohol and water and add to the powder. Filter, then mix the solution with 120 oz of syrup thoroughly.

Tincture of Carmine

Carmine	22 oz	Distilled Water, Enough to Make	500 oz
Strong Ammonia Water (Enough to Dissolve the Carmine)			

Dissolve the carmine in the ammonia water and add the distilled water.

Tincture of Indigo Carmine

Dissolve in distilled water any amount of the indigo carmine paste to get the desired strength or shade. Usually:

Paste	1 oz	Warm Water	20 oz

Dissolve, filter and add 5% alcohol.

Tincture of Annatto

One hundred ounces of annatto are extracted with 150 ounces of water containing 1½% potassium carbonate. Evaporating to 60 ounces, add 12 ounces of alcohol and filter.

This is used for coloring butter and cheese.

Sugar Coloring (Caramel Color)

Add 2 gallons of water to 10 gallons of caramel.

Coloring Materials for Liqueurs

Red 1 Steep 4 ounces of raspings of Red Sanders

	Wood in a pint of strong spirits of wine for 2 weeks. Strain and filter.
	2 Steep 3 ounces of Cochineal, finely powdered, in a pint of strong spirits of wine for 2 weeks; add 2 drams of powdered alum and filter clear.
Blue	Steep 4 drams of Indigo in a bottle with 2 ounces of Sulfuric Acid for several days, frequently putting the bottle into hot water, add ½ pint (8 ounces) of distilled water and filter.
Yellow	Steep 1 ounce of Saffron in ½ pint of spirits of wine for a week; filter.
Green	Mix equal parts of Blue and Yellow coloring.
Violet	Mix 1 part of Blue with 2 of Red coloring.
Pink	Steep 4 ounces of Cudbear in a quart of strong spirits for 2 weeks; filter.
Brown	Add 2 gallons of water to 10 gallons of Caramel.

Preparation and Use of Stabilizer Solutions in Whipped Cream

The following instructions are given for preparing and using stabilizer solutions on the basis of 10 gallons of cream-stabilizer mixture:

Sodium Alginate

Weigh out 2¾ ounces of sodium alginate. Place in 2 quarts of cold water and heat to 160 to 170° F by placing the container in hot water. Stir frequently during heating. Heat 80 pounds of 34% cream to 160 to 165° F and add the hot stabilizer solution to the hot cream. Mix well and cool to 40 to 50° F. Age 24 hours before whipping. The cream-stabilizer mixture will make 10 gallons of a product containing 32.3% butterfat and approximately 0.2% sodium alginate.

Gelatin

Weigh out 4¾ ounces of gelatin. Place in 2 quarts of cold water and heat to 140 to 143° F by placing the container in hot water. Stir frequently during heating. Heat 80 pounds of 34% cream to 140 to 143° F and add the hot stabilizer solution to the hot cream. Mix well and cool to 40 to 50° F. Age 24 hours before whipping. The cream-stabilizer mixture will make 10 gallons of a product containing 32.3% butterfat and approximately 0.35% gelatin.

Experimental work shows that the production of whipped cream for sale as such is entirely feasible. Whipped cream might advantageously be added to the list of other dairy products offered to the public by the milk dealer.

In a comparison of Guernsey and Ayrshire cream for whipping purposes, experiments show that Guernsey cream can be whipped in less time than Ayrshire cream. Also, the fat loss in the drainage from the whipped cream is considerably less with Guernsey whipped cream than with Ayrshire whipped cream. This may be explained on the basis of the average volume of the fat globules found in the two creams, those in the Guernsey cream having the greater average volume.

The temperature at which whipped cream is stored affects its stability and the amount of fat lost in the drainage. The whipped cream stored at 40° F had the best stability as evidenced by the amount of drainage taking place. However, little difference was noted between the samples stored at 40° F and 52° F. Samples stored at 60° F and 76° F showed considerable drainage, the greater amount taking place at the higher temperature.

Temperature had little effect on total fat loss when maintained at or below 64° F. However, considerable loss of fat occurred at 76° F. Also, whipped cream stored at 76° F has a tendency to become slightly sour at the end of 24 hours.

In so far as fat loss is concerned, whipped cream should not be stored at temperatures above 64° F. However, when considering stability and keeping quality, a storage temperature of 52° F or lower is to be preferred.

Good stability and little loss of fat in the drainage are desired in high quality whipped cream. A product, in which these qualities may be lacking under normal conditions, may be made to possess such qualities by the addition of a stabilizer to the cream prior to whipping.

Sodium alginate and gelatin have both proven satisfactory for increasing stability of, and decreasing fat loss in the drainage from whipped cream. The use of such substances is recommended in the manufacture of whipped cream in which it is impossible to obtain these qualities in any other way. However, prior to use, the fact must be established whether or not the use of such substances violates existing city or state laws governing the composition of whipped cream.

In comparing the relative advantages of the two stabilizers, the type of whipper must be considered. Whipped cream produced by means of the air whip and containing gelatin possesses greater stability with less fat loss in the drainage than that containing sodium alginate as the stabilizer. Whipped cream produced by means of the turbine whipper and containing gelatin possesses greater stability with a greater fat loss in the drainage than that containing sodium alginate as the stabilizer.

When sodium alginate was used as the stabilizer, the body of the whipped cream was apparently normal. When gelatin was used as the stabilizer, the body of the whipped cream was slightly soggy and sticky, this defect being more pronounced when the turbine whipper was used. This defect, however, was not sufficiently serious to be objectionable, except possibly to a very discriminating trade.

Differences noted on the various results of the use of sodium alginate and gelatin as stabilizers in cream for whipping are not sufficiently great to warrant recommending one or the other.

Both sodium alginate and gelatin increased the viscosity of the cream and reduced the overrun.

The type of whipper used is an important factor in obtaining the desired overrun. Under all conditions the overrun was approximately doubled by the use of the air whip compared to the turbine whipper. The body of the whipped cream produced by the air whip was more desirable even though it had a more open structure than the whipped cream produced by the turbine whipper. The air whip was found to be the better of the two types of whippers used for the production of whipped cream.

Stabilized Whipped Cream

(a) 40% Cream	1	gal	Hot Water (200° F)	1	lb
(b) High Grade Gelatin	1½	oz	Granulated Sugar	10	oz

Dissolve (b). Add while hot to (a) stirring constantly. Place in refrigerator over-night. Whip cream the following day. Then add 10 ounces granulated sugar. Flavor to taste.

This cream may be whipped either with a regular egg beater or on an air whip machine. If possible whip in a cool place.

Be sure the product for which this cream is intended is thoroughly cooled before using. When finished hold whipped cream products under refrigeration to prevent deterioration.

Whipped Cream Stabilizer

Water	1½	pt	Agar-Agar	½	oz
Sugar	1¾	lb	Egg Whites	1	pt
Corn Starch	1	oz			

Boil the water, sugar, corn starch and agar-agar for 3 minutes. Whip the egg whites with medium speed. When whites are up to full volume pour syrup slowly over whipping whites and whip until cool.

Blend with 2 quarts whipped cream.

Whipped Cream Stabilizer

Sugar	27	oz	Agar-Agar	1	oz
Corn Starch	8	oz	Salt	½	oz
Gelatin	1	oz	Vanilla Flavor	If desired	

The above ingredients are well blended, added slowly to 4 pints of boiling water, and cooked to 190° F. The mixture is cooled to room temperature until it forms a paste. One gallon of chilled cream (45° F, 38% fat) is then whipped at second speed until it holds the marks of the beater. One pint of the above paste is added to the cream, and the mixture heated for ½ to 1 minute. This whipped cream will not weep and will hold its shape.

Cremo Imitation Whipped Cream

This is made without cream for filling and covering cakes, filling sandwich sponge cookies and small sponge cakes.

Small Batch		*Large Batch*
2½ lb	Shortening (Preferably Half Butter and Part Plastic Coconut Fat)	25 lb
1½ lb	Sugar (Icing)	15 lb
2 lb	Nulomoline	20 lb
3¼ lb	Milk (Evaporated Basis)	32½ lb
¾ lb	Prepared Marshmallow	
As desired	Flavor (Vanilla)	As desired

The fats are creamed until thoroughly mixed and slightly lightened. The powdered sugar and nulomoline are added, and the mixture whipped lightly. During the whipping, the evaporated milk is added gradually. Just before the whipping is finished, the prepared marshmallow flavor is whipped in.

For banana fillings, about 4 ounces of Nulco Meloban for each ten-pound batch are incorporated in any convenient way, preferably being first blended with about an equal weight of sugar.

Cream Aid

This is a preparation to aid in whipping cream.

Lime water and sugar, or a saturated solution of calcium saccharate are added in the proportion of 1 teaspoonful to the pint of cream.

Cream Improver

Water	7 pt	Agar	1 oz
Sugar	3 lb		

Mix sugar and agar, and boil up with water for 3 minutes.

Water	1 pt	Corn Starch	4 oz

Cook this a day ahead and add to above. Mix and bring to boil.

Cream the above cold mix in the beater. Then add 1 pound meringue powder and beat until fluffy.

The addition of this finished mix, in the proportion of 1 gallon to 3 gallons of whipped cream, results in a product superior to that mixed with straight meringue. The cream should be whipped up first and then incorporated with the above improver.

Powdered and Flaked Egg Whites

The moisture content of powdered fresh egg whites is usually standardized at about 8½%, when ready for shipment, although flaked albumen may run as high as 12 to 18% in moisture. A pure egg white powder should be almost completely soluble in water, free from unwholesome odor and have a nice white color. Powdered dry whites are almost sterile as they come out of the drying equipment; the bacteria count is very low, and only in the presence of excess moisture and high humidity does it increase over a period of time. The whipping quality of the dried egg white is practically the same as that of the original egg albumen. The primary mechanical function of dried egg whites or albumen is aeration, i.e., incorporation of air, of the batch of cake or marshmallow so that it will be lighter in weight, shorter and smoother in texture. The syrup is often added hot to the beater to give the air cells in the egg albumen an opportunity to develop from the very start of the whipping. If the syrup is thick, heavy and cold, the volume of the batch will be materially reduced. The syrup must not, however, be heated to a temperature that will coagulate the egg white. The machine used for beating the egg white batch should be of the proper speed and design to provide air circulation; a machine equipped with whipping action is preferable to one that employs a beating action. The quality of egg white, the whipping ability of the machine and the quantity of syrup to be used determine the amount of egg white that should be used in a formula. When egg white or albumen is whipped into a syrup, the egg takes up moisture as the air cells form, causing the batch to become stiffer as the whipping progresses. It is important that a definite balance exist between egg, syrup, moisture and cooking temperature, in order to produce uniform results.

Marshmallow Cream

To make a firm, consistent marshmallow cream, 100 pounds of corn syrup, 9 pounds of water and 3 pounds of powdered, fresh egg whites are used. The syrup is cooked to 240° F. The body of this cream may be increased by the addition of a slurry of 1 pound of tapioca flour in 2 pounds of water to the sugar syrup after it has been boiled to 240° F. The marshmallow cream is made by adding the corn syrup solution to the egg whites dissolved in the water. After developing the volume, stability and texture that comes closest to expectations, cane sugar, invert sugar or cerelose may be substituted for corn syrup, pound for pound as given in the above formula. If further increase of the body of the marshmallow cream is desired, the addition of tapioca solution may be increased.

When powdered egg white is allowed to soak in water it deteriorates at room temperature just as rapidly as though it were fresh egg albumen. Therefore, it is quite important to be careful in this respect in order to avoid an unpleasant odor and taste. To make a rapid solution of powdered egg white, equal parts of water and whites are handwhipped until smooth and creamy. The balance of the required water is then added. This procedure may also be employed if an upright vertical type of machine is used, by using the low speed on the machine. Beaten egg whites may also be made from 2 to 4 pounds of cold corn syrup for each pound of powdered egg whites. The mixture is creamed at low speed for about 1 minute until completely dispersed; a required amount of water is added and the mixture whipped up at high speed to a good body consistency, after which the cooked syrup is added slowly, and the whipping continued to a sharp peak. The powdered egg whites are added dry to the 4 pounds of cold corn syrup.

When a horizontal type beater is used, part or all of the syrup is placed in the beater. The machine is stopped and the dry, powdered egg whites are added directly to the syrup along the backside of the beater. Any lids or covers are then closed and the machine is started; heat is applied for half a minute and the water slowly added; the beating is continued to a sharp peak. The egg white albumen must not be allowed to come in contact with copper, iron or grease, before or after dissolving it in water.

Nougat Cream and Marshmallow Topping

In making nougat cream, marshmallow topping, or other aerated confections, we are confronted with the delicate problem of creating tiny air cells, by the use of egg white or albumen, generally in the presence of cooked corn syrup, sugar and other ingredients that go into the finished batch.

Unless a chemical change takes place in the egg white, the air cells will collapse and cause the batch to break down and become soupy. The heat of the syrup appears to be the most important factor in inducing the chemical reaction necessary to stabilize the air cells. As a general rule, temperatures of 160 to 170° F in the batch will be found the ideal point at which to add the egg white and start the whipping or beating. Under some conditions even higher temperatures will be found advantageous, depending entirely on the type of beater and methods employed. Obviously, temperatures above 140° F will coagulate the egg white but whether coagulation or chemical reaction or both stabilize the air cells, is not definitely known. However, aside from the advantages of high temperatures in proper stabilization of the egg, it is not possible to build up an air cell structure in the presence of cold heavy syrup, due to the high viscosity of the syrup, which prevents the formation of air cells. The viscosity of the syrup must be low enough to permit free development of the air cells.

If the egg white is permitted to coagulate before the air cells are created, the albumen is rendered inert and will not form suitable cells. For this reason care must be exercised in adding the egg white, dry or in solution, to the batch and the temperature of the syrup must be regulated by whipping, to bring about proper stabilization and viscosity at the finish of the whipping process, which must be timed.

There are several ways of incorporating powdered egg whites into the batch. Each method offers definite advantages, depending on the type of beater or production system employed. A brief outline of various methods that may be used follows.

Upright Type Vertical Beater

The powdered egg white is added to an equal amount of water, by weight, and the mixture is creamed at low speed until the egg is thoroughly dissolved. This requires 1 to 2 minutes. As soon as a smooth pasty mass forms, the balance of the water specified in the formula for dissolving the egg is added and the whipping continued at high speed to a good peak. The cooked syrup, at 170 to 200° F, is added slowly and the desired consistency achieved by continued whipping.

Another very efficient method is to place 2 or 3 pounds of cold or warm corn or invert syrup, to each pound of powdered egg whites, in the bowl of the beater and add the dry powdered egg. The mixture is creamed at low speed for about 1 minute until the egg is dissolved, and a smooth, creamy mass forms. The water specified in the formula for dissolving the egg white is then added and the whipping continued at high speed to a good peak. The cooked syrup, at 170 to 200° F, is added, and whipping is continued to the desired consistency.

RECONSTITUTING DRY WHOLE EGG, DRY EGG YOLK AND DRY EGG ALBUMEN TO THE NATURAL STATE

1 lb of dry whole egg is equivalent to 4 lb liquid egg
1 lb of dry egg yolk is equivalent to 2½ lb of liquid egg yolk
1 lb of dry egg white is equivalent to 8 lb of liquid egg white

Whole egg contains 27.80% solids	pH 7.20
Egg yolk contains 43% solids	pH 6.50
Liquid egg white contains 11.85% solids	pH 8.62

Use of Glycerin in Food Products

The use of glycerin in food products, principally as a moistening or hygroscopic agent, has long been a common practice. However, in addition to its value as a hygroscopic agent, glycerin is sometimes used for its sweetening action. That glycerin is a normal product of the fermentation of sugars has long been recognized. Fermented beverages should not be classed as foods, but it should be recalled that bread, our most widely used food, is raised by fermentation and, consequently, contains some glycerin. In addition to the glycerin normally ingested in manufactured products, such as beverages, flavoring extracts and bread, glycerin, in combination with fatty acids, constitutes 10 to 12% of all fats and oils used as food.

There is no objection to the use of small quantities of glycerin in food products, provided it is of a purity suitable for food use and provided that its presence is plainly declared in the labeling when it is not a normal ingredient of the article. It should not, of course, be used to conceal damage or inferiority. In some products, e.g., vanilla extract, glycerin is regarded as a normal ingredient and it is unnecessary to declare it on the label.

CHAPTER II

MILK AND MILK PRODUCTS

Milk, when examined under a microscope, is seen as a transparent fluid which consists of many minute fat globules suspended, together with casein, the main protein of milk, in a solution of albumin, sugar and water. The fat globules in an average drop may number more than one hundred million.

Cow's Milk

Cow's milk is a white opaque liquid having a butter-like odor and a bland sweet taste. It has a specific gravity of about 1.030. On standing, a yellowish stratum of cream, which may constitute 2 to 7% of the total milk, rises to the top. Milk, in its raw state, contains all the known vitamins. The exact vitamin content is dependent upon the food consumed by the herd. The chemical composition of cow's milk varies. The water content may be 84 to 90%. The total solids average 10 to 16%, including from 2 to 7% fat, 2½ to 4½% casein, 2 to 6% lactose or milk sugar, a small percentage of albumin, a large quantity of calcium, and a small quantity of other minerals.

Butter is the fat extracted from milk. Casein, one of the chief components of milk, is used in the manufacture of cheese. The milk sugar, lactose, is a disaccharide and does not have a sweet taste. Lactic acid bacteria comprise the normal bacterial content of milk. They cause the souring of milk and are essential to the manufacture of all true cheeses.

Certified milk is raw milk produced under very strict regulations specified by the medical milk commissions. The consumer is assured of milk from clean healthy cows milked in clean, well-ventilated stables. This milk is promptly cooled, bottled and delivered, in the most modern sanitary manner, and with the greatest possible speed. This milk must also contain specified quantities of fat and other milk solids, and, when tested for bacterial content, must show no more than a certain number of bacteria per cubic centimeter.

Pasteurized milk is milk that has been heated to below the boiling point, for a length of time sufficient to kill, or at least to render harmless, practically all pathogenic bacteria. It is then immediately cooled to 50° F or lower, to retard the development of any remaining organisms. The chemical change taking place in the milk during this procedure is so slight as to be negligible. The advantages of pasteurized milk are that disease-carrying bacteria, which may have entered the milk from the cow or elsewhere, are killed, and that it remains fresh much longer than raw milk. If dairy-pasteurized milk is not obtainable, pasteurization can be accomplished by heating the milk in sealed bottles or other containers to a temperature of not less than 145° F nor more than 150° F, for 20 to 30 minutes, then cooling, as rapidly as possible, without breaking the container. It is then kept in a refrigerator or other cool place until used. The next best simple method of pasteurization is boiling the milk for a few minutes, cooling it rapidly, straining if necessary to remove clumps and then keeping it in a cool place.

Sterilized milk is milk that has been heated to the temperature of boiling water, or a little higher, and maintained at that point long enough to kill all organisms present. The term is also applied to milk that has been repeatedly pasteurized. These two processes rid the milk of the danger of being a disease carrier, provided that it is not subject to future contamination. Milk treated in this way is objectionable to many people because of the cooked flavor developed and the destruction of most, or all, of its Vitamin C content.

Dry, or desiccated, milk is a white powder obtained by the evaporation of wholly, partially, or completely skimmed milk. This product is largely used by bakers, chocolate manufacturers, ice cream companies, in baby food and for many other useful purposes. Dry whole milk, which is a soft cream color, averages about 28% fat, whereas the dry skimmed milk, a white or a light yellowish powder, usually contains an average of

½% fat. Whole milk powder, or skimmed milk powder, lends itself readily to enriching the diet by addition to bread, biscuit dough, soups, etc. Whole milk powder is reconstituted by adding 1 pound of the dry milk to 3¾ quarts of water; skimmed milk is reconstituted by adding 14 ounces of dry skimmed milk powder to 3 13/16 quarts of water.

Cream Tables

TABLE FOR MIXING CREAM AND MILK TO OBTAIN 100 POUNDS OF 20% CREAM

When Cream Shows Per Cent Butter Fat to be:	*Take* Cream, lb	*Take* Skim Milk, lb	*Take* Cream, lb	*Take* 4% Milk, lb
21	95	5	94	6
22	91	9	89	11
23	87	13	85	15
24	83	17	80	20
25	80	20	76	24
26	77	23	73	27
27	75	25	70	30
28	72	28	67	33
29	69	31	64	36
30	67	33	62	38
35	57	43	52	48
40	50	50	45	55

TABLE TO REDUCE 100 POUNDS OF CREAM TO 20%

To 100 lb Cream Showing: Per Cent	*Add* Skim Milk, lb	*Add* 4% Milk, lb	*To 100 lb Cream Showing: Per Cent*	*Add* Skim Milk, lb	*Add* 4% Milk, lb
40	100	125	26	30	37½
35	75	93¾	25	25	31¼
30	50	62½	24	20	25
29	45	56¼	23	15	18¾
28	40	50	22	10	12½
27	35	43¾	21	5	6¼

1 gal 30% cream weighs 8.3 lb 1 gal 40% cream weighs 8.2 lb

QUANTITY TO BE TAKEN FROM CREAM AND SKIM MILK TO MAKE 100 POUNDS CREAM TESTING

Cream Shows Per Cent Fat	8%	10%	12%	14%	16%	18%	20%
16	50.0	62.5	75.0	87.5	100.0	—	—
17	47.0	58.8	70.9	82.3	94.0	—	—
18	44.4	55.5	66.6	77.7	88.8	100	—
19	42.1	52.1	63.1	73.6	84.2	94.7	—
20	40.0	50.0	60.0	70.0	80.0	90.0	—
21	38.0	47.5	57.1	66.6	76.1	85.7	95.2
22	36.3	45.4	54.5	63.6	72.7	81.8	91.9
23	34.7	43.4	52.1	60.8	69.5	78.2	86.9
24	33.3	41.6	50.0	58.3	66.6	75.0	83.3
25	32.0	40.0	48.0	56.0	64.0	72.0	80.0
26	30.7	38.4	46.1	53.8	61.5	69.2	76.9
27	29.6	37.0	44.4	51.8	59.2	66.6	74.0
28	28.5	35.7	42.9	50.0	57.1	64.3	71.5
29	27.5	34.3	41.3	48.2	55.1	62.0	68.7
30	26.6	33.3	40.0	46.6	53.3	60.0	66.6
31	25.8	32.3	38.7	43.7	51.6	58.0	64.5
32	25.0	31.2	37.5	42.4	50.0	56.2	62.5
33	24.2	30.3	36.3	41.1	48.4	54.5	60.6
34	23.4	29.4	35.2	45.1	46.8	52.9	58.8
35	22.8	28.5	34.2	40.0	45.7	51.4	57.1
40	20.0	25.0	30.0	35.0	40.0	45.0	50.0
45	17.7	22.2	26.6	31.1	35.5	40.0	44.4
50	16.0	20.0	24.0	28.0	32.0	36.0	40.0

The figures 8, 10, 12, 14, 16, 18, 20 represent the per cent of butter fat desired.

The other figures represent the pounds of cream to be taken, and this number subtracted from 100 gives the pounds of skim milk required.

Example: Cream on hand shows 26% fat
Cream to be made shall show 20% fat

Find 26% in the left hand column and follow the horizontal line to 20% and you come to the number 76.9.

You must take 76.9 pounds of your 26% and add to it 23.1 pounds of skim milk to make 100 pounds 20% cream

$$\begin{array}{r} 100 \\ -\ 76.9 \\ \hline 23.1 \end{array}$$

Acidophilus Milk

The manufacture of pure acidophilus milk requires the greatest care, both in the preparation of the milk before inoculation and throughout the entire process of manufacture. It must be remembered that L. acidophilus is not a natural inhabitant of milk and must be acclimated to this medium. Even though it be acclimated to milk, its growth is, nevertheless, such that it is unable to overcome a slight contamination. Laboratory facilities, close supervision and adequate equipment are essential to the successful production of acidophilus milk.

The first step in the manufacture of acidophilus milk is to obtain, from a reliable source, a pure culture of the bacterium Lactobacillus acidophilus of proved therapeutic value. After a pure culture is obtained it must be kept pure as a starter. All chances of contamination must be avoided. Florence or Erlenmeyer flasks are the most suitable containers in which to carry the starter. The flasks should be about half filled with fresh skim milk and plugged with cotton. They are then sterilized in an autoclave at 15 pounds pressure for 20 minutes. The milk, after sterilization, has a slightly caramelized appearance. After the milk has cooled, a flame from a gas burner should be passed over and around the mouth of the flask, and the culture introduced into the flask after the mouth of the culture tube is passed through the flame. The cotton plug from the flask should be held in the hand during this operation and replaced in the mouth of the flask immediately after inoculation. The inoculated milk is incubated at 98 to 100° F until the milk has curdled. A starter should show a small quantity of whey on top of the curd which should be firm with no evidence of gas. The curd is broken up by rotating the flask vigorously. The starter should be of a creamy consistency and possess a clean acid flavor and a characteristic aroma.

This starter, or mother starter, is carried on by daily transfers from flask to flask. The same procedure as described above is followed, except that usually the first bit poured over the lip of the flask is discarded as it serves to wash off the lip. The care necessary to maintain a pure starter of L. acidophilus cannot be overemphasized. The starter should be examined frequently for purity either microscopically or by plates. Should it develop a bad flavor or odor, it must be discarded and a new culture obtained. In any case it is advisable to get a new culture at regular intervals. Kopeloff advises the use of a new fecal culture every 2 to 4 months to assure the therapeutic value of the product.

The bulk starter, which is the starter to be used to inoculate the batch, is prepared in the same way as the mother starter, and the same precautions against contamination must be observed. Florence or Erlenmeyer flasks of two-liter capacity or larger are most satisfactory in which to prepare this bulk starter. These flasks permit the autoclaving of the milk and are most easily handled under the conditions necessary to prevent contamination. The starter should be 18 to 24 hours old when used to inoculate the batch, and should possess all the qualities indicating purity previously mentioned for the mother starter.

Equally great care must be exercised in the treatment of the milk as in the preparation of the starter. Only fresh milk of the best quality should be used in the manufacture of acidophilus milk. Whole milk, milk which is partly skimmed, or skim milk may be used. The pasteurizer employed in the preparation of the product must be so fitted that the milk may be heated and cooled without removing it from the vat, which should be equipped with an efficient mechanical stirrer. The vat must be used not only as a pasteurizer but also as an incubator for the milk. The characteristics of the bacterium Lactobacillus acidophilus make it necessary that the bacteria in the milk be killed before the starter is added. The milk is heated to about 205° F for 1½ hours or longer and then cooled to 98° F. It should have a slightly caramelized appearance after heating. However, this, in itself, is not an indication that the necessary destruction of the bacteria has been accomplished. Bass recommends heating the milk to 190 to 195° F for 1 hour; then cooling it to 98° F, at which temperature it is held for 3 or 4 hours; again heating it to 190 to 195° F for 1 hour; then cooling it to 98° F and inoculating. After the milk has been cooled to 98 to 100° F it is inoculated with about 2% of the bulk starter. The greatest care must be observed in the inoculating process to prevent contamination of the batch. The mouths of the flasks should be passed through a flame before pouring the starter. The cover of the vat is lifted just sufficiently to permit the pouring of the starter and is replaced immediately. After inoculation, the milk is stirred for a few minutes to distribute the inoculum, and the batch is then allowed to incubate. The temperature of incubation, 98 to 100° F, should be maintained throughout the entire incubation period. The batch is incubated until a firm curd has been formed. This usually requires about 18 to 24 hours. The length of the incubation period is dependent upon the size of the inoculation, the temperature of incubation, and the activity of the starter. After the milk has curdled, it is broken up by stirring, cooled to room temperature, and bottled.

Kefir

Add ½ teaspoonful of sugar to a six-ounce or eight-ounce bottle of boiled and cooled water. Half a yeast cake is added to this sugar solution which is then set in a warm place overnight. This will give an active culture of the yeast and obviate the necessity of adding the yeast cake directly to the milk. This yeast culture should be ready at the time the buttermilk is received or, if made at home, at the time it is curdled.

One to 1½% sugar is added to the buttermilk. The extent of the alcoholic fermentation will depend on the quantity of sugar added to the buttermilk. Theoretically, about one-half of fermented sugar may be converted into alcohol. That is, milk to which 1% of cane sugar has been added may contain, after the fermentation, 0.5% of alcohol. The quantity of sugar added should be governed by the amount of carbon dioxide desired in the finished product. This should be sufficient to make the kefir distinctly effervescent and to impart to it the peculiar sharp taste of charged water. However, it should not be developed sufficiently to blow the fluid out of the bottles when the stoppers are removed. Experience shows that 1 to 1.5% of sugar will give the proper amount of gas. This may be approximated by adding sugar in the proportion of 2 even teaspoonfuls of sugar to each pint of milk. When the buttermilk and the yeast culture are ready, the sugar is dissolved in the buttermilk.

The yeast culture is added to the buttermilk in the proportion of 1 teaspoonful to 1 quart of buttermilk.

The buttermilk is mixed thoroughly and bottled. The bottles should be very strong, as sometimes sufficient gas pressure is generated to break ordinary bottles. The heavy bottles used for ginger ale or other carbonated drinks answer this purpose very well. They should be carefully cleaned and boiled or steamed before being filled and then tightly stoppered. The stoppers should be wired or tied securely in place.

The product is put in a cool place to ferment. If the fermentation is too active, the kefir will have a yeasty taste, and the curd is apt to become lumpy and filled with large gas bubbles. A temperature of 18 to 21° C (65 to 70° F) will be found satisfactory for kefir which is to be used on the third or fourth day. The floor of a cool cellar is a convenient place to ferment kefir made in the home. The bottles should be shaken as often as may be necessary to keep the curd in a finely divided condition. The finished product should be smooth and creamy, effervesce rapidly when poured from the bottle, and have the pleasant, acid taste of buttermilk, with the added sharpness caused by the gas and the trace of alcohol. Kefir, 2 or 3 days old, may have a yeasty taste, but if it has been properly made this will disappear as the fermentation of the sugar nears completion; made under these conditions, it should be used when 3 to 5 days old, but if put on ice it may be held for a week or even longer.

Determining the Vitality of Lactic Cultures

Cultures used for cheese making or cultured milk produce acid at different rates. The following formula is used to determine the vitality of lactic cultures. A pint of milk is placed in a sterile ground-glass stoppered bottle and heated to 100° F. Then 5 cubic centimeters of the lactic culture is added and 30 minutes later 1 cubic centimeter of rennet is added. One hour after adding the rennet, the curd is cut, and, 2 hours after cutting the curd, the whey is drained off. At periods of 2 and 3 hours after draining, the whey acid tests are made on the whey surrounding the curd. A slow culture will produce whey titrating 0.16 and 0.23% whereas an active culture will produce whey titrating 0.42 and 0.63% of acid at the two consecutive titrations. The titrations are made with tenth normal sodium hydroxide using phenolphthalein as an indicator. The result is expressed as lactic acid.

Soy Milk

If the dried beans, preferably yellow-seeded varieties, are soaked for a few hours, then finely crushed and boiled for about 30 minutes in the proportion of 3 parts of water to 1 part of mash, a milky emulsion is obtained, which is very similar in appearance and properties to animal milk. This liquid, separated out by means of a very fine sieve or cloth strainer, is the soybean or vegetable milk used so extensively in China. Soybean meal after the oil is extracted or whole soybean meal may be utilized quite as well as the whole bean. In the absence of animal milk, soybean milk is used extensively in the fresh state and as the basis of various kinds of vegetable cheeses in oriental countries. Soybean milk in the form of a powder is a commercial product in some European countries, and in parts of the United States it has been used in special feeding cases. The milk can be used successfully in numerous preparations, such as breads and cakes, in creaming vegetables, in milk chocolate, and in custards.

After separating the liquid from the solid material, the residue is still very rich in nutritive substances and can be dried and used for cattle feed or made into flour for human food.

Artificial Milk

Sugar	40 g	Dextrin	20 g
Soybean Powder	125 g	Egg Yolk, Liquid	50 g
Lactose	30 g	Calcium Lactate	6 g
Peanut Oil	20 g	Salt	2 g

Stir in water before use.

Soy Curd

The addition of magnesium or calcium salts or of rennet or lactic acid to soybean milk when hot precipitates some of the protein, forming a grayish-white curd which settles out, leaving a yellowish water liquid. This curd, after being drained and pressed, represents bean curd or tofu, which is extensively eaten and forms the basis of numerous fermented, smoked and dried cheeses in China and Japan. Bean curd is made fresh daily and is a staple article of diet among oriental peoples. In many cities of the United States having a large oriental population, fresh bean curd may be found in the Chinese and Japanese markets.

Cocoa Milk Mix

Cocoa	1¾	lb	Vanillin	1½ g
Cane Sugar	7	lb	Salt	1 g
Powdered Agar	1¼	oz		

Mix the above ingredients well and add to each gallon of milk in the pasteurizer at 185° F. Agitate for ½ hour.

Cocoa Malt Powder

Cocoa Powder	23	lb	Sodium Bicarbonate	2 oz
Fine Granulated Sugar	70	lb	Salt	8 oz
Malt Powder, Mild Flavor	20	lb	Vanillin	½ oz
Soluble Skim Milk	14	lb	Vanilla Extract	½ oz

Mix ingredients thoroughly and pass through a coarse sieve. This mixture can be packaged in cans, glass containers, or in 1¼ ounce envelopes for individual use.

Stabilized Chocolate Milk
Pat. No. 1,989,758

In carrying out the process of making the milk starch emulsion, the chocolate, sugars, when they are used, starch, and the gum may be introduced, as dry substances, into the milk. They should be thoroughly mixed, and the mixture heated to a temperature of 170 to 200° F, or higher if desired, although this is not necessary. This is preferable to temperatures approximating 240° F heretofore recommended, for periods from 20 to 30 minutes. Preferably, however, a syrup is first made of the chocolate and sugar, and this syrup, together with a preformed mixture, in proper proportions, of the starch and gum, is added to the milk, and the final mixture is agitated and heated as described.

As a matter of convenience to the beverage manufacturer, and in order to insure correct proportions between starch and gum, the starch and gum may be compounded together, and the compound delivered to the beverage manufacturer.

In making the compound, the agar-agar, for example, is preferably ground dry and screened to the same degree of fineness as the starch and is then thoroughly mixed with the starch in the proportions indicated by the specific examples given below. In such a mixture the agar-agar, although very small in quantity, approximately from 1 to 20 parts of agar to 100 parts of starch, will remain evenly distributed in the starch. It will not sift out. This novel mixture will disperse in the chocolate vehicle much more easily than if the ingredients were introduced into the liquid as separate substances. If the agar is not finely ground it will swell instead of dissolving, particularly at the low temperatures preferably used in compounding, with consequent loss of stabilizing power.

The following examples of typical mixtures, with preferred percentages of the ingredients, will serve to illustrate the character of the present invention. The percentages are by weight.

Formula No. 1

	Per Cent
Milk	90.48
Cane Sugar	4.82
Dextrose (Cerelose)	2.41
High Grade, Dark Cocoa	1.27
Raw Tapioca Starch (Scott Test 150)	1
Agar-Agar	0.02

Any suitable sugars may be used in the suspension or in the dry product or the sugar ingredient may be omitted if desired. The amount of the sugar ingredient may be varied to any extent. For any usable quantity the sugar does not add to the viscosity of the beverage. The amount of cocoa or chocolate may also be varied. Taste or economy will govern any increase or decrease. As much as 2.5% of cocoa may be used without changing the percentage of starch or gum. The starch ingredient may be increased to 2 or 3%. Experience has shown that 1% is near the critical lower limit. More than 2 or 3% gives too high a viscosity and is likely to give a distinct starch taste to the product. The agar-agar may be varied in amount from about 0.01 to 0.2%, but at the upper limit there is a strong tendency to segregation in jelly-like lumps.

Formula No. 2

	Per Cent
Milk	90.78
Cane Sugar	4.07
Cerelose	2.03
Cocoa (Cheaper Quality Than in No. 1)	1.673
Raw Corn Starch (Scott Test 100)	1.433
Gum	0.024

The first four items may be varied as indicated in No. 1.

The same quantity of modified corn starch may be used in place of the specified raw corn starch. The amount of corn starch may vary between 1 and 2%. Where raw corn starch is used the lower limit of the gum quantity should not be quite as low as in No. 1.

Formula No. 3

	Per Cent
Milk	91
Cane Sugar	4.01
Cerelose (Corn Sugar)	2.03
Cocoa	1.676
Wheat Starch (Scott Test 85)	1.2
Gum	0.024

The variations may be substantially the same as with No. 1.

The time of cooking with the raw corn starch should ordinarily be 25 to 30 minutes; with the modified corn starch 20 to

25 minutes; with the tapioca and wheat starches about 20 minutes.

Cocoa Milk

In this improved formula use:

Cocoa	20 lb	Skim Milk	90 lb
Sugar	90 lb		

To the above syrup add 2000 pounds of milk; heat to 143½° F and hold for 30 minutes. Homogenize the mixture at 2000 to 3000 pounds pressure while hot.

Cool and bottle.

Cocoa Milk, Non-Settling

Cocoa Powder	6 oz	Sodium Alginate	1 oz
Sugar	28 oz	Milk	15 qt

Mix together the cocoa, alginate and sugar. Heat the milk to 160° F, add the dry mixture slowly with constant stirring, for 30 minutes. Cool the batch to 45° F and hold for 2 hours before bottling in sterilized bottles. The cocoa powder can be of any fat content from 10 to 25%. The milk can be either whole milk or skim milk, or any mixture of each. Additional flavoring ingredients such as vanilla, malt, caramel, etc., may be added.

Cheese

Cheese is essentially a processed form of casein, or milk protein, which contains butter fat. The milk is generally acidified by the addition of lactic acid bacteria and the curd or casein coagulated by a soluble ferment called rennet. It is then separated from the whey which consists of water, milk, sugar, a little fat and minerals. One part of rennet may coagulate as much as 250 thousand parts of milk. Before the rennet is added the milk may be tested for ripeness for curdling. For this test, a teaspoonful of diluted rennet extract is added to 4 ounces of milk and, if the coagulation takes place within 45 to 60 seconds, the milk is ready for the addition of rennet.

Sour milk cheeses such as cottage, cream, Neufchâtel, Gervais and buttermilk owe their flavors chiefly to the lactic acid produced by the lactic acid bacteria. The characteristic flavors of other varieties of cheese are developed largely during later processing. These flavors result from exposure of the cheese to bacteria, molds, or yeasts that occur naturally or are added. The striking differences in the flavor and consistency of cheeses may also be due to the type of milk used, e.g., whole milk, skimmed milk, cream, or a combination of milks. The milk may be derived from cows, goats, sheep or it may be a mixture of these. The time spent in ripening and aging the cheese determines the definite characteristics, e.g., for cheddar types (ordinary American or store cheese) the lactic acid bacteria are permitted to develop in great numbers before the rennet is added, whereas in Swiss, Camembert or Limburger cheese, they are not. Also, in cheddar, the moisture content or whey, left in the curd is considerably less than that present in Camembert; this means that more fermentable milk sugar is present in Camembert. Emmenthaler (Swiss) curd is cooked much longer, and at a higher temperature, than cheddar. Camembert curd is not cooked at all. The amount of salt added to cheeses, whether it is incorporated before or after pressing, the difference in temperature during ripening and intensification of the activity of bacteria, e.g., in Roquefort cheese, all vary for different cheeses.

Great skill, acquired only after long practice based on years of experience, is required for the regularly repeated production of a specific flavor and consistency. These depend, not on one but on several, forms of minute microscopic life. For example, in Camembert, the chief biological factors are lactic acid bacteria, the ever-present milk mold, and a foreign Penicillium mold, the blue-green bread mold. If the lactic acid bacteria are permitted to develop in too great numbers before the curdling, the consistency of true Camembert cannot be attained. If the Penicillium overcrowds the milk mold, the consistency may be good, but the flavor is inferior.

Cheeses are commonly graded as special, fancy, good, prime and common. The terms used to denote size and weight are fanciful expressions such as, *Young America* for small cheese of 9 to 11 pounds, and *Long Horn, Daisy,* etc., for larger specimens.

Whole milk, or full cream, cheese is made of milk from which no portion of the fat has been removed. Cream cheese is made from milk and cream by a special process.

COMPOSITION OF MILK, MILK PRODUCTS AND BUTTER

Kind	*Per Cent Water*	*Per Cent Butter-fat*	*Per Cent Protein* (Casein and Albumin)	*Per Cent Milk Sugar*	*Per Cent Min-erals*	*Per Cent Total Milk Solids*
Liquid whole milk	88	3½	3¼	4½	¾	12
Evaporated whole milk	72	8	7¾	10½	1¾	28
Sweetened condensed milk	31	8	7¾	10½	1¾	28
(Cane sugar)	—	41	—	—	—	—
Powdered whole milk	1½	27½	27	38	6	98½
Liquid skimmed milk	91	—	3½	4¾	¾	9
Skimmed sweetened condensed milk	29	43	11	14½	2½	28
(Cane sugar)	—	—	—	—	—	—
Powdered skimmed milk	2½	1½	38	50	8	97½
Liquid 18% cream	75	18	2½	4	½	25
Powdered cream	½	72	10½	14½	2½	99½
Butter from fresh cream	13.03	83.75	0.64	0.35	0.14 2.09 Salt	—
Butter from ripened cream	13.78	82.97	0.84	0.39	0.16 1.86 Salt	—

American cheese, American cheddar, Store cheese — Made from sweet milk, whole or skimmed, in large molds, weighing from 10 to 60 pounds; of varying shades of yellow or uncolored; originally made in Cheddar, England, but now imitated in the United States; many varieties of packaged cheese made from it

Brick bat — An English cheese made from new milk and pressed into a brick-like form

Brie cheese — A richly salted cream cheese made in Brie, France

Camembert cheese — A finely flavored cream cheese, of a pale yellow color, made in France;

usually put up in small round wooden boxes and in round tins

Cottage cheese, Clabber cheese, Dutch cheese, Pot cheese, Smeerkaas — A pressed curd, into which cream is added, without rennet; cottage cheese of the best texture contains 70 to 75% of water

Cream cheese — A soft cheese made from unskimmed milk with or without the addition of cream; name does not imply that the cheese is made from cream

Dunlop cheese — A soft cheese similar to single Gloucester, made in Ayrshire, Scotland

Dutch cheese — a. A hard cheese made from skimmed milk pressed into round molds, the outside being colored a reddish tint with a preparation of madder
b. The same as cottage cheese

Edam cheese — A hard, mild cheese made in Holland; molded in round forms and usually coated with a dark red color or with paraffin; each cheese weighs from 3 to 4 pounds; packed 12 pieces to the case

Filled cheese — A kind of cheese made from skim milk, so called because a fatty filling such as lard or oleomargarine is used as a substitute for the cream

Gloucester cheese — A rich English cheese of mild flavor; called single or double Gloucester, depending on the amount of cream in its composition

Gorgonzola cheese — A rennet cheese made from fresh whole cow's milk in Northern Italy; takes its name from the village of

Gorgonzola, a few miles from Milan, but its manufacture has spread over a wide area; at its best, Gorgonzola is very similar to Roquefort

Gruyère cheese Schweitzer, Swiss or Emmenthaler cheese — A cheese made in France and Switzerland, intermediate between hard and soft, yellow, and of an open structure; characteristic holes of this cheese are caused by gas-producing bacteria during the process of ripening; a mild cheese, dry, and of a distinct, somewhat sweet flavor

A *Domestic Swiss* is made in the United States but is inferior to the imported cheese

Limburger cheese — A soft cheese originally made at Limburg, Belgium, and now in the United States; is kept until putrefaction begins; put up in one- and two-pound bricks; packed in full cases, weighing about 125 pounds, half and quarter cases

Lincolnshire cheese — An English cheese made in small molds from new milk and cream

Neufchâtel cheese — A delicate cream cheese made in small molds at Neufchâtel, France

Parmesan cheese — A hard, dry cheese of delicate flavor and colored with saffron; made in Italy; when broken, has a granular appearance; used grated for seasoning macaroni; long-keeping qualities

Pont l'Evêque cheese — A cream cheese similar to Neufchâtel cheese

Primost cheese — A cheese made from whey and brown

sugar, a by-product of American cheese; put up in two and one-half-pound pieces; especially liked by the Scandinavian people

Roquefort cheese — A hard cheese made at Roquefort, France, from ewes' milk; when almost solid and dry it is placed in caves, in which the temperature remains at 40° F, where it takes on a reddish color and develops the green mold around the bread crumbs used in preparing it

Sage or Green cheese — Regular American cheese to which leaves of the sage have been added to impart a green color and the sage flavor

Sapsago cheese — A hard skim milk cheese, made in conical molds, weighing about 6 ounces; contains a powdered herb of the clover type, which gives its characteristic flavor and green color; used for grating purposes mainly and especially liked by the German people who call it *Kraeuter-Kaese*

Slipcoat cheese — A rich cheese made from milk while it still retains its natural warmth; soft and resembles uncolored butter

Stilton cheese — A double cream cheese of strong taste originally made at Stilton, England

The government grades of most general interest to the consumer are:

1 U. S. Extra Fancy	Score 95 or above
2 U. S. Fancy	Score 92–94
3 U. S. No. 1	

Maximum Score	
Body and texture	40
Flavor	30
Finish and appearance	20
Color	10
Total	100

VARIATIONS IN COMPOSITION, AGING, ETC.

		Approximate Content			
Kind of Cheese	*Aging*	Per Cent Water	Per Cent Fat	Per Cent Protein Ash	*Milk Product Used*
American cheddar	Unripened to fully ripened	36	34	30	Whole cow's milk
American cheddar	Unripened to fully ripened	57	5	38	Skimmed cow's milk
Swiss	Ripened	33	30	37	Whole cow's milk
Cream	Unripened	39	47	14	Cream or enriched cow's milk
Cottage	Unripened	72	1	27	Sour skimmed cow's milk
Roquefort	Ripened	38	32	30	Cow's milk

Cheese Spread Manufacture

Cheese spreads are combinations of cheese with condensed whey, whey powder, condensed skim milk, skim milk powder or a combination of whey and skim milk. Avoid an excess of milk sugar in the manufacturing process. The cream curd used is usually made from a 6% fat milk which is pasteurized at 170° F for 30 minutes. Sometimes the addition of less than 1% of gelatin is used in this milk since it seems to hold the cheese curd together and also results in a smoother cheese spread. After the pasteurized milk has been held for 30 minutes, it is cooled to 145° F, homogenized under 2500 pounds per square inch pressure and allowed to cool to 60° F or even lower. Twelve to fourteen hours before the cream curd is to be used, the milk is once more brought up to 69 to 70° F, and a 5% starter added. This starter is a lactic acid culture and should give the milk a smooth curd and a rich creamy flavor.

Pimento Cream Spread

Cream Curd (From 6% Fat Milk)	252	lb
Pimento (Pass Through 5/16 in. Grinder Plate)	62½	lb
Vinegar-Sugar Solution	40	lb
Gum Paste	31	lb
Salt	5	lb
Certified Cheese Color	6	oz
Cayenne Pepper	1⅓	oz
Sugar	4	lb

Mix thoroughly before putting into cooking kettle. Heat to 170° F. Run through filling machine into sterilized glass containers of desired capacity. Use of emulsifier is optional but if it is used, add 1% sodium citrate or 2% di-sodium phosphate calculated on weight of batch mix.

Vinegar-Sugar Solution
(Called for in Cheese Spread)

Cider Vinegar	3 lb
Sugar	2 lb

Heat slightly and gently, if necessary, to hasten the solution of sugar.

Gum Paste
(Called for in Cheese Spread)

Gum	1 lb
Water	20 pt

The gum may be tragacanth, karaya or locust bean. The proportion of water used is not arbitrary. The mixture must not be too thick or too thin. An egg beater or similar device is helpful in getting the gum and water combined into a smooth-textured paste.

Relish Spread

Cream Curd (From 6% Fat Milk)	250 lb
Sweet Pickle Relish	60 lb

Vinegar-Sugar Solution	40 lb
Pimento (Through an ⅛ in. Grinder Plate)	27 lb
Condensed Whey or Whey Powder plus Water	30 lb
Gum Paste	40 lb
An Emulsifier (Optional)	

Mix constituents thoroughly, heat to 165° F or slightly higher, depending upon flavor desired, and run through filling machine into sterilized glass containers of desired capacity.

Special "Sharp" Spread

Smooth American Cheddar Cheese	340 lb
Salt	2½ lb
Di-Sodium Phosphate	6 lb
Condensed Whey	84 lb
Water	40 pt
Certified Cheese Color	To suit

If needed to firm up the batch, 1 to 2 pounds of sodium citrate or of sodium meta-phosphate may be added.

The American cheddar cheese should be partly broken down. One-half should be of clean flavor and aged.

Add cheese to heated kettle, slowly, as in making the limburger product. Add 15 pints of the water at the start. Do not add the condensed whey until after the batch has been heated to about 135° F. Finish by heating to 160° F and packaging in containers other than glass. Make into packages similar to those of processed cheeses.

Limburger Cheese in Glass

Limburger Cheese	180 lb
Cream Curd (From 6% Fat Milk)	60 lb
Water	10 lb
Di-Sodium Phosphate	10 lb
Salt	2 lb

Mix thoroughly and heat to 174° F. Run through filling machine into sterilized glass containers of desired capacity.

The preferred way of cooking limburger cheese, to assure

good spreading consistency, is to use 80 pounds of soft smooth limburger, 40 pounds of a firm cheese with a sharp limburger flavor and 60 pounds of limburger with a *clean* but slightly *off* flavor. Put 70 pounds of this limburger mixture in the cooking kettle with steam turned on and heat with stirring, until fat begins to separate sufficiently to give a slightly glistening appearance to the cheese which pulls together to form a mass.

Add 5 pounds of the phosphate and 1 pound of the salt with continued heating and stirring. Add another 40 pounds of the limburger mixture and stir until mass is smoothed out. Add 30 pounds of the cream curd and stir until smoothed out. Add remainder of limburger mixture. Heat and stir until smooth before adding the remaining 5 pounds of phosphate, 1 pound of salt and 30 pounds of cream curd.

Caution:

After the first charge of cheese is put into the kettle, additions of more than 60 pounds should not be made. Also, subsequent additions should not be made until the charge in the kettle has been thoroughly smoothed out by gradual heating and constant stirring by the mechanical agitator in the cooking kettle. The procedure followed in building up the batch in the kettle may be varied according to the operator's experience and the particular properties desired in the finished limburger cheese mixture.

Processed Cheeses

Processed cheeses offer to the manufacturer a new field for the sale of his milk products. Manufactured cheeses have always been an accepted commodity but processed cheeses offer a new way to make a variety of flavor and taste blends, which are meeting with increasing acceptance by the consumer.

American Processed Cheeses

For a white product, a white cheese should be used throughout; in making a yellow or orange product, the proper amount of annatto color should be added. The name cheese in the following formulas means cheddar cheese (American cheese, American cheddar cheese).

Formula No. 1

Green Young American Cheddars	6	
Sharp American Cheddar	1	
Salt	3	lb
Sugar	2	lb
Di-Sodium Phosphate	13	lb
Water	4	gal

1 The total weight of cheese equivalent to seven cheddars should be kept constant.

2 The ratio of sharp to young cheese may be varied to meet flavor requirements.

3 The water content should be adjusted so that the final product does not contain more than 40% moisture; 50% of the total solids should be butterfat.

4 It is suggested that half a cheddar of young cheese be left out in the above formula and replaced with any or all of the following:

a Previously prepared process cheese

b Swiss cheese

c Part skimmed cheese. The use of strong cheeses should be avoided as much as possible and the moisture limits and legal fat content should be watched closely. Stainless steel, or a similar unreactive metal, should be used for the stirrers, which are constructed for efficient scraping down of the sides of the kettle. The blades should be constructed to permit agitation on all levels; half of the blades should turn clockwise and the other half counterclockwise. The lid should be constructed of two tight-fitting halves with a draw-off at the lowest point in the kettle. In manufacturing processed cheese, about ¾ to 1 ground cheddar cheese should first be added to the kettle and melted down; water, sugar, salt, phosphate and, finally, the rest of the ground cheddars are added. The contents are heated to 170° F and held there for 10 to 15 minutes. The processed cheese is now drawn off into filling machines.

Modification of Above Formula

1 Brick blended with one young American cheddar and an equivalent amount of limburger added

2 Three or four cheddars and an equivalent amount of Swiss added; the rest of the ingredients added in the same proportions as in the above formula

Italian Type Process Cheese
Formula No. 1

	I	II	III
Green (Young) American Cheddar	1½ lb	—	—
Sharp (Aged) American Cheese	2 lb	1 lb	1 lb
Sbrinza (Argentina)	2 lb	4 oz	6 oz
Monterey (Argentina)	½ lb	—	—
Sodium Chloride	1 oz	½ oz	½ oz
Anhydrous Di-Sodium Phosphate	2 oz	—	—
Water	1¼ pt	½ pt	⅝ pt
American Processed Cheese	—	2 lb	2½ lb
Romano	—	⅜ lb	½ lb
Di-Sodium Phosphate (U.S.P.)	—	1½ oz	—
Sodium Citrate	—	—	2 oz

Italian Type Pimento
(Very Sharp with Red Pepper)
Formula No. 2

Sharp American Cheddar	3 lb	1 lb
American Processed Cheese	2 lb	3½ lb
Water	¾ pt	2 oz
Monterey Cheese	2 oz	—
Romano Cheese	4 oz	—
Red Peppers	¾ oz	¾ oz
Pimento Meat	4 oz	4 oz
Salt	½ oz	½ oz
Di-Sodium Phosphate (U.S.P.)	2 oz	2 oz
Pimento Juice	—	½ pt

Formula No. 3

	I	II	III	IV	V
Munster Cheese	7 lb	7 lb	7 lb	8 lb	7 lb
American Swiss Cheese	12 oz	12 oz	12 oz	—	—
American Cheddar	1 lb	1 lb	1 lb	3 lb	3 lb
Sodium Citrate	3 oz	—	4 oz	2½ oz	3 oz
Anhydrous Di-Sodium Phosphate	1 oz	4 oz	—	1 oz	1 oz

Sugar	4 oz	4 oz	8 oz	1 oz	1 oz
Whey Powder	2 oz	2 oz	2 oz	2 oz	2 oz
Water	⅝ pt	¾ pt	¾ pt	½ pt	½ pt

Cream Cheese

Sweet Cream (18% Fat)	100 lb
Locust Bean Flour	1½ oz (Optional)

Pasteurize at 160 to 180° F for 30 minutes. Cool, mix to 120 to 140° F. Homogenize at 3500 pounds, then cool to 72 to 75° F. The cooled homogenized cream is drawn into a coil vat where a 1% lactic culture is added with thorough agitation. Fermentation is allowed to proceed for 12 to 14 hours. After 0.7 to 0.8% titratable acidity has been reached, the curd is broken by rotating the coil. One to 2% salt is added, and the soured mix is heated slowly to 120 to 140° F and cooled immediately to below 50° F.

The cheese is drawn into bags, and each bag is tied with string. These bags are placed on wooden racks in an ice chest to drain. They should be reversed and repiled daily to facilitate drainage during the 2 or 3 days which are required to reach the proper condition.

The cheese usually is packed in special paper-lined boxes by a wormfed packing machine with a wooden block.

Cheese, Ice-Cream and Salad Stabilizer
U. S. Patent 2,007,218

Locust Bean Gum	65 oz
Powdered Irish Moss	35 oz
Karaya Gum	15 oz

When the stabilizer is used in the preparation of cream cheese, it is added at the time the curds are mixed with the cream, following the usual procedure for the manufacture of cream cheese, and in the proportion of about ½ of 1% by weight on a wet basis. The material is heated to about 165° F, homogenized, and then packed hot.

In ice cream, the stabilizer is used diluted with sugar, in the preferred proportion of 0.5%, on a wet basis. The stabilizer acts to prevent crystallization of ice particles and insures a fine, smooth texture and a body which will hold up under severe shocks, such as are encountered in transportation and handling. The use of it in ice cream usually results in more rapid freezing, especially in old-style freezers.

Cheese Emulsifiers
U. S. Patent 1,940,031

One to 4% of any of the following are used:

Sodium Mucate
Sodium Lactate
Di-Sodium Phosphate

Preservation of Rindless Cheese
British Patent 434,374

Bacterial action on the surface of rindless cheese is prevented by treatment with the following prior to heating to 65° F:

Hydrogen Peroxide (35%)	0.3%

A cheese low in fat content is first heated to 65° C, the peroxide is then added, the batch is mixed and reheated to 80° C.

Manufacture of Neufchâtel Cream Cheese

Cream cheese is a soft unripened cheese, usually made from cream testing from 12 to 20% fat, coagulated by the development of acidity or by the use of rennet, and then pressed in cloth bags. The product is creamy-white in color, of a fine smooth texture, and having a full rich creamy flavor. The fat content of the cheese varies from 30 to 40%. There is a corresponding variation in moisture.

Neufchâtel is a similar cheese, made from either whole milk or cream, and has a correspondingly lower fat content. The popularity of this and other soft cheeses has grown tremendously in recent years.

The Commercial Manufacture of Neufchâtel Cheese

This cheese, although commonly made from cream testing between 10 to 20% fat, may also be made from milk or cream of any fat content. Cream with a fat content of more than 25% is ordinarily impractical and uneconomical to use. If the fat content of the raw material is approximately equal to that of whole milk, the cheese is commonly called Neufchâtel cheese rather than cream cheese. Care should be taken in selecting the raw material. Milk or cream used in the manufacture of cream cheese should be fresh, sweet, clean, free from all foreign flavors and should contain as low a bacteria count as possible.

The cream is standardized to the desired fat content of 12 to 20%; the higher the fat content, the smoother, richer and better the keeping quality of the finished cheese. It should be heated to 145° F and homogenized at 2000 to 3000 pounds pressure, for 30 minutes, after which time it is cooled to 72° F and set with 0.10 to 0.25% of starter to give it an acidity of 0.6 to 0.7%, in 14 to 16 hours. The material may be set in shotgun cans, 10-gallon milk cans, or in any vat equipped with an agitating device.

Higher acidities make a sour cheese of poor quality, whereas lower acidities retard draining and make the body of the finished cheese rubbery or stringy. When the acidity is between 0.6 and 0.7%, the entire mass of curd is agitated and slowly heated to 120 to 125° F, or until definite whey separation is noticeable. The curd is then homogenized at 300 to 500 pounds pressure, cooled to 50 or 60° F, and placed in draining bags, 18 by 36 inches, of unbleached muslin, or similar material, that will allow free draining of the whey without loss of any curd. This last homogenization may be omitted if plant equipment or routine does not conveniently allow for it.

These bags, each holding about 10 gallons of curd and whey, are stacked in a *cooler* (40° F), either on a rack or a clean tiled floor, and frequently shifted to facilitate drainage. If the cheese is made in sufficient quantities to allow the bags to be stacked high enough, so that they will be pressed by sheer weight, no other arrangement is necessary to press the curd. If, however, the cheese is made in small quantities, as is the case in most instances, a press made from hard-wood slats 1½ by 1½ inches spaced from 2 to 3 inches apart, will serve the purpose. Under these circumstances it is only necessary to make two trays. The bags of curds are placed between these trays and additional weight, in the form of ten-gallon milk cans filled with water, is applied. The curd is pressed until it is firm enough to hold its shape. This usually requires 48 hours from the time the mixture is placed in the bags.

Salt may be added, either at the time the curd is placed in the bags, or after the curd is pressed. If added when the curd is placed in the draining bags, about 3 times as much salt is necessary, because more must be added to allow for the salt lost through whey drainage. The amount of salt used in either case depends on personal taste; 1 to 2% is usually sufficient, if added after draining. Cheese of this type is packed in one-quarter-pound cartons, glass or stone jars, metal foil packages, or in 3-or 5-pound loaves in wooden boxes. Any protection, such as that furnished by a container, a parchment wrapper, or by metal foil, which prevents the air from coming in contact with the surface of the cheese, will materially reduce spoilage. The packaged cheese should be kept as cold as possible without freezing. If properly made and kept under refrigeration, this cheese should keep for at least three weeks.

By adding various flavoring materials such as pimentos, olives, nuts, sweet pickle relish, honey, and similar materials a great many varieties of cheese and cheese spreads can be made. In making pimento cheese, one No. 2½ can to each 15 pounds of curd gives very satisfactory results. Olive nut cheese can be made by mixing 2 pounds of chopped olives and 1 pound of chopped pecans with 15 pounds of curd. Relish cheese is made by mixing 2 pounds of sweet pickle relish with 15 pounds of plain or pimento-flavored curd.

These cheeses find a ready market in most communities and afford an excellent outlet for milk which might otherwise become a burdensome surplus. Opportunities for developing this type of cheese business are practically unlimited and depend only on the ingenuity of the dairyman.

Another Method of Making Bakers' Neufchâtel and Cream Cheeses

The starting material may be skimmed milk, whole milk, or whole milk to which the cream from an equal volume of milk has been added. The resulting products will vary in flavor and in texture according to the amount of fat incorporated. Raw milk of good quality may be used, but pasteurizing the milk for 30 minutes at 143 to 145° F, and cooling promptly to 72° F, will tend to insure better cheese.

The starter is added at the rate of 1 teaspoonful to each quart of milk, or ½ pint to each ten-gallon can, or ⅗ pint to each 100 pounds.

Rennet extract, well diluted in cold water, is then added at the rate of 30 drops to each 100 pounds of milk. For small amounts of milk, ¼ of a junket tablet in 6 tablespoons of cold water may be used. Use 1 tablespoon of this solution to each quart of milk. The milk must be stirred well while the rennet solution is being added.

The milk is allowed to stand undisturbed at 72° F, until the curd is jelly-like and firm, and whey just starts to appear at the surface. A period of 12 to 15 hours is usually sufficient for setting.

While not imperative, cooling of the curd is beneficial, and stimulates drainage. This may be accomplished by setting the container in ice water until the temperature of the curd is down to approximately 50° F. Gentle stirring of the curd hastens the cooling. The addition, at this time, of a heaping tablespoon of salt for each 15 pounds of milk will also hasten drainage and will retard development of acid.

The curd is now dipped, or poured, into muslin draining cloths, or bags, and the bags hung in a cool place to allow the curd to drain. If clean, pure ice is available, small pieces may advantageously be added directly to the curd. The cloths must be opened occasionally and the dry curd scraped toward the center, with a large spatula, table knife, or spoon.

The curd is re-tied, and drainage continued until the whey is almost completely expressed. Pressure is applied, gently at first, because the soft curd tends to fill the meshes of cloth and interfere with the passage of whey. One method is to use a pail of water as a weight either on the end of a lever or set directly on a plate or board covering the cheese. The pressure may be readily varied by altering the amount of water in the pail. The drainage is slow, at best, and it is advisable to pack the curd bags in crushed ice during the pressing period, to check the development of acid.

Ability to judge when the curd is sufficiently dry comes with experience. The curd should be firm enough to show a clean break like cast iron or steel, and should be fairly resistant to pressure. The average yield is 14 to 16 pounds from each 100 pounds of whole milk, and 18 to 20 pounds from creamed milk.

When the curd has been sufficiently pressed, 1 to 2 pounds of salt to each 100 pounds of curd, or about 1 teaspoon to a pound, are added. The curd can then be smoothed by mixing, or by the use of a food chopper or cake mixer. Breaking and remixing the curd after pressing can be avoided by stirring the salt in, just before the curd is put to press. This method requires a greater amount of salt, to compensate for that which passes out in the whey, but reduces the labor, and keeps the cheese in cake form as it comes from the press.

Manufacture of Roquefort Cheese from Goat's Milk

The original Roquefort cheese process, as perfected in France, employs sheep's milk. In America, a similar cheese has been made successfully from cow's milk. Goat's milk, which is the milk used in the procedure that follows, is similar to sheep's milk in color and body, and, if manufactured with the usual dairy care, a cheese of excellent quality should result.

In the manufacture of this cheese the methods outlined below may be modified to meet specific conditions.

Fresh clean milk is necessary for making Roquefort cheese. The acidity should be between 0.16 and 0.18% for the best results. Over-ripe milk is likely to result in a gassy cheese and, sometimes, in an objectionable flavor. From 3 to 4% of good clean lactic starter, depending on the acidity of the milk and starter, should be added.

For sheep's milk, the temperature of setting used in France is from 76 to 84° F. Cow's milk should be set at 82 to 85° F, and goat's milk at 82° F. When goat's milk is set above this temperature, difficulty is encountered with cracking of the curd, but if 20% of cow's milk is added to the goat's milk this difficulty is overcome. When a combination of cow's and goat's milk is used, it is advisable to set the milk at 84° F. A lower temperature is apt to result in a soft, mushy curd which fails to drain properly, whereas a higher temperature may result in a tough curd.

The milk is heated in an ordinary cheese vat and stirred to insure even distribution of fat and heat. After the acidity has developed to 0.21 or 0.22%, commercial rennet, diluted from 20 to 40 times with pure cold water, is added to the milk, which is then agitated thoroughly.

The setting period should be approximately 1½ hours, but varies with the acidity, temperature and quality of the milk. Beads of whey, which gradually become a film, spread over the surface of the curd, indicating that it is ready for cutting. The curd is cut in two directions with a coarse, vertical knife with wires ⅝ to 1 inch apart. Soon after this, the curd slowly contracts and draws away from the sides of the vat. This contraction should not be allowed to go too far before dipping the curd.

A cheddar cheese vat similar in capacity to that in which the milk was set, makes an excellent drain vat, if it is provided with a false bottom covered with cheese cloth or muslin of good quality. The two vats should be placed near enough together to permit two or three workers to do the dipping. Large tin scoops, each with a capacity of 3 or 4 quarts, are suitable for this purpose.

The curd should be transferred to the drain vat rapidly and carefully. Two men should be able to take the curd from 1500 pounds of milk in 10 to 15 minutes. Raising the cloths from time to time aids in expelling the whey and in bringing the draining curd to the proper condition of dryness. The time required for draining is 20 to 40 minutes, depending on how often the cloths are raised during this period. The curd should be placed in the forms in a fairly moist condition. When it is too dry, the surface of the cheese becomes so open and rough, that undesirable organisms may have access to the interior. Greater losses occur in the curing process when such conditions exist.

The whey should be drawn from the drain vat and run through a whey separator. The average fat loss from goat's milk is 0.50 to 0.60%, whereas that from cow's milk is 0.20 to 0.30%. When 20% of cow's milk is added to goat's milk, the fat loss is reduced to about 0.40%. It is probable that a satisfactory reduction of loss would be obtained with somewhat less than 20% addition of cow's milk.

The hoops, or forms, for Roquefort cheese are made of galvanized metal with wire edges. They are round and open on both ends, and are 7⅜ inches in diameter and 6 inches high. They have six horizontal rows of holes, each row having twenty-five holes, ⅛ inch in diameter. The hoops are placed upon special reed mats, 10 x 24 inches, which, in turn, are placed upon boards of approximately the same size, each of which will hold three hoops. Cloths may be used instead of mats, but they are not as satisfactory. All hoops, mats and boards should be thoroughly cleaned and sterilized before using, to reduce the number of foreign molds which may appear later.

When it is put into the hoop, the curd is a white pulpy mass and should contain only a small excess of whey. The hoop is filled with four or five layers of curd, between which liberal applications of mold powder are sprinkled.

The drain room should have a temperature of 65 to 68° F and a relative humidity of 85 to 90%, to prevent the cheese from drying out. Higher temperatures favor the development of injurious molds. The cheese is left in the room for about 4 days. It is turned several times during the first day and twice daily thereafter, the object of the repeated turnings being to hasten drainage and insure a smooth, even surface. The hoops, which are allowed to remain around the cheese for the first 1 or 2 days, should be washed and sterilized before using them again. The boards should be washed every day. The cheese is scraped each day while in the drain room, in order to remove the coating of slime which forms over the surfaces.

The salting room should be dry and have a temperature of approximately 48° F. There should be sufficient ventilation to remove the excess of moisture drawn to the surface of the cheese by the salt. The cheese should be salted 2 or 3 times at two-day intervals, depending on the amount of salt desired and the conditions affecting its penetration. The cheese should be placed in the salting room several hours before the first salt is applied, in order to cool it to room temperature. Cheese salt is then rubbed vigorously over the surface, until no more will adhere.

After the cheeses are salted, they are piled two or three high, and the following day their positions reversed, with no addition of salt. On the third day they are again salted in the manner previously described. On the fifth day they may receive a light sprinkling of salt, if intended for markets preferring a highly salted cheese. The entire salting period should take from 10 to 14 days, depending on the type of cheese desired.

Some experience is required to salt Roquefort cheese successfully, the usual tendency being to undersalt or oversalt. Cured Roquefort cheese should contain at least 4% salt, an amount which is sufficient to prevent an excessive development of the mold, Oidium lactis. Most of the cheese made in the above way contained from 5 to 6%, which did not seem to affect the quality for most consumers, although some preferred a less salty cheese.

During the salting process a slimy coat, composed of softened cheese, whey, salt and micro-organisms, begins to form. This coating is scraped off with an ordinary case knife. Care should be taken not to scrape the cheese too deeply, since, at this period, a high percentage of salt on the surface is essential.

The cheese is punched immediately after the scraping. Oxygen is necessary for the proper growth of the Roquefort mold, and in order to let the air in, each cheese is punched with a needle in twenty to fifty places. The needles are about 1/8 inch in diameter and about 8 inches long. The cheese is placed on edge after the punching in order that the holes may not close, and that the maximum amount of air may come in contact with the mold.

Ripening or Curing

The curing room should have a temperature of 42 to 48° F, a relatively high humidity, about 92%, and moderate ventilation.

CONDITIONS FAVORABLE FOR CURING GOAT'S MILK ROQUEFORT CHEESE

Period	*Temperature in ° F*	*Per Cent Relative Humidity*	*Ventilation*
First 4 days	65–68	85–90	Slight
Next 10 days	46–48	80–90	Considerable
Third week	42–48	90–95	Moderate
Fourth and fifth weeks	42–48	90–95	Moderate
Second and third months	42–48	90–95	Moderate
Four and fifth months	46–48	80–90	Considerable

The humidity and ventilation indicated in the above table cannot be expressed with any great degree of accuracy at the present time. The table shows in a general way, however, the conditions that have proved most successful. With experience, they can be properly controlled.

In work done on this cheese, well-insulated curing rooms were used. The sides, tops and bottoms were insulated with 4 inches of cork laid in hot asphalt and covered with Portland cement plaster. Direct expansion ammonia refrigeration and brine coils were used in order to maintain the low temperature. A perforated water pipe was placed along the top of the room so that water could trickle down over a screen behind the shelves in order to maintain the proper humidity. Water, cooled to room temperature, is more economical than tap water, since the cost of refrigeration is lower. A vacuum cleaner motor was installed, and run continuously at low speed, in order to obtain the proper circulation of air. The air was drawn from the curing room and forced back again. Outside air was taken in only when the door was open.

During the first 2 or 3 weeks there should be a gradual development of a reddish slime upon the surface of the cheese, together with patches of white and green mold. In cases where the reddish slime fails to develop rapidly, scrapings may be obtained from other cheese. These, mixed with sterile water and then sprinkled on the cheese, give the desired growth without difficulty. This growth should be removed by scraping every three or four weeks in order to permit the air to reach the mold in the interior of the cheese. After this scraping, there is only a slight development of the white and green mold, but there is much greater development of reddish slime.

Within 10 days after punching, mold can be seen through the punched holes when the cheese is held before an electric light. Within 2 or 3 weeks or a month, there is probably as much green mold visible as when the cheese is cut several months later. The mold in the center of the cheese appears to develop more rapidly than that nearer the surface, probably because of the difference in concentration of salt. In the earlier stages of ripening, the cheese often tastes bitter; later this condition disappears and a sweet piquant flavor is noticeable with little or no bitterness. With the proper curing conditions, the texture of goat's milk Roquefort cheese can be made similar to that of the imported variety. In some cheeses, a typical peppery flavor, repellent to the taste, develops. This is probably due to an intense oxidation, evidenced by the production of formic acid, which occurs when the cheese is held for a long period at a high temperature.

High humidity prevents the cheese from drying and increases the rapidity of growth of the slime on the surface. No experimental data are available to show that the surface growth aids in the ripening of the cheese, but its presence is always associated with the best grades. This slime seems to check the growth of surface mold and to keep the cheese in a sweet, normal condition. Dryness causes the cheese to ripen slowly with a flavor lacking the characteristics of good Roquefort, and a texture which is too waxy.

The cheese may be taken from the curing shelves when about 3 months old, scraped for the last time, and wrapped in tin foil or aluminum foil. At this time there should be an abundance of mold and some flavor, but as a rule, the full flavor develops only after the cheese has been left in the foil for a few months.

Enclosing the cheese in the foil prevents drying, excessive oxidation, and the escape of the volatile substances. It also helps to develop the sweet flavor and reduces the formation of strong, biting, and soapy flavors which may occur when cheese is exposed too long to the air. Parchment paper is used to protect the foil. Paraffin cannot be used as a substitute for foil because of the moist condition of the surface of the cheese.

Roquefort cheese made from goat's milk requires from 4 to 7 months to ripen. When ripened it may be placed in cold storage at about 34° F, for eight months to one year, in order that a uniform supply of cheese may be placed on the market throughout the year. During this storage, the cheese should become sweeter and more fragrant.

Losses in Curing

In Roquefort cheese made from cow's milk, the average loss from scraping, brushing and handling is 7.4%, which seems to be less than the loss in sheep's milk Roquefort. The losses in goat's milk Roquefort are very small because of the firmness of the cheese.

The loss from shrinkage will vary with the moisture content of the cheese at the time it is wrapped in foil, the temperature and the time it is kept before shipment. Even under the best conditions, there is some shrinkage, due largely to evaporation and leakage of whey.

Yield

Goat's milk with an average butterfat content of 3.4% gives an average yield of 10.34 pounds of cheese for every 100 pounds of milk.

Preparation of Mold Powder

Mold powder for inoculation purposes is grown on ordinary white bread. Stock cultures of Roquefort mold should be obtained from a reliable laboratory several weeks before the manufacture of cheese is contemplated.

The mold powder is prepared in the following way. Fresh loaves of bread are sterilized by heating in a drying oven for 2 hours at 170° C (338° F), and then cooled to room temperature. Meanwhile, a suspension of mold is made by transferring spores from an agar culture of Roquefort mold, by means of a sterile platinum needle, to a flask of sterile water and shaking.

A sterilized 1 cc pipette is then filled with this suspension and the bread inoculated by allowing a small quantity of the suspension to flow into numerous holes, previously punched with a sterile glass rod or a sterile pipette. At least 1 ounce of the suspension should be used for each pound of bread, and the punctures must be well distributed to insure an abundant growth of mold in all parts of the loaf. The inoculated bread is put in a moist cold place (48° F) and left for several weeks to allow the mold to develop to the stage at which it is suitable for use in the cheese. This stage is indicated by the spread of the green mold throughout the bread.

The Roquefort mold develops more vigorously in the cheese and is less likely to be contaminated with foreign molds when grown at a low temperature. When there is evidence of foreign mold in the bread, the infected part should be removed or the entire loaf discarded. Since Roquefort mold grows rapidly at a low temperature in the presence of carbon dioxide gas, foreign molds cause but little trouble in the cheese. This is especially true with Roquefort mold grown in bread at a temperature of 48° F.

When the bread is completely invaded by the mold, it is thoroughly dried and then sliced and ground in a coffee grinder, or with a rolling pin, to a condition in which it is fine enough to pass through a salt shaker. It should be kept in a cool dry place to prevent the absorption of moisture. Before drying, the mold in bread is light green, but, upon drying, it darkens. It will keep for several months in a dark, dry place.

Methods of Packing

Roquefort cheese is marketed in wooden boxes (pine, white-fir, or spruce), holding 3, 6 or 12 cheeses each. The twelve-cheese boxes are 8 by 8 inches by 3 feet 9 inches inside dimensions. They are divided into 3 or 4 compartments to hold 3 or 4 cheeses each. The sides, top, and bottom, are 3/8 inch thick, and the ends and compartment pieces 11/16 inch thick. The six- and three-cheese boxes are made, proportionately smaller, from similar material.

The cheese is wrapped neatly and securely with foil. The dimensions of the foil are 22 by 13 inches and there are from 18 to 25 sheets to a pound. The foil is separated from the cheese by a layer of thin paper to which it is attached. Ordinarily, the trade name and design are printed in black upon the foil. When the cheeses are ready for shipment, they are wrapped in strong paper and tied with a cord. This holds them in shape and gives protection in transit. The net weight of cheese is marked on the paper.

The wrapped cheese is then packed in sawdust, pine shavings, or excelsior, to absorb any leakage brought about by a change from low to high temperatures. No difficulty has been encountered in the absorption of odors from the box or packing materials by the cheese.

Iowa Blue Cheese (Roquefort Type Cheese Made from Cow's Milk)

Iowa Blue cheese is a Roquefort type cheese made from cow's milk. Roquefort type cheese possesses a distinct peppery flavor due to the growth and activity of certain molds. Many types of cheese are known in which one of the major ripening agents is a mold, the growth of which throughout the cheese produces a distinctive and characteristic marbled blue color. This class of cheeses is referred to as blue vein, because of its appearance. In addition, the enzymes elaborated during the growth of the mold produce a typical, pleasing, peppery flavor, largely responsible for the high esteem in which Roquefort type cheese is held.

The present efforts of the Iowa State College of Agriculture to develop this type of cheese under the supervision of Goss, Neilsen, and Mortensen, were directed particularly towards producing a type of blue cheese which could be made by a simple, inexpensive and easily controlled process. The process described below is similar, in some respects, to the conventional French process.

Method of Manufacture

The procedure followed in the manufacture of Iowa Blue cheese is given in some detail below. The directions are given as precisely as possible at this time. The cheese maker must acquire some experience, which must serve as a basis for the successful control of certain physical properties of the curd which determine the quality of the final product.

The milk used in the experiments averaged 3.34% fat. Cream was added to it to raise the average fat content to 3.79%. A normal milk with 3.8 to 4.2% fat has also been used with success. Pasteurization of the milk for Iowa Blue cheese has not proved advantageous. Cheese made from pasteurized milk does not develop as much surface growth of bacteria and molds; less flavor is present, and the cheese does not become as sweet during curing, as the raw milk product. The pasteurized milk cheese carried 49.0% moisture in the 23-hour drained curd as compared with 47.4% moisture in the raw milk product. The poorer drainage properties of the pasteurized milk curd make it more difficult to obtain the right sort of body and texture. Losses of fat in the whey were 0.04% lower for pasteurized milk. This is so slight that it is not significant. More experience with this sort of cheese will be required before the modifications necessary to avoid certain ill effects of pasteurization can be specified. It seems at present that Iowa Blue cheese made from raw milk cures in a more normal and satisfactory manner than that made from pasteurized milk.

Ripening the Milk

After the milk has been placed in the usual American cheese vat, 1.5 to 2% of ordinary lactic cheese starter for ripening the milk is added. The acidity is permitted to increase until 0.19 to 0.20% has been reached, which usually requires from 1 to 1½ hours. At the end of the ripening period, the Marshall rennet test will show 1¼ to 1½ spaces. It is essential that an active and rapid lactic acid-producing starter be used. An inactive or slow culture will not facilitate satisfactory drainage, and the composition of the cheese cannot be controlled.

When cream is added to raise the fat content of the milk by approximately 0.4%, the increase in the ratio of fat to casein does not appear to cause any special difficulty in obtaining a sufficiently open cheese or satisfactory drainage. The setting temperature of 84° F is lower than that commonly used for most varieties of cured cheese, but this lower temperature seems to aid in the production of the desired body and texture.

Setting the Milk

When the acidity test shows 0.19 to 0.20% acid present, the rennet extract, which has been diluted with 20 times its volume of cold water, is added at the rate of 3 fluid ounces or 90 cubic centimeters per 1000 pounds of milk. Just before the rennet extract is added, the milk is set in motion. Thorough mixing should continue for 1 or 2 minutes to insure uniform distribution of the rennet extract. It will be noted that no color is added since it is desired to have the finished cheese as white as possible. The veining with the blue mold then shows up in sharp contrast against the uncolored cheese.

Curdling Period

A relatively long curdling period is used, usually 1¾ to 2 hours. During this time the milk thickens, then curdles completely and becomes progressively firmer, more brittle and less elastic. Beads of whey will stand out on the surface of the curd, gradually becoming larger areas until the entire surface is covered with a thin film of clear whey. The curd is now ready to cut. Sometimes, in the case of milk slightly riper than usual, the curd will shrink away from the lining of the vat before it is ready to cut.

Cutting the Curd

With ½ inch wire curd knives the curd is cut into cubes by means of a horizontal cut lengthwise, followed by a vertical cut across the vat, and finishing with a longitudinal cut with the same knife. Following this, the whey will separate rapidly and the curd will begin to settle out. An acidity test of the whey taken immediately after the cutting is completed will usually show 0.14 to 0.15% of acid. Lower acidities have frequently been followed by difficulty in drainage.

Firming the Curd

Following the completion of the cutting operation, the curd is left undisturbed for about 15 minutes. At the end of this period, it is stirred gently to prevent matting and the operation repeated at intervals of a few minutes for the next ½ hour. Fifty minutes after cutting, the temperature of the curd is raised 2 or 3° F, by dipping out a pail of whey and heating it to a temperature of 170 to 180° F, by injecting steam directly into it. Enough whey can easily be heated in this manner to raise the curd temperature sufficiently without danger of the curd matting in the bottom of the vat. It also avoids the application of too much heat, which might occur if steam were admitted into the jacket to heat the curd. The hot whey should be added carefully to the vat and then be thoroughly distributed with the wooden curd fork.

Dipping the Curd

The firming temperature of 86 to 87° F will slightly toughen the curd in a few minutes, and the acidity by this time will have reached 0.18 to 0.19%. This slight toughness prevents too much breaking up of the curd on the cloth, to which it is now to be transferred. When the toughening process has proceeded for about 15 minutes, some of the whey is drawn off through the vat gate, and one-half of the curd is transferred by means of curd scoops or flat-sided curd pails to a nearby drain vat. At the bottom of this vat is a rack over which two of the Swiss cheese cloths are spread. Drainage of the curd is assisted by raising and lowering the corners of the cloths occasionally, but too much manipulation will cause unnecessary breaking of the curd and loss of cheese solids. After 15 to 20 minutes the curd becomes a slightly tough, lumpy, porous mass. There will be very little free whey passing from it.

Hooping the Curd

The hoops used in this work are the so-called Muenster style mold, made of heavy tinned steel, round, 7½ inches in diameter, open at both ends, with six rows of twenty ⅛ inch holes drilled in the side for drainage. If the room is not sufficiently warm, it is well to dip the hoops in hot water just before using them. The curd may have cooled a degree or two during drainage before being placed in the hoops. The hoops may be placed on boards or directly upon a draining table, and each hoop filled one-fourth full with the pulpy, irregular curd.

Upon this first layer of curd is dusted a light layer of the prepared mold powder, using a pepper box. Following this, a second and third portion of curd is added, and upon each layer, mold powder is sprinkled lightly. The last fourth of the curd needed to fill the hoop is placed over the third layer of mold so that the hoop is filled with alternate layers of curd and mold powder. In the cheese made in this laboratory 0.06 to 0.07% of mold powder gave excellent results. Unless suitable laboratory facilities are available it will probably be more satisfactory to obtain the mold powder from a reliable laboratory than to attempt to prepare it in the dairy plant.

The curd is not packed in the hoop, nor is any follower or weight used. If the correct physical condition of the curd is obtained and the temperature is favorable, the surface of the cheese will close up reasonably well, but the cheese will remain rather open through the interior. The closing of the surface is desirable as it gives a smoother, cleaner appearance to the finished cheese and reduces losses during curing, but the open interior is necessary for proper mold growth. If difficulty is experienced in obtaining a smooth, well closed surface after draining for 20 minutes, the cheese may be surrounded with a small square of cheesecloth dipped in warm water. This procedure will insure a well closed surface, but this precaution has not generally been necessary in the college laboratory. Curd remaining in the whey still in the vat is now transferred to the cloths. This curd will have firmed somewhat during the draining and filling of the first lot of curd, so that drainage of the remainder will require less time. Fat losses in the whey average 0.25%.

Draining the Cheeses

After 20 or 30 minutes in the hoops, the cheeses are turned to smooth the upper surfaces and to facilitate drainage. A special draining room is not used in this work; the cheeses are placed on cloths in the drain vat in the mixing room and their temperature held between 75 and 80° F. Usually there are five turnings of the cheeses at increasing intervals of time during the afternoon. It is important that the temperature be held sufficiently high to facilitate drainage, but not so high that the open texture is lost. The high drainage temperature permits the curd to drain sufficiently so that it may be salted within 24 hours after it is made. During the drainage period the weight of curd decreases to 3.02 pounds of cheese per pound of fat, based on the records of approximately 8000 pounds of cheese. The cheese will show 45 to 46% moisture when it is ready to salt.

Salting

The salting of this type of cheese presents one of the major problems from the standpoint of producing uniform quality. A combination of brine and dry salting has been used to obtain the 4% (or more) of salt desired in the final product. After drainage, the cheeses are held at a temperature of 75 to 80° F overnight, and are then placed in a wooden salting tank containing a saturated solution of butter salt. The temperature is kept near 55° F. The cheeses are sprinkled lightly with dry salt on the exposed surface and the next day are turned over in the brine and the upper surface lightly salted. At the end of the second day in the brine, the cheeses are removed and dry salted rather liberally on all surfaces. This dry salting is repeated in 48 hours. At the end of another 48 hours the cheeses are rinsed off with clear water. The entire salting operation covers a period of 6 days; during the first 2 days the cheeses are kept in brine and during the last 4 days they are dry salted. During the salting period the surface of the cheeses becomes very firm but shows little evidence of surface growth. The hardness of the surface is not due to air-drying but to the effect of the salt. Samples of good quality imported blue cheese have been analyzed which showed as high as 5.7% salt. It is important that a carefully prepared sample show a salt content somewhat above 4.0%, upon completion of the salting process. During the salting period the cheese continues to lose weight, producing 2.89 pounds of salted cheese per pound of fat, as compared with 3.02 pounds of cheese per pound of fat at the beginning of the salting process. The acidity of the brine increases with use, and should be kept below 0.3% by the addition of proper amounts of milk of lime made from clean, unslaked, lump lime.

Punching the Cheese

At the end of the salting period excess salt is rinsed from the cheese surfaces and, when the surplus moisture has disappeared, the cheeses are skewered.

The proper growth of the flavor-producing mold culture requires that a certain amount of air, in addition to that present when the curd is placed in the hoops, be admitted during the curing process. This additional air is provided by skewering or pricking the cheese in about fifty places with a ⅛ inch diameter needle. These openings are evenly distributed over the flat sides of the cheese, permitting better growth of the mold in the many natural openings of the cheese, especially along the line of puncture. If large quantities of cheese are to be made, a special machine should be used for punching in order to economize on labor and produce more uniform results.

Curing

The cheeses are placed on edge on special racks for the curing period, which takes approximately 2 months. During this period the temperature of the curing room in which the cheeses are placed is held at 45 to 48° F and the humidity above 90% of saturation with slight or moderate air circulation.

The rooms used can be insulated with 4 inches of cork and provided with a floor drain and cold water connections; the cooling is effected by means of wall brine coils on one side of the room. High humidity and a slight air movement are produced by a fine water spray located at the top and discharging into a galvanized iron cylinder 6 feet by 1 foot 6 inches, open at both top and bottom, extending within a foot of the eight-foot ceiling and ending 1 foot above the floor not far from the drain. A fine spray of water passing downward through the cylinder humidifies and circulates the air, the excess water passing onto the floor and down the drain. Manual adjustment of a valve on the water line regulates the humidity and, to some extent, the air circulation. A difference of about 1° F between the wet and dry bulb thermometer readings should be maintained. The effectiveness of the curing conditions is also determined by the sequence of fermentations on the cheese surfaces, the appearance and stickiness being used as evidences of the desired curing room conditions.

During the first 3 or 4 weeks following the pricking of the cheese, a blue-gray mold develops in patches over the surface, sometimes almost covering the cheese. Some sticky surface growth is mingled with the mold, but the mold should predominate. At this stage the surface mold and bacterial growth are removed by immersion in cold water in a vat for ½ to 1 hour. This will soften the surface sufficiently so that a stiff brush may be used to render the cheese white, before replacing it in the curing room. During the second period of 3 or 4 weeks under curing room conditions, the reddish-brown sticky surface growth will almost entirely displace the mold, which predominated during the first month. This reddish-brown layer will probably have to be removed twice during the curing process. Its development is evidence of a normal cheese and, to some extent, of suitable curing conditions. However, if it becomes too vigorous, the surface of the cheese softens too deeply, and large losses result. High temperatures and infrequent removal favor too luxuriant growth. The removal of the surface growth with the brush causes some loss of weight during curing. The high humidity of the curing room considerably prevents evaporation of moisture from the cheese so that little loss of weight occurs on this account. The total loss of weight from the time the cheese is placed in brine until it is enclosed in foil amounts to 9.67%.

Placing in Foil

After 8 or 10 weeks in the curing room, the last removal of surface growth takes place and the white cheeses are enclosed in pure block-tin foil, as soon as most of the surface moisture has disappeared. The foil should be pressed upon the surface of the cheese with the hands so that there will be close contact between the foil and cheese. The foil should bear the brand and other information, in which case the sheets of foil will be hand-leaved with tissue paper. The cheeses can be handled with this tissue paper to keep the outside surface of the foil entirely clean. A yield of 2.72 pounds of cured cheese per pound of fat at the time the cheese was inclosed in foil can be obtained.

The cheese should be held for at least 1 month in foil at a temperature below 45° F, or at 32° F for a storage period of some months. During this ripening period in foil more of the characteristic, peppery flavor develops and the cheese becomes sweeter, losing certain raw curd characteristics which had not entirely disappeared at the time the cheese was inclosed in foil. The total time from making to selling for a milk-cured cheese will be 3 to 4 months.

Packing for Shipment

Absence of a definite rind and the friable characteristics of the cheese require precautions in handling and shipping to prevent injury. Just before shipment, each cheese is wrapped with white parchment paper and tied with a heavy cord around its circumference. Cheeses for express or freight shipment should be packed in ½ dozen lots, surrounded with clean excelsior in a special wooden box 7¼ inches wide, 10 inches deep and 22½ inches long (inside dimensions). The tops and sides are ⅜ inch, and the ends ¾ inch, in thickness. For a single cheese, a strong corrugated fiber box just large enough for the cheese may be used for express and parcel post shipments.

Yield

A yield of 2¾ pounds of Iowa Blue cheese per pound of fat can be obtained. This is the yield at the time the cheese is inclosed in foil. After the cheese is in foil the loss in weight will be relatively small, especially if it is stored for no longer than 1 or 2 months. The above yield would be at the rate of 11 pounds of cured cheese per 100 pounds of 4% milk.

The average composition of 9 cheeses from representative lots made by the method outlined contained 34.5% fat, 36.66% moisture and 4.79% salt. There was 54.5% fat in the dry matter.

Lactic Acid Prepared from Milk

There are three grades of lactic acid, made from calcium lactate of different degrees of refinement: (1) the commercial acid is made from concentrated crude calcium lactate liquor from the vacuum pan after the first evaporation, (2) the edible grade from the centrifuge after the first crystallization and (3) the water-white U.S.P. grade from the purest grade of calcium lactate.

Concentrated, crude calcium lactate as it comes from the vacuum pan is decolorized in a wooden mixing tank, treated with sulfuric acid and separated from the calcium sulfate and other precipitates by filtration. The concentration of the lactic acid coming out of the filter-press is about 22%; this product is suitable for sale as commercial acid. In manufacturing an edible grade of lactic acid, the process of refinement is repeated from the standpoint of using decolorizing carbon, etc. The liquor coming out of the filter-press is subsequently concentrated in stainless steel evaporating pans. The acid sold to the trade, known as edible lactic acid, averages about 50% by weight. This acid is almost as clear as water, free from high ash and unpleasant odor. One pound of lactose theoretically produces 1.17 pounds of 50% lactic acid.

Lactic acid is finding increasing favor in the manufacture of ice-cream, sherbets, soft drinks, and fruit syrups. It is extensively used in the salt brine used in olive processing, to adjust the acidity and to retard the growth of bacteria, which may occur when grinds are too low in acidity.

Butter Manufacture

Butter can be made from the milk of sheep, goats, buffalo and other animals, but the product that we know and consume is usually made from cow's milk. The old-fashioned method was to allow the milk to stand for about a day and then place the cream, which had risen to the top, in earthen crocks, to be held there until sufficiently ripened or soured by the action of the lactic acid bacteria. There are many pitfalls in this process of manufacture, especially if the atmosphere surrounding the cream is infected with the wrong kind of bacteria during the ripening process.

When the creamery method is used, the cream is generally separated from the whole milk, while it is still sweet, by running it through centrifugal separators. It is pasteurized, then treated by the addition of a lactic acid starter, which consists of pasteurized skimmed milk previously curdled by the addition of pure cultures of lactic acid bacteria. Butter made from sweet cream, without the aid of a starter, is flat in flavor and lacking in aroma. Proper use of a good starter results in the butter, made from sweet cream or neutralized sour cream, having a distinct character, good flavor and pleasant aroma.

The next step in the process is churning, which brings the fat globules into contact with each other so that they unite into granules of butter. The granules are drained of the remaining liquid, which is the buttermilk. They are then washed and salted, and finally worked into the finished homogenous product.

Butter should not have a lardy, oily or greasy flavor. It should be neither oily nor dry in appearance, nor cloudy nor streaked. It should be free from holes or crevices, as these enclose moist air and favor fermentation. Renovated, or processed butter is produced by reworking low grade or slightly deteriorated lots. The butter is melted and allowed to settle; then the froth and curd are skimmed off and the mixture freshened by strong currents of air. Freshly inoculated milk cultures are added. The butter is then churned rapidly, cooled and worked to remove any excess of milk. However, there are differences between fresh and renovated butter and the latter should not be sold as fresh butter.

Fruit-Flavored Milk Beverages

The production of this type of beverage is possible only when certain considerations and regulations are complied with. The chief considerations in the production of fruit-flavored milk beverages are the prevention of curdling, control of acidity, sweetness and production of a true fruit flavor. The curdling of milk is the coagulation of the casein; this coagulation starts at a pH of 5.5 and ceases at a pH of 4.7. The acidity of the milk increases as the pH decreases. From this it becomes evident that the progressive addition of an acid syrup at first causes a slight precipitation of casein and, when the acid is sufficiently concentrated, the curd is formed, with whey as the by-product.

Acidity, therefore, is an important consideration and must be very carefully controlled to regulate, or to completely prevent, even a trace of curd. The amount of sugar added for sweetening purposes is important because addition in too large quantities offsets the fruit flavor. The addition of a large quantity of fruit flavor often causes the milk to curdle. Unless there is a sufficiently strong fruit aroma, the milk beverage may be very unsatisfactory, since milk requires a powerful flavor to mask its own aroma and taste.

The suggested fruit flavors for making milk beverages are:

Strawberry	Black currant
Raspberry	Loganberry

The most satisfactory proportion of milk to syrup is 100 parts of milk to 20 parts of fruit syrup. The syrups listed, when added to milk, introduce a characteristic flavor without curdling.

Flavor	*Per Cent Sugar*	*pH of Syrup*	*Per Cent Acid*
Strawberry	66	5.85	0.4
Raspberry	64	4.09	0.62
Loganberry	65	4.32	0.92
Black currant	65	4.70	1.5

The addition of these fruit-flavored syrups in the proportion of 20 parts to 100 parts of milk will regulate the finished milk beverage so that the pH will average about 5.5 and the acidity will be regulated so that a curd will barely form or not at all.

CHAPTER III

CONFECTIONERY, COCOA, COFFEE AND TEA

Candy Products and Confectionery

Chocolate Peanut Bars

Sugar	3 lb	Water	1 pt
Corn Syrup	3 lb		

Cook to 240° F. Then add 8 pounds of washed peanuts and cook for 10 minutes. Remove from fire, and roll out on a slab. Cut into small pieces and dip in chocolate.

Coffee Caramels

Sugar	4 lb	Cream	½ gal
Corn Syrup	3 lb	Good Java Coffee	½ gal

Cook the above to 238° F. Then add 1 quart sweet cream and cook to about 242° F. Remove from fire, pour on greased slab.

Butterscotch Squares

Sugar	10 lb	Water	3 pt
Corn Syrup	4 lb		

Cook to 285° F. Add 4 ounces of butter, ½ teaspoonful of salt. Stir and cook to 295° F. When cooled down flavor with oil of lemon. Pour on greased slabs.

Everton Toffee

Sugar	4 lb	Cream of Tartar	½ tsp
Water	3 cups	Butter	2 oz

Put the water, sugar, and cream of tartar into a pan and stir until they boil, but no longer. Add the butter after the other comes off the fire, but do not stir it in.

Licorice Drops

Sugar	12½	lb	Oil of Wintergreen	30	min
Water	3	pt	Oil of Anise	20	min
Cream of Tartar	1	dr	Powdered Charcoal	1	dr
Powdered Extract of Licorice	1	oz			

Cook the sugar, water, and cream of tartar to 340° F. Pour out on an oiled slab. Add the other ingredients, and fold and knead until well mixed. Then stamp into drops.

Fruit Jelly Candies

Fruit Juice	2 qt	Corn Syrup or Invert Syrup	3 lb
Sugar	3 lb		
Commercial Pectin Syrup	1 qt		

Mix fruit juice and pectin syrup. Add the sugar and corn, or invert, syrup. Boil to 222–223° F. Remove from fire. Pour into starch molds or oiled pans, about ½ inch deep. Allow to harden 24 hours. Remove from molds and dust with powdered sugar or dip in chocolate.

Fruit Caramels

Sugar	2	lb	Ground Dried Fruit	1	lb
Chopped Walnuts	½	pt	Butter	2	oz
Corn Syrup	1¼	lb	Whipping Cream	2	qt

Cook sugar, corn syrup, and one-half of cream to 238° F. Then add one-half of the remaining cream and cook to 242° F. Add remaining cream, dried fruit, butter, and chopped nuts and cook to 248° F. Pour into an oiled pan or on a slab to cool and harden.

Fruit Marshmallows

Powdered Edible Gelatin	4 oz	Dried Apricots, Figs, Prunes, Pears, Chopped or Whole Raisins	3 lb
Hot Water	1 pt		
Corn Syrup	2½ lb		
Powdered Sugar	2½ lb		

Dissolve the gelatin in hot water. Cook corn syrup to 250° F. Beat into it the dissolved gelatin and powdered sugar until the mixture is light. Flavor with vanilla. Then add the dried fruit. Beat a short time and pour on oiled paper or a slab to harden. Three-quarters of a pint of fruit syrup may be used instead of dried fruit.

Molasses Wafers

Sugar	5 lb	New Orleans Molasses	1 pt
Corn Syrup	1 lb	Butter	¼ lb
Water	1½ pt		

Stir and cook to 290° F. Roll out on a greased slab.

Maraschino Cherries

Place firm ripe Royal Ann Cherries in a paraffin-lined barrel, a crock, or a bottle, depending upon amount. Do not remove the stems. Cover the cherries with the solution made as follows:

Water	1 gal	Citric Acid	½ oz
Bisulfite of Soda	1 gal	Salt	6 oz

Dissolve the above ingredients. Let stand until cherries are bleached, which may take 2 weeks or longer. Then discard the solution. Stem and pit the fruit. Boil the fruit in five or six changes of water until free from sulfur taste and tender. Prepare a 40% sugar solution, and for each 10 gallons of cherries dissolve ⅙ ounce of certified red color, Ponceaux 3R. Also add ½ teaspoonful citric acid. Now cover cherries with this colored sugar solution, and cook for 10 minutes. After 24 hours, cook cherries and syrup again, after adding 10 pounds of sugar. Repeat this again on third day. Then allow to remain in crock for 2 weeks. Now add ¼ ounce of bitter almond oil dissolved in a little alcohol, for flavor. Instead of the red, a green food color may be used, and instead of wild cherry, mint flavor may be added.

Wild Cherry Drops

Sugar (Crystal A)	4 lb	Extract of Bitter Almonds	3 dr
Water	1 pt	Powdered Orris	3 dr
Cream of Tartar	24 gr	Red Color	Sufficient

Cook the sugar, water, and cream of tartar to 335° F. Pour out on an oiled slab. Let it cool a little. Add the extract, the orris and the color. When cool enough, work thoroughly and stamp into drops.

Aniseed Drops

Sugar (Crystal A)	5 lb	Oil of Anise	40 min
Water	22 oz	Red Color	Sufficient
Cream of Tartar	40 gr		

Put the sugar in a suitable container. Add the water and the cream of tartar. Cook to 335° F. Pour out on an oiled slab. As it cools, add the oil of anise and the color. Fold over until cool enough to handle. Then work thoroughly and stamp into drops.

Nut Glazes

Sugar	10 lb	Water	3 pt
Corn Syrup	2½ lb		

Cook sugar and water to 290° F. Shut off heat and drop in nuts to be sugar coated or glazed. Remove with fork.

Revivifying Dry Popcorn

Popcorn which pops poorly because of having become too dry may be restored to good popping condition by the following method:

Put 40 pounds of corn into a 10-gallon can. Add 1 to 3 pounds of water, according to the dryness of the corn, as indicated by the way it pops. If its popping yield is less than one-third the normal yield of the variety, add 3 pounds of water; if it is two-thirds normal, add only 1 pound. For intermediate degrees of popping add intermediate quantities of water. For different varieties of popcorn the normal popping yield varies from 15 to about 30 volumes.

Put on the cover, using a rubber gasket, and clamp it down tightly. Shake thoroughly. Let it stand 2 days or longer before popping.

The poor popping quality of some lots of corn is due to other causes than lack of moisture. In such cases the popping cannot be improved by adding water.

The above method of restoring poppability is applicable only to small quantities of corn. When it becomes necessary to increase the moisture content of a large quantity of popcorn, it should be done by storing the corn in a cool, damp place, as, for example, in a shed outside during winter, or in a basement room during summer.

Popcorn Fruit Crisp

Sugar	3 lb	Butter	3 tbsp
Corn Syrup	1½ lb	Chopped Dried Fruit	¾ lb
Water	1 pt	Salt	Sufficient to flavor
Popped Popcorn	4½ qt		

Cook sugar, corn syrup and water to 285° F. Add the butter, salt and fruit, and stir. Now stir in the popcorn. More or less popcorn may be added to suit.

Salted Peanuts

Heat coconut oil in a kettle on a gas flame to about 300° F. Put shelled Spanish peanuts into a basket. Immerse this basket in the hot oil, and roast until brown. Now pour peanuts over a fine sieve, and when the oil has drained off add the salt.

Unwrapped Licorice Caramels

Cane Sugar	60 lb	Sweetened Whole Condensed Milk (8%)	3 gal
Corn Syrup (43° Bé)	50 lb	Table Salt	6 oz
Cream (40%)	4 gal	Fat (140° F m.p.) (Hydrogenated Peanut or Cottonseed Stearine)	2 lb
Fresh Unsweetened Whole Condensed Milk (10%)	5 gal	Licorice Syrup	8 lb

Put all ingredients except licorice in kettle. Cook to a soft ball. Add licorice syrup and black color. Cook batch to hard ball in cold water. Pour on slab and let stand overnight. Cut to size desired. No wrapper is required for this excellent "stand-up" caramel. Put in waxed paper cups.

Licorice Taffy

Cane Sugar	15 lb	Cocoa Butter	1 lb
Licorice Syrup	8 lb	Corn Syrup (43° Bé)	13 lb

Place in kettle. Cook to soft ball in cold water and add licorice syrup. Cook to hard ball, add black color, and spread out on marble to desired thickness. Let cool and cut to suit. Wrapper should be about 20-pound white waxed.

This is a splendid number for wax-wrapped five- and ten-cent bars. May also be used to increase variety and give novelty to chocolate and bonbon centers. Makes an excellent center for pan goods.

Licorice Jelly Gum Drops

Gum Arabic	50 lb	Corn Syrup (43° Bé)	5 lb
Cane Sugar	80 lb	Water	4 gal
Licorice Syrup	12 lb		

Soak gum in hot water, using 1 pound of gum to 1 quart of water. Let stand in hot room overnight. Skim off and strain through very fine sieve. Cook sugar, corn syrup, and water to 250° F. Add licorice syrup and black color and continue to cook to 255° F. Add gum and cast in starch. Put in hot room for 48 hours. Take out and let stand until cool. When cool, remove from starch and place in crystal pans. Use crystal cold, cook to 232° F. Leave in crystal overnight.

The smaller sizes of licorice jelly gum drops make excellent centers for pan work. For this work they are, of course, not crystallized. Put them in the pan after removing them from starch and handle them the same as you would jelly beans or cordial drops. Do not color the syrup but finish white.

Chocolate or bonbon-coated, this luscious jelly gum drop makes an excellent addition to the conventional cordial box or summer package.

Hard Licorice Drops (Pastilles)

Gum Arabic	60 lb	Licorice Syrup	12 lb
Cane Sugar	36 lb	Corn Syrup	6 lb
Glycerin	3 oz		

Melt gum in the proportion of 1 pound to 1 quart of hot water, and let set overnight in hot room. Bring to a boil next day and strain through a very fine sieve. Cook sugar and corn syrup to 280° F. Add gum, licorice syrup, and black color, and put in hot room overnight. Next morning, skim off and cast in starch.

Place in hot room again and allow to set for three days. Temperature of hot room should be maintained between 130 and 140° F. After removing from the hot room, let stand for about 24 hours to cool. Clean off starch and return to hot room until product is hot, rub in refined petrolatum (white vaseline) until the desired shine is obtained.

Wrapped Licorice Caramels

Corn Syrup (43° Bé)	10 lb	Cream (40%)	4 gal
Cane Sugar	30 lb	Licorice Syrup	6 lb
Cocoa Butter	3 lb		

Put all ingredients except black color and licorice syrup in kettle. Cook to a soft ball. Add licorice syrup and black color.

Cook to a hard ball in cold water. Pour on slab and allow to stand overnight. Cut and wrap.

Licorice Marshmallow Caramels

Cane sugar	60 lb	Sweetened Whole Condensed Milk (8%)	3 gal
Corn Syrup (43° Bé)	50 lb	Table Salt	6 oz
Cream (40%)	4 gal	Licorice Syrup	8 lb
Fresh Unsweetened Whole Condensed Milk (10%)	5 gal		

Put all ingredients except color and licorice in kettle. Cook to a soft ball. Add licorice syrup and black color. Cook batch to a hard ball in cold water. Turn out ⅛ inch thick. On this lay a sheet of sheet marshmallow. Put top piece of licorice caramel on when batch is cool enough to handle.

Cut and pack in waxed paper cups, straight-packed or in assortments, alternating rows with plain licorice caramels.

Licorice Caramel-Nougat Center for Chocolates or Bonbons

Corn Syrup (43° Bé)	10 lb	Cream (40%)	4 gal
Cane Sugar	30 lb	Licorice Syrup	4 lb
Cocoa Butter	3 lb		

Mix equal quantities of this licorice caramel and a vanilla nougat. Do not add any additional color. The resulting batch will be lighter than licorice caramel but of good color.

The licorice-caramel-nougat mixture makes a particularly fine eating piece of candy and is confidently recommended to those who seek something distinctive in the way of chocolate or bonbon centers.

Licorice Turkish Paste

Cane Sugar	75 lb	Cream of Tartar	6 oz
Corn Syrup (43° Bé)	15 lb	Licorice Syrup	20 lb
Cooking Starch	9 lb	Water	4 gal

Put sugar, corn syrup, and water in kettle and bring to boil. Add cooking starch which has previously been soaked in 1 gallon·of cold water. After batch has boiled 10 minutes, add cream of tartar which was previously melted in 6 ounces of water. Next, add licorice syrup and continue boiling until batch is finished. Cook until it drops easily from a knife.

This constitutes a licorice confection of mild character and velvety texture.

Licorice Strings, Opera Drops or A. B. Gum-Work

Cane Sugar	50	lb	Special Thick-Boiling Starch	20	lb
Corn Syrup (43° Bé)	135	lb	Licorice Syrup	12	lb
Cream of Tartar	2½	oz	Water	8	gal
Tartaric Acid	1½	oz			

Put sugar and corn syrup in kettle with water. Bring to a boil. Add starch, which has been previously dissolved in 3 gallons of cold water, and cream of tartar. Cook 1 hour. Add licorice syrup. Cook for another ½ hour, making a total cooking time of 1½ hours. Turn off steam. Add vegetable black color, and finally add acid which has been melted in a little hot water.

Cast in blood-warm starch (98–100° F). Leave in hot room, at 140° F, for 3 days. Take out of hot room and let stand until cold. When cold, sugar up and let stand in hot room or a very dry room for 24 hours.

Cook crystal to 233° F and apply, while hot, to goods. Leave goods in crystal for 10 hours. Drain off crystal and empty crystal pans into boards. Be sure goods are thoroughly dry before packing.

This formula makes a tender and delicious string or drop. It may also be used for A.B. gums, if desired.

Licorice Paste for Bag Work

Powdered XXXX Sugar	10	lb	Well Beaten Whites of Eggs	8	oz
Licorice Syrup	1	lb			

Beat egg whites to a stiff fluff. Add sugar and licorice syrup, and sufficient black vegetable color to get the desired shade. Keep this paste in a covered container and use as needed. It sets up brittle and crunchy, like a biscuit. It is a deliciously flavored paste for making *dominoes, clubs and spades, Hallowe'en witches, black cats,* and all sorts of holiday bag work and bag work novelties.

Licorice Bonbon Fondant
(Natural Color)

Cane Sugar	80	lb	Water	2½	gal
Egg Whites	1	pt	Licorice Syrup	8	lb

Cook sugar and water to 242° F. Pour on ball beater and let cool. Beat egg whites to a stiff fluff, add to batch, and start beater. When thoroughly mixed, add licorice syrup. Use no color. When fondant is finished, put it in a cream tank. Do not use a wet cloth to cover fondant; use a metal or wood cover instead.

This is a mild flavored bonbon coating of natural light brown color. It adds a new piquant note to the character of bonbons and advances but does not dominate the flavor of the center.

Cocoanut Center for Licorice Bonbons

Fresh Grated Cocoanut	55	lb	Cane Sugar	15	lb
Corn Syrup (43° Bé)	35	lb	Table Salt	¼	oz

Put all of the above ingredients in steam kettle and cook to a soft ball. Add 10 pounds of nougat cream and 4 ounces vanilla extract. Pour on marble. Let cool. Cut or roll to desired size. Fork-dip in licorice bonbon coating.

This is a delicious fresh cocoanut bonbon for the high-class retail trade. A less perishable center can be made using desiccated instead of fresh cocoanut.

Low-Priced Wrapped Licorice Caramels
(Natural Color or Colored)

Corn Syrup (43° Bé)	60	lb	Whole Sweetened Condensed Milk	25	lb
Cane Sugar	40	lb	Vanilla Extract	4	oz
Cocoanut Butter (92° F)	8	lb	Licorice Syrup	8	lb
Full Cream Caramel Paste	25	lb			

Put all the ingredients, except licorice syrup and vanilla extract, in kettle. Cook to a soft ball. Add licorice syrup and vanilla extract. Add black color, if desired. Cook to a hard ball in cold water. Pour on slab and allow to stand overnight. Cut and wrap.

This is an inexpensive caramel of excellent stand-up and eating qualities.

Licorice Wrapped Kisses
(Semi-Nougat Type)

Corn Syrup (43° Bé)	20 lb	Whole Sweetened Condensed Milk	1¼ lb
Cane Sugar	10 lb	Licorice Syrup	1½ lb
Cocoa Butter	¾ lb	Color	Sufficient

Put all materials together in kettle. Cook to 255° F. Pour on slab and let cool. Pull for five minutes. Spin out and cut to suit.

This is a pleasing departure from the ordinary run of wrapped kiss formula. This licorice kiss is inexpensive to make, will stay soft, and is easy to eat.

Almond Licorice Nougat

Cane Sugar	80 lb	Licorice Syrup	10 lb
Corn Syrup (43° Bé)	70 lb	Water	8 gal
Almonds	15 lb		
Egg Albumen (Good Beating Quality)	1¼ lb		

Soak egg albumen in water overnight. Then put in nougat kettle and beat 5 minutes. Cook sugar, corn syrup, and water to 240° F. Take 3 quarts of this mixture and beat into egg fluff very slowly. Cook balance of sugar and corn syrup batch to 300° F. Add slowly to batch in nougat kettle until all are thoroughly mixed. Now add licorice syrup, almonds, and sufficient vegetable color to get a good black color.

If a short nougat is desired, the batch should be beaten about ½ hour. If a chewy nougat is preferred, 15 minutes beating will be sufficient.

Turn nougat out into steel or wood trays and cut the following day.

This makes a high-class, cut nougat which is without equal for adding color contrast and flavor variety to pallid vacation numbers, summer packages, and straight nougat assortments.

Genuine Licorice *Anthracite*

Licorice Syrup	12	lb	Whole 8% Sweetened Condensed Milk	5	gal
Granulated Cane Sugar	50	lb	Salt	½	lb
Corn Syrup (43° Bé)	60	lb	Cocoanut Butter (92°F m.p.)	6	lb
Hard Fat (Hydrogenated Peanut or Cottonseed Stearine)	2	lb			

Put all raw materials in kettle, with the exception of licorice syrup, and cook to a crack. Add licorice syrup and black vegetable color to suit. Cook to a hard crack. Pour on marble slab and allow to cool. Cut into irregular chunks resembling coal.

A rich, delicious piece of the butter-toffee variety. Has the novel appearance of anthracite and a savory background of licorice.

Genuine Licorice Black Birds

Cane Sugar	12	lb	Corn Syrup (43°Bé)	30	lb
Refined Dextrose (Corn Products Cerelose or Equivalent)	20	lb	Thin Boiling Starch	9	lb
			Gelatin	3¾	lb
			Licorice Syrup	5	lb

Dissolve cane sugar, dextrose, and corn syrup in 4½ gallons of water. Cook as for a regular high-cooked gum drop. Suspend 9 pounds of thin boiling starch in 2 gallons of water and add gradually to the boiling batch. The cooking is continued until a high-cooked gum drop results.

Dissolve 3¾ pounds of gelatin in 5½ quarts of water and warm up. Mix this in with the remainder of the batch toward the end of the boil. Finally, add licorice syrup and black vegetable color to suit. It is customary to give this piece a high gloss after taking it out of starch.

This is the popular chewy black licorice drop in the appealing all-licorice flavor. Those desiring a still chewier drop will find that they can vary the above formula by adding to the gelatin solution an equal amount of gum arabic dissolved in water.

Sugar Coating for Candy

Cocoa Butter Substitute	35	lb	Powdered Sugar	50	lb
Powdered Skim Milk	15	lb	Salt	2	oz
			Vanillin	½	oz

This may be colored and flavored with oil-soluble colors. After melting the cocoa butter substitute, add the other materials gradually in a dough mixer and finally run over finishing rolls at a temperature of 110° F. In using, run on enrober at 104° F.

Chocolate Coating

Cocoa Butter Substitute	28	lb	XXXX Powdered Sugar	52	lb
Cocoa Powder	10	lb	Vanillin	½	oz
Skim Milk Powder	10	lb	Salt	2	oz

Melt fat and mix thoroughly. This coating may be used at a temperature of about 110° F. For summer use, add 2% cottonseed stearine (140° F m.p.).

Vanilla Marshmallows

Water	3	qt	Corn Syrup	15	lb
Gelatin	15	oz	XXXX Sugar	15	lb

Dissolve gelatin in 3 quarts of hot water and set aside. Heat the corn syrup to 250° F in a kettle. Remove from heat and, while mixing, add 15 pounds of XXXX sugar. Now pour into the batch, through a strainer, the 3 quarts of gelatin. Work the whole lot together and flavor with vanilla. Transfer the batch to the marshmallow beater, and beat until very stiff. Then pour it on a slab previously powdered with XXXX sugar. Let stand 4 to 5 hours. Then cut into marshmallow size. Press knife down and cut.

Agar Jelly Gum Drops

Agar-Agar	10 oz	Water	5 gal
Sugar	18 lb	Corn Starch	1 lb
Corn Syrup	8 lb		

Soak 10 ounces of agar in 4 gallons of water, overnight. In the morning cook up the agar for several minutes, and then strain into the sugar and corn syrup. Now bring the batch to a boil, while stirring. When the batch comes to a boil, add the 1 pound of corn starch made into a paste with 1 gallon of water. Cook the entire batch up to about 230° F or to a stiff string and then cast in starch. Add color and flavor, after removing batch from the heat. For fruit flavors, use 1 or 2 ounces of citric acid.

British Unwrapped Caramels

Condensed Whole Milk		Corn Syrup	20 lb
(40% Sugar)	50 lb	Nulomoline	5 lb
Salted Dairy Butter	5 lb	Brown Sugar	10 lb
Hard Coconut Butter		Rice Flour	5 lb
(88–90° F m. p.)	10 lb	Salt	3 lb

Flavor with a combination of:

Vanilla Flavor	Strong
Violet Flavor	Trace
Maple Flavor	Moderately strong

Place the condensed milk, butter, coconut butter, corn syrup, and Nulomoline into a kettle, heat, and mix until the fats are melted. Add the sugar and rice flour, mix, and boil the batch to a firm ball. Now add the salt and flavors, mix well and spread the batch out to cool, then size and cut.

North Carolina Caramels
(For Caramel Twirls)

Condensed Skim Milk (40% Sugar)	35 lb	Fine Granulated Sugar	20 lb
Corn Syrup	50 lb	Soft Winter Wheat Flour	10 lb
Coconut Oil (76° F m.p.)	5 lb	Salt	6 oz
Nulomoline	10 lb	Basic Fondant	15 lb
		Flavor and Color	As desired

Heat the Nulomoline, corn syrup, milk and coconut oil, then add the sugar, which has been mixed with the flour, and stir. Cook the batch to a hard ball. Add the fondant, salt, flavor and color, and mix well. Then spread the batch out to cool.

Roll the batch into thin sheets, placing a layer of nougat (New York Nougat) on the caramel. Then roll it up in much the same manner as jelly rolls. Spin the caramel into strips about 1 inch in diameter and later cut into pieces of the desired size.

Basic Fondant

Fine Granulated Sugar	80 lb	Water	To Dissolve the Sugar
Corn Syrup	20 lb		
Nulomoline	10 lb		

Cook together to 244° F.

Smyrna Caramels

Corn Syrup	10 lb	Salt	4 oz
Nulomoline	10 lb	Ground Figs	30 lb
Coconut Oil (76° F m.p.)	5 lb	Red Color	
Sugar	15 lb	Roman Punch Flavor	
Condensed Whole Milk (40% Sugar)	20 lb		

Heat the corn syrup, Nulomoline, coconut oil and condensed milk together. Add the sugar. Continue to stir and bring the batch to a boil. Now add the ground figs and cook to a firm ball. Then add the salt, flavor and color. Spread the batch out to cool. Cut into squares and wrap in cellophane or coat in chocolate or pack in parchment paper cups.

Swiss Caramels
(Unwrapped)

Sugar	45 lb	Egg Whites (Fresh or Frozen) or 2 lb of Albumen Dissolved in 4 lb of Water	10 lb
Nulomoline	5 lb	Salt	2 oz
Corn Syrup	40 lb	Coffee Flavor	
Cream (20% Butter Fat)	100 lb		

Mix the egg whites or albumen solution with the cream. Meanwhile, place the sugar, Nulomoline and corn syrup in a kettle. Add enough of the cream to dissolve the sugar and bring the batch to the boiling point. Gradually add the balance of the cream and egg mixture and boil the batch to a medium firm ball. Add the salt, flavor, and color if wanted, and spread the batch out to cool. Different flavors and nut meats can be added to this batch to produce a variety of caramels.

Butterscotch
(Squares and Slabs)

Refined Sugar	10 lb	Lecithin	½ oz
Brown Sugar	10 lb	Special Caramel Paste	10 lb
Corn Syrup	10 lb	Frappe No. 1	2 lb
Nulomoline	5 lb	Salt	3 oz
Water	6 lb	Vanilla Flavor	Strong
Salted Dairy Butter	3 lb	Bergamot Oil	Trace
Coconut Oil (76° F m.p.)	2 lb	Nutmeg Oil	Trace

Melt the butter and coconut oil, add the lecithin and stir until it is melted, then add the sugars, corn syrup, Nulomoline and water and boil together to 280° F. Gradually add the caramel paste. Mix well and continue to cook the batch to the first crack. Turn off the heat. Add the frappe, salt and flavors and spread the batch out between iron bars so that it will be approximately ½ inch thick. Mark with a cutting frame. Then cut into bars. This makes a good five-cent item for wrapping in waxed paper or cellophane.

Special Caramel Paste

Corn Syrup	35 lb	Coconut Oil (76° F m.p.)	10 lb
Condensed Whole Milk (40% Sugar)	45 lb	Granlated Sugar	20 lb
		Salt	½ lb

Heat the corn syrup, coconut oil and condensed milk. Add the sugar and salt and heat until the sugar is dissolved. Then cook to a soft paste and spread out to cool.

Frappe No. 1

Nulomoline	50 lb	Albumen	3 lb
Corn Syrup	50 lb	Cold Water	2 lb

Boil the corn syrup to approximately 245° F. Shut off the steam. Add the Nulomoline cold and stir until it is melted. Place the batch in a marshmallow beater. Start the beater. Then gradually add the albumen, dissolved in the water, and beat until light.

French Chewing Candy
Mixture No. 1

Corn Syrup	25 lb	Albumen—Dissolved in	3 lb
Nulomoline	25 lb	Water	6 lb

Boil the corn syrup to 245° F. Add the Nulomoline and stir until it is melted. Place the batch into a beater. Start the beater. Then gradually add the aqueous albumen solution and beat until light.

Mixture No. 2

Sugar	30 lb	Cream (20% Butter Fat)	10 lb
Nulomoline	5 lb	Coconut Oil (76° F m.p.)	2 lb
Corn Syrup	25 lb		

Boil altogether to 290° F. Pour this into 20 pounds of Mixture No. 1, adding 6 ounces of gelatin that has been previously soaked in 6 ounces of hot water, and beat until quite light. Now add 10 pounds of roasted cashew nuts, 2 pounds of unsweetened chocolate and vanilla flavor. When the batch is almost cold, spin it into round sticks about 1½ inches in diameter. Then cut these sticks or rolls into pieces approximately 6 inches long and wrap in cellophane or heavy waxed paper. This may also be cut into blocks or bars.

Japanese Strips
(Coconut Pieces)

Corn Syrup	20 lb	Coconut Oil (76° F m.p.)	3 lb
Water	8 lb	Desiccated Coconut	25 lb
Brown Sugar	5 lb	Salt	2 oz
Nulomoline	5 lb	Red Color	
Condensed Whole Milk		Oil of Cinnamon (Ceylon)	
(40% Sugar)	10 lb	Currants	7 lb

Heat the corn syrup, water, Nulomoline, sugar, condensed milk and coconut oil to the boiling point. Add the coconut, continue to mix and cook to a medium firm paste, then add the color, salt, and flavor, mix well, then add the currants. Mix and spread out to cool on an oiled slab dusted with XXXX sugar. Then roll the batch down and, when cold, size, cut and roll in sugar.

Mandarin Slices
(Coconut Cut Pieces)

Sugar	15 lb	Desiccated Coconut	25 lb
Water	10 lb	Kumquat Pieces	10 lb
Nulomoline	10 lb	Preserved Ginger	4 lb
Corn Syrup	25 lb	Yellow Color	

Heat the sugar, water, Nulomoline and corn syrup to the boiling point. Then add the coconut and stir and cook to a medium firm paste. Add the kumquat pieces, ginger and color, and mix well. If the fruit contains an excess amount of water, cook the batch to a firm paste.

Spread the batch out on an oiled slab dusted with XXXX sugar and, before the batch is quite cold, form into rolls about 1 inch in diameter. When cold, moisten the rolls with simple syrup or thin icing and roll at once in sesame seeds. The rolls may later be cut into slices and placed in paper cups.

Mont Royale
(Fudge Coconut Kisses)
Mixture No. 1

Desiccated Coconut (Fancy Shred)	50 lb	Nulomoline	5 lb
		Hot Water	5 lb

Mix the Nulomoline with the hot water, then mix it through the coconut and allow to set for ½ hour before using.

Mixture No. 2

Condensed Whole Milk (40% Sugar)	35 lb	Winter Wheat Flour	5 lb
Corn Syrup	25 lb	Basic Fondant—made from:	100 lb
Nulomoline	15 lb	80 lbs Sugar	
Brown Sugar	15 lb	20 lbs Corn Syrup	
Coconut Oil (76° F m.p.)	9 lb	10 lbs Nulomoline	
Salt	8 oz	Water to Dissolve the Sugar	

Cook to 244° F.

Heat the condensed milk, corn syrup, Nulomoline and coconut oil together. Add the sugar and flour and continue to stir and cook the batch to a firm ball. Then add the salt, fondant and vanilla flavor and mix well. Now add the coconut (Mixture No. 1). Mix well and spoon the batch out into roughly formed kisses, or it may be formed into bars or egg shapes and coated in chocolate.

Sicilano Coconut Cuts
(Licorice Flavor)

Sugar	12 lb	Desiccated Coconut	25 lb
Nulomoline	8 lb	Gelatin (Soaked in 8 oz of Water)	8 oz
Corn Syrup	20 lb	Salt	1 oz
Water	8 lb	Black Color	
Coconut Oil (76° F m.p.)	2 lb	Anise Flavor (Optional)	
Licorice Paste (Properly Dissolved)	3 lb	Almonds	4 lb

Heat the sugar, Nulomoline, corn syrup and water to the boiling point. Add the properly dissolved licorice paste, then the coconut and stir and cook the batch to a medium firm paste. Now add the soaked gelatin, salt, coconut oil, color, flavor and almonds and mix well. Spread the batch out on an oiled slab dusted with XXXX sugar that has been colored black. Roll the batch down to the desired thickness using colored XXXX sugar, and when cold cut into pieces.

Massachusetts Molasses Mint Cream Centers
(Short-Cast Cream Centers)

Mixture No. 1

Basic Fondant—made from:	100 lb	Nulomoline	10 lb
		Water	To Dissolve the Sugar
Sugar	80 lb		
Corn Syrup	20 lb		

Cook together to 242° F.

Mixture No. 2

Dark Grandma's Molasses	25 lb	Frappe No. 1	15 lb
Sugar	20 lb	Convertit	2 oz
Water	5 lb	Mint Flavor	Sufficient

Place the fondant (Mixture No. 1) into a cream-melting kettle and heat the batch to approximately 120° F. Meanwhile, boil the molasses, sugar and water to 234° F. Then pour this at once on top of the fondant and mix thoroughly. Now add the frappe and convertit, mix well, then work in the flavor. The casting temperature is approximately 140 to 145° F.

Frappe No. 1

Nulomoline	50 lb	Albumen	1 lb
Corn Syrup	50 lb	Cold Water	2 lb

Boil the corn syrup to approximately 245° F. Shut off the steam, add the Nulomoline and stir until melted. Place the batch in a marshmallow beater. Start the beater, then gradually add the albumen, in aqueous solution, and beat until light.

Pennsylvania Pudding
(Cast Whipped Creams—Chocolate Flavor)
Mixture No. 1

Sugar	80 lb	Nulomoline	10 lb
Corn Syrup	20 lb	Water	20 lb

Boil together to 242° F. Then cool to 110° F and beat into fondant in the usual manner.

Mixture No. 2

Corn Syrup	50 lb	Soaked in Cold Water	3 lb
Nulomoline	50 lb	Albumen	1½ lb
Good Grade Gelatin	1½ lb	Soaked in Cold Water	3 lb

Boil the corn syrup to 245° F. Add the Nulomoline and stir until it is melted. Place the batch in a marshmallow beater, add the gelatin solution. Start the beater, then gradually add the albumen solution and beat until quite light.

Melt 100 pounds of fondant (Mixture No. 1) to 150° F. Add 5 pounds of unsweetened chocolate, 2 ounces of salt and vanilla or rum flavor, and mix well. Then add 30 pounds of frappe (Mixture No. 2) and mix well. Cast the batch into starch at 140 to 145° F. This is suitable for small or large pieces.

Virginia Variety Creams
(Crystallized Creams)

First prepare a fondant as follows

Sugar	80 lb	Water	To Dissolve the Sugar
Corn Syrup	15 lb		
Nulomoline	5 lb		

Boil all together to 246° F and cream as usual.

Prepare a bob as follows:

Sugar	80 lb	Water	To Dissolve the Sugar
Corn Syrup	10 lb		
Nulomoline	5 lb		

Boil together to 245° F.

Place 100 pounds of the fondant in a clean cream-melting kettle. Turn the steam on partly and heat the fondant to approximately 125° F. Then add color and flavor and mix well. Now gradually add the boiled bob and mix thoroughly.

Have the hopper of the depositor heated to approximately 125° F.

Cast into fancy shaped starch impressions and later crystallize the creams using a syrup of 34° Bé at a temperature of 75° F.

Italian Cream Fudge

Mixture No. 1

Sugar	80 lb	Cream (26% Butter Fat)	20 lb
Nulomoline	20 lb		

Place the sugar, Nulomoline and 15 pounds of the cream into a kettle, stir the batch constantly until it boils, then wash down all grains of sugar that adhere to the interior of the kettle. Now add the remaining cream and boil the batch to 240° F. Pour the batch onto a cream cooler and do not disturb until it has cooled to approximately 110° F. Then beat into fondant in the usual manner.

Mixture No. 2

Sugar	50 lb	Hard Coconut Butter (86° F m.p.)	14 lb
Corn Syrup	20 lb	Salt	10 oz
Condensed Whole Milk (40% Sugar)	40 lb	Frappe No. 1	10 lb
		Vanilla Flavor	

Place the condensed milk, corn syrup and coconut butter into a kettle and heat until the fat is melted. Then add the sugar and stir. Cook to 240° F. Turn off the heat. Add the salt, flavor, nuts as desired, and frappe. Add 100 pounds of Mixture No. 1, and mix thoroughly. Then spread the batch out on oiled paper to set.

Frappe No. 1

Nulomoline	50 lb	Albumen	1 lb
Corn Syrup	50 lb	Dissolved in Cold Water	2 lb

Boil the corn syrup to approximately 245° F. Shut off the steam. Add the Nulomoline, cold, and stir until it is melted. Place the batch in a marshmallow beater. Start the beater. Then gradually add the albumen solution and beat until light.

Georgia Gum Drops
(Starch Jelly Gums)

Sugar	40 lb	Corn Starch (No. 50–60)	10½ lb
Nulomoline	10 lb	Cold Water	100 lb
Corn Syrup	50 lb	Cream of Tartar	2 oz

Place the water, sugar, starch and 30 pounds of the corn syrup in a kettle equipped with a good stirring device. Turn the steam on partly, and allow the batch gradually to reach the boiling point. Add the cream of tartar. Turn the steam on full and boil the batch as rapidly as possible, using 80 to 100 pounds steam pressure. Cook the batch to a string. Turn off the heat. Add the balance of the corn syrup and the Nulomoline, then color and flavor. Mix thoroughly and cast the batch into starch. The temperature of the starch into which the jellies are cast may range from 65 to 175° F.

If a continuous starch drier is available, heat the starch to 160 to 175° F. Then cast the batch into the starch and leave the jellies in a room at normal temperature for at least 10 hours. They may then be removed from the starch, moistened with steam vapor and sanded with fine granulated sugar.

In the absence of a continuous starch drier the jellies may be placed in a drying room, heated to 140° F, and allowed to dry for 24 to 48 hours.

Moscow Jellies
(Cut Agar-agar Jellies)
Mixture No. 1

Sugar	30 lb	Japanese Gelatin (Agar-Agar No. 1)	1 lb
Nulomoline	20 lb		
Water	40 lb		

Mixture No. 2

XXXX Sugar (Berry Sugar)	225 lb	Green Coloring	
		Peppermint and Caraway Oils	Combination

Mixture No. 1: Soak the agar-agar in the water for several hours, preferably overnight. Then boil the agar-agar and water for several minutes, stirring occasionally. Strain the agar solution through a No. 30 sieve. Add the sugar and Nulomoline and boil to 222° F. Allow the batch to cool to 140° F.

When the batch has cooled to 140° F, add the color, flavors and XXXX sugar and mix well, pouring the batch into trays which have first been lined with ordinary wrapping paper.

When the jelly has set, sprinkle the surface with XXXX sugar, remove the paper from the jelly sheets, sprinkle them well with XXXX sugar and cut into pieces of the desired size. Then sand well and let them dry before packing.

Swan Marmalade (Cast)
Apricot Starch Jellies

Dry Apricots (Soaked in Water until Soft)	20 lb	Starch (No. 50–60)	10 lb
		Cream of Tartar	1 oz
Sugar	50 lb	Water	90 lb
Nulomoline	10 lb	Red Coloring	
Corn Syrup	40 lb		

Rub the apricots through a No. 20 sieve to remove the skins. Then place the apricot pulp in a kettle. Add the sugar, Nulomoline, corn syrup, starch, water and cream of tartar, and boil all together to a good string. Add the red coloring. Mix well and cast the batch into warm dry starch.

Place the jellies in a dry room at 130° F for 24 hours and, when cold, remove from the starch, run them through a sanding machine and sand them in granulated sugar which has been colored yellow.

New Jersey Jelly
(Agar Jelly for Cutting or Casting)

Sugar	50 lb	Japanese Gelatin No. 1 (Agar-Agar)	1¾ lb
Corn Syrup	25 lb	Water	75 lb
Nulomoline	25 lb	Flavor and Color	As desired

Soak the gelatin in the water for at least 2 hours, preferably overnight. Place the soaked gelatin in a kettle and boil for several minutes, stirring occasionally, then strain through a No. 30 sieve. Add the sugar and Nulomoline to the strained gelatin and boil the batch to 224° F. Turn off the heat. Add the corn syrup. Mix well, and, when the batch has cooled to approximately 125° F, add flavor and color. The batch may now be poured into trays that have been lined with wrapping paper or it may be cast into warm dry starch.

If the jelly is poured into trays, allow it to remain undisturbed until set, then turn the jelly sheet upside-down. Moisten the paper with water. Peel off the paper and roll the jellies in XXXX sugar. Then cut and roll again in XXXX sugar and let the pieces dry before they are packed into containers.

If aerated jelly is desired, as soon as the batch has boiled to 224° F, place it at once in a marshmallow beater. Start the beater. Then gradually add ½ pound of albumen, dissolved in 1 pound of water, and beat the batch until white.

Acid or flavors containing acid should be added when the batch has cooled to 125° F or less.

Turkish Paste
(Starch Jellies—Cast into Starch)

Sugar	55 lb	Water	100 lb
Nulomoline	45 lb	Cream of Tartar	2 oz
Thin Boiling Starch No. 50–60	12 lb	Color and Flavor	As desired

Place the sugar, Nulomoline, water, starch and cream of tartar into a kettle and cook and stir the batch occasionally until it reaches the boiling point. Continue to stir and boil the batch to a good string. Turn off the heat. Add color and flavor and cast the batch into dry starch.

Place the jellies in a drying room overnight. Then remove the jellies from the drying room and clean well. Run them through a sanding machine. When the jellies are quite dry, they are ready to be packed.

Delaware Desserts
(Plain Marshmallow)

First Grade Gelatin	2½	lb	Corn Syrup	40	lb
Water	32	lb	Nulomoline	10	lb
Fine Granulated Sugar	60	lb			

Mix the gelatin with 17 pounds of cold water. Place the remaining 15 pounds of water in a kettle and add the sugar. Heat and stir until the sugar is dissolved. Do not boil. Then add the soaked gelatin and stir until it is dissolved. Now add the corn syrup and Nulomoline and beat the batch until 1 gallon measure filled with the marshmallow weighs approximately 4 pounds. Add the flavor. Mix well and place the batch into the hopper of the depositor, which should be heated to approximately 100° F.

Cast the marshmallow into starch, containing not more than 8% moisture and heated to 90 to 95° F. Place the trays containing the marshmallow in a room at normal temperature for 10 to 20 hours.

After a slight crust has formed on the marshmallows, remove them from the starch. If they are to be packed plain, dust them with a combination of 9 pounds of dry powdered sugar to 1 pound of dry starch. Let them remain covered with the starch-sugar mixture for several hours before packing.

These marshmallows can also be placed in a revolving pan and coated with colored and flavored jelly (made by formula for New Jersey Jelly). After the marshmallows are well coated with the jelly, add desiccated coconut or colored sugar and turn them out to set.

Maryland Marshmallow
(Toasted Coconut Cut Marshmallow)

Granulated Sugar	25 lb	Water	18 lb
Nulomoline	15 lb	Good Grade Gelatin	2½ lb
Corn Syrup	60 lb		

Soak the gelatin for at least 15 minutes in 10 pounds of cold water. Meanwhile, heat the sugar with the remaining 8 pounds of water until it is dissolved. Do not permit the batch to boil. Turn off the heat, add the Nulomoline and soaked gelatin and stir until the gelatin is completely dissolved. Then add the corn syrup. Place the batch in a beater and beat until quite light. Add flavor and color as desired. Spread in paper-lined or greased trays, and immediately sprinkle the surface with toasted coconut. When the marshmallow has set, cut it into squares or bars and roll in toasted coconut. Permit the bars to dry for several hours before packing.

Danske Nuga
(Fluffy Short Nougat)
Part No. 1

Albumen	12 oz	Tapioca Flour	12 oz
Dissolved in Water	2 lb	Grandma's Molasses	4 lb

Place the albumen solution, tapioca flour and Grandma's Molasses in an upright beating bowl and beat until rather stiff.

Part No. 2

Sugar	35 lb	Nulomoline	5 lb
Corn Syrup	15 lb	Water	8 lb

Boil all together to 245° F. Add 2 gallons of this syrup to Part No. 1, beating at second speed. Boil the remainder of the syrup batch to 250° F. Gradually add this to the beaten batch, and continue the beating until quite light. Add 2 ounces of powdered sugar to start the grain when making the first batch.

When grain becomes visible add 5 pounds of powdered

skimmed milk, 3 pounds of hard coconut butter (88° F m.p.), 2 ounces of salt, vanilla flavor and broken walnut pieces.

Spread the batch out on an oiled and dusted wood or metal slab and, when it has set, cut into bars of the desired size, and coat with chocolate or icing.

Montelemar Nougat
Mixture No. 1

Honey	10 lb	Albumen	3½ lb
Corn Syrup	10 lb	Dissolved in Water	7 lb
Nulomoline	10 lb		

Heat the Nulomoline, honey and corn syrup until they are melted. Place this mixture in a beater. Gradually add the dissolved albumen, and beat until quite light.

Mixture No. 2

Corn Syrup	40 lb	Orange Flower Flavor	
Sugar	60 lb	Vanilla Flavor	
Nulomoline	10 lb	Pistachio Nuts	10 lb
Water	15 lb	Blanched Almonds	15 lb
Coconut Oil (76° F m.p.)	2 lb	Lightly Roasted Almonds	10 lb
Salt	4 oz		

Boil the corn syrup, sugar, Nulomoline and water to 280° F. Pour this mixture gradually into Mixture No. 1, beating well. Add the flavors, salt, nuts and coconut oil and mix well. Then spread the batch on an oiled slab and, while warm, cut into bars.

New York Nougat
Mixture No. 1

Albumen	2 lb	Dissolved in Water	4½ lb

Beat until light, then place in a cold kettle.

Mixture No. 2

Sugar	75 lb	Water	16 lb
Corn Syrup	50 lb		

Cook all together to 245° F. Draw off about 25 pounds of the syrup, and pour it slowly into the beaten albumen, stirring constantly.

Cook the remainder of the batch (Mixture No. 2) to 270° F. Add this to the beaten batch and beat until grain is visible, then add Mixture No. 3.

Mixture No. 3

Nulomoline	10 lb	Corn Syrup	10 lb

Boil together to 265° F. Pour this into the beaten batch. Add 3 pounds of coconut butter, 4 ounces of salt, flavor, nuts or diced fruits, and mix well. Spread the batch out on an oiled slab to set.

This makes an ideal layer to be used between caramels or coconut work. It may also be wrapped plain or coated in chocolate.

English Toffee Nougats
(Chewing)

Corn Syrup	38 lb	Brown Sugar	15 lb
Nulomoline	10 lb	Fine Granulated Sugar	20 lb
Dairy Butter (Salted)	5 lb	Salt	5 oz
Coconut Butter (86° F m.p.)	10 lb	Flavor with a combination of Clove Oil, Almond Oil and Vanilla	
Condensed Whole Milk (40% Sugar)	40 lb		

Place the following ingredients in a caramel cooking kettle: Nulomoline, corn syrup, butter, coconut butter and condensed milk. Give the steam valve a few turns and allow the batch to mix until the fats are melted and mixed with the other ingredients. Then add the brown and white sugar, turn the steam on full and cook the batch as rapidly as possible to a firm ball.

This toffee can be cooked quite successfully, using as much as 110 pounds of steam pressure. After the batch has been cooked to the desired consistency, turn off the heat. Add the salt and flavor and mix thoroughly. Spread the batch out at once to cool. Then gradually fold it together so as to secure an even temperature throughout the batch. It is then ready to pass through drop rolls.

After the toffee nuggets are cold, place them in a revolving pan. Spray them well with melted hard coconut butter. Then add powdered sugar that has been flavored with mint oil. Allow the candies to revolve until they are well coated, spread them out to set. Then pack.

Rum-Butter Toffee (Wrapped)

Condensed Whole Milk (40% Sugar)	40 lb	Fine Granulated Sugar	30 lb
Corn Syrup	30 lb	Hard Coconut Butter—No. 90	15 lb
Nulomoline	10 lb	Salt	5 oz
Grandma's Molasses	5 lb	Rum and Butter Flavor	

Place the condensed milk, corn syrup, Nulomoline, molasses and hard coconut butter in a caramel cooking kettle equipped with double action mixers. Turn the heat on partly and mix the batch until the fat is completely melted. Then add the granulated sugar. Mix well. Turn the steam on full and boil the batch as rapidly as possible to a medium firm ball. Turn off the heat. Add the salt and the rum-butter flavor, and mix thoroughly.

Spread the batch out to cool. Then cut into pieces of the desired size and wrap.

Barcelona Brittle (Chocolate Flavored)

Sugar	15 lb	Coconut Oil (76° F m.p.)	7 lb
Water	4 lb	Unsweetened Chocolate	2 lb
Nulomoline	3 lb	Lecithin	2 oz
Corn Syrup	2 lb	Salt	2 oz
Salted Dairy Butter	3 lb	Ground Roasted Almonds	5 lb
Condensed Whole Milk (40% Sugar)	4 lb	Ground Roasted Filberts	5 lb

Place the coconut oil and chocolate into a kettle and heat until melted. Add the lecithin and mix until it is dissolved. Then add the sugar, Nulomoline, corn syrup, water, butter and condensed milk and cook the batch to 300° F. Now add the ground roasted nuts. Mix thoroughly and turn the batch out on an oiled slab. Then roll as thinly as possible and cut into squares.

It is also possible to grind this brittle roughly and work it into ice cream or use it as a topping for sundaes. It can also be worked into chewing nougat to replace fruits, nuts, etc.

Glasgow Crackle

Sugar	15 lb	Water	2 lb
Nulomoline	3 lb	Ground Roasted Peanuts	Maximum amount
Salted Dairy Butter	2 lb	Salt	2 oz
Coconut Oil (76° F m.p.)	8 lb	Rum Butter Flavor	
Lecithin	1½ oz		

Melt the dairy butter and coconut oil. Add the lecithin and stir until it is melted. Then add the sugar, Nulomoline and water and boil all together to 310° F. Now add the maximum amount of ground roasted peanuts, the salt and flavor, and mix well.

Spread the batch out on an oiled slab. Roll into a thin sheet and cut with a roller knife.

This piece may be coated in chocolate or wrapped in cellophane or waxed paper, and used either for packages or bars. It is also possible to grind this candy roughly and work it into ice cream or use it as a topping for sundaes. It may also be worked into chewing nougat to replace fruits, nuts, etc.

Chewing Gum

Chewing Gum Bases

a Bubble Gum Base

Washed Pontianak Gum	425 lb	Washed Gutta Katian	400 lb
		Washed Gutta Soh	75 lb
		Candelilla Wax	10 lb

The mixed gums and wax are heated until the total batch contains only 8 to 9% moisture.

b Stick Gum Base			
Pontianak Gum	425 lb	Gutta Katian	400 lb
		Gutta Soh	75 lb
		Candelilla Wax	60 lb

Ball Gum

Base *b* (above)	22 lb	Wax	1½ lb
Corn Syrup	48 lb	Caramel Paste	2½ lb
Sugar	117 lb	Flavor	2⅔ oz
Chicle	3 lb		

Penny Stick Gum

Base *a* (above)	40 lb	Sugar	140 lb
Corn Syrup	40 lb	Flavor	30 oz

Bubble Gum

Base *a* (above)	35 lb	Sugar	115 lb
Pontianak Gum	5 lb	Flavor	28 oz
Corn Syrup	45 lb		

Ice-Cream Powder

Ice-cream powder is the perfect smoothener for ice cream and other frozen preparations. It consists of:

Powdered Sugar	48 lb	Gum Karaya	50 lb
White Gum Acacia	2 lb		

It is a pure vegetable compound which has the following advantages:

1 It can be used to advantage in the manufacture of the best grade of sherbets and ice cream, as well as cheaper grades.
2 Although it contains no gelatin, it can be used to replace gelatin entirely.
3 It will cause milk and cream to beat up to more than double their original volume.
4 It will make ice cream stand up, preventing it from melting quickly and giving it a firm substance. It prevents ice cream from becoming snowy or rough and from shrinking in volume while being stored or shipped.

5 It renders all water ices smoother and richer.
6 It is an incomparable thickener.

Directions for Use

Straight Cream	2 oz to 10-gal freezer
¼ Milk and ¾ Cream	4 oz to 10-gal freezer
⅔ Milk and ⅓ Cream	8 oz to 10-gal freezer
Straight Milk	12 oz to 10-gal freezer

The specified quantity of ice cream powder should be thoroughly incorporated with the amount of sugar which is used with the batch of 10 gallons of ice cream. Into this mixture sufficient straight milk, or cream and milk combined, should be stirred to make a smooth paste. To this paste, the balance of the milk and cream should be slowly added. Stir and mix constantly until the whole is dissolved. When dissolved, freeze the batch in the usual way.

Ice-Cream Preparation
(Freeze-It)

Albumen	10 oz	or:	
XXXX Sugar	80 oz	Dried Milk	50 oz
Whole Dried Milk	10 oz	XXXX Sugar	50 oz
Indian Gum	5 oz	Cream of Tartar	4⅖ oz
Vanillin	1½ oz	Sodium Carbonate	2 oz
		Vanillin	0.06 oz

Package 4½ ounces = 127.5 grams

Stir contents of the package into ½ pint (1 cup) cold milk. Stir until dissolved. Whip ½ pint (1 cup) chilled whipping cream. Add the dissolved Freeze-It to the whipped cream and mix. Freeze.

Chocolate Ice Cream

Milk	32 oz	Eggs	4 oz
Sugar	16 oz	Cream	32 oz
Flour	2 oz	Vanillin	¼ oz
Salt	⅛ oz	Unsweetened Chocolate	4 oz

Heat milk and add flour, salt, and sugar. Stir thoroughly in double boiler for 20 minutes after batch is brought to a boil. After the mass thickens, add the beaten eggs and cook for 5 minutes longer with constant stirring. Cool, add cream which has been whipped into a stiff paste, and then add the flavoring. Add the melted chocolate which was previously mixed with a little sugar and warm milk to form a paste. Put in a refrigerator or pack in ice and salt until frozen.

Ice Cream Without Gelatin

Butter Fat	12	lb	Milk Serum Solids	11¾	lb
Granulated Sugar	12	lb	Vanilla Flavor	To suit	
Cerelose (Corn Sugar)	4	lb			

Water Ices and Sherbets

The figures are given on the basis of 100 pounds of mix, which is about 10½ gallons.

Water Ice

Cane Sugar	25	lb	Water, Fruit, Fruit Acid, Flavor and Color	67⅖	lb
Corn Sugar	7	lb			
Agar (3.2 oz or 90.6 g)	⅕	lb			
Gum Tragacanth or Galagum C (6.4 oz or 181.2 g)	⅖	lb			

Overrun 20 to 25% Total yield 13 gallons

Sherbet Using Milk

Cane Sugar	25	lb	Gum Tragacanth or Galagum C (3.2 oz or 90.6 g)	⅕	lb
Corn Sugar	7	lb	Whole Milk	50	lb
Agar (3.2 oz or 96.6 g)	⅕	lb	Water, Fruit, Fruit Acid, Flavor and Color	17⅗	lb

Overrun 25 to 30% Total yield 13.5 gallons

Sherbet Using Ice-Cream Mix

Cane Sugar	25 lb	Ice-Cream Mix, without Sugar or Gelatin	10	lb
Corn Sugar	7 lb	Water, Fruit, Fruit Acid, Flavor and Color	57⅗	lb
Agar (3.2 oz or 90.6 g)	⅕ lb			
Gum Tragacanth or Galagum C (3.2 oz or 90.6 g)	⅕ lb			

Overrun 25 to 30% Total yield 13.5 gallons

Orange Water Ice
(For 10 Gal Batch)

Granulated Sugar	21 lb	Orange Juice (Or Its Equivalent in Orange Flavor)	1 gal
Corn Sugar	7 lb	Citric Acid	To suit
Galagum C	3 oz		

Make up to 10 gallons with water. Takes no overrun.

Orange Sherbet
(10 Gal Mix)

Cane Sugar	22½ lb	Orange Concentrate	4 oz
Cerelose (Corn Sugar)	7½ lb	Citric Acid	To suit
Milk	4 gal	Color	To suit
Gelatin	11 oz		

Make up to 10 gallons with water.

Cocoa Junket

Cocoa	2 oz	Junket Tablets	2
Boiling Water	4 oz	Cold Water	1 oz
Sugar	4 oz	Vanilla Extract	¼ oz
Milk	32 oz		

Cook mixture of cocoa and water in double boiler for 5 minutes. Add sugar, stir until dissolved, and then add milk which has been preheated to 100° F. Add vanilla extract and heat to 120° F. Stir in junket tablets which are dissolved first in a little water. Pour into containers immediately, let stand until set.

Cocoa and Chocolate Products*

Cacao (cocoa) is derived from the seeds of Theobroma Cacao. The cacao, often called cocoa, tree is indigenous to Brazil, Venezuela, Ecuador, and Mexico, thriving only in a tropical climate in localities that have well-watered, well-drained soils, and plenty of rainfall. It is a medium-sized tree, 20 to 30 feet high. The fruit is a pod, having a hard, thick shell, and is from 8 to 10 inches long and from 3 to 4 inches thick. There are forty or more beans in each pod.

Different Kinds of Cocoa Beans

Porto Cabello or Caracas

These beans are large, but not very uniform in size. They are covered with a reddish, sometimes silver-gray dust. The husks are thicker than on ordinary beans and are of dark red color inside. Ninety-nine hundredths of 1% is theobromine. The small Caracas bean has less fat and is therefore cheaper.

Trinidads

These beans are nearly flat. The husk is dark gray, and the kernels are almost black. Ninety-two hundredths of 1% is theobromine. This is a good bean for the price.

Guayaquils (Ecuador)

These are large, flat, irregular-shaped beans. The husk and kernel are brown. They have a fine strong flavor.

Cubans

These are egg-shaped, flat beans. The husk is mauve, and the kernel is dark brown.

Paras (Brazil)

These beans are irregular in size, and long and flat on one side. The husk is gray or reddish, and the kernel is light brown. They are rich in fat, and good for a filler.

---

*Courtesy of B. J. Zenlea, *Food Industries.*

Grenadas

These are similar to Paras.

Javas

These are oval, slightly flattened beans. The husk is thin and easily peeled. The kernel is light brown, of bitter, but agreeable, taste. One and twenty-two hundredths per cent is theobromine. They have a fine flavor, but are high in price.

Bahias

These beans are flat, triangular in shape and rough. The yellow-looking bean is the best, the grayish-red one, inferior. The kernel is mauve in color and produces a dark liquor. This is a very popular bean in the United States.

Test for Ripeness of a Cacao Bean

Cut the bean with a knife. If it slices and no shell falls off, the bean is unripe.

The (Cocoa) Cacao Bean and its Products

Cocoa has been cultivated since ancient times, but this plant and its products were unknown to Europe until the beginning of the sixteenth century, when Cortez and his soldiers became acquainted with it upon their landing in Mexico. From then on the use of cocoa spread very rapidly throughout the civilized world.

Cocoa is grown almost entirely south of the equator, the largest crops coming from the west (Gold Coast) of Africa. The next largest crop comes from Ecuador. The cocoa from each country has a distinctive character and flavor of its own. The cocoas coming from Venezuela are the finest. The cheapest are from San Domingo.

Cocoa beans are the seeds of a tree that grows in most tropical countries. The seeds grow in pods very much like the seeds of the pumpkin and squash, and hang from both the branches and trunks of the trees. When ripe they are picked from the tree and gathered into large piles. They are then cut open and the seeds and pulp scraped out.

The seeds with their surrounding pulpy mass are placed into large vats or pits and allowed to ferment. The time of fermentation varies according to the kind of cocoa. The sweeter variety requires a shorter fermentation while the coarser or more ordinary type requires a longer fermentation.

During the fermentation starch is converted by the diastase present into soluble sugars, which are further fermented into alcohol, and finally converted to acetic acid. The pulpy mass is decomposed and flows away as a vinegar-smelling fluid. During the fermentation, care must be taken that sufficient air is circulated through the beans, otherwise, butyl alcohols are formed which give the cocoa a bad odor. To prevent this the beans are turned over every day. Fermentation changes the color of the bean from a light to a dark, rich, red-brown, liberates the alkaloid theobromine, and lessens its bitter and astringent taste.

When the fermentation is completed the beans are separated from any remaining pulp, dried, and sometimes polished. The drying must be done very carefully, as improperly dried beans are apt to become moldy. The best cocoa is obtained by drying in the sun, although methods using steam- or water-heated pipes are also in use.

Upon importation into the United States, the beans are inspected and are rejected if they contain more than 5% of either wormy or moldy beans.

The raw cocoa beans are cleaned by passing them over a shaking-table covered with a metal screen which has holes smaller in size than the beans. Sand, dust, and other foreign materials are screened out. At the end of the shaking-table, the beans drop into a chute located at the bottom of an air duct. The cleaned beans are carried up by air suction, and deposited into a conveyor. Stones, metals, and other heavy materials drop to the bottom of the chute, which is counterpoised by a weighted lever. When sufficient heavy material has gathered, the bottom opens and the foreign material drops to a receiving pan.

The clean raw cocoa beans are run into a roaster of approximately 600 pounds capacity, of the revolving drum type, which is heated by coal fire using forced draft, or by gas or superheated steam. A roast usually requires 45 minutes, depending on the type of cocoa beans, and the kind of roast desired. The final flavor of the cocoa powder and chocolate depends a great deal on the proper roasting of the cocoa beans.

The cocoa beans attain a temperature from 240 to 300° F at the end of the roast, are quickly discharged into a cooling truck with a perforated metal bottom, and are quickly cooled by cold air suction.

The cooled beans are carried to a cracker and fanner device, which cracks up the beans and blows the shells aside. The cocoa germs and fines are separated from the broken nibs by sieving them through a rotating cylinder covered with screening of increasing mesh. The clean nibs are carried to hoppers which feed a battery of triple stone mills, and are ground into a heavy, viscous fluid known as chocolate liquor, which is the starting point in the manufacture of cocoa powder and chocolate.

In the manufacture of cocoa powder the chocolate liquor is transferred to hydraulic presses, and pressed from 8 minutes to 1¾ hours, depending on the amount of fat that is to be left in the cocoa powder. The cocoa press cake is cooled, and broken up by a cracking device, passed over a magnetic belt to remove stray iron, and then put through a micro-pulverizer. The ground press cake is cooled on a conveyer, and is refined by being passed through silk bolting cloth used as a sieve; or a current of air is played on the particles of cocoa in such manner as to separate the fine cocoa from the coarse. The coarse cocoa, or cocoa tailings, are returned to the grinders and reground over and over again, until the particles are fine enough to pass through the sieve, or can be separated by air flotation.

There are many kinds of cocoas, but they can be divided into two main classifications: (a) natural process cocoa and (b) Dutch process cocoa.

Natural process cocoas are cocoas to which no alkalies or *dutching agents* have been added during manufacture.

Dutch process cocoas are cocoas to which an alkali has been added. In preparing Dutched cocoas, the clean nibs are boiled in a solution of alkali, such as potassium, sodium or ammonium bicarbonate, or carbonate. The water is evaporated and the

Dutched nibs dried in a roaster. The nibs are milled, the liquor pressed, and the cocoa refined in the same manner as natural process cocoas.

Dutched cocoas are darker in color, less acid, and less bitter than natural cocoas. They are more economical to use, and, for certain purposes, cannot be replaced by natural cocoas.

The various types of cocoas can be described as follows:

1 Semi-cocoas, cocoa nuggets, cocoa crumbles, cocoa chips, etc., are prepared by crumbling press cake, or by mixing chocolate liquor with a quantity of cocoa to reduce the fat. In the latter case, the product is molded into thin cakes, and then broken into small chips. Because of the high fat content, 32 to 45%, and chocolate flavor, they are used extensively for fine quality ice cream, and for fancy icings.

2 High-quality cocoas usually contain from 25 to 32% cocoa butter. They are made from select quality cocoa beans and are carefully blended. These cocoas are either Dutched, natural, or semi-Dutched. They are used for fancy ice cream, for drinking, for icing, as a filler between layers of cakes, and for syrups.

3 Breakfast cocoas contain not less than 22% cocoa butter. The natural process and semi-Dutched breakfast cocoas are used for drinking, syrups, and general household use. The Dutched type breakfast cocoas are used for chocolate biscuits and crackers, for devil's food cakes, and for home made candy, such as chocolate fudge.

4 Semi-dry cocoas usually contain 15 to 18% fat. The natural and semi-Dutched cocoas are used for malted milk mixtures, cocoa milk drinks, chocolate flavored desserts, chocolate pies and chocolate-flavored syrups. The percentage of fat they contain is just right for use in the products mentioned.

5 Dry cocoas contain 10 to 12% fat, and are used for miscellaneous purposes. The natural process cocoa is used for milk drinks, cocoa skim milk mixtures, sweet cocoas, icings, etc. The Dutch process cocoa is used for cocoa extracts, for baking, and devil's food mixtures. These cocoas

are the most common type found on the market, and are used where low cost is of importance.

6 Extra dry cocoas contain 8 to 9% cocoa fat. They are used for thin syrups and other mixtures where any excess fat is not desired.

7 Very dry cocoas contain 6½ to 7¾% fat, and this fat content is about the limit for cocoas made by pressing the fat from the chocolate liquor. They are used for making substitute or imitation chocolate coatings by incorporating sugar and another fat, such as cocoanut oil. They are desired for their low fat content, because of the difficulty of blending cocoa butter and cocoanut oil. This cocoa is used as a dusting powder in confectionery.

8 Solvent-extracted cocoas, or defatted cocoas, contain from ½ to 3% fat. They are usually manufactured from cheap low-fat cocoa cake, by extracting the remaining fat by means of a solvent. Solvent-extracted cocoas have almost no chocolate flavor. Some have a faint odor and taste of the solvent used for their extraction. They are usually high in moisture content, which is, at times, as high as 10%. Solvent-extracted cocoas are used as dusting powders in confectionery, and for imitation chocolate.

9 By-product cocoa and cocoa cake usually contain 5½ to 11% cocoa fat. This product is obtained by passing cocoa fines and dust through the Anderson expeller, which removes a portion of the fat. The by-product cake has a limited use as a fertilizer ingredient, as raw material for theobromine extraction, and for cattle feed for which it has not been very successfully used.

10 Prepared cocoas are cocoa mixtures, such as cocoa and sugar, cocoa skim milk and sugar, and cocoa malt skim milk and sugar. They are used for instant chocolate and milk drinks in ice cream parlors, and are also put up in envelopes of the proper weight sufficient for one cup.

Manufacture of Chocolate

Chocolate, because of its very fine flavor, its concentrated food value, pleasing appearance, good keeping qualities, and convenience in handling, is a food of exceptional value to all.

The definition of chocolate, technically speaking, is the solid or plastic mass obtained by grinding cocoa nibs free from shell or husk and germs, and is the product commonly known as chocolate liquor. By common usage, however, the word chocolate is usually applied to sweet chocolate, milk chocolate, and chocolate covered centers.

Although chocolates may be similar in composition as to percentages of chocolate liquor, sugar, added cocoa butter, and milk solids, there is a great deal of difference in their flavor and texture.

Of the many varieties of cocoa beans, approximately thirty are used in the larger chocolate factories. The cocoa beans vary in color, chocolate character, pungency, acidity, mildness, bitterness, and creaminess. The delicate flavors of the cocoa beans are developed by careful roasting, cleaning, and milling. The flavor of the resulting chocolate depends on the skill used in blending, milling, and conching the various varieties of cocoa nibs with the proper amount of sugar and cocoa butter, with or without milk solids, and flavors.

The possibility of developing blends of chocolate is unlimited. Most formulas are limited to blends of one to six varieties. The number of combinations that can be made with thirty varieties of cocoa beans, blended with varying percentages of sugar, cocoa butter, milk solids, and flavors, and milled to various degrees of fineness, and adjusted to various viscosities and colors, is astounding.

There are many varieties of chocolates on the market, but they can be classified into the following general types: bitter, bittersweet, dark sweet, medium dark sweet, light sweet, milk chocolate, ice-cream coating, skim milk, buttermilk, malted milk, imitation, and lecithin coatings.

It is difficult to state exactly what type of chocolate coating is best suited for a particular center. It depends on a number of factors among which are sweetness, texture, and blending properties of both the center and coating. As a rule, chocolate liquors and bittersweets are used for every sweet center. Dark sweets and medium dark sweet coatings are used for moderately sweet centers. Milk chocolates are used for caramels, nougatines, marshmallows, etc. Light sweets are used for not-too-sweet centers and for candy bars.

The chocolate liquor of commerce is also known as chocolate, bitter chocolate, chocolate liquor, chocolate paste, bitter chocolate coating, and unsweetened chocolate. It is the product obtained by grinding cocoa nibs free from shell or husk and germs, usually through triple stone mills, Bausman disc, or similar devices.

The above product comes in two types, natural process and Dutch process. The Dutch process type, or alkalized type, is prepared in the same manner as described for cocoas except that none of the fat is expressed. The natural chocolate liquor has a pH 5.1 to 6.0. The Dutched chocolate liquor has a pH 6.0 to 7.8, depending on the original pH of the nibs, and the amount and kind of alkali used. If more than 1½% alkali is used the liquor has a tendency to develop a saponified taste.

Chocolate liquors are molded into ten-pound cakes for commercial use, and in one-quarter- to one-pound cakes for household use. Chocolate liquor is used for coating very sweet centers, such as hand roll centers and butter creams. It is also used as an ingredient for ice cream, candy, culinary products, and chocolate syrup. The Dutched type of chocolate liquor is used for flavoring tobacco, and for general uses where a milder and darker product is desired. Chocolate liquor is often used in place of cocoa powder when a creamier and richer product is desired. The high fat content, 50 to 56%, however, makes its use difficult without such aids as homogenizers or emulsifiers.

Chocolate liquor is considered pure if it contains not less than 50% cocoa fat, and, on the moisture and fat-free basis, not more than 8% total ash, not more than 0.4% ash insoluble in hydrochloric acid, and not more than 7% crude fiber. The standards for Dutched liquor are the same as for natural process liquor, after allowance is made for the kind and amount of alkaline substance added. The amount of alkaline substance permitted is not more than 3 pounds of potassium carbonate, or its neutralizing equivalent in other alkaline substances to each 100 pounds of cocoa nibs.

Sweet chocolate, sweet chocolate coating, is chocolate mixed with sugar and/or dextrose with or without the addition of cocoa butter, spices, or other flavoring ingredients. This definition applies to all sweet coatings regardless of the amount of sugar and chocolate liquor they contain. The phrase, other flavoring ingredients, is somewhat vague, and has led to some misunderstanding. The definitions for food products are so framed as to exclude anything not mentioned by name, but the strict interpretation of flavor would include any and all flavoring ingredients such as small amounts of malt, malted milk, skim milk, dairy butter, essential oil, etc.

Although dextrose, corn sugar, is permitted in chocolate, either alone or mixed with sucrose (cane sugar or beet sugar), it has been used only to a very limited extent, and has not been studied sufficiently to have its virtues discussed.

Bittersweet coatings contain from 80 to 95% chocolate liquor and 5 to 20% sugar. Since the cocoa butter content is very high, 40 to 50%, and the viscosity low, the addition of extra cocoa butter is not required. These coatings are prepared by adding sugar to chocolate liquor with or without flavor and refining. Bittersweet chocolates are dark in color and are used as couvertures for cordials, peppermint creams, fudge, whipped cream, Italian creams, hand rolls, butter creams, and marshmallow centers.

Mild bittersweet coatings contain from 50 to 75% chocolate liquor, from 25 to 38% sugar, and from 38 to 42% cocoa butter. As the sugar content increases and the chocolate liquor content decreases, it becomes necessary to add additional cocoa butter to obtain the proper viscosity. Mild bittersweet coatings are used for practically the same type of centers as bittersweet coatings, and are used to some extent for covering marshmallow crackers.

Dark sweet chocolates contain from 25 to 45% chocolate liquor, from 40 to 50% sugar, and vary in total cocoa butter content from 34 to 42%, depending on the desired viscosity of the finished coating. Dark sweet chocolates are used for enrobing such centers as nougats, cordials, peppermints, fudge, and nut-flavored cream fondants.

Medium dark sweet coatings are the most popular used, to a great extent, for molding chocolate bars and a variety of solid chocolate items, which are neither too sweet nor bitter for the average consumer's taste. They contain from 20 to 40% chocolate liquor, from 45 to 55% sugar, and from 32 to 42% total cocoa butter. The commercial product is sold in molded ten-pound cakes, and is used for enrobing such centers as caramels, nougats, fruit-flavored cream fondants, jelly centers, cocoanut cream fondants, vanilla cream fondants, nut clusters, raisin clusters, dates, figs, marshmallows, nut-flavored cream fondants, cocoanut, sponge, chip and marshmallow cracker.

Light sweet coatings are of two types, the expensive type and the cheaper type. The expensive types are prepared from blends of light-colored and mild but chocolate-like cocoa nibs such as Java, Ceylon, Maracaibo, etc., and a moderate amount of sugar, usually 45%. These coatings are very expensive, and are used for enrobing moderately sweet centers of the better sort. They are finely milled, conched, delicately flavored, and are representative of the chocolate expert's skill in the art of fine chocolate making.

The cheap type of light sweet chocolate coatings owe their light color to the small amount of chocolate liquor and the large amount of sugar used in their manufacture. They usually contain from 5 to 15% chocolate liquor, from 55 to 68% sugar, and a total cocoa butter content of 31 to 39%, depending on the fineness of grinding and their ultimate use. These coatings are used for special purposes, and for enrobing nut rolls, nuts, nut clusters, nut chews, raisin clusters, and for candy rolls containing peanuts, marshmallows, caramels, fudge, and other ingredients. The light sweet coating, although lacking in chocolate flavor and excessively sweet, nevertheless, blends in nicely with the types of centers mentioned.

Milk chocolate, sweet milk chocolate, is the product obtained by grinding chocolate with sucrose and/or dextrose, with the solids of whole milk, or the constituents of milk solids in the proportion normal for whole milk, and with or without cocoa butter, and/or flavoring material. It contains not less than 12% milk solids. Milk chocolates usually contain 7 to 17% chocolate liquor, 35 to 55% sugar, and a total fat content of 28 to 39%. There are many types of milk chocolate, varying in their degrees of milkiness, chocolate character, color, fineness, and viscosity. Milk chocolates are prepared from fresh whole milk which is first evaporated, and from milk powders, both spray and roller process types, and from skim milk powder to which has been added an amount of milk fat either in the form of dairy butter, or butter fat, necessary to convert the skim milk solids to whole milk solids.

In converting skim milk solids to whole milk solids by the addition of dairy butter, it is necessary to bear in mind the fact that dairy butter contains 80 to 84% milk fat, and approximately 16% water. The difficulties of removing the water and of blending the dairy butter with the skim milk solids, and the fact that 100 pounds of dairy butter yields approximately 80 pounds of milk fat, make this method of milk chocolate preparation undesirable, and often more expensive than the method using whole milk solids.

Milk chocolates owe their popularity to their delicate milk flavor, and their lack of a harsh or bitter taste. They are particularly liked by women and children. The manufacture of milk chocolate with a very fine milky flavor is a difficult art, and is a secret closely guarded by chocolate manufacturers. Milk chocolates are sold over the counter of the ice-cream parlors and candy stores as broken milk chocolate, and in the form of bars which vary from a fraction of an ounce to a pound. Milk chocolate coatings are molded in ten-pound cakes and are used for enrobing such centers as nougats, vanilla cream fondants, nut rolls, nuts, nut clusters, and nut chews.

Ice cream coatings are chocolate coatings of high cocoa butter content containing from 50 to 60% fat. Their high fat content is necessary to obtain a very thin running chocolate, which will quickly drain off the solid ice cream center when dipped. Ice cream coatings may be of the dark sweet, medium dark sweet, light sweet, and milk chocolate types. Allowance should be made for the high fat content. As their name implies, they are used exclusively for dipping ice cream in the manufacture of ice cream pops and Eskimo Pies. Since ice cream coatings are frozen instantly on the ice cream, and the ice cream is kept at temperatures below freezing, the user of ice cream coatings is not concerned with exact enrobing temperatures, nor with fat or sugar bloom. His chief problem is to combat the thickening-up of the chocolate due to moisture absorption from the melting ice cream. This difficulty is often remedied by the incorporation of $^3/_{10}$ of 1% lecithin in cocoa butter.

Skim milk, buttermilk and milk chocolate coatings are sweet chocolate coatings containing not less than 12% of, respectively, skim milk solids, buttermilk solids, or malted milk and milk solids. These coatings are not defined in F.I.D. 191, or later additions to the Federal Pure Food Laws. Buttermilk coatings, and skim milk coatings, have become popular in the last few years. They are a cross between milk chocolate coatings and light sweet coatings. Skim milk and buttermilk cut down the excessive sweetness of a light sweet coating without sacrificing the light color. These coatings are used for enrobing the same types of centers as light sweet coatings, and in some cases, milk chocolate coatings. These coatings, in the absence of a published definition for them, should be labeled and sold for exactly what they are, namely, skim milk, buttermilk, or malted milk chocolate coatings.

Modified chocolate coatings are sweet chocolate coatings of various types, such as dark sweets, light sweets, and ice cream coatings, to which has been added a small amount of skim milk, buttermilk, whole milk, or malt sugar. The amount added is usually 3 to 5%. These ingredients are added as flavoring ingredients, and to cut down the excessive sweetness of a light sweet coating, and also to cut the bitterness or sharpness of a dark sweet coating. The above ingredients are used to some extent in ice cream coatings for the reasons mentioned, and for obtaining a softer eating chocolate of less brittleness than the ordinary chocolate coating, when it is frozen on the ice cream.

Imitation chocolate coatings are mixtures of cocoa, sugar, and a substitute fat such as cocoanut oil, cocoanut stearines, illipe oil, etc., with or without flavoring ingredients. They are used mostly for enrobing cheap centers, and for medicated products where a higher melting point chocolate is desired than is obtained with cocoa butter. Cocoanut oil coatings are used to some extent for ice cream. Imitation coatings are what their names imply, imitation chocolate coatings often lacking the true flavor and taste of pure chocolate.

Lecithin chocolate coatings are chocolate coatings of all types to which has been added a small amount of lecithin, usually not exceeding 4/10 of 1% lecithin in cocoa butter. Lecithin modifies the physical properties of chocolate to some textent. Coatings can be made with 3 to 8 pounds less of added cocoa butter per 100 pounds of chocolate, making them less fatty. Lecithin coatings resist thickening-up due to moisture, allow greater flexibility in enrobing temperature, and cover centers more uniformly. Lecithin stabilizes chocolate coatings so that the viscosity is only slightly changed on aging, heating, or mechanical manipulation.

Cocoa butter, is the only fat common to chocolate, and is the edible fat obtained from sound cocoa beans either before or after roasting. It is a yellowish-white solid having a faint, agreeable odor, and a bland chocolate-like taste, and is usually brittle at temperatures below 25° C (77° F). It is slightly soluble in alcohol, but is soluble in ether, chloroform, petroleum benzin, benzene, and in many other solvents. It has a specific gravity of approximately 0.8580 at 100° C (212° F), and 0.9100 at 40° C (104° F), when compared with an equal volume of water at 25° C (77° F). Cocoa butter melts between 30 and 35° C (86 and 95° F), has a refractive index of approximately 1.4578 at 40° C (104° F), a saponification value of 188 to 195, an iodine value of 33 to 38, and a free fatty acid content of 0.70 to 1.4% calculated as oleic acid.

Cocoa butter is obtained by pressing chocolate liquor in the manufacture of cocoa powder, by passing cocoa fines through the expeller, and by solvent extractions from low-fat cocoa press cake and cocoa fines. Cocoa butter is used mostly for the manufacture of chocolate and confectionery, and to some extent for the manufacture of suppositories, creams, lotions, ointments, and soaps. It is also used for the treatment of burns, sunburn, and for massaging.

Chocolate, when properly molded, possesses a gloss which,

if not protected, soon becomes dull in appearance. Both heat and moisture rapidly destroy the luster of chocolate, and cause a mottled gray surface commonly known as bloom.

Chocolate must be carefully melted, carefully tempered, and carefully cooled when manufactured, otherwise bloom may develop in a short time. The handling of chocolate is a fine art and requires very careful attention to details. If chocolate is cooled too rapidly, the outer surface will solidify, trapping the latent heat within, which will gradually find its way to the surface and cause fat bloom. The proper temperature for storing chocolate is 55 to 65° F, and the proper relative humidity 60 to 65%.

USES OF COCOA

Uses	*Semi-Cocoas, Cocoa Nuggets, Cocoa Crumblers, Cocoa Chips, 32–45% Fat*		*High Quality Cocoas, 25–32% Fat*		*Breakfast Cocoas Not Less Than 22% Fat*		*Semi-Dry Cocoas, 15–18% Fat*		*Dry Cocoas, 10–12% Fat*		*Extra Dry Cocoas, 8–9% Fat*	*Very Dry Cocoas, 6½–7¾% Fat*	*Defatted or Solvent Extracted Cocoas ½–3% Fat*	*By-Product Cocoa and Cocoa Cake, 5½–11% Fat*	*Prepared Cocoas Sweet Cocoas*
	Natural	*Dutched*	*Natural*	*Dutched*	*Natural*	*Dutched*	*Natural*	*Dutched*	*Natural*	*Dutched*					
Baking					●	●	●	●	●	●					
Beverages			●	●	●	●	●	●	●	●					
Cake fillers			●	●											
Cattle feed														●	
Chocolate biscuits						●		●		●					
Chocolate crackers						●		●		●					
Choc. flavored desserts					●	●	●	●	●	●					
Chocolate pies					●	●	●	●	●	●					
Cocoa, malt, sugar mixt.							●	●	●	●					
Cocoa milks							●	●	●	●	●	●			
Confectionery, fudge			●	●	●	●	●	●	●	●					
Devil's food cake						●		●		●					
Dusting powder												●	●	●	
Fertilizers														●	
Household use					●	●			●	●					
Ice cream	●	●	●	●											
Icings	●	●	●	●											
Imitation choc. coatings									●		●	●			
Medicinal powders or tablets					●	●	●	●	●						
Sweet cocoas					●	●	●	●	●	●					●
Syrups			●	●	●	●	●	●	●	●	●				
Theobromine													●	●	
Thin syrups									●		●	●			
Tobacco			●	●	●	●	●	●	●	●					

USES OF CHOCOLATE

	Chocolate Liquor	*Bittersweet*	*Dark Sweet*	*Medium Dark Sweet*	***Light Sweet***	*Milk Chocolate*	*Ice Cream Coating*	*Skim Milk, Buttermilk*	*Imitation Coating*	*Cocoa Butter*
Butter creams	●	●	●							
Candy bars				●	●	●		●		
Caramels				●	●			●		
Chocolate bars				●	●	●				
Chocolate ice cream	●									
Chocolate pops							●		●	
Chocolate syrup	●									
Cocoanut creams				●					●	
Cordials		●	●							
Culinary products	●									
Frappé		●							●	
Fruit-flavored creams				●					●	
Fudge	●	●	●		●			●		
Hand rolls	●	●								
Ice cream bars							●		●	
Italian creams		●								
Jelly centers				●					●	
Lotions										●
Marshmallows, marshmallow crackers		●	●	●	●			●		
Nougats			●	●	●	●		●	●	
Nuts, nut clusters, nut rolls				●	●	●		●		
Nut-flavored creams			●	●					●	
Ointments										●
Peppermint creams	●	●	●							
Raisins, dates, figs				●		●		●		
Sponge centers				●					●	
Suppositories										●
Tobacco	●									
Vanilla-creams				●		●		●	●	

If chocolate is taken out of a cold room and placed in a warm or damp room, a film of moisture forms on the surface. The fine sugar particles dissolve and form a thin layer of syrup. On evaporation of the moisture, the sugar crystallizes in minute form, giving the chocolate a rough, grayish appearance, sometimes mottled by red-brown spots due to the solution of the coloring matter of the chocolate in the moisture. This phenomenon is known as sugar bloom.

If chocolate is stored in a warm place or exposed to heat, the cocoa butter softens or melts, and expands, forcing its way through the spaces between the sugar and cocoa particles to the surface. On cooling, the cocoa butter crystallizes, forming a grayish-white film which is greasy to the touch. This type of bloom is known as fat bloom, and is a source of trouble during the summer months. Although the flavor is usually not impaired, the grayish appearance of the chocolate is considered by the public to be a sign of staleness, and this results in great economic loss to the manufacturer. Chocolate picks up foreign

APPROXIMATE COMPOSITION OF CHOCOLATE COATINGS

	Chocolate Liquor	*Sugar*	*Total Fat*	*Whole Milk Solids*	*Skim Milk or Buttermilk Solids*
Bitter	100	0	50–56	0	0
Bittersweet	80–95	5–20	40–50	0	0
Mild bittersweet	50–75	25–38	38–42	0	0
Dark sweet	25–45	40–50	34–42	0	0
Medium dark sweet	20–40	45–55	32–42	0	0
High grade lt. sweet	15–25	40–50	34–39	0	0
Low grade lt. sweet	5–15	55–68	31–39	0	0
Sweet milk	7–17	35–55	28–39	12+	0
Ice cream bar	9–29	29–39	50–60	12+	0–6
Skim milk, buttermilk, malted milk	7–17	35–55	30–40	0	12+

Small quantities of flavoring materials such as vanillin, coumarin, coffee cinnamon, peppermint, orange, lemon, anise, and synthetic cocoa bean flavor may also be used.

odors readily, and should be stored in a clean room with free circulation of air.

Cocoa shells are a by-product of the cocoa and chocolate industries. They are used chiefly as a fertilizer ingredient, and for the extraction of their fat and theobromine content.

Coffee

Coffee is the fruit of Coffea arabica, indigenous to tropical Africa, especially Abyssinia, from where its growth undoubtedly spread across the continent to Liberia, where it grows wild and is now extensively cultivated. The Liberian coffee plant is a distinct species, which seems to be less subject to disease and which has been successfully introduced into the East Indies. The cultivation soon spread to the Western world, Cuba, Mexico, Central and South America. At the present time Brazil produces about three-quarters of the whole world market.

In its wild state, the coffee tree attains a height of from 20 to 30 feet and looks somewhat like a cherry tree. In cultivation, it is trimmed down to the height of about 6 feet. The coffee tree develops best in rich soil and in a tropical climate where the rainfall exceeds 75 inches per year.

The fruit is an oval, deep purple, two-seeded drupe. The parchment-like endocarp incloses two plano-convex seeds, which are placed together by their flat sides and constitute the raw coffee of commerce. There are two general varieties:

I Brazil coffees
II Mild coffees

The Brazil coffees include the Santos and Rio coffees. The mild coffees include the Java, Mocha, Bogota, Guatemala, Porto Rican, Mexican, Maracaibo and a few others.

Santos coffee is shipped from the port of Santos. It is mild and sweet. Santos coffee is usually blended with Bogota to obtain a good flavor and color.

Rio coffee is shipped from the port of Rio de Janeiro. It is rather pungent.

The roasting of coffee results in a notable reduction of some of the constituents, especially the caffein, fat and sugar. When roasting is properly conducted, the total loss in weight amounts to from 12 to 18%, of which about 8% is moisture.

Coffee is sometimes glazed with sugar before roasting. According to Hilger and Juckenack, glazed coffee must be heated to a much higher temperature. This results in about double the loss of caffein and fat. French roasted coffees belong to this class.

Tea

The tea plant is an evergreen shrub, which grows to a height of from 12 to 15 feet in the wild state, but under cultivation is kept about 4 feet high. Tea will grow in a variety of climates, but a sub-tropical climate, especially in sections where the rainfall is about 50 inches annually seems to be the best. There are three different kinds of tea, depending on the method of curing.

I Green or unfermented tea
II Black or fermented tea
III Oolong or semi-fermented tea

All of these teas can be produced from leaves picked from the same bush.

Green Teas

A Japan Green Teas
  1 Basket-fired green tea (long leaves)
  2 Pan-fired green tea (short leaves)
B China Green Teas
  1 Gunpowders (small round rolled leaves)
  2 Imperials (large round rolled leaves)
  3 Young hysons (long rolled leaves)

There is no difference in the drinking qualities of these subdivisions, only in appearance. China green teas are generally known as Souchong.

Oolong Teas

Oolong teas are produced on the Island of Formosa. They are fired when the leaves are only half fermented.

Black Teas

A Black teas produced in China are called English breakfast tea or Congou tea.
B Black teas produced in India, Ceylon, Java, and Sumatra are classified according to the size of the leaf. The three young shoots at the tip are known as Orange Pekoe and Pekoe, which contain the least fiber, most juice, and produce the finest grade of tea. The leaves immediately below the Pekoe are named Souchong, a tea of popular price. Pekoe and Souchong are sometimes mixed and sold as Pekoe Souchong.

The difference between green and black tea is due to the fact that the green tea is dried shortly after gathering, and then rolled and carefully fired, whereas black tea is made up into heaps, which are exposed to the air for some time before firing and allowed to undergo fermentation, resulting in the conversion of its original olive green color to a black color.

CHAPTER IV

DESSERT POWDERS AND PUDDINGS

Gelatin Dessert Powders

Essentially, a gelatin dessert powder represents a mixture of gelatin and cane sugar, alone or combined with dextrose, powdered gelatin, natural or imitation fruit flavor, fruit acid such as citric or tartaric, and certified food color.

Gelatin desserts are more popular today because of the constant striving of the manufacturer to improve the quality of the fruit flavor and to use a quality of gelatin as free as possible from bacteria, off odor and poor taste. The use of natural fruit flavor alone is incapable of producing that individual fruity taste, and, therefore, the addition of an imitation fruit flavor is necessary. To make a free-flowing gelatin dessert powder (one which will remain that way in the package) from 5 to 15% corn sugar must be used, to retard serious lumping. In the manufacture of pudding powders, the finest quality starch, low in moisture and bacteria content, should be employed. All products should be sifted after mixing and before final packaging.

Gelatin Dessert Powder
Formula No. 1
(For Bulk Sales)

Granulated Sugar	84	lb	Natural Fruit Flavor	1½ lb
Gelatin Powder	12	lb	Imitation Fruit Flavor	To suit
Citric Acid Powder	2	lb	Certified Food Color	To suit

Add 5 pints of boiling water to 17 ounces of above mixture. Stir until dissolved.

Formula No. 2
(For Package Sales)

Granulated Sugar	71 lb	Natural Fruit Flavor	3 lb
Gelatin Powder	12 lb	Imitation Fruit Flavor	To suit
Corn Sugar	13 lb	Certified Food Color	To suit
Citric Acid Powder	2 lb		

A 3¼ ounce-package of the above mixture added to 1 pint of boiling water, will make a gelatin dessert. If fruits, nuts, etc., are to be used, add after gelatin dessert is cold and set sufficiently to hold them in place.

Flavors for Gelatin Desserts

Amount of Color and Flavor for a 100 lb Batch

Strawberry

True Fruit Flavor	6 oz	Amaranth Color Powder	4 oz
Imitation Fruit Flavor	1/7 oz	Orange I Color Powder	1 oz

Raspberry

True Fruit Flavor	6 oz	Amaranth Color Powder	6 oz
Imitation Fruit Flavor	⅓ oz		

Cherry

True Fruit Flavor	6 oz	Amaranth Color Powder	6 oz
Imitation Fruit Flavor	1 oz		

Pineapple

True Fruit Flavor	6 oz	Tartrazine Color Powder	4 oz
Imitation Fruit Flavor	1 oz		

Lemon

Lemon Oil	4 oz	Tartrazine Color Powder	3 oz

Orange

Orange Oil	4 oz	Orange I Color Powder	1 oz

Lime

Lime Oil	4 oz	Light Green S.F. Powder	3 oz
Tartrazine Color Powder	½ oz		

Dissolve the color in a minimum amount of hot water. Spray color and flavor on gelatin batches during mixing operation.

Pudding Powders

Chocolate Flavor
Formula No. 1
(Average Quality)

Granulated Sugar	60 lb	Salt	1 lb
Cocoa Powder	15 lb	Vanillin	3 oz
Corn Starch	25 lb		

Formula No. 2
(Better Quality)

Granulated Sugar	53 lb	Ground Sweet Chocolate	9 lb
Corn Starch	22 lb	Salt	1 lb
Tapioca Starch	6 lb	Vanillin	2 oz
Cocoa Powder	9 lb		

Formula No. 3

Granulated Sugar	60 lb	Salt	1 lb
Corn Starch	23 lb	Vanillin	3 oz
Cocoa Powder	20 lb		

The above puddings are weighed up into 4 ounce-packages, cooked with one pint of water or milk, and stirred constantly until they reach the proper thickening point.

Vanillin Flavor
Formula No. 1

Granulated Sugar	64 lb	Vanillin	3 oz
Corn Starch	27 lb	Certified Food Egg Color	If desired
Skimmed Milk Powder	8 lb		
Salt	1 lb		

Weigh the above into 4 ounce-packages. Stir mixture into 1 pint of boiling water or milk.

Formula No. 2

Granulated Sugar	70 lb	Salt	1 lb
Corn Sugar	15 lb	Vanillin	2 oz
Corn Starch	30 lb		

Use 4 ounces to 1 pint of boiling milk.

Formula No. 3

Granulated Sugar	55 lb	Salt	1 lb
Corn Sugar	15 lb	Vanillin	2 oz
Corn Starch	30 lb		

Use 4 ounces to 1 pint of boiling milk.

Cocoanut Flavor Pudding

Granulated Sugar	40 lb	Powdered Gelatin	2 lb
Corn Starch	38 lb	Fine Cut Cocoanut	10 lb
Skimmed Milk Powder	10 lb	Vanillin	2 oz

Use 4 ounces of the above mixture in one pint of boiling water or milk. Thicken in double boiler.

Maple-Flavor Powdered Pudding

Light Brown Sugar	65 lb	Powdered Caramel Color	1 oz
Corn Starch	25 lb	Maple Flavor	1 oz
Salt	2 lb	Skimmed Milk Powder	6 lb

Use 4½ ounces to 1 pint of boiling milk. Cook over moderate heat, to gloss (185 to 190° F).

Caramel, cocoanut and butterscotch-flavor pudding can be manufactured by a similar method.

Custard Pudding
Formula No. 1

Granulated Sugar	71 lb	Turmeric Powder	1 lb
Corn Starch	20 lb	Oil Bitter Almond	1½ oz
Gum Tragacanth	3 lb	Lemon Oil	⅛ oz

Use 4 ounces of the above to 1 pint of boiling milk. Then pour into dishes, and bake as custard.

Formula No. 2

Granulated Sugar	71	lb	Turmeric Powder	1	lb
Corn Starch	10	lb	Oil Bitter Almond	½	oz
Arrowroot Starch	10	lb	Oil Nutmeg	¼	oz

Use in same manner as formula No. 1.

Formula No. 3

Granulated Sugar	70	lb	Salt	½	lb
Corn Starch	20	lb	Vanillin	2	oz
Whole Egg Powder	6	lb			

Use as above recipe.

Custard Cream
(French Style)

Granulated Sugar	60	lb	Shortening	4	lb
Corn Starch	18	lb	Salt	½	lb
Whole Milk Powder	18	lb	Vanillin	2	oz
Powdered Egg Yolk	4	lb			

Use 4 ounces of the above to 1 pint of boiling water. Then add to it 3 pints of cold water and whip up until mixture thickens.

Custard Filling for Pies
Formula No. 1

Granulated Sugar	70	lb	Powdered Whole Egg	5	lb
Whole Milk Powder	8	lb	Salt	1	lb
Corn Starch	16	lb	Vanillin	3	oz

Stir 2½ pounds of above in 1 pint of water. Pour this into 7 pints of boiling water and stir until it thickens. Use when cold.

Formula No. 2

Granulated Sugar	50	lb	Tapioca Starch	15	lb
Skimmed Milk Powder	20	lb	Vanillin	1	oz
Powdered Whole Egg	10	lb	Oil of Mace	¼	oz
Shortening	5	lb			

Whip 2 pounds of the above into a smooth paste with 1½ pints of water. Add to 5 pints of boiling water and heat until it thickens. Use when cool.

Custard Pie—Rich and Non-Watering

Whole Milk Powder	30	oz	Powdered Whole Egg	15	oz
Granulated Sugar	43	oz	Salt	½	oz
Corn Starch	10	oz	Nutmeg	5	g

Work the above ingredients into a smooth batter with about 2 pints of water. Pour this into 8 pints of boiling water, stir until it thickens. Then bake in individual pie plates at 425 to 450° F, with good bottom heat.

Dry Vanilla Cream Powders
Formula No. 1

Granulated Sugar	30	oz	Corn Starch	11	oz
Powdered Egg Yolk	1	oz	Color and Vanilla		
Salt	¼	oz	Flavor	If desired	

Add the above to 1 quart of milk and mix until smooth. Then add gradually to 4 quarts of hot milk. Boil to 190° F, or until thick and glossy. A better product is obtained upon addition of 4 ounces of butter.

Formula No. 2

Granulated Sugar	41	oz	Corn Starch	16	oz
Powdered Skimmed Milk	27	oz	Color and Vanilla		
Powdered Whole Egg	4	oz	Flavor	If desired	
Salt	1	oz			

Add the above to 2 quarts of water. Mix into smooth paste. Then add gradually to 3 quarts boiling water. Boil to 190° F, or until thick and glossy.

CHAPTER V

SAUCES, DRESSINGS AND PICKLES

Sauces

Sauces are blended extracts of condiments, vegetables, fruits, vinegar, etc., for use with meats, fish, soups and various other foods. They have been employed for culinary and table purposes since the time of the ancient Romans. Sauces of the Worcestershire kind, if of good quality, generally contain soy as their chief character ingredient. A typical formula of Worcestershire style includes, in addition to soy and vinegar, lime juice, onion, tamarinds, garlic, fish, such as anchovies or pickled herring, red chilies and spices. Sauces are really built-up foods and not simply a mixture of water, vinegar, some spice and a little sugar.

To manufacture a good sauce, with definite distinctive flavor, it is necessary to give special attention to the selection and blending of the raw ingredients used.

The process of manufacture generally means the boiling together of the various ingredients, pulping or finishing, and then bottling. However, there is a great deal more than just this if a good product is desired. For example, one must consider the preliminary treatment of the spices or fruits, which often have to be boiled in vinegar in order to bring out the flavor. Very often it is necessary to thicken the sauce to give it more body; gums, starches, flour, pectin, or purified Irish moss may very safely be used for this purpose. Thickeners should always be added at the end of the manufacturing process, except in those cases in which a sauce is made entirely without heat or cooking. If the thickener is not added at the right time to cooked sauces, the acid tends to break down the thickener to simpler elements, through hydrolysis.

Even with thorough boiling, solid pieces, such as herbs, spices, skins, pits, etc., sometimes remain in the sauce. Unless these are properly removed, they detract substantially from the quality of the product. The best way to accomplish this is to pass the sauce through a pulping machine or force it through a fine sieve. The sauce can also be greatly improved by running it through a colloid mill or homogenizer, just before transferring it into the filling machine. This tends to give it a closer texture, and the sauce is less apt to settle out on standing.

One of the problems the sauce manufacturer has to cope with is that of fermentation in the bottle. Fermented merchandise is generally returned to the manufacturer, involving financial loss and other inconveniences. This condition can be corrected by heating the sauce and filling the bottles while it is still hot. The sauce is much easier to pour while hot and, at the same time, the chances of outside contamination are greatly reduced.

In order to obtain a mellow flavor, the sauce should be pumped into huge aging or mellowing vats, and then allowed to mature in bulk, rather than in bottles. Frequent stirring with wooden paddles during the maturing process is necessary in order to prevent caking or settling of solid particles. The best sauces are those which have been allowed to age. Of course, it is assumed that perfect ingredients were used to begin with. The use of glacial acetic acid, which is sometimes added to sauces, is prohibited by law.

Tough substances should be soaked in vinegar, water or sugar syrup before use, to facilitate the diffusion of flavor later on in the main mixing vats. Where it is necessary to chop up or pulp the fruit, hard fruits may be steamed and allowed to soak overnight in hot water; they can then very easily be put through a food chopper or pulper.

In some instances, separation seems unavoidable in the finished product but this can be corrected by addition of a thickener, stabilizer, or by putting the sauce through a homogenizer.

It would be well to consider the importance of using sterilized spices in the manufacture of sauces. As shown in the table below, fruits and vegetables, cereal flours, and similar sauce materials are carriers of bacteria, molds and yeasts and serve as favorable media for their proliferation. These spice ingredients, even though used in small quantities, may be the cause of popping and fermentation in the finished product. During

TOTAL BACTERIA COUNTS ON SAMPLES OF UNTREATED SPICES

Spice	*Bacteria Per Gram*	*Molds and Yeasts Per Gram*
Allspice (ground)	214,000	2,300
Cinnamon (ground)	110,000	22,200
Coriander (ground)	400,000	—
Coriander (whole seed)	190,000	4,640
Garlic (ground)	5,775	—
Ginger (ground African)	2,100,000	—
Nutmeg (ground)	10,000	110
Paprika (ground)	1,500,000	140
Pepper (whole black)	2,770,000	90
Pepper (whole Carolina red)	20,000	560
Pepper (whole Cayenne red)	2,047,000	100
Pepper (whole Japanese finger)	4,000	40
Pepper (ground fine, red)	16,000,000	2,500
Pepper (ground white)	158,000	2,220
Turmeric (ground)	1,835,000	100

TOTAL BACTERIA COUNT ON SAMPLE OF UNTREATED FOOD INGREDIENTS

Food	*Bacteria Per Gram*	*Molds and Yeasts Per Gram*
Acacia (gum) powder	3,330	494
Apricots (dried)	195	75
Broccoli	32,800	120
Cocoa	627	10
Corn flour	18,000	110
Corn flour (processed)	5,000	10
Karaya (gum) powder	19,350	1,387
Prunes (dried)	49	28
Raisins (dried)	82	60
Soy flour	3,800	—
Spinach	1,020,000	1,570
Strawberries (whole)	172,800	2,600
Tragacanth (gum) powder	14,540	70
Tragacanth (gum) ribbon	640	160
Wheat flour	13,000	—

recent years much progress has been made in the treatment of raw ingredients with the aim of reducing the microorganism content of food likely to cause spoilage in the finished package. Spices and flavoring herbs have generally been considered to be free of contamination and to possess antiseptic properties, but this has long ago been proven false.

TOTAL REDUCTION BACTERIA COUNT IN SPICES TREATED BY HEAT

Test No.	*Spice*	*Untreated Count Per Gram*	*Treated Count Per Gram*	*Per Cent Kill*
1	African ginger	2,100,000	80,000	96.19
2	Fine red pepper	16,000,000	2,400,000	85.00
3	White pepper	80,000	12,200	84.75
4	Coriander	400,000	30,000	92.50
5	Nutmeg	10,000	Sterile	100.00
6	Cinnamon	110,000	7,000	93.64
7	Allspice	214,000	21,500	89.95
8	Paprika	1,500,000	300,000	80.00

Spiced Sauces

Barbecue Sauce

Formula No. 1

Powdered Garlic	1/5	oz	Cane Sugar	6	oz
Powdered Onion	1/5	oz	Tomato Puree	6	oz
Butter	4	lb	Prepared Mustard	8	oz
Vinegar (100 Grain)	2	pt	Worcestershire Sauce	2	oz
Salt	3	oz	Tabasco Sauce	1	oz
Paprika	2	oz	Locust Bean Kernel Gum	8	oz
Black Pepper	3/4	oz	Water	36	lb
Red Pepper	3/4	oz			
Chili Powder	2	oz			

To the hot water, with the aid of a high speed agitator, add the locust bean kernel gum. When it is dispersed add the balance of ingredients. Heat to boiling point. Fill bottles while sauce is hot, with straining and then pasteurize.

Formula No. 2

Worcestershire Sauce	8 oz	Paprika	1½ oz
Tomato Juice	3 lb	Cayenne	¼ oz
Corn Oil	8 oz	Distilled Vinegar	3 pt
Pepper	1½ oz	Minced Garlic	¼ oz
Salt	2 oz	Grated Onion	1 lb
Dry Mustard	1½ oz		

Mix all of the above ingredients together.

Formula No. 3

Butter	7½ lb	Chili Powder	½ oz
Hot Water	22½ lb	Sugar	¼ lb
Vinegar	2 lb	Prepared Mustard	½ lb
Salt	⅛ lb	Worcestershire Sauce	½ oz
Black Pepper	⅓ oz	Tabasco Sauce	⅓ oz
Red Pepper	⅓ oz	Chopped Onion	⅖ oz
Paprika	½ oz	Garlic	⅒ oz

Mix ingredients. Heat to boiling point.

Brown Sauce

Beef Fat	1 lb	Flour	1½ lb
Minced Onion	½ lb	Beef Stock	8 lb
Parsley	1 oz		

Melt fat over low heat, add vegetables, fry gently for 10 minutes. Now add flour. Cook over medium heat, strring until flour turns a deep brown. Add the beef stock, stirring until smooth and thickened. Salt, Worcestershire, etc. may be added for taste. Strain the sauce and, while it is hot, fill bottles and then pasteurize.

Beef Steak Sauce

Tamarinds	30 lb	Cayenne	8 oz
Onions	4 lb	Caramel	4 oz
Garlic	½ lb	Soy Sauce	5 gal
Sugar	10 lb	Mushroom Catsup	5 gal
Salt	5 lb	Malt Vinegar (24 Grain)	30 gal
Shredded Horseradish	3 lb		
Mustard	3 lb		

Remove seeds from tamarinds by simmering for 15 minutes in the vinegar and passing through the pulping machine.

Soak the ground spices in water for 48 hours. Peel and chop the garlic and onion fine. Make the mustard into a thin paste with water.

Place all ingredients in a steam-jacketed kettle and simmer for 1 hour.

Fish Fruit Sauce
(Thick)

Water	3	pt	Unsweetened Apple Pulp	2	pt
Distilled Vinegar (100 Grain)	6	pt	Tomato Puree	5	pt
Invert Sugar Syrup	1¼	lb	Garlic Powder	½	oz
Ginger Syrup	½	pt	Shallots	5	oz
Ground Dates	2½	lb	Tamarinds	2	lb
Sweet Orange Marmalade	2	lb	Anchovy Essence	½	pt
White Chopped Raisins	2	lb	Caramel Color	½	pt
			Salt	¼	lb
			Tragacanth	1	oz

The above ingredients are mixed and boiled. The tragacanth is added last. Then add the following:

Mace	½	oz	Cinnamon	⅛	oz
Cayenne Pepper	¼	oz	Lactic Acid (50% Strength)	¼	lb
Mixed Spice	½	oz	Corn Starch	¼	lb
Cloves	¼	oz			

Simmer entire recipe for ½ hour. Make paste of corn starch, etc.

Soak tamarind, just to loosen fiber. Crush apples. Use whole or powdered spice, etc.

Italian Tomato Sauce

Olive Oil	1	lb	Bay Leaves	2	oz
Minced Garlic	1	oz	Thyme	1	oz
Minced Green Pepper	8	oz	Salt	6	oz
Chopped Celery	1	lb	Cayenne Pepper	½	oz
Chopped Onion	2	lb	Whole Cloves	5	oz
Parsley	2	oz	Cane Sugar	12	oz
Chopped Carrots	1	lb	Italian Tomato Paste	4	lb
Canned Tomatoes	20	lb			

Heat oil and add vegetables. Fry gently and do not allow vegetables to brown. Add canned tomatoes, spices, and simmer for 45 minutes. Strain. Add tomato paste and, if a hotter sauce is required, add tabasco sauce. For a thicker sauce, add flour. Cook until it thickens. Fill bottles while sauce is hot and pasteurize.

Meat Sauce

Onions	10	oz	Beef Stock	20	lb
Carrot	20	oz	Bay Leaf	3	oz
Garlic	¾	oz	Whole Cloves	2	oz
Fat	2½	lb	Thyme	1	oz
Flour	1½	lb	Pepper	2	oz

Melt fat over low heat. Then add vegetables which have been put through a meat grinder. Fry gently for 10 minutes with stirring. Now add flour. Cook with stirring until flour turns dark. Add spices and continue cooking until thick. Season to taste. Strain. Fill bottles while sauce is hot and pasteurize.

Mushroom Anchovy Sauce

Chopped Anchovies	12	lb	Madeira Wine	3	qt
Mushroom Catsup	1	gal	Cayenne Pepper	½	lb

Boil catsup, chopped anchovy and pepper for 30 minutes and pass through sieve. Add wine when quite cold and stir well.

Pungent Sauce

Malt Vinegar	1½	pt	Tamarinds	1	lb
Curry	2	oz	Water	½	pt

Soak tamarinds for 12 hours. Pass through sieve, and add vinegar. Add curry powder while stirring. Boil rapidly for 5 minutes. Pour, while hot, into bottles.

Mustard Sauce
Formula No. 1

Mustard Flour	20 lb	Ground Mace	4 oz
Tapioca Starch	2 lb	Distilled Vinegar (100 Grain)	6 gal
Powdered Garlic	2 oz	Powdered Turmeric	Enough to color
Powdered Onion	2 oz	Water	16 gal
Sugar	20 lb		
Salt	6 lb		

Soften the mustard and mace overnight in about 5 gallons of water. To the balance of water, in a steam-jacketed kettle, add the sugar. Bring to a boil. Add the starch, which has been suspended in a little water. Stir the mixture vigorously, with steam going until the starch becomes translucent. Shut off steam. Add the balance of ingredients, stirring well, and boil gently for about 15 minutes. Pass the entire batch through a fine sieve and fill at a temperature higher than 180° F.

Formula No. 2

Table Mustard	20 lb	Sodium Alginate	6 oz
Onions	2 lb	Ground Mace	4 oz
Garlic	2 lb	Vinegar (24 Grain)	25 gal
Sugar	20 lb	Yellow (Certified Color)	As desired
Salt	7 lb		

Soak the gum in 2 gallons of water for 24 hours.

Soak the mustard and mace in 2 gallons of water for 24 hours. Peel and chop the onions and garlic. Rub mustard and gum through a fine sieve.

Place all ingredients in a steam-jacketed kettle and boil gently for 15 minutes. Add yellow color. Pass batch through a fine sieve. Fill container at a temperature higher than 180° F.

Sea Food Sauce

Catsup	4 lb	Pepper	½ oz
Lemon Juice	6 oz	Worcestershire Sauce	6 oz
Tabasco Sauce	½ tsp	Dry Mustard	1 oz
Celery Salt	3 oz	Minced Parsley	1½ oz
Salt	1½ oz	Minced Onion	1½ oz

Mix the above ingredients thoroughly and keep in cold place.

Tomato Sauce (Good Quality)

Tomato Puree	10 gal	Allspice	1¼ oz
Vinegar	1½ gal	Ground Cloves	¼ oz
Sugar	15 lb	Ground Mace	¼ oz
Salt	3 lb	Cayenne Pepper	⅛ oz
Powdered Onions	½ lb	Powdered Garlic	1/10 oz
Ground Paprika	¼ lb		

Cook to a final volume of 12 gallons. Add most of sugar, salt, paprika, onions and garlic to batch when about half finished. Pass through finisher or smoothening machine.

Worcestershire Sauce

	Good Quality	Medium Quality
Vinegar (100 Grain)	12 gal	14 gal
Malt Vinegar	5 gal	
Water		
Mushroom Catsup	10 gal	
Walnut Flavor	8 oz	1¼ gal
Soy Sauce	6 gal	3½ gal
Sherry Wine or Flavor	5 gal	
Blackstrap Molasses	2½ gal	3 gal
Salt	11 lb	7½ lb
Brown Sugar	20 lb	15 lb
Tamarinds	10 lb	
Onion Powder	2 lb	
Garlic Powder	¼ lb	8 oz
Ground Anchovies	21 lb	
Tragacanth	½ lb	
Cayenne Pepper	1 lb	3½ lb

Black Pepper	1 lb	
Lovage Root	⅛ oz	
Allspice	1 lb	
Cloves	⅛ oz	3 oz
Cinnamon	⅛ oz	
Mace	7 oz	
Nutmeg	3 lb	
Ginger	1 lb	
Lemon Peel	2 lb	

Prepare the spices in advance by allowing them to soak in vinegar. All of the ingredients should be agitated for at least several hours. Run through strainer, to catch large pieces, before filling bottles or barrels.

Hot Curry

Coriander	1½ lb	Mustard	1 oz
Cumin	½ lb	Fenugreek	1 oz
Turmeric	1 lb	Cayenne Pepper	1½ oz
Ginger	2 oz		

Table Sauce

Allspice	2 oz	Brown Sugar	3 lb
Black Pepper	1 oz	Salt	1 lb
Cloves	1 oz	Vinegar	4 gal
Mustard	8 oz		

Mix and place on the fire and simmer for about 2 hours. Then strain through a cloth and add enough sugar color to give it the desired dark shade. Let stand for 2 weeks before using. Kept in a keg, it will improve with age.

Chili Sauce

Ripe Tomatoes (Chopped Fine)	5 lb	Red Pepper	1 dr
		Salt	2 oz
Vinegar	2 pt	Lemon Juice	5 oz
Garlic	1 oz		

Boil for an hour. Then force through a colander and bottle while warm. Cork tightly.

Table Mustard

Mustard Flour	1 lb	Jamaica Pepper	1 dr
Vinegar	1 qt		

Boil the vinegar and pour over the mustard. Stir until all lumps are gone. Add the pepper and let stand several days, well covered. Add salt to suit. Force through a colander, if necessary.

French Mustard

The same as table mustard with the addition of 8 ounces of sugar and a trace of powdered cloves.

Sausage Seasoning

Sage	½ lb	Cumin	1 oz
Pepper	1 lb	Cassia	1 oz
Allspice	1 lb	Nutmeg	2 oz
Salt	1 lb	Pimento (Allspice)	6 oz
Or		Black Pepper	8 oz
Cayenne Pepper or Capsicum	1 oz	Salt	8 oz

Soup Spice

Lemon Peel	½ lb	Dried Parsley	1 lb
Thyme	1 lb	Celery Seed	¼ lb
Sweet Marjoram	½ lb	Sweet Basil	¼ lb
Summer Savory	½ lb		

Curry Powders

Coriander	300 lb	Celery Seed	7½ lb
White Pepper	30 lb	Cloves	7½ lb
Red Pepper *	30 lb	Or	
Yellow Mustard	30 lb	Coriander	3 lb
African Ginger	15 lb	Black Pepper	1 lb
Mace	7½ lb	Cayenne Pepper	¼ lb
Cumin Seed	60 lb	Yellow Mustard	1 lb
Fennel Seed	7½ lb	Jamaica Ginger	1 lb

* If a milder curry is desired, leave out the red pepper.

Cardamom	30	lb	Cumin Seed	¼	lb
Turmeric	75	lb	Cardamom	½	lb
Foenugreek	30	lb	Turmeric	3	lb
Saigon Cassia	25	lb	Cloves	¼	lb
Mint	18	lb	Allspice	½	lb

Sweet Sauces

Butterscotch Sauce

Brown Sugar	8	lb	Butter	1	lb
Corn Syrup	3	lb	Vanilla Extract	1½	oz
Boiling Water	2	pt	Salt	½	oz

Heat sugar, corn syrup and water to 230° F. Shut off heat. Add butter and when cool add salt and vanilla flavor.

For butterscotch cream sauce use above recipe, using cream instead of water. Cook sugars and cream together until sugar is melted and mixture is just cooking. Stir constantly. Add flour when cold.

Caramel Sauce

Corn Syrup	4	lb	Butter	½	lb
Brown Sugar	3	lb	Sweet Cream (18% Fat)	2	lb

Add the brown sugar to the warm corn syrup and heat until completely dissolved. Add the butter and stir until melted. Now add the cream. Stir and heat to about 230° F.

Chocolate Sauce

Bitter Chocolate	6	lb	Sugar	25	lb
Powdered Skimmed Milk	3	lb	Butter	3	lb
Water	2	gal	Vanilla Extract	3	oz
Flour	3	lb	Salt	2	oz

Dissolve the sugar in the hot water. Add flour and water as a thin paste, heat until the flour just gels. Add the chocolate, butter and mix. Now add the milk powder, which is made into a thin smooth paste with water. Cook while stirring constantly until smooth and thick. Flavor with vanilla. This sauce may be used cold or warm.

Fudge Sauce

Bitter Chocolate	1½ lb	Sugar	8 lb
Butter	8 oz	Salt	½ oz
Boiling Water	4 pt	Vanilla Extract	1 oz
Corn Syrup	1½ lb		

Add butter to the melted chocolate and, when dispersed, add the boiling water with stirring. Continue cooking. Add corn syrup and sugar. Boil for about 5 minutes. Cool, then add salt and vanilla flavor.

Maple Sauce

Maple Syrup	16 lb	Chopped Nuts (Walnuts, Pecans, or Almonds)	4 lb
Sweet Cream (18% Fat)	6 lb		

Cook syrup and cream to 230° F, mixing thoroughly. The chopped nuts may be omitted.

Mocha Sauce

Bitter Chocolate (Small Pieces)	6½ lb	Sugar	40 lb
Strong Hot Coffee (Prepared with 4 lb of Coffee and Water)	2 gal	Corn Syrup	10 lb
		Butter	3 lb
		Vanilla Extract	8 oz
		Salt	4 oz

Dissolve the sugar in the hot corn syrup. Add the butter with stirring, then the chocolate. Mix until completely dissolved. Add the hot coffee, salt, and cook to 230° F. Upon cooling, flavor with vanilla.

Mint Sauce

Granulated Sugar	120 lb	Water	16 gal
Vinegar	16 gal	Spearmint Oil	4 fl oz

Sugar and water must be mixed and agitated previous to the addition of the vinegar and spearmint. Final batch must stand overnight before bottling.

Soy Sauce

There are no definitions or standards of identity for so-called genuine soy sauces, however, manufacturers of this ingredient have developed their own special methods, and the following is a composite analysis:

Composite Soy Sauce

Water	56.6 to 75.2 %	Protein (N x 6.25)	1.9 to 8.0%
Sodium Chloride	13.7 to 20.5	Density	1.15 to 1.27
Reducing Sugars	2.0 to 13.3	pH	4.4 to 4.7

Staley Soy Sauce

Moisture	60.0 %	Sugars	11.25%
Protein (N x 6.25)	7.5	Specific Gravity	1.25
Sodium Chloride	19.0		

Chinese Soy Sauce

Moisture	57.7 %	Sugars	11.25%
Protein (n x 6.25)	4.6	Specific Gravity	1.25
Sodium Chloride	20.1		

Syrup for Pickling Fruits

Various fruits, particularly figs and peaches, can be packed in a heavy vinegar syrup as fruit pickles. The fruit is prepared as for canning and cooked in syrup to about 60% sugar.

Syrup

Dry Ginger Root	12 oz	Cider Vinegar (40 Grain)	3 gal
Whole Cloves	14 oz	Sugar	125 lb
Stick Cinnamon	18 oz	Water	20 gal

Place spices in cheesecloth bags. Cook fruit in syrup. Then allow to stand overnight. Remove spices. Pack in vacuum sealed jars and pasteurize at 175 to 180° F for 30 minutes.

Mayonnaise

Aside from its popularity as a bread spread appetizer and salad dressing, mayonnaise is greatly appreciated for its food value. In food manufacturing it is very conveniently used as a base for vegetable salads, other salad dressings, etc.

The manufacture of a high-grade mayonnaise requires the use of good quality eggs, oil, vinegar, spices and the use of superior equipment. A good mayonnaise is a semi-solid emulsion of edible vegetable oils, egg yolk or whole egg, vinegar and/or lemon juice, salt, seasoning, sugar and/or dextrose.

Mayonnaise
Formula No. 1

Ingredient	Amount	Unit	Ingredient	Amount	Unit
Egg Yolk (Fresh or Frozen)	15	lb	White Pepper	1	oz
Sugar	3	lb	Salad Oil	12	gal
Mustard Flour	1¼	lb	White Vinegar (100 Grain)	4	qt
Salt	1	lb	Water	2	qt
Paprika	¾	oz			

Beat the yolks at high speed until they are thoroughly broken, which may require about 1 minute. Now add the sugar, salt and spices and beat for another 2 to 5 minutes until the mixture is well creamed.

Now start adding the oil slowly, at the rate of a gallon per minute, until 8 gallons have been added. Change the speed of the beater to first speed while adding 1 quart of vinegar and 1 quart of water; the mixture should be thoroughly agitated. Turn the beater on high speed and add the remainder of the oil at the rate of 1 gallon every 2 minutes. Then shift the beater to low speed and add the remainder of the vinegar and water.

The beater should now be shut off to scrape down the sides of the bowl. Then mix the entire batch at low speed for at least another 2 minutes taking care not to overbeat as this tends to make the mayonnaise thin.

Formula No. 2

Whole Eggs	24	Salt	11 oz
Salad Oil	12 qt	Mustard	3½ oz
Vinegar	1¼ qt	Pepper	1¼ oz
Sugar	1¼ lb		

Dissolve the sugar and salt in the vinegar. Put the eggs in the bowl and beat for 10 minutes at high speed. Now start the oil slowly. Gauge the rate of flow of the oil so that it will take about 12 minutes to add the first one-third of the oil. Stop the machine. Introduce the dry ingredients, mustard, etc. Start the machine and start adding the oil. Increase the rate of flow of the oil so that the balance of it will be added in about 12 minutes. As soon as all of the oil is in, add the vinegar. It is not a good practice to add the vinegar through the oil dropper; it is much better to pour it from a glass or porcelain pitcher. All of the vinegar should be added in 2 or 3 minutes. Then stop the machine and do not beat further.

For a mayonnaise in which only yolks are used, its is ordinarily necessary to add a part of the vinegar before all of the oil is introduced. Otherwise the products become too heavy and separate. Add only enough vinegar to prevent separation. It is not necessary to take much time for making mayonnaise which is for relatively immediate use. The eggs and dry ingredients can be put in the bowl at the same time and beaten for only a minute or two before the oil is started. The rate of flow of the oil can also be increased somewhat so that the whole operation can be completed in 10 or 15 minutes. However, it is best for the maker to take plenty of time until he has acquired experience in handling.

It is possible to make a mayonnaise with considerably fewer eggs than required according to these recipes, but the recipes given here give a product which is satisfactory to most people and which has good stability. In case a batch does separate, as it sometimes will, the material can be reclaimed by starting a new batch and then adding the separated materials before putting the vinegar in.

In general, the amount of vinegar used varies with the number of eggs used and, to lesser extent, with the time of beating. With the proportion of one whole egg to each pint of oil there should be about one-tenth as much vinegar as oil. If two yolks to each pint of oil are used, about one-eighth as much vinegar as oil is right.

Cooked corn starch is very commonly used as a stabilizing material. The corn starch is cooked to the consistency of a jelly and is added cold. As much as 2% of corn starch is used. Gum tragacanth or alginates are also satisfactory.

Formula No. 3

Egg Yolk (Fresh)	12 lb	Salad Oil	13	gal
Sugar (Sucrose)	2½ lb	White Vinegar (100 Grain)	4	qt
Salt	1 7/16 lb	Water	2	qt
White Pepper	¾ oz			

For this mayonnaise always use fresh egg yolks. If you cannot purchase fresh yolks, break fresh eggs, separating the yolks.

The yolks are thrown into the beater bowl. Set beater at high speed. Beat the yolks until they are well broken up. This requires about 1 minute. Add sugar and salt and beat until creamed. This will require from 2 to 5 minutes, depending on the amount of egg white which adheres to the yolk. Too much egg white is undesirable and necessitates a longer beating time.

Now start adding the oil, slowly at first, and then at the rate of 1 gallon per minute until 9 gallons have been added. The beater is then run on low speed while 1 quart of vinegar and 1 quart of water are added and thoroughly mixed in.

Now run beater on high speed and, slowly, add the remainder of the oil at the rate of 1 gallon every 2 minutes. Shift beater to low speed and then add the remainder of the vinegar and water.

Shut off beater. Scrape down sides of bowl. Mix entire batch for 2 minutes longer at slow speed. Long beating of the entire mix tends to make a thinner mayonnaise.

Formula No. 4

Egg Yolk	5 lb	Mustard Flour	4 oz
Oil	2¼ gal	Salt	6 oz
Vinegar (45 Grain)	1¼ qt	White Pepper	½ oz
Sugar	10 oz	Paprika	½ oz

This formula yields 25 pounds of mayonnaise showing 70% oil and 20% egg yolk.

Formula No. 5

Frozen Egg Yolk	120 lb	Vinegar *	160 lb
Winter Oil (Cottonseed Oil)	330 lb	Salt	200 lb
Spice Mixture (As in Formula No. 1)	24 lb	Ground Mustard (No. 1)	100 lb

* Vinegar: Use cider vinegar. Dilute 9 gallons of vinegar with 1 gallon of water.

Mix the frozen egg yolk, which must have been taken out of the refrigerator the night before making the batch, with a part of the vinegar and the spices. Be sure to keep some vinegar to be added last or if the batch breaks. Allow the oil to flow in slowly.

This is a high quality mayonnaise, sharp in taste, and is good for grocery trade.

Formula No. 6

Frozen Egg Yolk	7½ lb	Spice Mixture (As in Formula No.1)	2 lb
Winter Oil (Cottonseed Oil)	33 lb	Vinegar (As in Formula No. 1)	10 lb

This is a thicker mayonnaise than that of Formula No. 5. It is milder in taste, has good keeping-qualities and is cheaper.

Formula No. 7

Cottonseed Salad Oil	70¼ lb	Salt	1½ lb
Egg Yolk	10 lb	Sugar	3½ lb
Vinegar (50 Grain)	10 lb	Mustard	1 lb
Water	4 lb	White Pepper	⅒ lb

The egg yolk can be reduced to 7½ pounds. The sugar can be left out.

Formula No. 8

Frozen Egg Yolk	4½	lb	Mustard No. II	8	oz
Winter Pressed Cottonseed Oil	75	lb	White Vinegar	1	qt
Salt	13	oz	Water	2	qt

This is a cheap product of good keeping-qualities and very popular in restaurants.

Modified Mayonnaise

Egg Yolk	2	lb	Salt	4¾	oz
Oil	21	lb	Mustard Flour	2½	oz
Pectin Powder (#100 Grade)	1½	oz	Lemon Juice	3	oz
Sugar	4¾	oz	Water	14	oz
			Vinegar (100 Grain)	20	oz

Salad Dressing

Add the spices and the egg yolk, beat at high speed and add half of the vinegar slowly. The oil is then added in small quantities with constant beating. As the mixture gradually thickens, add the rest of the vinegar and oil and continue beating until a creamy texture is obtained. Now add the cooled corn starch paste and beat at high speed until creamy.

Salad Dressing
Formula No. 1

Corn Starch	5	oz	Salt	5	oz
Water	90	oz	Sugar	10	oz
Cottonseed Oil	51	oz	White Vinegar	42	oz
French Mustard	25	oz			

Mix the corn starch with a little cold water and add to the remaining water which has been heated. Stir and boil until it stiffens. Set aside to cool. Put the mustard into a bowl and gradually add the oil, stirring all the time. Next add sugar. salt, vinegar, and then the cooled corn starch.

Formula No. 2

Tapioca Starch	7 oz	Salt	2 oz
Water	30 oz	Cane Sugar	4 oz
Corn or Cottonseed Oil	40 oz	White Vinegar (5%)	17 oz

Formula No. 3

Corn Starch	5¼ oz	Cayenne	Dash
Water	80 oz	White Vinegar (100 Grain)	20 oz
Oil	80 oz	Egg Yolk	14 oz
Mustard Flour	25 oz		
Paprika	Dash		

To make starch paste, add to the steam-jacketed kettle all the required water minus a small portion with which to make a cold starch paste. Bring this water to a boil and add the cold starch paste with stirring, and cooking until transparent.

Allow this starch paste to cool to room temperature before manufacturing it into a salad dressing.

Formula No. 4

Water	3 qt	Pepper	½ oz
Corn Starch	12 oz	Egg Yolks	2¼ lb
Sugar	6 oz	Oil	1½ gal
Salt	3 oz	Vinegar	¾ qt
Mustard	3 oz		

First dissolve the corn starch in a part of the water. Put the balance of the water in a double boiler and add the dissolved starch. Cook until the starch clears and forms a jelly.

During the time that the starch is cooking rub the sugar, salt, mustard and pepper into a paste with a part of the oil. This may be done in the mixing bowl with the flat beater.

After the starch has formed a jelly put it, while still hot, in the mixing bowl. Add the egg yolks, or whole eggs, and one-fourth of the vinegar. Now run the machine in third speed, whipping while it is cooling. This operation will take from 3 to 5 minutes.

Now start the oil and gauge the flow so that the oil will enter the batch in a period of 12 to 15 minutes. As soon as the oil has been incorporated add the rest of the vinegar and allow the machine to operate just long enough to thoroughly incorporate it.

French Dressing
Formula No. 1

Salad Oil	4 oz	Pepper	1 gr
Cider Vinegar	2 oz	Paprika	1 gr
Salt	6 gr		

Combine the above ingredients.

Formula No. 2

(a)		Salt	13 lb 13 oz
Water	1 gal	Black Pepper	9⅗ oz
Cider Vinegar	8 gal	Sugar	20 lb
Worcestershire Sauce	2 qt		

(b)		(c)	
Mustard Flour	2 lb	Olive Oil	8 gal
Paprika	5 lb	Cottonseed Oil	2 gal 2¾ qt

Mix (b) into (a) by stirring until the salt and sugar are dissolved, which may require 5 to 10 minutes. Then add (c) and agitate vigorously until there is an emulsion formed before filling bottles.

List of spicy seasonings, which can be used in manufacturing French Dressing

Garlic	Tomato Catsup
Celery Seed	Curry Powder
Horseradish	Chopped Chutney
Onion Juice	Worcestershire Sauce
Dry Mustard	Tabasco Sauce
Tarragon Vinegar	Cayenne
Chili Sauce	

Russian Salad Dressing

Mayonnaise	25 lb	Drained Chili Sauce	2 gal
or		Tabasco Sauce	⅙ oz
Salad Dressing	25 lb	Minced Green Pepper	1 lb

Mix the above ingredients.

Thousand Island Dressing

Mayonnaise	25	lb	Minced Green Peppers	1½	lb
Drained Chili Sauce	2	gal	Minced Onion	⅓	lb
Chopped Olives	3	lb	Minced Pimento	⅓	lb

Mix the above ingredients.

Salad Dressing A

Starch Paste

Water	16½	lb	Starch	3½	lb
Vinegar (50 Grain)	10	lb	Sugar	3	lb

The vinegar and 10 pounds of water are brought to about 160° F in a steam-jacketed kettle. Meanwhile, blend the starch to a smooth free-flowing suspension with the remaining water. Add the starch suspension, only as rapidly as the agitator will disperse it. Continue agitation at about 30 r.p.m. until the paste is clear and thick.

As soon as cooking is complete, add the sugar and mix thoroughly. Cool quickly at least to 80° F with slow agitation. It is not wise to prepare starch pastes too far in advance. Best practice is to cool to 80° F and use immediately.

Emulsion Base

Frozen Egg Yolk (10% Sugar)	1½	lb	Dry Mustard	4	oz
Sugar	3	lb	Water	12	oz
Salt	1¼	lb	Vinegar (50 Grain)	4	oz
			Salad Oil	10	lb

Place the defrosted yolks in the bowl of the mixer with the sugar, salt and mustard sifted together. Blend. Add about one-third of the water and whip until light. Begin adding oil and water or vinegar alternately. First add oil in a thin stream for 30 seconds while whipping at high speed and continue whipping until a uniform mass is obtained. Next add a small amount of water, then more oil, working in each addition well. The quantity of oil added at one time may be gradually increased. Regulate liquid (water or vinegar) additions so that there is some available to finish the process. Temperature for emulsifying should be 43 to 45° F.

Final Blend (20% in Product)

Emulsion Base	17 lb	Starch Paste	33 lb

Blend emulsion and paste in a power mixer until a fine glossy uniform mass is obtained.

Salad Dressing B

For a 35% oil dressing the same starch paste should be used with a higher fat emulsion base.

Frozen Egg Yolk (10% Sugar)	2¾ lb	Dry Mustard	4 oz
		Water	1½ lb
Sugar	1½ lb	Vinegar (50 Grain)	8 oz
Salt	1 lb	Salad Oil	17½ lb

Manufacture of Prepared Mustard

Prepared mustard is a paste composed of a mixture of ground mustard seed, mustard flour or mustard cake, with salt and vinegar, with or without sugar or dextrose or other condiments. The above definition naturally permits the utilization of a wide range of spices and vinegar. The amount of spices, vinegar and salt determines the quality and taste of the finished product.

In practically all types of prepared mustard, a blend of different varieties of mustard seed is used, the exact proportion of each variety depending partly upon the color and flavor desired in the finished product, and partly upon current market conditions.

Practically all formulas call for either or both white (or yellow) mustard seed and black (or brown) mustard seed. All varieties of mustard seed contain a bland fixed oil. The seed of the white mustard, Sinapis alba, contains little or none of the hot, volatile mustard oil, allyl isothiocyanate. This seed is the main ingredient in the mild, yellow, or French style mustard. Black mustard seed, from Brassica nigra, or related varieties, contains appreciable quantities, not less than 0.6% of the volatile mustard oil. Brassica juncea, also extensively used, is an oriental species with a yellow seed, containing appreciable quantities of volatile oil.

Distilled (white or spirit) vinegar is used, cut to a strength of 35 to 40 grain. Enough vinegar must be used to give the proper consistency and flavor, and to assist in the preservation of the product. Too much vinegar should not be used because a pronounced acid flavor is not desirable. The spices used depend to a great extent upon the market to be supplied. Some sections of the country prefer a mild smooth flavor, while others demand a hot, highly spiced product. Among the spices used are cinnamon, cloves, nutmeg, ginger, allspice, black, red, cayenne, and tabasco peppers, bay leaves, celery seed, turmeric, curry powder, thyme and powdered garlic.

The mustard seeds are crushed, the vinegar diluted to the proper strength, the spices finely ground, the salt freed from lumps, and the ingredients for a given formula are weighed or measured, and blended in a large vat before milling. A wet grinding process is used, the blended ingredients being run through buhrstones and finished in a second set of stones. The stones must be carefully and frequently dressed by an experienced person and must be loosened and readjusted to the proper pressure after each stopping of the mill. A good finish depends upon the cut of the stones, the pressure or set of the stones, the speed of rotation, and the rate of flow of the mix.

A prepared mustard, to be made up in smaller quantities for immediate use, may be made by stirring boiling vinegar into mustard flour. Salt and spices or spice oils should be added to the vinegar before blending with the mustard flour. Cheap imitation mustards are on the market, in which more or less mustard bran (ground mustard hulls) is used as a filler.

Equipment required in a mustard plant includes mixing tanks, pumps, grinding and finishing mills.

Tomato Catsup

Catsup is a descriptive term for a number of different products which consist of the pulp, strained and seasoned, of various fruits. The variety made from tomatoes is the most popular condiment in this country. A good catsup is judged by flavor, consistency, uniformity and attractive color.

In the manufacture of tomato catsup, the tomatoes are scalded with steam, and then converted into pulp. They are passed through a pulping machine or finisher, in which they are forced through a fine-mesh cylindrical screen. This removes skins, seeds, and other undesirable parts of the tomato. The pulp is then pumped to the catsup boiling tanks, where it is mixed with salt, vinegar, sugar, and various spices. A common proportion is 70 pounds of salt to 500 pounds of sugar, and sufficient vinegar to yield 1.6% acetic acid in the final product.

When a batch of catsup reaches the proper consistency in the boiling tank, it is emptied through a finisher, similar to those described for tomato juice. This serves to smooth the catsup and to eliminate any tendency on its part to be curdy. From a receiving trough under the conditioner, the catsup is pumped to supply tanks over to the bowls of the bottle-filling machines. If the catsup is placed hot in the bottles at a temperature around 205° F, it is not necessary to pasteurize it for 45 minutes at 185° F.

Manufacture of Concentrated Tomato Pulp and Catsup

The open kettles used for concentrating tomato pulp may be of wood, copper, block tin or Monel, or glass lined. The open kettles are often not steam-jacketed, but are heated by closed copper coils, known as flash coils, 3 inches in diameter. Under normal conditions it takes 35 to 50 minutes to concentrate 50 gallons of pulp to one-half the volume with the use of flash coils.

Where wood tanks are required, they are made of cypress

wood. Unless these tanks are always kept very clean and free from mold growths, a musty or moldy flavor may be imparted to the tomato pulp. Copper kettles are more expensive and have only one disadvantage: the copper may spoil the color of the pulp.

Cooking the Pulp

While the pulp is being concentrated, there may be considerable foaming, which can be prevented by adding a small amount of cottonseed oil. The kettle is filled with the tomato pulp, just enough to cover the coils or steam-jacketed part of the kettle. Steam is admitted and the boiling is started. The concentration must be accomplished rapidly to retain the bright color. Boiling for 30 minutes is usually sufficient in a tank having a good flash coil and enough high-pressure steam. The finishing point of the tomato pulp, or puree, is determined by a specific gravity which should be 1.050.

Spices and Condiments Used

The cloves should be the headless type. Penang cloves are the most popular. The cinnamon used is the broken stick form, preferably Saigon cinnamon. The mace is received from the Penang or Banda districts. Cayenne and onions should be selected for pungency. Hot onions are preferable to mild ones, the same is true of the cayenne. Paprika is sometimes used to intensify the red color of the catsup, but adds very little flavor.

Extraction of Spices

Spices are usually added to tomato catsup as a vinegar extract, prepared by adding the spices to distilled vinegar and steeping at about 185° F in a covered wooden or glass-lined tank for 2 to 3 hours. The special vinegar is now separated from the spices and added to the pulp in the catsup kettle. Sometimes onions, garlic, and spices are tied in a bag and suspended in boiling catsup for extraction. In this case, the spices are used a second time by replenishing with fresh spices.

Typical Catsup Formula

Heavy Concentrated Tomato Pulp (Puree) (Sp. gr. 1.06)	100	gal	Cloves	15	oz
Salt	28	lb	Allspice	15	oz
Sugar	200	lb	Cayenne	3½	oz
Chopped Onions	25	lb	Chopped Garlic (Optional)	4	oz
Cinnamon	25	oz	Ground Paprika	2	oz
Mace	3½	oz	Vinegar (100 Grain)	15	gal

Prepare the spice extract with vinegar, but leave the onion, garlic, and paprika out. Add these three separately. The puree is now boiled quickly, usually in an open tank. Add the sugar and, towards the end, add the salt. Near the end of the cook add the spice vinegar extract.

Catsup

Tomatoes (Skins, Seeds, and Green Spots Removed)	30 lb	Redistilled Vinegar *	2¼ to 3⅓ cups
Salt	1¼ to 1½ cups	Sugar	5 to 8 cups

* If redistilled vinegar cannot be obtained, use 4½ to 6⅓ cups of white vinegar, 5% strength, or 6 to 8 cups of cider vinegar, 4% strength.

Spice Formula No. 1
(Mild-Spiced Catsup)

For a mild-spiced product, add to the above catsup mixture the following:

Celery Seed	3 tsp	Paprika	3 tsp
Mustard	3 tsp	Cayenne Pepper	2 tsp
Cinnamon	2 tsp	Onions	4

Spice Formula No. 2
(Heavily Spiced Catsup)

For a heavily spiced product, add to the catsup mixture the following:

Mace	2	tsp	White Pepper	4	tsp
Cassia	2	tsp	Celery Seed	2	tsp
Chillies	2	tsp	Grated Garlic	¼	tsp
Cayenne Pepper	2	tsp	Mustard	2	tsp
Paprika	4	tsp	Onions	8	

The spicing of catsup is largely a matter of individual taste and can, therefore, be varied at will, but at least the minimum quantities of vinegar, sugar, and salt given in the recipes must be used. The higher quantities mentioned give a heavier catsup, which does not require as much boiling to concentrate to a desired consistency. In some of the popular commercial catsups as much as 8 cups of sugar, 1⅔ cups of salt, and 4½ cups of vinegar are added to every 30 pounds of tomatoes.

To produce a catsup of proper consistency, 30 pounds of tomatoes should be boiled down to give 16 pounds of catsup. However, depending upon whether the low or high quantities of sugar are used and upon the amount of water in the tomatoes at the start, boiling may have to be continued until the material is reduced to 13 pounds. The consistency of catsup, like the flavor, is largely a matter of personal preference.

To determine the point where boiling should be discontinued, the kettle and contents should be weighed before and after boiling, e.g.:

Before boiling	
Weight of kettle	2 lb 8 oz
Weight of tomato juice	30 lb
Total weight	32 lb 8 oz
After boiling	
Weight of catsup	16 lb
Weight of kettle	2 lb 8 oz
Total weight	18 lb 8 oz

Manufacture of Tomato Juice

After inspection and grading, the tomatoes are dumped into sluices containing running water. They are soaked and washed, and then conveyed to an elevator which carries them to a rotary washer. This washer is an inclined cylinder made up of closely spaced bars and tubes down which the tomatoes tumble, under a spray of wash water.

From the washer, the tomatoes are passed to a conveyor belt which moves down the middle of a sorting and trimming table. Here the tomatoes are carefully picked and trimmed, only the best ones being diverted to the juice plant.

In the juice plant the specially selected tomatoes are passed through a chopper. The chopped pulp is then heated by means of a tubular heater, from which it is passed to a finisher, where the juice is forced through a very fine mesh cylindrical screen to remove skin, seeds and other large pieces of solid matter. The juice is then allowed to flow to surge tanks equipped with small agitators to insure uniformity of the juice. A small quantity of salt is added to the juice in these tanks, usually about 2 pounds for each 50 gallons of juice.

From these tanks the juice is pumped into large tanks, which act as reservoirs for the homogenizers and can fillers. The homogenized juice is heated to about 200° F, by means of a tubular heater, before it is piped to the bowls of the filling machines, from which it is packed into bottles.

The bottles of juice are then sealed under vacuum and passed through water at 200° F to insure sterilization. They are cooled and packed into cases.

Tomato Juice Cocktail
Formula No. 1

Pure Fresh Tomato Juice	100	gal	Fine Granulated Sugar	18	lb
Fine Salt	12	lb	Finely Ground White Pepper	¾	oz

Add the salt and sugar directly to the tomato juice with minimum agitation. The salt and sugar are fine enough to be readily soluble. The pepper should be sifted in evenly and smoothly, to prevent the formation of clots or clumps, which are very difficult to break up.

With the above recipe as a base, any number of interesting spiced flavors can be produced. For example, any number of varieties of pepper, sweet or hot types, can be employed. Also, various types of seeds, such as caraway and coriander, can be used.

Formula No. 2

Tomato Juice	10 pt	Tabasco Sauce	1 cc
Celery Salt	1 oz	Sugar	2 oz
Lemon Juice	7 oz	Onion Juice	2 oz
Worcestershire Sauce	2 oz		

Mix the above ingredients. Fill bottles and pasteurize.

Sauerkraut-Tomato Juice Cocktail

Sauerkraut Juice	1 pt	Worcestershire Sauce	1½ oz
Tomato Juice	3 pt	Salt	¾ oz
Onion Salt	⅓ oz	Pepper	¾ oz

Mix above ingredients. Fill bottles and pasteurize.

Pickles

Pickles are vegetables or fruits preserved, singly or mixed, in vinegar, mustard, brine, and variously flavored. Among the most popular are cucumbers, gerkins, mixed pickles, chilies, mango melons, cauliflower, onions, watermelon rinds, green and red tomatoes, etc.

Pickles should always be kept in a tightly covered receptacle and stored in a dark, dry place. Excessive cold and heat are harmful. Although pickles in brine solution do not freeze easily, pickles in vinegar freeze almost as quickly as water. Pickles should be stirred from time to time as both brine and sweet vinegar have a tendency to sink to the bottom, leaving a dilute liquid on top. Any scum forming on the surface of the pickle brine must be removed, since it has a tendency to soften the pickles.

General Classification of Pickles

All manufactured pickles may be logically grouped in three general classes.

- I Dill Pickles
  - A Fermented dill pickles
    - a Genuine dill pickles
    - b Genuine Kosher dill pickles
    - c Overnight, or fresh fermented, dill pickles
    - d Overnight, or fresh fermented Kosher, dill pickles
    - e Polish dill pickles
  - B Unfermented dill pickles made directly from fresh cucumbers
    - a Fresh or pasteurized dill pickles
    - b Iceberg or quartered dill pickles
  - C Dill pickles made from salt stock
    - a Processed imitation, or summer, dill pickles
    - b Processed imitation, or summer, Kosher dill pickles
    - c Pasteurized processed dill pickles
    - d Pasteurized processed Kosher dill pickles
- II Sour Pickles
  - A Sour pickles
    - a Plain sour pickles
    - b Sliced or hot sour pickles
  - B Mixed sour pickles
    - a Mixed unspiced
    - b Spiced or hot mixed pickles
    - c Mixed chutney
  - C Relish, chow-chow, etc.
    - a Chow-chow
- III Sweet Pickles
  - A Plain sweet pickles
    - a Standard sweet pickles
    - b Midget sweet pickles
    - c Burgherkins

- d Slice, chips or wafers
- e Candied chips
- f Sweet dill pickles
- g Bread-and-butter, or old-fashioned, pickles
- h Peeled pickles

B Mixed sweet pickles

- a Plain mixed
- b Mustard pickles
- c Jamaica pickles

C Relish, or chopped sweet, pickles

- a Plain relish
- b Spread relish
- c India relish
- d Piccalilli
- e Fruit relish
- f Mexican relish
- g Vegetable relish

Dill Pickles

A general definition of dill pickles includes pickles, either fermented or unfermented, to which have been added dill flavoring, in the form of fresh or cured dill weed or oil of dill, salt, vinegar or lactic acid and mixed spices. The lactic acid may be produced either by fermentation of the pickles or by the addition of edible lactic acid.

Genuine Dill Pickles

Genuine dill pickles are pickles produced by the natural fermentation of fresh cucumbers to which have been added from 9.54 to 10.6% of salt (36° to 40° salometer brine), 10 to 15 pounds of cured dill weed per barrel, mixed spices, about 1 pound per barrel, and 1 quart of 100 grain vinegar. The vinegar may or may not be added.

It requires from 3 to 6 weeks for the pickles to cure thoroughly and for the fermentation to be completed. When properly made, genuine dill pickles should contain 7 to 12 grains of acid, expressed as lactic acid, 4.25 to 5.3% salt (18 to 20°

salometer). There should be a primary flavor of dill and a secondary flavor of the mixed spices. To retain the flavor of the natural fermentation, genuine dill pickles should be marketed in the original brine. This can be done only by selling them in casks, barrels or glass, in which they are packed in the original brine, a practice which has been discontinued to a great extent because of the growth of the scum yeast, Mycoderma, which oxidizes the acid and leads to a loss of flavor and softening of the pickles. Genuine dill brine may also become cloudy from the presence of microorganisms, soil particles and other debris present.

There are several ways by which the cloudiness may be overcome. First, by filtering the brine, or by packing the pickles in colored glass containers and stating on the label that the pickles are packed in the original milky brine in order to retain the original flavor. Another and less satisfactory method is to discard the original brine and make up a synthetic brine simulating the original brine in acidity, salinity and spicing and adding it to the genuine dill pickles. This is an easy way out, but not too satisfactory as this synthetic brine lacks the real flavor of the original brine and, therefore, detracts from the flavor of the pickles. This condition doubtless accounts for the decline in the popularity of genuine dill pickles and is to be deplored. The hazards attending the making and marketing of genuine dill pickles have led to the many substitutes for them.

Fabian has shown that scum formation on the surface of genuine dill brine can be prevented by the addition of a few drops of mustard oil or by covering the surface with paper or some other material impregnated with mustard oil. If this were done, it would still be possible to market genuine dill pickles in open casks or barrels as was formerly the practice.

Polish Dill Pickles

Polish dill pickles are pickles made from fresh cucumbers which are fermented in a low salt brine. The fresh cucumbers are placed in 5.3% salt solution (20° salometer) at first, rather than in 40° salometer brine as in genuine dill pickles. The finished pickles contain about 2% salt and test 4 to 6 grains of acid. The spicing is about the same as for genuine dills except for the addition of onions, garlic and red peppers. The better brands of Polish dills are completely fermented. Some of the cheaper brands are not.

Some type of treatment is necessary to preserve them, as the salt and acid content is low. Sodium benzoate is used in some cases; in others, pasteurization at 165° F for 30 minutes is employed. The latter method is preferable since the presence of benzoate, and acid, results in a burning sensation in the mouth.

Polish dills were very popular when introduced into this country. This doubtless stimulated American manufacturers to simulate them, leading to the greater variety of pickles, some of which are superior to many of the Polish dills.

Overnight, or Fresh Fermented Dill Pickles

Overnight, or fresh fermented, dill pickles are produced by the natural fermentation of fresh cucumbers to which have been added 5.3% salt (20° salometer brine), 10 to 15 pounds of cured dill weed and 1 pound of mixed spices, with or without 1 quart of 100 grain vinegar per barrel.

The length of the fermentation period varies from several days to 1 week. Frequently the pickles are placed in a cold place or in cold storage, where some fermentation takes place. Usually, only partial fermentation takes place before consumption. When they are removed from the cold and placed in the store in a barrel, fermentation continues at a more rapid rate unless sodium benzoate is added. The short fermentation period leads to the name overnight dill pickles, although some people prefer to call them fresh fermented dill pickles. The finished pickle has about 2% salt and an acidity between 3 and 5 grains.

Since the salt and acid content is so low, it is obvious that this type of pickles is very perishable. They must be sold within a short time if marketed in barrels; if canned, they must be heated. The safe pasteurizing temperature is 165° F for 30 minutes; the addition of 0.1% sodium benzoate cannot be relied upon to preserve them.

Kosher Dill Pickles

The term Kosher, as commonly used, simply implies a more highly spiced pickle, especially with respect to garlic, onion and peppers. These should be labeled *Kosher style* pickles to be correct. The term Kosher should be applied to dill pickles of any type, genuine, overnight, fresh fermented, pasteurized, etc. which contain garlic and other spices and which have been prepared according to the Hebrew ordinances. Thus, we may have Kosher style genuine dill pickles, which would differ from genuine dill pickles only in the flavoring. A Kosher genuine dill pickle would differ only from the Kosher style genuine dill pickle in that the former had been prepared according to the Hebrew ordinances with the assistance of a Rabbi whereas the latter had not.

Fresh or Pasteurized Dill Pickles

Fresh or pasteurized dill pickles are unfermented pickles made by placing fresh cucumbers in a weak brine (20° salometer), containing 5.3% salt and about 8 to 10 grains of acetic acid. They are flavored with dill oil and essential oils. Oil of garlic and pepper may be added, if these flavors are desired, also, from ½ to 3% sugar. Cured dill weed and whole spices may be used, if desired. It is quite common to find a sprig of dill weed, mustard or pepper seed present, and the other spices added in the form of oils. A small red pepper may be added for flavor and to improve the attractiveness of the jar. The addition of other whole spices is not recommended, as they detract from the appearance and flavor of the pickles.

Since these pickles are unfermented, it is necessary to preserve them. Chemicals, such as sodium benzoate, are unsatisfactory for this purpose. Pasteurization at 165° F for 30 minutes is the most satisfactory method to use.

In commercial practice, fresh cucumbers, ranging in size from 3 to 5 inches, are washed and packed tightly inside glass jars. The glass jars are then filled with the brine, capped and pasteurized. The pickles are not cured and therefore have the flavor of fresh cucumbers, in addition to the flavor of dill and other ingredients present. They are very palatable and appetizing.

Since the curing and flavoring of the fresh cucumber is all done inside the jar, it is very important that the ratio of the brine to the cucumbers be kept practically constant. The amount of salt, acid and spicing is also important in this connection, if the flavor of the finished pickle is to be satisfactory.

Iceberg or quartered dill pickles are the same as fresh or pasteurized dills except that the ends of the fresh cucumbers are cut off, and the cucumbers are quartered before being packed in the jars. Larger sizes are cut into six or even eight pieces. Twenty degree salometer brine, containing from 8 to 10 grains of acetic acid and the spicing in the form of oils, is poured over the pickles in the jars. A sprig of dill weed, mustard seed and one small red pepper are added. The jars are then pasteurized at 165° F for 30 minutes. The pickles should contain from 2 to 2¼% salt and 4 to 5 grains of acid when finished.

Processed Dill Pickles

Processed dill pickles, imitation or summer dills, are pickles made from cured salt stock by the addition of a brine, simulating dill brine, which is made by heating dill weed and whole spices in the brine, or by the addition of the essential oils to the brine, or both. There is considerable variation in the analysis of the finished pickles, the salt content ranging from 3 to 4¼% and the acidity from 5 to 10 grains.

In making this type of pickle, the salt stock is freshened and processed with turmeric and alum. Then the processed pickles are placed in a 20° salometer brine containing about 17 grains of acetic acid, dill weed or dill oil and whole spices, or dill-flavored brine, prepared as previously described. The pickles are allowed to stand in the brine for several days or longer.

It has been found that the cloudiness, which sometimes appears in the brine after the pickles are placed in jars, is due to the presence of micro-organisms. There are three principal causes of growth of micro-organisms in this type of pickle. One cause is low brine acidity which permits the growth of microorganisms. Dirty barrels, or barrels containing sugar, sometimes are at fault. Frequently, the condition has been eliminated by using the oil flavors instead of dill weed and whole spices, to eliminate the possibility of contaminated spices. Any one, or combination, of the above conditions is almost certain to lead to cloudiness.

While processed dill pickles are not fermented, as other types of dill pickles are, they are made from salt stock which has previously been fermented and cured. In this way they differ from all the other types of dill pickles previously discussed.

Processed, imitation, or summer Kosher, dill pickles are processed dill pickles which have been treated in the manner previously described under Kosher dill pickles.

Pasteurized Processed Dill Pickles

Pasteurized processed dill pickles are pickles made from salt stock pickles by freshening and processing them in the usual manner. They are then placed in a weak brine containing 5.3 per cent salt (20° salometer) and about 8 to 10 grains vinegar and are flavored with dill oil and 4 to 6 other essential oils. Oil of garlic, onion and pepper may be added if desired, also from ½ to 3% sugar, if a slightly sweet taste is preferred. A sprig of dill weed and a small red pepper may be placed in the jar when the pickles are packed to add to their attractiveness. Whole spices have not proved satisfactory. The filled jars should be pasteurized at 165° F for 30 minutes to insure lasting qualities.

There has been such a demand for fresh, or pasteurized, dill pickles that it has been difficult to make enough of this type of pickle from fresh cucumbers during the short cucumber season. For this reason, some pickle packers have been using salt stock pickles, instead of fresh cucumbers, to meet the demand after the supply of pickles, made from fresh cucumbers, has become exhausted. The pasteurized processed dill pickles, however, lack the flavor of fresh cucumbers but are otherwise satisfactory.

Pasteurized processed Kosher dill pickles are the same as pasteurized processed pickles except that they are treated in the same manner described under Kosher dill pickles.

Sour Pickles

Sour pickles are pickles containing from 2 to 4% salt and 18 to 20 grains of acetic acid. They are prepared from salt stock by freshening and processing cucumbers which have previously been fermented and cured in salt brine. The final salt concentration of the brine stock at the end of the curing process ranges from 13.25 to 18.55% salt (50 to 70° salometer). In practice, the salt is withdrawn from the pickles by placing them in fresh water until the desired amount of salt has been removed. The amount of salt left in the pickles is sufficient so that, when they are subsequently soured (by placing them in 50 grain vinegar), from 2 to 4% salt will be present.

Spiced (or hot sour) pickles are made in the same manner as sour pickles except that mixed pickling spices and Chili peppers are added to them.

Sour mixed pickles are pickles containing freshened and processed salt stock pickles, cauliflower, onions, sweet peppers and spices. They are made from the cured salt stock of all the above ingredients and contain the same amount of salt and vinegar as sour pickles.

Sour spiced, or hot mixed, pickles are the same as sour mixed pickles except that spices and hot peppers are added to them. The spices may be added whole, or in the form of the oils.

Chutney

Mixed chutney consists of cured and freshened cut pickles, cauliflower and pearl onions, seasoned with lemon peel, sliced lemons, coriander, celery seed, Japanese chilies, oils of cassia and cloves. It should test 21 to 22 grains of acetic acid, 18 to 20° Bé and 2% salt. Sometimes cut mangoes and Burr pickles are added to the above ingredients, especially if the product is to be sold in cans.

Chow-chow

Chow-chow is a combination of freshened and processed salt stock pickles, cauliflower and onions. The above ingredients are cut and covered with a chow sauce, consisting of yellow and brown mustard seed, turmeric, garlic, cinnamon, cloves, ginger, nutmeg, cayenne, black and white pepper and vinegar.

Sweet Pickles

Plain sweet pickles are pickles made by freshening and processing cured salt stock and then adding sugar, vinegar and spices to the freshened pickles. The sugar content of the finished pickles ranges from 12 to 22° Bé (21.7 to 40° Brix), the acidity from 20 to 28 grains, calculated as acetic acid, and the salt from 1 to 2%. Spices may be added either whole or in the form of the oils.

Sweet pickles containing less than 20 grains of acid will spoil unless sodium benzoate is added to them or they are pasteurized. The amount of sugar added, within the given ranges, is purely a matter of cost and flavor, although the general tendency is to lower the sugar content of pickles in order to lower costs. Only minute traces of spices are necessary to produce the best flavored pickles. The combination and manner of incorporating the spices are of prime importance.

Standard sweet pickles are pickles ranging in size from 10,000 to 2700 per forty-five-gallon barrel. They conform to the general specifications for plain sweet pickles just given.

Midget sweet pickles are pickles which are smaller than the 10,000 size, but otherwise conform to the general specifications for plain sweet pickles.

Burgherkins are a species of the cucurbit family, which are processed and finished in the same manner as the pickling variety of cucumber and which conform to the general specifications for plain sweet pickles. However, they must be sweetened more slowly and over a longer period of time than ordinary pickles, because of the shriveling action which accompanies all sweetening.

Slices, chips or wafers are plain sweet pickles that have been cut crosswise in thin rings. Considerable loss results from the marketing of this type of pickles due to shrinkage in packing and the extra labor involved, unless costs are carefully figured.

Candied chips are sliced sweet pickles containing 11 to 16 grains of acetic acid, 24 to 28° Bé of sugar (44 to 52° Brix), 1.5% salt and suitable mixed spices, either whole or in the form of the oil. Due to the low acidity and high sugar content, this variety is sweeter than ordinary sweet pickles. The high concentration of sugar helps to preserve these pickles, so that it is safe to have low acidity without pasteurization.

Sweet dill pickles are the same as candied chips except they are made from genuine or processed dill pickles, instead of salt stock pickles, and are sliced lengthwise, instead of crosswise. Sometimes they are made from salt stock pickles with the addition of dill weed flavor.

Spiced Pickles

Spiced peeled pickles are made by peeling large salt stock pickles and freshening and processing the peelings. They are then treated with vinegar and sweet spiced liquor. The finished product is similar in taste to spiced watermelon or cantaloupe.

Bread-and-Butter Pickles

Bread-and-butter (or country-style) pickles are made from fresh cucumbers and onions. They are washed, sliced into chips and placed in a 25° salometer brine for 4 to 12 hours. They are then placed in jars, to which is added a heavy spiced syrup containing mustard and celery seed; the jars are then heated to 170° F for 25 minutes. The finished product usually tests from 10 to 16° Bé (18 to 29° Brix), 8 to 17 grains acidity, and from 1 to 2.5% salt. There are other methods used in making this variety, such as cooking the fresh cucumbers in a weak sweet liquor and then placing them in jars, with the addition of a heavy sweet liquor, and heating further to prevent spoilage.

Plain mixed sweet pickles are a combination of freshened and processed salt stock pickles with cauliflower, onions, and sweet peppers. They contain from 21 to 22 grains of acid, 18 to 20° Bé of sugar (22.6 to 36.3° Brix), 1.5% salt and spices as desired. Due to the cut surfaces of the ingredients, this variety should contain more acetic acid than whole sweet pickles do in order to keep them from spoiling.

Mustard Pickles

Mustard pickles, or sweet chow-chow, are a combination of cured, freshened, processed and sweetened cut pickles, cauliflower, onions and sweet peppers. They contain 21 to 22 grains of acetic acid, 18 to 20° Bé of sugar (32.7 to 36.4° Brix), 1.5% salt, suitable spices and yellow ground mustard.

Jamaica Pickle

Jamaica pickle is a type of mixed pickle made from preserved ginger, cut cauliflower, cucumbers, onions, Valencia raisins, lemon, orange and citron peel, brown sugar and vinegar. To this mixture is added a cooked sauce composed of ground cauliflower, tamarinds, tomato pulp, dates, Canton soya, old orange marmalade spiced with powdered spices such as mace, cardamom seed, celery seed, ginger, curry, and flavored vinegars such as cider, garlic and shallott vinegar. It is then sweetened with brown and granulated sugars, colored with caramel and thickened with gum arabic. The finished product is bottled and processed at 180° F for 15 to 20 minutes.

Relish

Plain Relish

Plain relish is a combination of cured, freshened processed pickles, green tomatoes, cauliflower stocks, red peppers and white onions; cabbage and/or green tomatoes are sometimes added. The finished product contains from 22 to 23 grains of acetic acid, 18 to 20° Bé of sugar, 1.5 to 2.5% salt and spices.

The acid content of sweet relish should be 2 to 3 grains higher than for uncut materials to prevent spoilage, due to the cut surfaces.

Spread Relish

Spread relish is a plain relish which is soured to 27 to 30 grains acid. This is mixed with salad dressing or mayonnaise, so that 40 to 60% of the sandwich spread is made up of relish. Relish used for salad dressing should test from 27 to 30 grains acid, 18 to 21° Bé and 1.75 to 2.75% salt. The excess juice should be drained from it in order not to dilute the salad dressing. If these precautions are not taken, fermentation results.

India Relish

India relish is a combination of cured, freshened and processed pickles, green tomatoes, cauliflower stocks, white onions and red bullnose peppers. The finished product contains 22 to 23 grains of acetic acid, 18 to 20° Bé sugar, 1.5 to 2.0% salt, and suitable spices (usually ground ginger, nutmeg, celery and yellow mustard seed).

Fruit Relish

Fruit relish is a relish used for stuffing fruits and melons. It is made from plain, or any other type, relish by addition of currants, citron, orange and lemon peel, and sugar. This mixture is cooked for 15 to 20 minutes and is then ready to use.

Piccalilli

Sweet piccalilli is made from sliced green tomatoes, onions and sweet pickles, to which a sweet spiced liquor consisting of sugar, vinegar and spices is added. The mixture should be pasteurized at 165° F for 30 minutes.

Vegetable Relish

Vegetable relish is a mixture of plain relish, ground fresh cabbage, red peppers, additional turmeric, mustard and celery seed, regular pickling spices and sufficient sugar to test 12 to 15° Bé (21.6 to 27.1° Brix), acid to test 12 to 15 grains and 1.5 to 2.5% salt. The product must either be pasteurized or sodium benzoate must be added.

Mexican Relish

Mexican relish consists of freshened red chili peppers, chopped processed pickle nubs and crooks, cauliflower stocks, chopped ripe olives and green tomatoes, to which a sweetened liquor made from sugar and distilled vinegar is added. All ingredients except the peppers are cooked. Sodium benzoate is added and it is bottled hot.

Pickle Spicing

Practically all manufacturers have their own spicing formula. In some cases, as in genuine dill pickles, whole spices are used. Not infrequently, a combination of whole spices and oils is used, as in fresh dill pickles, processed dill pickles and some types of sweet pickles. Again, only the oils and oleoresins (resins of the spices), are used, as in certain types of sweet pickles.

The oils and oleoresins of spices are practically insoluble in water; therefore, it is necessary to put them in solution with some substance which is soluble in the pickle liquor. Alternatively, an emulsion of the oils and resins with some emulsifying agent, such as gum tragacanth or arabic, can be made and the spices added to the pickles in this form.

Spicing is important in the pickle industry if a well-flavored pickle is to be produced, but space does not permit going into detail on this phase of the subject. However, some idea may be gained of the number of spices used in a formula from the following mixed spice for pickles: spices used are; allspice, coriander seed, broken Batavia cassia No. 2, Zanzibar cloves, broken Jamaica ginger, broken bay leaves, Penang mace, cardamom seed, E. D. long peppers, yellow mustard seed, black pepper No. 2, Japanese chilies No. 1.

The proportion of each of the above spices in the formula is as important as the individual spice. Predominance of any one of the spices gives an undesirable flavor. Equally important is the amount of spices added to the pickles, since too much spicing will ruin the finest pickles produced.

One thing is certain about any good spice formula; it must contain some of the heavy oil spices, such as cloves, cassia, allspice or pimento, and some of the light oil spices, such as nutmeg, ginger, thyme, cardamom, orange, lemon, etc. A proper blending of the light with the heavy oil spices gives the ideal spicing mixture, whether it consists of whole spices or oils.

CHAPTER VI

FISH, FRIED FOODS, HEALTH FOOD, NUTS, BREAD, SOY PRODUCTS, BAKING POWDER

Fresh and Salt Water Fish

The annual catch of fish in the United States averages more than three million pounds, most of which is consumed in this country. The loss in dressing, throwing away unused parts, may vary from 15 to 50% of the total weight. In addition, this country imports roughly two to three hundred million pounds of fresh, salted and canned fish.

Fish is a food rich in proteins and fat, has excellent nutritive values and is easily prepared and cooked. Due to the abundance of the many varieties of salt and fresh water fish, it is also an inexpensive food. Both salt and fresh water fish are equally wholesome and nutritious, and distinctively flavored. Fish is very similar to lean beef in its food composition. Among the fish most abundantly supplied with fat are butter fish, gray fish, herring, mackerel, salmon, shad, trout and tuna. The digestibility of fish varies with the different kinds. It is the general opinion that fish of a fine texture are more easily digested and that fresh fish are more easily assimilated than smoked or dried. Canned, uncured fish correspond very closely in digestibility to fresh fish of the same variety. Freshness is essential; the flesh should be firm, the skin and eyes bright, and the meat should not be so soft that the pressure of the finger leaves a mark. Unless the fish is used immediately it is kept best by submerging in finely cracked ice. For commercial shipment and storage, fish should be quickly frozen to about 5° F or lower, if possible, and held at the lowest possible point, between 0 and 15° F.

Dried, salted, smoked and pickled fish should be kept out of the sun and as cool as possible. Pickled fish should always be kept submerged in the pickling brine. If the brine dries out or leaks away, it should be replenished with new brine.

The advertisement of the use of principal food fishes is continually being conducted in the United States, under the jurisdiction of the United States Bureau of Fisheries, assisted by the State Fisheries Commission, sportsmen's associations and clubs. The United States Bureau maintains hatcheries along the coast, in the chief shad and salmon rivers and at various points along the Great Lakes. Subsequently, the young fish are distributed in natural spawning grounds of the immediate vicinity and great quantities are shipped to all parts of the country, for replenishing of ponds, rivers and lakes. A small percentage is brought to maturity for breeding purposes. The young spawn fish are called fry while the larger ones, still small, are called fingerlings. In distributing the fry and fingerlings overland, the fishery bureau operates many specially constructed railway cars equipped with air pumps and other required appliances for transportation of living fish.

SALT WATER FISH

Anchovies
Bass
Blackfish
Bluefish
Bream
Cod
Eel
Gurnet
Haddock
Hake
Herring
John Dory
Lafayette
Ling
Kingfish
Mackerel
Marlin
Mullet, Gray
Mullet, Red
Pollack
Porgy
Red Drum
Sheepshead
Smelts
Snapper
Sprat
Tuna
Weakfish
Whitebait
Whiting

FLAT SALT WATER FISH

Brill	Plaice
Flounder	Skate
Fluke	Sole
Halibut	Turbot

FRESH WATER FISH

Carp	Pike (Pickerel)
Bass (Striped)	Roach
Bass (Otsego)	Salmon
Bass (Black)	Sturgeon
Bass (Oswego)	Tench
Perch (White)	Trout

SALT FISH

Anchovies	Haddock
Bloaters	Herring
Cod	Kippers

SHELL FISH

Clams	Oysters
Cockles	Prawns
Crabs	Scallops
Crayfish	Shrimps
Lobster	Winkles
Mussels	Whelks

CANNED FISH

Bonito	Salmon
Caviar	Sardines
Crab	Shrimp
Lobster	Tuna

Fish with less than 2% of fat are known as white fish and are suitable for invalids.

Among these fish are:

Cod
Flounder
Haddock
Ling
Pike
Pickerel
Plaice
Skate
Sole
Turbot
Whiting

Fat or oily fish with more than 5% of fat include:

Eel
Halibut
Herring
Mackerel
Salmon
Trout

It is very important to determine whether fish are fresh or not. The following are the criteria of fresh fish:

1 The eyes are bright.

2 The gills are red.

3 The flesh is firm and odorless. The flesh springs back when pressed with the finger.

Put fish in water and if it sinks it is fresh. If it floats it is a sign that it is not fresh and should certainly not be used.

Salt Water Fish

Anchovy

This is a small bony fish of the herring family about 6 inches long, found in European waters, mainly in the Mediterranean Sea.

Bass

The sea bass, found in the bays as early as May, is a small fish, averaging ½ to ¾ pound. It is dark brown in color. In June the larger bass are caught. They are generally dark blue or indigo with black predominating in a checker effect. They reach 4 to 5 pounds.

Blackfish

A name given to different fish, such as the New England tautoc, the sea bass and some others. Blackfish are found from the Jersey coast to South Carolina. They average from 3 to 8 pounds; occasionally larger ones are caught. They are in season in the spring and fall.

Bluefish

These are also called snapping mackerel, tailor, greenfish, fat back and snapper. They are common along the Atlantic coast from Nova Scotia to Brazil. Bluish in color, they range from 6 to 15 pounds and are voracious. They are in season from July to November. They have excellent food value.

Bream

This fish is allied to the carp.

Cod

The cod inhabits the northeastern seas and is also found in enormous quantities off the coast of Newfoundland and the shores of New England. They grow to 40 to 50 pounds in weight and are very prolific. A single fish produces from 2 to 9 million eggs. The small or baby fish are called codling and are found close to the shore.

Cod are named from the place where they are found, the nature of their food, etc., as shore cod, bank cod, rock cod and George's cod (caught in the vicinity of George's banks, a fishing ground off the coast of Maine). Cod come into season about the beginning of October, but are considered at their best about Christmas time. They are out of season by the middle of March.

Eel

This is a fish distributed in almost all fresh waters and seas of the temperate and tropical zones, valuable for its nutritive properties. Eels bite best in spring and fall. When the weather is mild, they are caught as late as December, but with the approach of cold weather they burrow into the mud, where they stay until spring.

The Conger eel is well known on all rocky parts of the British Isles. Some of them are very large specimens, measuring 10 feet in lenth, 18 inches in girth and weighing over 100 pounds.

Gurnet or Gurnard

This constitutes several species of a spiny fish inhabiting the North Atlantic coasts. The red or cuckoo gurnard frequents the English shores.

They are known by their scaly body, compressed laterally. The head is very much broader than the body, the upper jaw extending beyond the lower. The eyes are near the top of the head, large and prominent.

The gurnet is a nicely flavored fish with firm white flesh.

Haddock

The haddock inhabits the Arctic seas to about the Bay of Biscay on the European coast and to Cape Hatteras on the American side. They do not reach a very large size, 8 pounds being about the biggest. They have a long body, the upper part of which is a brown color. They have a silvery belly and a black spot on each side just behind the head.

Either fresh or smoked haddock forms an excellent table dish. Its flesh is firm and sound, white in color, with a creamy curd between the flakes. The larger the fish is the firmer the flesh it has. If it can be boiled in sea water the flavor is perfection. The haddock is in season nearly all year, but it is at its best in March and April. The finest, come from Finnan in Kincardineshire, Scotland. They are generally recognized by their odor and creamy dark yellow color.

Hake

This fish belongs to the codfish group. In shape it is not unlike a pike and on account of its voraciousness it is called the sea pike. The hake is sometimes 3 or 4 feet in length, weighs about 5 pounds and is of a whitish color, grayish on the back. The fish is rather migratory and found in all temperate and moderately cold waters. Frequently it is sold under the name of rock salmon.

Its flesh is very solid and is better baked or fried than boiled. Hake is in season during the summer months.

Herring

This is one of our most abundant fish, found in all northern seas not far from land. The common herring is found throughout the North Atlantic, and the California herring in great numbers in the North Pacific. The herring is distinguished by its brilliant silver color, small fins, and deeply forked tail. It arrives at our shores about the end of May, and averages about 1 pound; in October a larger herring, weighing as high as 2½ pounds, arrives and often stays until December. This is the hickory shad, better known as fall herring or shad herring. Herring travel in large shoals and their appearance is marked by certain signs, such as the number of birds which follow the shoals to prey on them.

The herring is an important food supply and herring fisheries constitute a valuable industry.

John Dory

In the United States and Canada this name is applied to the wall-eyed pike.

In European seas, especially off the Devonshire and Cornish coasts, the name is applied to a fish of a yellowish-golden tint. The name is derived from the French words *jaune doré,* meaning yellow gilt. It is not a very common fish, and consequently rather expensive. Its flesh is not unlike that of the lobster's claw. The best fish are those weighing between 5 and 6 pounds. They should be boiled on account of the great quantity of oil they contain.

Lafayette

Also called spot, goody or oldwife, this is a bottom fish resembling the perch, and rarely weighs more than a few ounces. It abounds from Cape Cod to Texas. It is of a bluish-silver hue with about fifteen wavy transverse bands.

The name is also applied to the butterfish, harvest-fish and dollar-fish.

They generally arrive in July and stay until October.

Ling

This fish belongs to the family of the cod and is a native of the northern seas of Europe. It is not unlike the cod in form but is more slender. It grows to a large size, varying from 5 to 7 feet in length. Its liver yields more oil than the codfish. The flesh is firm and very white. It is better fried or baked than boiled. In the west of England very good pies are made with the ling.

In the United States the name ling is applied to a hake. They average about 2½ pounds. Ling are most plentiful in spring and fall, but can be caught all year. They come into shore at night to feed along the surf.

Kingfish

This is a fish caught on the Atlantic coast where it is quite common, from the Chesapeake to Virginia. It is generally small, rarely exceeding 12 inches in length, weighing about 1 pound, but occasionally may be 3 to 4 pounds.

Kingfish are beautifully marked in a sort of two-tone brown with wormlike dark markings on the back and sides. They are excellent food fish and have a delicate flavor. They are in season in June, July and August.

Mackerel

Mackerel are distributed in practically all parts of the world, and seem to be as abundant in one part as another. Their habits are similar to those of the bluefish. They move in large shoals, and although not the wanton killers that bluefish are, they do destroy a large number of fish. The mackerel is well known for its beauty, brilliancy of color, elegance of form, and its value as an article of food. They vary in length from 14 to 16 inches and in weight from 1 to 3 pounds, but very often larger fish running from 3 to 6 pounds are caught.

They are best when caught in the months of May and June, when they are considered superior in flavor to those caught in spring or autumn.

Mackerel must be eaten only when very fresh, because they become unfit for human food more quickly than any other fish. Those with the least amount of roe and of middle size are the best in flavor. The mackerel fishery of the world ranks in importance with the cod and herring.

Marlin

This is also known as spearfish, and is a large fish of the swordfish family.

Mullet

(Gray) This fish is most abundant on the sandy shores of England, ascending the rivers for many miles.

The best are medium-sized, about 9 inches long. The flesh is exceedingly delicate and has a nice flavor of the sea. The gray mullet, unlike the red mullet, must be scraped, emptied, and thoroughly cleaned. It should be purchased as fresh as possible.

The red mullet has been a favorite from time immemorial. The ancient Romans used to pay an extravagant price for it. It frequents the shores of England, particularly the Cornish and south coasts. This fish is always in great demand, the flesh being firm, white, and well flavored. They are considered in perfection when the roe is forming and that is generally in the summer. Unlike other fish, they are dressed without being gutted, and after being kept for a day or two their flavor is so improved that they are often termed sea woodcock.

Pollack

This fish is of the cod family. It is also known as coal fish or green cod, and is found in the North Atlantic. The pollack, like the codfish, reaches great size. Specimens 4 feet long have been recorded.

Porgy

Also called scup, scuppang, this is a small fish, rarely averaging more than ¼ pound, and resembling the sheepshead. They are found on the Atlantic coast from Cape Cod to Florida. Porgies arrive in June and are caught until October. They are highly valued for food.

Red Drum

Also known as channel bass, redfish, drum, red bass, rief bass, red horse, these fish range along the Atlantic coast from New Jersey to Texas. They belong to the same family as the weakfish, but do not resemble them. They run along shore in July and August, and reach weights up to 60 pounds. The black drum is related to the red drum, but is not as handsome as the latter. They reach weights up to 100 pounds.

Shad

This is a small fish belonging to the herring family, but grows from 20 to 30 inches in length.

Sheepshead

These fish resemble the blackfish. They sometimes reach a weight of 15 pounds, but average about 3 pounds. They are now scarce on the Atlantic coast north of Virginia, while in the waters of Florida sheepshead are abundant.

Smelt

This is a genus of fish belonging to the salmon family. They inhabit the salt water about the mouths of rivers from August until May, and return to the sea after spawning in March. The American smelt is found on the coasts of New England. It does not exceed 12 inches in length. When freshly taken from the water there is a strong cucumberlike smell from this fish. The smelt is eaten in enormous quantities in Paris, as they are abundantly taken from the river Seine. The French word for smelt is *éperlan*.

Snapper

This is the young bluefish.

Sprat

This is a fish of the herring family, rarely exceeding more than 4 or 5 inches in length. Very abundant on the coasts of Great Britain, they visit the shores after the herring and other fish have departed. They are dried and salted, and sold in the market as kilkies, also as Norwegian anchovies.

Tuna or Tunny

This fish, belonging to the mackerel family, sometimes attains a length of 10 feet and a weight of over 1000 pounds.

On the Atlantic coast this fish is known as the great horse mackerel, on the coast of California as tuna, and in the Mediterranean, off Spain and Italy, where they are very abundant, as tunny.

July and August are the best months for tuna fishing. At this time, they may be caught at almost any point between Montauk Point and Cape May from 6 to 30 miles off shore.

The albacore is shaped like a tuna, its body marks differing.

A relative to the albacore is the Arctic bonito. Practically like it, it weighs up to 20 pounds, has dark red meat and a strong taste.

Weakfish

Also called sea trout, these are caught in large numbers from May to October along the shores of the Atlantic coast from Cape Cod to Florida. They run from ½ to 10 pounds.

Whitebait

A small fish, 2 to 5 inches long, this is abundant in the river Thames during spring and summer, and is much prized by Londoners.

Whiting

Also known as winter weakfish, frost fish and silver hake, these fish are called whiting because of their pearly silvery whiteness. They arrive in October or November in shoals that generally contain a few hundred fish. The usual weight of a whiting is from 1½ to 2 pounds, inshore, but those found in deep water have scaled as much as 6 to 7 pounds in weight. They are considered a very fine table fish, and, because they are easily digested, are good for invalids. They are free from oily matter, and are very nutritious.

Flat Salt Water Fish

Brill

This is a European fish, found in most parts of the British coasts, but more plentiful in the English Channel. The brill is similar to the turbot in form, but inferior to it in size and quality. Its skin is perfectly smooth, of a pale, yellowish-brown color with reddish-orange spots and covered with medium-sized scales.

Fluke

This is a species of flatfish, much like the common flounder. From June to November it is found almost everywhere from Cape May to Montauk Point. They average about 2 pounds, but fish running from 5 to 15 pounds are sometimes caught.

Flounder

This is a small flatfish, with an oval, strongly compressed body. The adult has both eyes on the same side of the body. The so-called winter flounder, dark brown in color, is caught on the Atlantic coast in winter and early spring. It is inferior to the brill and plaice, but is easily digested when fresh. Flounders rarely exceed 20 inches in length and 5 pounds in weight.

Halibut

The name *halibut* is derived from holy butt meaning holy flounder, because this fish is eaten particularly on holy days.

This is the largest of all flatfish, some of them turning the scale at 120 pounds. The best for food purposes are, however, those weighing from 30 to 40 pounds. The halibut is found along both shores of the northern Atlantic and northern Pacific. It has an elongated, thick body with a broad head, both eyes on the upper side, and a large mouth. The upper surface is covered with small dark brown scales; the under surface is smooth and white. The flesh is delicate, very wholesome, and resembles the turbot. In fact, it is the poor man's turbot.

Plaice

This is a valuable European flatfish found in the Baltic and Mediterranean, and mainly in the North Sea.

The eyes of the plaice are on the right side, the skin grayish-brown, and, when it is perfectly fresh, large orange-colored spots are found on the skin. As the fish gets stale, the spots darken until they become black when the fish is bad. The plaice usually measures less than 2 feet and weighs from 8 to 10 pounds.

Skate

This is a fish of the ray tribe, extremely abundant and widely distributed over temperate seas, especially in the north. The common skate of Europe often attains a weight of 200 pounds. The barn door skate of the United States is about 4 feet long.

The flesh is thick and very nutritious, particularly appreciated by the French, who value it as a very nutritive fish. This fish varies in quality more than any other fish. The finest are those which have broad and thick backs. The dark side should be brown and healthy looking and the bottom is a delicate creamy-white. The young skate are called maids and are very tender and delicate, but the larger having more flesh, are more profitable.

Sole

The sole holds a prominent position among the flatfish. It is second only to the turbot in point of excellence and is highly esteemed for its firm, white and delicate flesh. Sole, the European flounder, is distinguished by its eyes, which are on the right side and by its oblong form with a rounded muzzle. They vary in size from the little delicate morsels called slips to the larger ones weighing from 4 to 7 pounds. Those that are caught in the shallow waters are the smallest and are really the best in flavor.

The color of the sole depends greatly upon the ground where they feed. If light in color, the skin is light and they are called lemon soles, but if they feed on muddy shores, they are called black soles.

The flesh of the sole is easily digested.

Turbot

This fish may be termed the king of flatfish. Epicures have appreciated this fish from the remotest time. No classic feast in ancient Rome or Greece was complete without the turbot. The adult turbot is about 24 inches in length and has been known to weigh 30 to 40 pounds. The largest supply comes from the North Sea. An American species, the spotted turbot or windowpane, is common on the New England coast.

The turbot is distinguished by having the eyes on the left side, a broad diamond-shaped body, no scales and sandy-colored skin, beneath which lie little bony tubercles.

Fresh Water Fish

Bass

One of many species of valuable food fish, both fresh and sea water fish, allied to the perch, and abundant in the inland waters of the United States is the bass.

The varieties include:

- Striped bass
- Otsego bass (Otsego whitefish)
- Black bass (Oswego bass)
- Grass bass
- Red-eye bass (Rock bass)
- Calico bass

Carp

A long, large-scaled fresh water fish, native of China, with a compressed body and a rather small mouth, furnished with four barbels, this fish is largely used as food and bred in ponds. It breeds rapidly, grows to a large size and lives to a great age. The carp is generally dark brown in color, from 12 to 17 inches long and weighs about 20 pounds, often more.

Perch

The common perch, a fresh water fish with powerful fins and sharp spines, is found in clear rivers and lakes throughout the whole of temperate Europe. In America, the yellow perch is best known. It is an excellent food fish.

The name, perch, however, is given to many other salt or fresh water fish, such as:

Black perch (Black bass)

Gray perch (The fresh water drum)

Red perch (The rosefish of the north Atlantic or the Garibaldi of California) and others.

Pike

This is a voracious fresh water fish found throughout the northern hemisphere. It has a narrow body, 20 to 45 inches in length, and weighs from 2 to 20 pounds. In America, the pike is represented by several fish commonly called pickerel.

Roach

This is a small European fresh water fish. In America, the name is applied to various fish which resemble the roach, such as the sunfish, the Lafayette, etc.

Salmon

Justly called the king of fish and ranking foremost among the food fish of this and other countries, the salmon is found in all the northern climates of America, Europe and Asia. It attains a length of from 3 to 4 feet and an average weight of from 12 to 30 pounds; specimens have been known to weigh over 70 pounds. The adult fish is a blackish-blue on the back and the bottom has a glistening silvery-white luster. The flesh, when fresh, is of a bright orange color, and has the best flavor when taken after sea feeding. In fall, the salmon ascends the rivers, sometimes as far up as 70 or 100 miles. Here it deposits its spawn in gravel or sand. The spawn lies buried until the middle or end of March, when the young come forth. For 2 years they gradually increase in size, until they attain the smolt stage. The young fish are then supposed to leave the river and make their first migration to salt water. The smolts grow rapidly in the ocean and return in about 2 months to fresh water in the form of grilses, weighing from 3 to 4 pounds. In the grilse stage they are capable of depositing eggs and then return to the sea again. After their third stay in the ocean, they return after 3 months' absence as adult salmon.

Particular care should be taken that, no matter in what form salmon is cooked, it is always well done, because if this fish is eaten when insufficiently cooked, it is both unwholesome and indigestible.

Sturgeon

Commonly termed royal sturgeon, this is a large fish found in northern Europe, especially Russia, in the Black, Caspian and Mediterranean Seas, in the Don, Danube and other large rivers, and in the lakes of North America. The Pacific coast has a large and important species in the white or Columbia River sturgeon, weighing from 300 to 600 pounds. Sturgeon grow to a very large size, from 10 to 12 feet in length and from 500 to 600 pounds in weight. Those that inhabit the Caspian and Black Seas sometimes attain a length of 20 to 24 feet and a weight of 2000 pounds.

There are many species of sturgeon, and all pass the greater part of the year in the sea. Like the salmon, they regularly ascend the large rivers in the spring and deposit their spawn. A few species are confined to fresh water. Its flesh is considered very delicious. The commercial value of the sturgeon, in Russia alone, is enormous, including the isinglass (the air bladder) and the caviar, which is the roe of the sturgeon.

Tench

This is a fresh water fish of the carp family, found in standing water all over Europe and remarkable for its tenacity to life.

Trout

The brook or speckled trout is easily distinguished by its red, round spots, which cover the whole of the fish as far as the belly. It does not exceed more than 12 or 14 inches in length, or more than ½ to ¾ pound in weight. It inhabits clear running water with a pebbly bottom, swims very fast and can jump like a salmon, to whose family it belongs. The flesh of the trout is of a delicious flavor and, from the time of the Romans, has been considered a delicacy. The colder the water is from which the trout comes, the better its flavor. Trout are found in mountain streams all over Europe and America.

The lake trout exceeds all others of the trout family, reaching a weight of over 100 pounds.

Specifications of Canning Trade for Fish *

Anchovies

Anchovies are put up both in brine and in olive oil.

In Brine

Size of container	½ Kg.	1 Kg.	2 Kg.	8 Kg.	12 Kg.
Approximate net weight of fish content	5 oz.	12 oz.	2¼ to 2¾ lb.	9 lb.	17 lb.

In Olive Oil

Size of container	Small can	3-ring bottle	4-ring bottle	6-ring bottle
Approximate net weight of fish content	2¼ oz.	2½ oz.	3¼ oz.	5¼ oz.

Anchovies in olive oil are packed both plain and rolled in capers.

Bloaters

Bloater has two meanings

1. A fat herring, salted and smoked.
2. A kind of cisco, called also the lake herring, very abundant in the Great Lakes of the United States.

Cod

Codfish appear in the market as

* Reprinted, by permission, from *Canning Trade Almanac,* Baltimore, Md., (1938).

1. *Codfish Tablets*

These are strips of boneless codfish, cut 6 inches long, 3 inches wide and about 1¼ inches thick. The tablets are usually made up of 2 or 3 pieces and come in 1-pound boxes, packed 12 to 24 in a case.

2. *Codfish Strips*

Skinned codfish, usually the whole fish, made up in a roll, varying in weight from 2 to 4 pounds.

3. *Codfish Middles*

The thickest and choicest part of the codfish. Packed in 40-pound boxes.

Haddock

A cured and smoked haddock is offered in the market as "Finnan Haddie" (See Haddock under Salt Water Fish).

Herring

A. *Norwegian Herring*

2K Herring—small herring, running from 650 to 700 to a 220 barrel (100 Kg.).

4K Herring—large herring, running from 400 to 450 to a 220 barrel (100 Kg.).

B. *Holland Herring*

Holland herring are packed in:

100 Kg. barrel
90 Kg. barrel
½ barrel
¼ barrel
Keg (about 5 Kg. or 11 lb.)

Specialties Made with Herring

Marinated Herring

Herring pickled in vinegar with lemon slices and onions, bay leaves, allspice, black pepper, mustard seeds, cloves (whole), and small Japanese peppers.

Roll Marinated Herring

Marinated herring, split in halves, boned and rolled around a piece of pickle. Fastened together with a toothpick.

Kippers

Any fish, also mostly herring, may be called "kippered" when split, cleaned, dried, and rubbed with salt and pepper.

Shell Fish

Clams

A name for almost any edible bivalve mollusk other than a scallop, mussel, or oyster.

Two kinds are generally found in the market:

1. The hard or round clam (known as "quahaug"), contained in heavy triangular shells, and
2. The soft clam (also called surf or hen clam), in thin elongated white shells.

Cockles

A small edible mollusk related to the American quahaug (clam). In Europe much used as food.

Crabs

The blue crab is found from Cape Cod to Florida. The fiddler crab is found in great numbers on the Atlantic coast.

The crab offers a much larger proportion of non-muscular material than the lobster. It has more flavor, is more easily digested, and is cheaper and preferable to the lobster as a food.

Crayfish or Crawfish

Closely related to the lobster, inhabiting rivers and ponds.

Lobsters

Lobsters were known and eaten by our ancestors 500 years ago. Those of greatest commercial importance are the common lobster of eastern North America and its ally, the European lobster.

It may be that the lobster and crayfish are the least desirable as food, as they so frequently cause indigestion. Nevertheless, they are extremely popular as a luxury and change. Their flesh contains little or no oil, therefore they should be taken with plenty of oil and vinegar, which assists the solid fiber of which they are composed.

Mussels

A name given to a number of sea and fresh water mollusks. The common edible mussel is found on the shores of North America and in Europe. They may be obtained nearly all the year round, are very inexpensive, and full of flavor.

Oysters

A well-known edible mollusk. There are over 100 known species of oysters. It is the most extensively eaten shellfish and, next to the herring, the most valuable commercially of all food obtained from the sea.

The largest known oyster ground in the world is Chesapeake Bay, with Long Island second. In Europe the "Whitstable Native" and the "Colchesters" are considered the best.

The American oyster cannot be transplanted to European waters.

Prawns

A small shellfish of the shrimp family, prized for food.

Scallops

A kind of shellfish with a shell like a comb.

Shrimps

A small shellfish allied to the lobster, crayfish, and prawn, found on sandy shores all around North America.

Removing the shells of shrimps and canning them forms an extensive industry.

Winkles, Whelks

A large class of mollusks of marine gasteropods like snails, slugs, limpets, etc.

Canned Fish

Bonito

The bonito resembles the mackerel in appearance and the tuna in flesh, although the color of the flesh is reddish.

The bonito is canned in the same way as the tuna.

Caviar

Caviar is the salted roe of the sturgeon. In the United States eggs of fish other than sturgeon are prepared as caviar. The product must, however, be labeled with the name of the fish used.

Grading

Whole Grain Salted.	Very large eggs (similar in size to a large buckshot), containing 8–10% salt. Gray in color.

Other Grades

Caviar Paste, Pressed Caviar, Small Grain.

Crab

The domestic crab meat is obtained from the blue crab caught along the Virginia coast. Its meat is flaky and of grayish white color. The Japanese crab meat is white and similar to that of the lobster. The meat of the domestic crab is superior in flavor to the Japanese crab meat.

The Japanese Canned Crab Packers' and Explorers' Association prescribes for:

Fancy Crab Meat	Chiefly large leg meat of beautiful appearance and perfect as to container and processing. By rule, in order to grade Fancy, the pack must run 70% leg meat. Most Fancy packs, however, average better than 80% leg meat. In order to grade strictly Fancy, the leg meat must be large.
Choice	Chiefly composed of leg meat somewhat smaller than appropriate for Fancy pack, or a pack in which slight imperfection in technique of packing or processing occurred, which would otherwise grade Fancy.
Fair	Slender pieces of leg meat or a Choice pack in which defects are apparent.
Passed	Flake meat, chiefly composed of body meat with or without slender leg meat.

All crab meat is packed in tin cans, lacquered inside, with a parchment paper lining to prevent the meat from turning black.

Packed in cans	½s Flat	1s Flat
Net weight of contents	6½ oz.	13 oz.

Lobsters

Lobsters are canned in Maine, Nova Scotia, and on the northeast coast.

Packed in cans:

Size of container	¼s Flat	½s Flat	No. 1 Flat
Net weight of meat	3 oz.	6 oz.	12 oz.

Grades

Fancy (Basis of No. ½ size can.)
Should show the whole claw (3 or 4) carefully sandwiched at top. Meat of claws should be white, retaining its bright red skin. Under the claws should appear the tails (3 or 4), carefully encircled so that the tip of one tail will overlap the butt of another. Tail meat should be white, retaining its bright red skin. Within the circle of the tails should appear the fine arm meat in quantity not much over 1 ounce. The entire package should be practically free from blood (blood appears as white curd adhering to the meat). The strength of the sauce water should not be over 2½% saline, and in quantity about 2 ounces in weight. The out turn of meat should be 6 ounces exclusive of liquor. The No. 1 Flat Cans and the No. ¼ Flat Cans should contain a proportionate amount of claws, tails, and arm meat, and the out turn of meat for each size should be 12 or 3 ounces, respectively.

Oysters

Oysters were first canned in this country in Baltimore and developed a large market for the famous Chesapeake Bay oyster. On account of the large demand for raw oysters, the supply of northern oysters for canning became limited and canning has largely been confined to the south Atlantic and Gulf coasts (Virginia, Carolina, Florida, Georgia). Oysters are packed in cans.

Size of container	4 oz.	5 oz.	6 oz.	8 oz.	10 oz.	12 oz.
Approx. number of oysters per can:						
Standard	45 or less	60 or less	....	90 or less	120 or less	....
Select	15/20	15/20	25/28	....	....	55 or less
Extra Select	....	....	18/22	....	....	40 or less

Grades

Extra Selects and Selects	Large, firm oysters, practically uniform as to size, uniform color, liquor should be tasty, bright, and approaching sky blue in color.
Extra Standard	Firm oysters, reasonably uniform as to size. Practically uniform color. Liquor should be bright and tasty.
Standard	Firm oysters, ungraded for size, reasonably uniform color.

Salmon

The Chinook, King, or Spring salmon is the largest of the salmons, and its flesh, when canned, readily separates into very large flakes. It is usually very rich in oil and is commonly sold in several grades, according to the color, which varies from a deep red to white. The large flakes and abundance of oil gives this fish a soft texture, and it is more suitable for salads than for cooked dishes.

The Red, Sockeye, or Blueback usually has a deep red color, a small flake, and a considerable amount of oil. It is firm in texture and is particularly well adapted to salads and to eating as it comes from the can. It may also be used for sandwiches and cooked dishes, but other kinds, costing less, are equally suitable for these purposes.

While the two kinds of salmon already described may mature in four or five years, or even longer, the Medium Red, Coho, or Silver salmon generally does so in three. It is smaller than the King, but larger than the Red. Its flake is large and the color distinctly red, but averaging lighter than the Red salmon. It is an excellent fish suitable for all purposes.

The Pink salmon is the smallest of all the salmons, and is more finely textured than the rest. The flakes are small and range from a light to a deep pink in color. The Pink salmon has a distinctive delicate flavor and is particularly well adapted to sandwiches, chowders, and various cooked dishes. Over half of the entire American pack consists of this species, and its abundance brings the price very low.

The Chun, or Keta salmon, while highly esteemed by the fishermen as fresh fish and by the Indians for drying, becomes, when canned, the cheapest of them all. It is a rather large and somewhat coarsely textured fish with little color; however, it can form the basis for many nourishing and palatable cooked dishes, bringing the healthful qualities of the sea to homes that cannot afford the other kinds costing more.

Salmon comes packed in cans.

Size of container	½s Flat	No. 1 Flat	No. 1 Tall
Net weight of contents	8 oz.	16 oz.	16 oz.

Canned salmon should be labeled with the name of the species of fish contained in the can (U. S. Food Insp. Decision No. 105).

The can should show about ¼ inch head space, not more than 2 or 3 pieces of fish in the can. The long axes of the pieces should parallel the long axis of the can. Ends of pieces should be clean cut, not jagged, free from skin on the end of the cylinder of fish. Free from pieces of gills, fins, and intestines. Free from all clotted blood that can be washed out without tearing the fish. Blood sometimes settles behind the backbone or a fish is bruised before the blood has congealed. These blood clots cannot be removed and their presence does not signify poor cleaning. Ribs should be loose, flesh should not be torn, vertebrae should break easily between the fingers. Ribs should be brittle, breaking without much bending. In the majority of cases an 8-inch or more vacuum is satisfactory. (Determined with a compound vacuum pressure gauge equipped with a hollow piercing point and a rubber gasket.)

Sardines

According to a definition by the Bureau of Fisheries, United States Department of Commerce, any small fish of the herring family, properly prepared, are entitled to be called sardines.

In Europe the pilchard, sprat, and bristling are canned as sardines.

In Maine the young herring and in California the Pacific sardine, a species of the pilchard.

Domestic Sardines

Size of container	¼s square	½s square	½s oval	No. 1 oval
Net weight of contents	3¼ oz.	7½ oz.	7½ oz.	15 oz.
Style of packing	Olive Oil Cottonseed Oil	Olive Oil	Tomato Sauce Mustard Sauce	Tomato Sauce Mustard Sauce (pickled)

Imported Sardines

Norwegian

Size of container	⅟₁₆s	⅛s	¼s	⅛s	¼s	¼s high	½s
Appr. net weight of contents	1½ oz.	2½ oz.	3¾ oz.	2¼ oz.	4 oz.	5 oz.	8 oz.

Grades

Bristling	22/24—18/20 fish per can
Mussq or Sild	16/18—12/14—8/10 per can

French

Size of container	⅛ club	¼s	¼s high	½s
Appr. net weight of contents	3½ oz.	4¼ oz.	6¾ oz.	8 oz.

Grades

The French and Portuguese packers offer sardines under the self-explanatory grades of

Small Count, High Count, Boneless (with tails), Boneless and Skinless (without tails).

Requirements for All Varieties

Fish should be of uniform size, firm but tender. Vertebrae should crush easily when pressed between thumb and finger. Can should show a liberal quantity of oil or sauce.

Shrimps

The canning of shrimps is carried on in the South Atlantic and Gulf States, where two styles of pack are put up, the "wet pack" and the "dry pack."

For wet pack shrimp some brine is added, and the label should show the drained weight. They are packed in cans and in glass.

Size of container	No. 1 Tin	No. 1½ Tin	5¼ oz. Tumbler	6½ oz. Tumbler
Net weight, dry pack ...	5 oz.	8¼ oz.	......	......
Net weight, wet pack ...	5¾ oz.	9¾ oz.	5¼ oz.	6½ oz.

Tuna and Tonno

There are four varieties of canned tuna:

1. The Albacore Tuna or Long Fin Tuna—meat white.
2. The Yellow Fin Tuna—meat slightly pink.
3. The Blue Fin Tuna—meat pink.
4. The Striped Tuna or Skipjack—meat brown.

They are packed in cans.

Tuna

Size of container	¼s Flat	½s Flat	No. 1 Flat
Net weight of contents	4 oz.	7 oz.	13 oz.
Net weight drained	3½ oz.	5¾ oz.	11½ oz.

Tonno

Size of container	¼s Flat	½s Flat
Net weight of contents	3½–4 oz.	7 oz.
Net weight drained	3–3½ oz.	5 oz.

The net drained weights shown above for tuna (not tonno) are those prescribed by the United States Department Agriculture Bureau Chemical Circular, April 29, 1925.

Grades

Standard White Meat	Packed from albacore (long fin tuna), perfectly processed, not less than 80% steak, meat white. No discolored meat, perfectly cleaned, a quantity of oil sufficient to thoroughly season the meat (approximately 1 ounce to a No. ½ can at time of filling).
Yellow Fin	Packed from yellow fin tuna, perfectly pressed, not less than 80% steak, meat slightly off to pink from white, no discolored meat, perfectly cleaned, a quantity of oil sufficient to thoroughly season the meat (approximately 1 ounce to a No. ½ can at time of filling).
Blue Fin	Packed from blue fin (leaping tuna), perfectly pressed, meat practically the same as yellow fin, slightly darker, well cleaned, not less than 80% steak, no discolored meat, a quantity of oil sufficient to thoroughly season the meat (approximately 1 ounce to a No. ½ can at time of filling).
Striped	Packed from striped tuna (skip jack), perfectly processed. This is the darkest meat of the four grades of tuna, having a decidedly brownish-pink cast. Not less than 80% steak, well cleaned, practically free from discolored meat, a quantity of oil sufficient to thoroughly season the meat (approximately 1 ounce to a No. ½ can at time of filling).
White Meat Flakes or Pieces	Packed from the small pieces or flakes that are accumulated when cutting fish for the standard white meat pack. No

mixed varieties, good color, reasonably free from discolored meat. Also the same amount of oil is used in preparing this pack as in the regular pack; the can, when cut, will not show any considerable quantity of "free oil" as the flaked condition of the meat promotes greater absorption than in the "steak" pack.

Light Meat Pieces or Flakes — Packed from the small pieces or flakes that are accumulated when cutting the regular packs of yellow fin, blue fin, and striped tuna; may contain these three varieties singly or mixed, free from badly discolored meat, also the same amount of oil is used in preparing this pack as is used in the regular pack, the can, when cut, will not show any considerable quantity of "free oil" as the flaked condition of the meat promotes greater absorption than in the "steak" pack.

Albacore (long fin tuna) is the only variety that may be labeled white. The three other varieties may be labeled light meat or California tuna.

All labels must bear the legend "packed in cottonseed oil" or an equivalent phrase, such as "in salad oil made from cottonseed" when this variety of oil is used.

Tonno

Italian style in olive oil.

Packed in Kanopen cans, may be either lithographed tins or labeled.

Packed from yellow fin or blue fin or striped tuna, but no mixed varieties, all steak, no flakes, perfectly pressed, meat brownish-pink, well cleaned, no discolored meat, liberal quantity of olive oil (about 2 ounces to a No. ½ can when packed).

Smoked Fish

It is hardly possible to furnish directions for smoking all species of fish, under the various weather conditions which will be encountered during the changing seasons. Only the general methods used on a typical variety under average conditions can be given here. This is intended as a guide, not an infallible recipe. To smoke fish successfully, experiment and use intelligence, altering the method (amount of salt and smoke flavor) according to the preference of markets, the variety of fish and weather conditions.

There are two general methods of smoking fish—hot smoking, or barbecuing, and cold smoking.

Any fish may be hot-smoked, or barbecued, but the following varieties are some of those to be preferred:

Butterfish
Kingfish
Mullet
Sailfish
Spanish mackerel
Shad
Grouper

Sturgeon is always hot-smoked.

Because of the keeping qualities of cold-smoked fish, certain varieties offer market possibilities for quantity production, such as:

Alewife or river herring
Shad
Drum
Mullet
Red snapper
Redfish
Grouper
Kingfish
Robalo or snook
Squeteague (Spotted trout)
Spots

In the first method, the fish are laid 3 or 4 feet above a fire and cured at temperatures from 150 to 200° F. The fish are wholly or partially cooked by this method, and therefore, no matter how carefully prepared, or how long smoked, will keep for periods of from a few days to several weeks. If fish is to be preserved for any period of time, the cold smoking method should be used. In this process the fish are cured over a low, smoldering fire at a temperature of 90° F, or less. The efficiency of the process depends on the drying action of the fire, and must be carried on at a temperature that will not cook the flesh. Fish may be given a short cold smoke, if preservation is intended for a few days only, or cured for several days if they are to be kept for some time. This product is comparable to ham or bacon and should be cooked before using. The same general principles governing smoking, handling, and storing of cured meats should be followed in smoking fish.

A smokehouse for curing small lots of fish may readily be made. Obtain a box or make one, about 6 by 3 by 3 feet. One end, that resting on the ground, should be removed. About 12 inches above this end a false bottom with auger holes at two-inch intervals is built. This end of the box is set over a pit 2 feet wide by 18 inches deep. A trench about 1 foot wide by 1 foot deep is dug from this pit for a distance of about 10 feet. The fire pit, a hole 3 feet wide by 3 feet long by 18 inches deep, is dug at the end of this trench, which is then covered by sheets of galvanized iron, forming a chimney for the smoke from the fire pit to the smokehouse. If a more permanent house is desired, terra cotta drain or sewer pipe may be used to connect the fire pot with the smokehouse. Cleats are nailed inside the box on the sides, the first set about 12 to 14 inches below the top. The trays for holding the fish, or the ends of the smoke sticks, rest on these cleats. A few holes should be bored for ventilation in or near the top of the house.

If mullet or Spanish mackerel are to be smoked, the following process is recommended.

The fish should be split along the back just above the backbone, almost to the tail, so that it will lie flat in one piece, leaving the belly portion solid. Clear out all traces of intestines, black skin, and blood, taking special care to remove the coagulated blood and kidney just under the backbone. The head may or may not be removed, depending on the individual. If the head is cut off, the hard bony plate just below the gills should be allowed to remain, as it will be needed to carry the weight when the fish are hung on rods. If it is cut off the fish often pull loose and drop from the sticks.

After splitting and cleaning, the fish should be dropped in a brine made by adding 2 cups of salt to 4 gallons of water. They are left in this brine 30 minutes to soak out blood diffused through the flesh. At the end of this time they should be taken out, rinsed, and freed from any remaining traces of blood or other offal. Drain for a few minutes, then drop each fish singly in a shallow box of fine salt, dredging it about, then pick it up with as much salt as will cling to it, and pack the fish in even layers in a tub or box.

The fish should be left in salt from 1 to 3 hours, depending on weather, size of fish, fatness, and length of time for which preservation is desired. The exact length of time must be determined by the smoker. When the fish are taken out of salt they should be rinsed in brine, scrubbing off all visible particles of salt or dirt. The fish should then be laid on chicken wire drying racks kept out of the direct rays of the sun, but located where a good breeze can reach them. Wire drying racks are desirable as the fish can dry on both sides. If laid on boards, one side will remain wet. The fish should be given about 3 hours' drying, until a thin film is formed on the surface, before putting the fish in the smokehouse. If put in immediately after taking out of salt, the fish will be too moist, will require longer smoking, will not color and dry as well, and will not have as good a surface.

The fish may be placed in the smokehouse on wire mesh trays, or hung on sticks or iron rods. In no case should two fish touch, as this will prevent the drying and penetrative action of the smoke. If hung on rods, more fish may be smoked at one time, and they will smoke better, with a clearer color. Trays, of course, give less trouble. Rods are run through the fish just under the hard, bony plate at the neck, one rod on each side. In this way, each fish hangs from two rods. Twelve or fourteen fish may be hung on a set of two rods 3 feet long.

The fire should be started an hour or two before the fish are put in the house. It should be low and smoldering. Almost any hardwood or wood other than pine may be used for fuel. Pine or other pitchy woods will give the fish a bitter taste. Some of the woods that may be used in the Southern States are scrub oak, live oak, hickory, sweet bay, river mangrove, palmetto roots, button wood, and coconut husks. In smoking any one kind of fish, such as mullet, variety of flavor may be obtained through the choice of wood used in smoking. In addition to the woods listed above, orange wood gives a particularly pleasing flavor. Cypress may also be used. The fire should not give off too much smoke during the first 8 to 12 hours. A dense cloud of smoke should be built up for the balance of the process. The fire must be small and steady. Two short chunks of wood, about 2 feet in length and the thickness of a man's arm, are usually sufficient. The fire pit is kept covered with a sheet of metal to drive as much smoke as possible up into the smokehouse, and to keep the fire from burning rapidly. The fire must not be allowed to blaze up. The air in the smokehouse should not feel warm to the hand. The fish should be smoked for 24 hours, if they are to be kept for a few weeks, and for 4 or 5 days if they are to be kept for some time. The fire should not be allowed to die out at night, nor should it be built up too large the last thing at night.

After taking the fish out of the smokehouse dry for an hour or two in the air, then wrap in sheets of waxed paper, sprinkling a little fine table salt on each one, and store in tin or wooden boxes. Keep in a cool, dry place. If signs of mold appear, sponge off with vinegar and give the fish a short smoking for from 3 to 6 hours.

Hot Smoking—German Method

The following method is recommended if one wishes to prepare a hot smoked fish that can be used immediately without cooking. It will keep without molding or souring longer than other hot smoked fish.

Split, clean, and soak the fish to remove blood as previously instructed. Then prepare a brine with the following: 2 pounds of salt, 1 pound of sugar, ½ ounce of saltpeter, 1 ounce of crushed whole black peppers, 1 ounce of crushed cardamom seeds. Make this up into a 90° brine, that is, one that will float a potato with a 10-penny nail stuck in it. Increase the amount of ingredients according to the quantity of brine you wish to make. The number of spices used can be increased in variety and amount. Various spice mixtures are used.

Put the fish in this brine for a period varying from 2 to 4 hours, depending on the size and thickness of the fish, amount of fat, and the taste of the individual. Some people require a less salty taste than others. The exact length of time must be determined by experiment. Rinse the fish off in fresh water, and place on drying racks outside in a cool, shady, breezy place for about 3 hours before putting in the smokehouse.

For the first 8 hours that the fish are in the house, give them a cool smoke in a dense cloud of smoke. Then increase the fire until the temperature is between 130 and 150° F for 2 or 3 hours, or until the fish have a glossy, brown surface. This partially cooks, or hot smokes, the fish. Wipe any moisture off the fish, and cool for a couple of hours before storing. Wrap in waxed paper, and store in a cool, dry place. Do not store them in a wet cold place or allow them to come into contact with ice.

In some cases, the fish are brushed over lightly with vegetable oil, usually cottonseed, either just after finishing the cold smoking part of the process, or on taking out to cool. Another method of handling this fish after smoking is to cut the flesh up into fingers the length of a No. 2 can or pint glass jar. Skin and pack into the can or jar. Then add vegetable oil, cottonseed, or olive oil if you have it, until the spaces between the pieces of fish are filled and there is a layer of oil up to within ⅛ inch of the top. Seal the cans or jars and store in a cool place, such as an icebox, until used. Under such conditions, it should keep almost indefinitely. As this product is not sterilized, the cans or jars should be thoroughly scalded before use. In some cases the oil is poured in hot, and the containers sealed immediately.

Lake Herring and Whitefish

The processes of smoking lake herring and whitefish are identical. If the fish are frozen when received at the smokehouse, they are thawed in the open air, or better, by immersing and stirring them in a barrel of water of medium temperature. After thawing, they are split down the belly to the vent, eviscerated, washed thoroughly, and pickled in butts or barrels. About 4 pounds of fine salt to 100 pounds of fish are scattered among them and sufficient brine of 90° salinity to cover them. Either dry salt or brine alone may be used, the former being preferred in warm weather and the latter during the winter. In case brine alone is used, some dry salt should be placed on top to strengthen the weak pickle floating at the surface. After remaining in the pickle for 10 to 16 hours, according to the strength of the pickle and the flavor desired, the fish are removed and strung on the smoke rods, 10 to 20 fish to each rod, according to its length and the size of the fish.

In stringing, some curers pass the rod through the body immediately below the nape bone, effectively preventing the fish from falling down in smoking, but also marring its appearance somewhat. A more usual way is to pass the stick in at the right gill-opening and out of the mouth. Some people pass the rod through the head near, or through, the eyes, others pass it immediately back of the throat cartilage. The latter way leaves a neat appearance, but permits more fish to fall in the smoking process than when the rod is passed through the head or the shoulders. In some houses, the smoke-stick is not passed through the fish, instead a stiff iron wire, curved in *S* shape, is used to attach the fish to the stick, one end of the wire passing through the fish at the head or beneath the nape bone and the other hung over the smoke-stick. At Grand Haven, and to some extent in Chicago, Milwaukeee, and one or two other places, the fish are secured by having stout smoke-sticks, about 1½ inches thick and 2½ inches wide in the top of each, and about ¼ inch from the edge, a row of tacks or small wire nails at intervals of about 3 inches is driven, projecting about ½ inch above the surface. Ordinary cotton wrapping cord is tied to the wire nail at the end of each stick, and by means of this cord passing around each nail, a single herring is held in place between each two nails throughout the length of the stick, the fish being placed with the back of the neck against the stick and the cord passing from one nail around the throat of the fish, entering under the gills on each side, and then around the next nail, and so on to the end. By having the stick of sufficient width, a row of small nails may be placed on each edge to enable attaching a row of fish at each side. This removes nearly all risk of the fish falling, and their appearance is not marred by holes through which the smoke-stick has been passed.

Some markets prefer the herring well smoked on the inside, and to accomplish this, the sides of the abdominal cavity are stretched open by means of small wooden sticks or toothpicks, one or two sticks to each fish. This permits the smoke to permeate the stomach cavity better and results in a more durable article. In general, the western trade prefers the stomach cavities stretched open, while the eastern markets prefer them without the sticks; but there are exceptions. The smoked lake herring sold in Washington are mostly extended by means of a small stick, or, in case of large fish, by two small sticks.

The fish, attached to the sticks, are dipped in fresh water to remove surplus or undissolved salt, loose scales, etc., (unless they have been rinsed before stringing) drained, and suspended in the smokehouse 4 to 8 feet above the floor, and subjected to a gentle smoke for 4 or 5 hours. The door or damper is then closed, the fires spread or built up, and the fish cooked for 1 or 2 hours according to the amount of fire, the height of the fish, and the particular cure desired. After cooling, which is accomplished either by opening the doors of the smokehouse or by removing the fish to the outside, they are ready for the trade. One hundred pounds of round fish, or 85 pounds of dressed, yield about 65 pounds of smoked. Ordinarily these fish keep 1 or 2 weeks, and even longer.

Lake Trout and Carp

Smoked lake trout and carp are prepared to a small extent in the manner already described for lake herring or whitefish.

Alewives or River Herring

River herring or alewives are smoked in a number of localities, but principally in Maryland and Virginia.

In preparing these fish in the Chesapeake region they are washed in vats and scaled with a knife as soon as practicable after removal from the water. They are next immersed overnight in strong brine, containing 12 to 14 pounds of Liverpool salt to each 100 pounds of fish, with some dry salt on top to strengthen the weak pickle that rises to the surface. The following morning the round fish are strung on smoke-sticks, the stick usually being inserted at the left gill-opening of each fish and ejected at the mouth, as in case of hard herring or bloaters on the New England coast. The strings of fish attached to the stick are then dipped in fresh water to rinse them off and, after draining and drying for a few hours, are suspended in the smokehouse about 6 or 8 feet above the fire, and exposed to a dense but cool smoke made of pine shavings or similar material for about 2 or 3 days. Care must be taken to prevent the fire from becoming too hot, causing the fish to crack at the lower end or to fall from the sticks to the floor. Prepared in this manner, the river herring will usually keep in good condition in the Chesapeake region for 30 days during the spring and for a somewhat shorter period in the summer. As the fish are not eviscerated before smoking the decrease in weight is small, 100 pounds of round fish yielding about 85 pounds of smoked. The wholesale price is about 20 to 22 cents per dozen, according to the size and condition.

In Washington, Baltimore and one or two other places the river herring are prepared in the following manner.

The fresh herring are scaled with a knife, gibbed like the pickled herring of Scotland, washed, and pickled for 3 hours in brine, about 20 pounds of Liverpool salt being used for each 100 pounds of fish. On removal from the pickle, they are strung on small iron rods, the rod passing through the eye sockets, drained for an hour or so, and hung in the hogshead smokehouses, in the bottom of which a fire has been made of equal quantities of oak and hickory wood. The fish are dried for a few minutes and then the tops of the hogsheads are covered with old sacks or other suitable material. From time to time the fire is sprinkled with water to produce a vapor exposing the fish to heat, smoke and steam for about 3 hours, after which they are removed and cooled and are then ready to be eaten. Only oak and hickory should be used as fuel; other materials do not produce the proper flavor. If the fire becomes too warm it should be smothered with oak or hickory sawdust.

The process of smoking alewives commonly employed in the New England states differs from the Chesapeake process in a few minor particulars. The smokers are usually not so careful about removing the scales with a knife, depending generally on the frequent handling of the fish to scale them, if cured soon after removal from the water. It is also customary, in salting the fish, to permit them to make their own pickle, the fish remaining in the pickle for 3 to 5 days. On removal, they are soaked in fresh water for 5 to 6 hours and strung on hardwood sticks, the stick entering through the left gill-opening and coming out at the mouth. They are then rinsed, drained and dried for a short while. They are then suspended in the smokehouse, where they are exposed to a smoldering fire of hardwood and sawdust for 3 to 4 days. After cooling, they are ready for sale.

Shad

In the Chesapeake region and at various points along the coast, small quantities of shad are smoked, usually in precisely the same manner as already described for river herring or alewives.

Catfish

Intended as a substitute for sturgeon, catfish are smoked in identically the same manner as are sturgeon. The fish as received at the smokehouse are usually beheaded and eviscerated. They are skinned and cut into small pieces, weighing about 1 or 1½ pounds each, and are pickled for 6 or 8 hours in tight barrels. This may be accomplished by rubbing the pieces with salt and placing them in the barrel either with dry salt scattered among them, or simply by placing them in the barrel with dry salt or with strong brine. On removal from the brine, the pieces are rinsed by dipping in fresh water, to remove slime, surplus salt, etc. They are then attached to the smoke-sticks and drained for an hour or so, then placed in the smokehouse, where they are smoked for 7 or 8 hours in the same manner as are sturgeon. One hundred pounds of dressed catfish yield from 65 to 70 pounds of smoked, and the product sells usually at about 15 or 16 cents per pound. The total annual product of smoked catfish in the United States does not exceed 50,000 pounds, and its sale is confined principally to those who are willing to accept a substitute because of its cheaper price.

At several points in the Mississippi Valley, the small catfish are smoked whole, like lake herring. They are split to the vent and eviscerated, the head and, in some instances, the skin being left on, struck with salt in tight barrels and smoked for a few hours in the manner described for lake herring.

Eels

Generally the eels are received at the smokehouse fresh, directly from the fisheries, but some are received frozen from cold storage. In the latter case they are thawed by immersing them in water a few hours or by exposure to the open air. Some smokers slime the eels with salt, that is, they rub the skin with a small quantity of fine salt to remove the slime. In dressing, the fish are split from the head to the vent and the viscera removed. It is desirable to continue the splitting down to the end of the tail sufficiently deep to remove the large vein along the backbone, but sometimes this may be pulled out without splitting the fish more than inch or two beyond the vent. Few smokers, however, give attention to this item. The eels are immersed in strong brine from 1¾ to 7½ hours, according to strength of brine, size of fish, and the desired flavor. This brine should be quite strong, about 20 pounds of Liverpool or other good salt being required for each 100 pounds of fish.

In New York the eels are usually pickled for 2 hours, whereas on the Great Lakes the length of time is generally about 7 hours. On removal, the fish are washed, bristle brushes being used by some smokers, whereas others simply dip the fish in water to remove the slime and surplus salt. A few smokers throw them in a tub of water and beat them with a net for several minutes to accomplish the same purpose. The eels are next strung on iron or steel rods 1/3 inch in diameter. The rod passes through the head of each eel, or through the throat cartilage and comes out of the mouth. They are then hung in the open air for a few hours to dry. But if the atmosphere is moist, or the saving of time necessary, they may be placed in the smokehouse at once.

In New York, where small brick ovens are used, the fish are subjected to a mild smoke for about 4 or 5 hours until they have acquired the proper color. Then the fires are gradually increased and the fish are hot smoked or cooked for 30 or 40 minutes. At Buffalo and some of the other Great Lakes ports, the smoking is usually continued at an even temperature throughout for 6 or 8 hours. Mahogany or cedar sawdust is used in New York for making the smoke, and hickory or white oak wood is used for cooking, the latter being preferred. In Washington, the eels are suspended in the hogshead smokehouses over a fire made of oak and hickory wood and dried for 20 minutes, when the hogshead is covered with sacking. In this way they are hot smoked for 3 or 4 hours, the fires being sprinkled with water from time to time to produce a hot vapor. The smoking must be carefully attended, for if the heat becomes too great the fish will curl up out of shape. A good test to determine whether the cooking is sufficient, is the ease with which the skin may be separated, or peeled, from the flesh when the eel has been split.

The decrease in weight by dressing and smoking is about 35%, 100 pounds of eels yielding 65 to 75 pounds of smoked. When eels have been pickled 6 or 8 hours they ordinarily keep 10 or 12 days; but when the salting has been only 2 hours, as is usual in New York, they are apt to mold after 5 or 6 days. Smoked eels keep a shorter length of time than almost any other smoked fish.

Eels are sometimes skinned before being smoked, the process being the same as described above, except that less salting and smoking is required, and it is very difficult to keep them from falling down off the rods in the smokehouse.

Salting (Including Corning) River Herring

The fish are usually taken from the boats on the day on which they are caught, but in some cases not until the third or fourth day. All handling of the fish is with scoop nets. When taken from the boats, they are spread upon the wharf for cutting. Sitting on a low inclined seat with his knees on the wharf, the cutter removes the head and belly and scrapes out the roe and viscera. The cut fish are placed in a basket, and the roe in a bucket. The fish are then dumped into the washing vats. These are 12 feet long by 6 feet wide by 3 feet deep of two-inch pine. In some, the bottom is inclined about 30° to one side, with a horizontal false bottom of slats above the incline. Scales, dirt and other washings settle down in the deep angle of the bottom and are drawn off with the wash water through two flood gates, without loss of time. Others still employ flat-bottomed vats with resultant loss of time in cleaning.

The fish are agitated in the vats, which are kept filled with water, for about 10 minutes to thoroughly wash them and then scooped out with dip nets into slat cars holding about 1200 fish. Here the fish drain as they are transported to the salting vats. The latter, built of two-inch Virginia pine, are 10 feet long by 6 feet wide and 24 to 30 inches deep. The salting vats contain saturated brine to a depth of 4 inches. As each car of fish is dumped into the brine, additional salt is added, the amount depending upon conditions of temperature of fish, etc. When full, the vats contain from 12,000 to 15,000 fish (about 4000 pounds). The fish should be roused once each day while striking. After each rousing, the fish are tamped down lightly and top dressed with a thin layer of salt.

Corning

Early in the season, most of the packers in the lower Potomac corn their herring for immediate consumption. This method is usually followed for about 6 to 10 days from April 1. The earliest caught fish are kept in the brine from 12 to 48 hours, according to temperature. Fish brined 12 hours when the temperature is from 40 to 50° F should keep for 10 days. After brining, the fish are taken from the vats and spread on the floor, covered with salt, and the salt and fish thoroughly mixed, after which they are packed in sugar barrels and immediately shipped to the trade. No fish are corned after the temperature rises above 60° F.

Hard Cure or Tight Pack

Herring intended for storage are kept in the brine for 7 to 10 days, according to temperature. At temperatures from 50 to 60° F, 9 to 10 days is sufficient; at from 60 to 70° F, 7 to 9 days will cure them satisfactorily. After the fish are cured, they are taken from the brine and piled on the draining floor to a depth of from 1 to 4 feet, according to available space, and allowed to remain there from 4 to 10 days, according to the demand for the space. The fish are then weighed or counted (weighing is most accurate) and packed in the barrels, the first layer backs down, the balance backs up, with from 2 to 2½ pounds of salt to the layer. A properly packed barrel should contain 160 pounds of fish and 40 pounds of salt.

Salted Fish

Considerable trouble has been experienced in salting fish in warm climates. The methods followed commercially in other regions have not produced a product of good quality, and the directions generally given for salting small quantities, or for the home curing of fish, have not always proved satisfactory.

If attempts are made to preserve fish by pickling, or curing in brine, in a warm climate, the product will either turn rusty and sour, spoiling in a short time, or, if the quality is good at first, the fish will soon deteriorate. The best method for curing fish in this region is dry salting. That is, a combination of salting and drying. If the fish are handled carefully, and the directions given below closely followed, a high-quality product that will not spoil nearly as rapidly as the salted fish now prepared can be produced. But if instructions are not followed, it is useless to expect much.

In the first place, the fish must be absolutely fresh. Do not try to save a fish that may be stale, by salting. The fish should be bled, when caught, to drain out all blood as well as possible. Blood decomposes much more easily and quickly than flesh. Fish will keep longer if blood is not diffused through the flesh. They should be thoroughly cleaned as soon as possible. Fish should not be handled roughly in taking it out of the net or while in the boat. If fish are piled in heaps, walked on, or forked roughly, they will be of inferior quality and spoil much more readily than they would otherwise. Fish should not be left under the direct rays of the sun in an open boat. A tarpaulin should be rigged above the fish.

Mullet and Spanish mackerel are among the best fish for dry salting, for many reasons, a few of which are: they are split more easily and the loss of weight is less in splitting and cleaning; they are two of the commonest southern fish, and obtained more easily and cheaply. Using this outline as a guide, however, many other varieties of fish, such as grouper, sheepshead, alewives or river herring, spot, croaker, and drum, may be cured successfully, with the resultant product of good quality.

Most fish should be split along the back, just above the backbone on which no flesh should be left. The fish are split, mackerel style. That is, they must lie flat in a single piece, containing the backbone. When the knife is drawn toward the tail it must not go clear through the skin, so that the fish will be in two pieces near the tail. The head may or may not be removed. In splitting Spanish mackerel and other fat fish, the backbone is cut out nearly to the tail, where it is broken off. In cleaning, remove all traces of blood from under the backbone and clear away all the black skin. A wire brush should be used for the blood. *Black skin* is best wiped out by a piece of canvas or gunny sack. If the head is left on, clean out all traces of gills. All cleaning must be done thoroughly and carefully.

When the mullet or mackerel are cleaned they should be rinsed, then dropped in a tub of light salt brine (2 pounds of salt to 5 gallons of water). The fish should be left here to soak 30 minutes. The principal object of brining is to remove traces of blood from the cut flesh. It also cuts slime, and is better for washing than water. Never use sea water from around a fish house, dock, or near shore. It is invariably contaminated and increases the possibility of spoilage.

Score with a knife under the backbone and then longitudinally through the flesh on the other side. After the fish have soaked 30 minutes take them out, making sure that each one is properly cleaned. Drain them for 15 minutes. If salted at once the excess moisture will require more salt.

Use a very finely ground mined salt. Ordinary sea salt is more apt to cause reddening. Coarse salt is not as good as fine salt. Pour the salt into a shallow box about 2 feet square. Dredge each fish in this salt, rolling it about 2 or 3 times and rubbing salt into the slashes. Pick it up with as much salt as will stick to it. Scatter a thin layer of salt on the bottom of the tub or box used for salting. Then lay in the fish in an even layer, flesh side up. Be sure that no two pieces of fish touch without salt between. Scatter a little salt on top. Continue this until all the fish are in salt. Each layer should be laid in at right angles to the preceding layer. The top layer should be weighted down, to keep the fish under the surface of any brine formed. The top layer should also be packed skin side up. Use about 1 part of salt to 3 of fish.

The salting shed should be light, open, airy and cool as possible. The mullet will have absorbed enough salt for curing purposes in about 36 hours. Mackerel should be in salt about 48 hours. At the end of this time take the fish out of the salt and scrub them in a brine of the same strength as that used in cleaning to remove all excess salt and dirt. No traces of salt should be visible on the surface. After draining 15 to 20 minutes, the fish are ready for the drying racks. These are frames of wood, covered with chicken wire and standing on legs 3 or 4 feet high.

The drying racks must be placed on dry ground, preferably covered with gravel. Oxidation, or rusting, sets in immediately if drying is carried on under the direct rays of the sun. If fish are kept shaded in a breezy location they will dry well with a clear color. For this reason, drying is best done in the shade under a roof without walls, so located that as much of a current of air as possible will pass over the fish. The fish are laid out skin side down but are turned 3 or 4 times the first day.

The fish are gathered up and placed under shelter at night to prevent spoilage through dampness. If left spread out in the open at night, they will sour and mold. The time required for drying depends on weather conditions during the drying period, and on the size of the fish being cured. The exact time must be determined by the person curing the fish. For mullet it should average about 4 days; for Spanish mackerel, 5 days. The more the fish are dried, the less danger there will be of reddening or rusting. When the surface looks dry and hard, and if the thumb can be pressed into the thick part of the flesh leaving no impression, the flesh can be considered as cured.

In weather in which air-drying is impossible, or in climates too humid for this process, the following method may be used. When the fish are struck through, or have absorbed enough salt for curing purposes, they should be taken out of salt, scrubbed off in brine, then piled in stacks, flesh side down. These stacks should be heavily weighted down in order to press moisture out of the fish. After 10 to 18 hours in the stack, the fish should be repacked in dry salt with the top weighted down, and put in storage in a cool dry place.

Store the fish in wooden boxes lined with waxed paper. Scatter a little dry salt between each layer of fish, (about 1 pound of salt to 10 pounds of fish). Store in as cool and dry a place as possible. If signs of rust or mold appear, scrub the fish off in brine and dry in the air for a day or two.

Reddening of salted fish is a form of bacterial spoilage caused by the salt used in curing. Contrary to popular belief, salt is not strictly an antiseptic, and certain types of bacteria live and thrive in a salt medium. Salt most apt to be contaminated is that obtained by evaporation of sea water. Several types of salt used extensively in fish curing are apt to be so contaminated. In salting fish, every effort should be made to use a salt as pure and high in grade as possible. It is advisable to heat salt and bake it thoroughly before using. If, however, reddening appears at any time, all tables and other equipment used in salting should be thoroughly disinfected. Unless every effort is made to keep the salting equipment clean, the use of sterilized salt or other precautions will be useless, as the fish can be contaminated through unclean equipment. After curing, the fish should be stored in the coolest place possible, as the salt reddening bacteria grows best at a warm temperature. At first signs of reddening the fish should be removed, washed thoroughly in pure salt brine, and given a few hours' careful drying and repacked with a thin layer of dry salt between each layer of fish, using from 10 to 15 pounds of salt to 100 pounds of fish. Reddening is most apt to appear in fish stored in pickle (brine) and held in a warm place. It will remain in good condition longer if packed in dry salt and held in as cool a storeroom as possible.

Canning Alewives or River Herring Roe and Buckroe

The following method of canning alewives has proved quite satisfactory. The fish are cut, washed, and placed in the salting vats in the same manner as for salt curing. After 12 to 14 hours they are removed from the vats and washed in an abundance of lukewarm fresh water. During the washing they are trimmed, the balance of the fins and scales being removed. They are then cut to can size and placed in the cans, after which they are processed for 55 minutes at 244° F for No. 1 cans and 60 minutes for No. 2 cans.

Herring roe intended for canning is collected in buckets as the fish are cut and washed in fresh water in special trays, blood and adhering particles of entrails being removed. The roe is then put in the cans. As it swells considerably in processing, the cans must not be entirely filled. If of the sanitary type, the cans are filled to within about ¾ inch of the top with roe and then filled to the edge with cold salt brine. About 1 pound of salt to 8 or 10 gallons of water is used to make the brine. The brine is added solely for seasoning. The cans are capped immediately and placed in the processing baskets. If solder-top cans are used, the filled cans are placed in the exhaust box. Upon removal from the exhaust, the necessary air space is provided-for by pressing the roe down with a plunger. Material clinging to the groove where the solder is to be applied is removed with a brush, and the cans are capped and tipped. The canned roe is processed in a closed kettle for 45 to 55 minutes at a temperature of 240 to 245° F. The milt roe may be canned in the same manner as the roe except that the cans may be more completely filled, as this product does not swell in the processing. As the quantity of brine used in this case will be somewhat less, it should be made correspondingly stronger.

Note: In canning, the fish should be drained of superfluous water before they are placed in the cans, and no water added to can contents. That the fish may retain their shape in the can and stand transportation, the cans should be well filled. The shrinkage of the fish in processing must be taken account of in filling the cans.

Canning Clams (Alaska)

The first operation is the removal of the clams from the shells. This is done by immersing them in boiling water, either in vats especially designed to receive the wire baskets in which the clams are placed, or the clams are passed through the water on an endless belt. After remaining in the water several minutes, they are thrown on a table and the shells fall away from the meat. The clams are then passed on to workers, who open the stomachs and necks, remove the sand and sediment, and sever the black part of the neck. The cleansing process is continued by placing the meat in a cylindrical perforated washing machine, which revolves automatically half a turn both ways in a tank filled with water. Any sediment that may have remained after the hand operations were completed is removed in this way. The clams are now ready to be canned and are taken directly to the filling tables if whole clams are packed, or to the grinder if the minced variety is desired. The cans are filled by hand with both meat and juice, after which they pass through the topping and sealing machines and are sealed. The process is completed by cooking the canned product in retorts at a temperature of about 245° F from 1 to 1½ hours, depending upon the size of the container used. The juice which is thrown off in the process is used in preparing the finished product, the surplus being sealed in cans.

Anchovy Paste from Sprats

Anchovy paste from sprats may be made as follows. Two pounds of common salt, 3 ounces of bay salt, 1 pound of saltpeter, 2 ounces of prunella and a few grains of cochineal are pounded well together in a mortar. (These quantities are sufficient for a peck of sprats.) Into a stone jar place first a layer of fish, then a layer of the pounded ingredients, and so on until the jar is filled; press them down hard and cover closely. After 6 months they will be ready for use.

Note: Persons using such preservatives as saltpeter should consult the Bureau of Chemistry, Washington, D. C., to determine whether they are using an amount in excess of that held to be proper under existing law.

Anchovy Butter

Take 1 part of anchovies which have been beaten to a paste, and pass through a sieve; add 2 parts of butter, and spice to suit. Cayenne pepper or paprika may be used to advantage.

Anchovy Essence

Anchovy essence can be made with either canned or bottled anchovies. Take the fish, and rub to a pulp in a mortar, and then pass through a fine sieve. To ¼ pound of anchovies add ¼ pound of water; boil 15 minutes, and strain; then add ½ ounce of salt and ½ ounce of flour, and the pulped anchovies. The mixture is allowed to simmer over the fire for 3 or 4 minutes. After the preparation is cool, add 2 ounces of strong vinegar. The product should be bottled in small bottles and tightly corked and covered with bottle wax.

Anchovy Paste

Take 1 pound of anchovies, 1 pound of water, 2¼ ounces of salt and 2¼ ounces of flour. Add a small quantity of cayenne pepper (say 1/10 ounce), a small quantity of grated lemon peel, and ½ ounce of mushroom catsup.*

Anchovy Sauce

Take 3 or 4 anchovies, and chop them fine; add 3 ounces of butter, 2 ounces of water, 1 ounce of vinegar, and 1 ounce of flour. Melt the butter over a water bath, add the water and the vinegar, and finally the flour and the anchovies. Stir until the mixture is thick, then rub through a wire sieve. This preparation should be kept on ice. It will not keep indefinitely.

Christiana Anchovies

In the preparation of Christiana anchovies many methods and flavoring ingredients are used, depending on the skill and ideas of the curer and the markets for which the preparation is intended. The following is one of the most popular processes.

The fresh sprat or anchovies are immersed in brine for 12 or 18 hours, 15 pounds of Liverpool salt being used for each 100 pounds of fish. On removal, the fish are drained in a sieve and then packed loosely in a barrel, with the following ingredients, which have previously been finely crushed and well mixed: 4 pounds of Luneburg salt, 6 units of pepper, 6 units of sugar, 6 units of English spices, 1 unit of cloves, 1 unit of nutmeg, and 1 unit of Spanish pepper. The anchovies remain saturated with these ingredients for 2 weeks, when they are repacked tightly in kegs or barrels, being carefully arranged in layers, with the backs downward. A quantity of the above mentioned ingredients is sprinkled over each layer, with the addition of a few cut bay leaves or cherry leaves. At the bottom and the top of the package are placed two whole bay leaves, but, before the top leaves are laid on, brine is poured over the fish. The barrels or

*** Mushroom Catsup**

Upon a suitable quantity of the fresh mushrooms, about 1 to 4, sprinkle salt, and after 3 days squeeze out the juice. To every gallon of juice, add black pepper, ginger and cloves, ½ ounce of each, 2 ounces pimento, 2 ounces mustard seed, and a sufficient quantity of salt. Boil for 5 minutes and set aside to settle. Strain after 7 days.

kegs are then coopered and rotated daily for the first few days, and after that every other day for 2 or 3 weeks.

The following process is also used to some extent.

The fish are salted for 24 hours and then immersed in sweetened water, 20 parts of water to 1 part of sugar being used. The fish are then packed with a mixture of Luneburg salt with 90 units or parts of allspice, 60 units of pulverized sugar, 19 units of whole peppers, 15 units of cloves, an equal quantity of nutmeg or mace and of hops (Origanum creticum), and some bay leaves.

The following is a choice method of preparing *Matjeshering* in Germany.

Fresh full herring, both spawners and milters, are well washed, and the gills, stomach, and intestines are removed in such a way as not to necessitate cutting the throat or abdomen. This is accomplished by pulling them through the gill flap. The fish are next immersed for 12 or 18 hours in a 7% solution of white-wine vinegar, from which they must be removed before the skin becomes flabby. They must be wiped dry and covered with a preparation composed of 2 pounds of salt, 1 pound of powdered sugar, this quantity being sufficient for 75 herring. The fish are then packed in a barrel, which is sealed. When there is not sufficient brine to fill the barrel, more should be made of 1 part of the above mixture and 4 parts of water which has been boiled.

Spiced herring, *Gewürzhering,* are prepared in Germany in the manner above described, with the addition of spices mixed with the salt. The spices commonly used consist of 1 part of Spanish pepper, 5 parts of white pepper, 4 parts of cloves, 2½ parts of ginger, an equal quantity of mustard, and a particle of mace and of Spanish marjoram, with a few bay leaves scattered between the layers.

Fried Foods

Fried noodles, pies, doughnuts, crullers, potato chips, etc. are becoming increasingly popular. The frying of foods in fat is an art which can be acquired after a little experience. Properly fried foods have a special delicate flavor which can be obtained when they are manufactured under the strictest regulation of temperature and time control. The success of fried foods depends primarily on the use of proper equipment, appropriate fat, the correct methods of preparation and the care taken of the fat frying equipment.

The quality of fried foods depends on the quality of the fat, which should meet requirements, such as having a low absorption value, i.e., the fat should be able to permeate and fry food easily, a high smoking point, a long frying life. It should leave the food with a dry, non-greasy surface and impart good flavor. It must be digestible, stand up against rancidity and not give off disagreeable fumes under normal frying conditions.

In the frying of food products, fresh fat and oil are continually added to replace the absorbed fat. If the consumption of frying fat is rapid, the level of the fat must not be allowed to get too low. All frying fats eventually break down when used at usual frying temperatures. The most stable frying fats break down the least. When a large amount of frying fat is present in the kettle, an excess of frying fat breaks down at the frying temperature. It, therefore, follows that, when the amount of food that has to be fried is small in proportion to the total fat in the kettle, there is likely to be more than the usual breakdown of fat. To obtain the best results in frying foods, a kettle of appropriate size (not too large) and the smallest quantity of fat possible, should be used. As the fat is being used up, burnt food particles collect, which, if not removed, may cause smoking, and will shorten the frying life of the fat. A procedure for removing burnt food particles is to strain the fat through several layers of cheese cloth. If this method is unsatisfactory, the fat can be conditioned by drawing it off into a separate receptacle and cooling it to 212° F, after which the addition of hot water drags the burnt particles down to the bottom of the receptacle. The water and sediment can then be drawn off by means of a spigot on the bottom of the container. The proper time to replace all of the frying fat is determined by tasting the fried food daily; if the product has a noticeable flavor of cooked fat, the fat should be discarded.

Potato Chips

The manufacture of potato chips offers an easily sellable product. The best results are obtained from second crop potatoes grown in a dry, sandy soil, although Colorado potatoes are also very good. Regardless of variety, however, it is very important to pre-determine the proper soaking time and frying temperature. The preparation of the potato for frying involves peeling, by hand or machine, cutting out the eyes and slicing by machine into a tank of cool water. The chips are allowed to soak for ½ to 2 hours, the exact time to be determined from experience with a particular type of potato. The sliced potatoes are then removed from the water, allowed to drain for a few minutes and then fried.

A satisfactory fried chip is obtained by frying at a temperature of 200 to 230° F. The chips are usually lowered into the frying oil by means of wire baskets. One pound of sliced chips to every 25 pounds of fat roughly represents the average quantity of chips for one batch. The average frying time is from 3 to 5 minutes, with allowance for the degree of browning desired. After frying, the chips must drain thoroughly, the wire basket being hung either above the frying kettle or removed to a metal draining shelf. The chips should be kept warm during the draining procedure. Salt is then applied at any time after the chips are free from excess fat, the amount added depending upon individual taste. The chips are then thoroughly cooled before packaging, to prevent sweating and sogginess in the containers.

Nuts, Fried and Salted

The popular demand for fried salted nuts in this country is steadily increasing. Nuts are important from the standpoint of nutrition, since they possess concentrated energy in the form of protein and fats. Only good sound nuts should be used for frying. The most desirable nut is one that has had a chance to dry out, since high moisture may result in toughness, whereas low moisture renders a crisp tender product. The storage temperature of nuts until ready for frying should be from 50 to 70° F.

Nut meats are specially prepared and can be purchased ready for frying. Peanuts may be fried with or without the skins, while almonds and Brazil nuts are always fried without their skins. The same equipment can be used for frying nuts as for frying other foods, with the exception that the wire basket must be of a finer mesh. The determination of correct frying temperature for nut meats depends on the type of nut, the size and the price obtainable for the finished product. Fat absorption in the nut meat should be held to a minimum from the standpoint of cost. Small nuts, such as unblanched Spanish peanuts, should be fried at high temperatures, whereas larger nuts should be fried at lower temperatures, resulting in a more uniform color as well as increased yield.

The frying of nuts is simple, about 1 pound of nuts for every 15 pounds of fat in the frying kettle representing a normal batch.

Unblanched Spanish Peanuts

These nuts should be fried as rapidly as possible at 360 to 380° F to a cream color. The frying time is 16 to 18 minutes. The yield is 98 to 99%.

Blanched Spanish Nuts

The nuts should be continually stirred at a frying temperature of 350° F, until the desired color is obtained, with a yield of 98 to 99%.

Blanched Jumbo Virginia Peanuts

To obtain a uniform color, a frying temperature of 300° F is usually about right. The time required should be around 12 minutes, with a yield of 100%.

Unblanched Jumbo Virginia Redskin Nuts

These nuts are fried at 300° F for 20 to 25 minutes. This should give a product of an even cream color, with a yield of 101 to 103%.

Cashew Nuts

The cashew is a very delicate nut and since it colors rapidly, should be fried very slowly. The frying should be started at 200° F, with a gradually increasing temperature to 275° F. Uniform results are obtained by frying for approximately 35 to 40 minutes. To make sure that the cashews are fried throughout, the temperature should be increased slowly at the start. The yield is 100 to 106%.

Pecans

The pecan is a very delicate nut and is one of the most difficult to fry. It has a high natural oil content and its softness causes shriveling and darkening if the temperature of the frying fat is not properly regulated; if it is underfried, the nut is soft and lacking in crispness. A frying temperature of 175° F with a gradual increase to 240° F is most desirable, for a period of 15 to 18 minutes. The yield should be 112 to 114%.

Almonds

This nut meat should be fried slowly with continual stirring and must not be allowed to color very deeply. The starting temperature of 160° F with a gradual increase to 240° F, should prove satisfactory, if applied for approximately 25 minutes. Before the almonds are fried, they should be soaked in water, blanched and then dried in a heated cabinet with careful stirring. Overheating must be avoided or premature rancidity may result.

After frying, nut meats should be cooled as rapidly as possible. They should be spread in thin layers on appropriate pans and cooled, by blowing cool air over and under them by means of an electric fan or other methods. The nuts should be salted with a very fine salt while warm enough to cause even adherence of the salt particles, two pounds of salt usually being required for salting 100 pounds of fried nuts. The nuts should be strained with a 10% gum arabic solution in water before salting.

Nut meats fried in fat with a high melting point have a drier surface than those fried in liquid oils, while nut meats subjected to cold temperatures may become dull or lack gloss. These nuts may be reconditioned quickly by rapidly frying, cooling and then sprinkling them with a liquid oil, before salting while the nuts are still warm. One to 2 pounds of liquid oil per 100 pounds of nut meat will bring about a new gloss.

The development of foam in frying oil baths after several days of frying is the result of the impurities coming out of the nut meats. These impurities are almost gelatinous in nature, representing gums, resins, etc. These contaminating ingredients retard normal filtration since they quickly clog straining cloths, small filters, etc. It is, therefore, suggested that the impurities be removed from the frying oil at the end of each day, to clear the surface of the oil. Foaming does not always mean that the fat has broken down. Removal of these impurities in most cases eliminates foaming and the fat can then be used for further frying.

Fried Soybeans

The soybean has, for centuries, been an important source of food in the Far East and is becoming increasingly important in this country. Production has expanded tremendously and considerable research work is constantly being done on the soybean. The fried soybean is a pleasing nut-like confection, which will prove a product of sufficient interest to many lovers of nut meats. There are many varieties of beans, some of which are better than others for fried food purposes. Experience has shown that the Illiné variety is preferable, although other varieties can be fried satisfactorily, e.g., Rohusum, Mammoth Yellow, Dixi, Easy Cook, and Hohto. By writing to the National Soybean Oil Manufacturers Association, Box 388, Taylorville, Illinois, all information on the bean and sources of supply can be acquired.

Before frying the soybean, it should be washed in water to remove particles of dirt that may be adhering. In order to obtain a good product with a nut-like texture, the bean is allowed to swell by soaking in boiling water for a period of 10 minutes. The water is drained off and the beans are cooled and fried. The beans are then placed in a proper wire basket, with 5 pounds of fat, previously heated to 370 to 380° F. The addition of the bean to the fat will cause a drop in temperature to about 340 to 355° F, at which point most of the frying takes place. Temperature control is important since frying at high temperatures produces a bean with a scorched flavor, while frying at too low temperatures produces a product with a beany instead of a nutty flavor. After frying, the beans are drained, salted and cooled in the same manner as other fried nut meats.

Fried Noodles

There are a number of methods for making fried noodles. One is to roll out the noodle dough very thin, roll up the dough to prevent drying out, cutting it into thin slices and frying at a temperature of 365 to 375° F. Noodles made this way are very superior and delicious for serving with chop suey, chow mein and other fine food dishes requiring a crisp brown noodle.

A second method for making fried noodles is to take the dried noodle and boil it in water until soft and tender. These noodles are then fried at a temperature of 365 to 375° F. They have a distinctly different flavor from the ones made by the first method. These noodles, after frying, should be salted, cooled and served with cooling drinks and salads. Fried noodle products are perishable, being salable over a 3 to 5 day period, or for a longer time, if properly packaged.

Fried Noodles

Flour	5 lb	Eggs	1 lb
Water	1 lb	Salt	2 oz

Mix the above ingredients together to form a smooth dough, which is then rolled out, partially dried and then rolled up and cut into thin slices. The cut noodles are allowed to dry and then fried.

Cheese Sticks

These newly developed fried products have a pleasant taste, are appetizing and make ideal additions to salads, drinks and lunches. The method for making the sticks is to roll out the dough to a thickness of ¼ inch, cut into strips, roll into sticks and fry at a temperature of 370 to 380° F. The sticks are then removed, allowed to drain and salted; cheese sticks can also be baked if desired.

Cheese Sticks

Flour	1 lb	Pepper	1/16 oz
Grated Cheese	1¼ lb	Eggs	6 oz
Baking Powder	½ oz	Paprika	1/16 oz
Salt	¼ oz		

The above ingredients are mixed together with a sufficient quantity of milk to make a stiff batter.

Health Foods

Natural, vitamin-laden, health-building foods and drinks are essential for the building of healthy body tissue. The best diets are those containing plenty of fresh, crisp, green vegetables, fresh berries, fruits or their juices, honey, grapefruit, marmalade, raspberry and strawberry jam, sun-dried bananas, nuts and cheeses.

Nuts and soybeans are good substitutes for meats, fish, eggs and sea foods. Pistachio nuts are non-acid-forming and are rich in organic salts and vitamins. They are also rich in easily digested fats and carbohydrates. Their high percentage of alkaline minerals makes them an alkaline-yielding food. Next to the pistachio nut, both in nutritional value and non-acid-forming properties, is the coconut. While not very high in total mineral content, it is rich in alkaline minerals, and its carbohydrates are easily digested. When finely ground, its fibrous cellulose supplies very good bulk. In its fresh state, both its water and its soft meat are delicious and nourishing. Properly ripened in its fresh state, it has vitamins A and B. The next nut in order of nutritional value and non-acid-forming properties is the cashew nut, which is rich in proteins and low in carbohydrates. Next in order of the least acidity is the pecan nut, which is also rich in mineral salts. Next to the avocado pear, it is the best kind of emulsified fat to eat.

The sweet almond, of which there are many varieties, is rich in vitamin A and fairly rich in vitamin B. Although it is fairly rich in alkaline minerals, though relatively low in potassium, on account of its excessively high phosphorus content which is the highest known in the vegetable kingdom, according to latest authoritative analyses, it is decidedly acid in reaction. In making butter or meal, the skin should never be used. Blanching is very easily done by throwing the almonds into hot water, and after a minute the skin separates. The almonds should then be removed and the skin wiped off. The meal is rich in protein, sugar, and minerals. The bitter almond is 10% higher in average protein content and 10% lower in fat content, than the sweet almond. But because of the high content of amygdalin, a poisonous glucoside, bitter almonds are not edible, because of danger of hydrocyanic acid poisoning. Although amygdalin is present in the seeds of the apricot, cherry, prune, peach, etc., it is especially high in the bitter almond, varying from 2.5 to 3.5%. It is split up by various reagents and by enzymes in the seed into benzaldehyde, glucose and hydrocyanic acid. Benzaldehyde is the chief constituent of the essential oil of bitter almonds, prepared by distillation from the seed.

The oil of sweet almond is sweet and bland and is nourishing. It is often adulterated with similar oils from peach, apricot or plum kernels, which so resemble it as to render detection difficult or uncertain.

The next nut in order of its nutritional value and acid-forming properties is the paradise (sapucaia) nut, a native of Brazil. In appearance and general structure it very much resembles the

Brazil, or Para nut, but it is higher in protein and fat, and lower in mineral content. It is considered superior to the Brazil. The Brazil nut comes next in order in the scale of nuts with respect to their nutritional value and acid-forming properties. However, both are acid-forming, but both are rich in vitamin B and have some A and C.

Next comes the filbert, a native of Italy, which is the best of the hazelnut genus. It is also grown in England and the United States, where it is known as the hazelnut, but this is not as good as the native Italian filbert. Although the protein content is a little lower than that of the Brazil, it is a more complete protein. In all other respects, the filbert is inferior to the Brazil. It is also less acid-forming than the Brazil and has the same vitamins, A, B, and C, but it is not as digestible as the Brazil.

Next in order comes the hickory nut. Although its exact chemical properties have not been established in detail, it is considered superior to butternut on account of its lower and better quality protein content. It is lower in mineral content than the butternut. It also has vitamins A and B, in which it is very rich.

Next in order is butternut. As stated, it is richer in minerals than the hickory nut but both are acid-forming on account of their high acid-forming minerals. Butternut is especially recommended for those who want to avoid carbohydrates, as it is lowest of all nuts in carbohydrates, but in fat it is nearly as high as the hickory. Its vitamin content is unknown.

Next in order comes the walnut, which really belongs to the same genus as hickory and pecan, but, on account of its inferior quality and quantitatively higher protein content it ranks much lower than pecan and hickory. It is also lower in mineral content than the hickory and is more acid-forming. It is high in fat and starch content and has a small amount of vitamin B with traces of vitamin A. Of its many varieties, the black is the best, because of its lower carbohydrate and considerably lower fat content. Next in order comes the Japanese walnut, and the English walnut, grown in the United States ranks last. The analyses available indicate that black walnut, Japanese walnut, and butternut are characterized by relatively high protein and low oil; English walnut, hickory and pecan, by relatively low protein and high fat content. Of the three, the pecan is highest in fat and lowest in protein.

Pine Nuts

The various species of the pine family nuts differ greatly in their composition. The oil content usually exceeds that of all other constituents and is often several times the protein content. Starch is a minor constituent, at least in the species growing in Europe and the United States. Analyses of the Brazilian pine nut are not available. The European and Brazilian nuts are longer than those growing in the United States, which have the further disadvantage of being thick-shelled. The European varieties are, therefore, more popular and are used in this country, where they are known as pignolia. Italy and Spain are the chief sources of supply.

The pine nut or pignolia is the highest of all nuts in protein. The only other vegetable protein that excels it is the soybean. It ranks low in carbohydrates, and its mineral content is highest of all nuts. The Italian varieties, in this respect, rank above the pistachio. On account of its high protein and phosphorus content, it is not alkaline. It has some vitamin B. The protein content of pine nuts is incomplete. This distinguishes them from genuine nuts.

Beechnut ranks high as a fat-containing food. It is quite high in protein and minerals. It has a very high phosphorus content, which makes it an acid-forming food.

The oil, which is expressed on a small scale in Europe, is used for foods and lighting. Beechnut cake is a distinctly European product, as the American nut is not gathered on a commercial scale. The nut is pressed, either whole or after shelling, and the cake utilized as a cattle food. The cake has been said to be injurious to horses. No evidence is available that the nut kernel is injurious to human beings. A toxic substance, fagin, was claimed to be an ingredient of it. Later analyses isolated cholin and showed it to be identical with fagin.

Peanuts, erroneously regarded as nuts, are legumes and belong to the same family as soybeans, etc. Their protein content is high. They are also high in fat and carbohydrates, as well as in mineral salts, of which phosphorus predominates and excels potassium, which is their highest alkaline element. The next highest alkaline element is magnesium, which excels sulfur. On account of their incomplete protein, partly digestible fat, high purine content, as well as phosphorus and sulfur, they are a high acid food. They have some vitamin A and a considerable vitamin B content. Originally a native of Brazil, the peanut is now grown throughout the tropics and warmer regions of the temperate zone. Large-seeded peanuts, commonly called the Chinese variety, are grown in Virginia and Georgia in large quantities. The small-seeded Spanish and African varieties are superior to the others.

Peanut butter is manufactured in immense quantities in the United States, but it is usually made of roasted and salted peanuts. The butter made of the Spanish peanut is superior in flavor as well as in nutritional value.

Soybean

The soybean is justifiably called the king of the legumes, not only because it is lowest in starch, but also because it is the most complete and best protein for both growing children and adults. It is unusually high in mineral content, especially in potassium, calcium and magnesium. It also contains some sodium, iron, manganese and copper. Although it has a high phosphorus content it is an alkaline food. It is also rich in vitamin B, and has a considerable vitamin A content. Its fat content is quantitatively high and of good quality. Starch may be found in unripe seeds, but does not occur in the mature seeds of most of the varieties. It is also rich in enzymes and coenzymes, especially in its germinated state.

A native of the Far East, the soybean has been cultivated since the dawn of civilization in China and Japan, where the seeds furnish millions of human beings with food. From the seeds are prepared soy cheeses (Tofu, Nattomiso), soy milk (also butter, spread, loaf, and bologna), and soy sauce, the latter being used in chop suey. Because of the absence of starch in many varieties, soybean flour has come into use in the Occident as a diabetic food. Soybean oil is of growing industrial importance.

Chinese Bean (Choy Sprouts)

This is a tiny, round, green, pea-shaped bean high in carbohydrates. It is also very rich in alkaline minerals and, therefore, is a non-acid-forming food. Its vitamin content is unknown. It requires no soaking and little cooking.

Chick Pea (Garbanza Bean)

This comes next in order of the legumes. Although rather high in carbohydrates, it is relatively low in protein. It is exceptionally high in magnesium, fairly high in potassium and has some calcium, iron and sodium. Although also high in phosphorus, it is regarded as a non-acid-forming food. It has a considerable vitamin B content and some vitamin A. It is one of the few highly nourishing legumes and vegetables. It is best when eaten in its green state.

String Beans

Fresh string beans are rather low in protein and carbohydrates but high in potassium, calcium and magnesium. They also contain a considerable quantity of sodium and some iron. They have a much lower phosphorus content, 5 times the amount of sulfur and 6 times the amount of chlorine than soybeans, yet they are a non-acid-forming food. They are rather high in vitamin B content. When not overcooked, they are one of the most nourishing legumes.

Kidney Beans

The kidney bean in the green state is superior to that in the dried state being low in protein and carbohydrates, high in potassium, calcium, and containing considerable sodium, magnesium, and some iron. It is not very high in phosphorus but considerably higher in sulfur and chlorine than that in the dried state. On account of its high phosphorus content and lower potassium and sodium in the dried state, it is acid-forming when dry but not so in the green state. The red kidney bean is richer in minerals than is the white, and is therefore preferable to the latter. The French green kidney bean, known as the flageolet, is a better one than both in every respect.

Green Peas

These come next in order for their low protein and fat content. They are not very high in mineral content, of which the potassium content is highest, next phosphorus and sulfur, and then magnesium, calcium, sodium and some iron. They are best in the raw, tender, fresh condition, being very high in vitamin C and rich in vitamins A and B. In the dried condition vitamin C almost disappears and the other vitamins are considerably reduced, although vitamin G, or B_2, appears in considerable quantities.

Lentils

There are several varieties of lentils, of which the Egyptian is the best; next comes the Russian, with the Chilean and German last. Rather high in protein and carbohydrates, it is low in fat and high in minerals, of which potassium is highest. Then come sodium, calcium, magnesium and iron. Although its phosphorus content is high, it is not an acid-forming food as far as the Egyptian and Russian varieties go. It is also high in vitamins B_1 and A, and, in the sprouted form, in vitamin C.

Lima Bean

This bean is best in its green state. It is very high in potassium, sodium, magnesium, and has a considerable calcium content and some iron. It also has a considerable phosphorus content, sulfur and some chlorine, but, on account of its high alkaline salts and low protein, is an alkaline food. The dried lima bean has almost 3 times the amount of protein and carbohydrates, and, although its mineral proportion is relatively higher, it is lower in potassium as well as every other mineral. Vitamin content is unknown. Many commercial sources claim that the dried lima bean is alkaline. This is not true on account of its high protein and carbohydrate content. The green one is alkaline. The small, baby, lima is best.

Fava bean, of which the Canary Mexican is the best, comes next and, in some respects, is even superior to the domestic lima bean.

Horse bean is very similar to the lima bean, except that it has considerably less fat as well as alkaline minerals, and is much higher in phosphorus. Although lower in carbohydrates, it is still an acid-forming food. Vitamin content is unknown.

Acorns

These are very high in carbohydrates and fat. They are fairly high in minerals, potassium being the highest, phosphorus next. Calcium, magnesium, sulfur, and small quantities of chlorine, iron, and sodium are also present. Low in protein, acorns are of an inferior quality and, on account of their high carbohydrate content, highly acid. It has small amounts of vitamins A and B. It is fit only for animal consumption.

Chestnut belongs to the same oak family and resembles the acorn, but it has a sweet taste whereas the acorn has a bitter taste. This is due to the sugar content of the chestnut. It is higher in protein and carbohydrates and much lower in fat than the acorn. It is also a little higher in mineral content, but much higher in sulfur and therefore considerably more acid-forming. It has only a small amount of vitamin B.

Cereals

Breadfruit

This is a tropical starchy fruit which heads the list of cereals, strictly speaking. It is best when baked on hot stones. It is also edible raw when thoroughly ripe and the starch is nearly totally converted into sugar. It is rich in minerals and some vitamins.

Banana is not strictly a cereal, but being a typically starchy food, like the breadfruit, is included. When ripened, at least partly, on the stalk, it is lower in starch content and is non-acid-forming. There are many different kinds, sizes and shapes. The most delicious and richest is the large Samoan variety which is seldom exported. Those ordinarily imported are picked too green to ever ripen properly. The longer the banana remains on the stalk before cutting, the rounder it becomes. The tropically ripened sun-dried banana is preferable to those ripened here. The tropically ripened banana is naturally rich in vitamins and minerals, and is of great nutritional value. In powdered form, it is the best baby food, and is a good building food for underweight adults. (See sun-dried bananas.)

Potato

Potato, strictly speaking, is a vegetable of the bulb family but, because of its starch content, can be classed with the cereals. On account of its high alkaline mineral content, potassium particularly, it is not only a valuable and economical food but is also non-acid-forming. The most nourishing part of the potato is found near and in the skin, the part usually peeled off and thrown away. It is best when eaten in the steamed or baked form, with the skin. The many varieties of sweet potato vary in their relative nutritional properties, but they are superior to the Irish potato. This is also true of yams of some varieties.

Wild Rice (Indian, Water or Black Rice and Water Oats)

This stands next in the list of cereals for its nutritional value, especially in minerals and some vitamins, and for its relatively low starch content. The Indians parch it and eat it in that state, but when cooked it is a good dish, especially when prepared with some vegetables like onions, etc. It need not be soaked or cooked as long as ordinary rice on account of its low starch content. From 15 to 25 minutes' steaming in a waterless steamer is all that is needed. In its sun-dried, unparched, state it has even higher food value.

Barley, Whole

Whole barley is relatively low in protein and fat and high in carbohydrates. It is rich in potassium and magnesium and has a considerable quantity of calcium and sodium. Although rather high in phosphorus, it is low in the other acid-forming minerals and, therefore, the least acid-forming of the cereals named below. It contains vitamin B_1 and some vitamin A.

Buckwheat, Whole

This is only a little higher in protein, fat, carbohydrates and minerals than barley. It is a little lower in potassium, magnesium, phosphorus, sulfur and silicon.

Corn, Whole, Unbolted

This product is high in carbohydrates and fat, but relatively low in protein. It also contains phosphorus and is rich in vitamins B_1 and A. The yellow corn is best and is most nourishing.

Wheat, Whole

This product is relatively high in protein, fat, and carbohydrates. It is also relatively high in minerals, being rich in potassium and magnesium, and rather high in phosphorus. It is also rich in vitamins A, B, and E. Of the various wheats, the western hard spring wheat is the best.

Wheat Germ

This is far superior to any of the cereals mentioned because of its high mineral and vitamin content and its low starch content. It is very rich in vitamins B_1 and E and has a considerable quantity of B_2 and A. Cooking destroys nearly all of its nutritional value. It may be sprinkled on salads or other foods.

Rice, Whole, Unpeeled and Unpolished

This comes next to wheat in order as a cereal. It is rich in vitamins B and E. It is rapidly spoiled by storing.

Rice Polishings

Rice polishings stand next in order to wheat germ as superior to any cereal, particularly on account of their high vitamin B content. Like wheat germ, it should not be cooked. It may be added to wheat germ or to any other food, cooked or raw. It may be used as a thickener for soups.

Rye, Whole

This comes next in order. It is lower than wheat in protein, fat, carbohydrates, and minerals. It is rich in vitamin B and has considerable quantities of vitamins A and E.

Oats, Whole

This is the fourth highest of all cereals in proteins. Wheat, rye, and corn are higher. It is highest in fat, relatively low in carbohydrates and high in minerals, especially potassium and phosphorus. It also has considerable quantities of magnesium, calcium, and sodium, and some iron. It is highly acid-forming. It contains a considerable quantity of vitamin B, some E and G, and traces of A. It also has some zinc, manganese, copper, aluminum, and arsenic. Steel-cut oats are hard to digest, and therefore unsuitable for raw food diet.

Millet (Proso)

This belongs to the same family as the sorghum and is closely related to it, but its better varieties are somewhat superior to sorghum.

Sorghum (Indian Millet)

This also includes durra, and comes next in order. It has a small quantity of vitamin B. It is highly acid-forming.

Sugar- and Starch-Free Food Formulas

Vanilla Dessert

Formula No. 1

Skimmed Milk Powder	77 oz	Vanillin	Enough to flavor
Powdered Agar-Agar	10 oz	Saccharin	Enough to sweeten
Tartrazine Color	Trace		

Use 3½ ounces to 1 pint of hot water or milk, or a mixture of both.

Formula No. 2

Skimmed Milk Powder	70 oz	Tartrazine Color	Enough to color
Soluble Casein Powder	10 oz	Vanillin	Enough to flavor
Powdered Agar-Agar	10 oz	Saccharin	Enough to sweeten

Use 3 to 3½ ounces to 1 pint of hot water or milk, or a mixture of both.

Formula No. 3

Skimmed Milk Powder	60 oz	Tartrazine Color	Enough to color
Powdered Gum Acacia	20 oz	Vanillin	Enough to flavor
Powdered Agar-Agar	8 oz	Saccharin	Enough to sweeten

Use 3 to 3½ ounces to 1 pint of hot water or milk, or a mixture of both.

Chocolate-Flavored Desserts

Formula No. 1

Skimmed Milk Powder	65 oz	Vanillin	Enough to flavor
Cocoa Powder	20 oz	Saccharin	Enough to sweeten
Powdered Agar-Agar	15 oz		

Use 3 to 3½ ounces to 1 pint of hot water or milk, or a mixture of both.

Formula No. 2

Skimmed Milk Powder	70 oz	Powdered Agar-Agar	15 oz
Cocoa Powder	10 oz	Vanillin	Enough to flavor
Powdered Bitter Chocolate	5 oz	Saccharin	Enough to sweeten

Use 3 to 3½ ounces to 1 pint of hot water or milk, or a mixture of both.

Formula No. 3

Skimmed Milk	12 oz	Powdered Agar-Agar	12 oz
Cocoa Powder	35 oz	Vanillin	Enough to flavor
Powdered Gum Acacia	40 oz	Saccharin	Enough to sweeten

Use 3 to 3½ ounces to 1 pint of hot water or milk, or a mixture of both.

Desserts may be made with fruit flavors such as raspberry, cherry, strawberry, lemon and lime by rubbing the flavor with gum acacia, drying, and then using the same ingredients as suggested above..

Manufacture of Sugarless Jams

Raspberry, strawberry, grape or pineapple marmalade can be made by following the procedure suggested below.

Formula No. 1

Fruit	50 oz	Water	35 oz
Powdered Agar-Agar	3 oz	Saccharin	Enough to flavor
Gelatin	2 oz	Sodium Benzoate (1⁄10%)	If desired
Glycerin	5 oz		
Powdered Gum Acacia	10 oz		

Dissolve the gelatin in 5 parts of hot water. Mix the agar-agar and acacia and let soak overnight in 25 parts of water. Now add to gelatin solution, bring to boil. Add fruit and just bring to boil. Then add the glycerin. Add more water if jam seems too thick. Fill bottles while hot. Cap, and pasteurize for 35 minutes at 185° F.

Sugarless Marmalade for Diabetics

Lemons	1½	Water	7 oz
Peel of 1 Large Orange		Gelatin	¼ oz
Saccharin	5 gr		

Wash the orange and lemons. Peel the orange, avoiding white pith, and chop the skin up small. Add the juice and pulp of the lemons. Put into saucepan and cover with the water. Bring to boiling point and simmer for 2 hours, adding water when necessary to keep to stated amount. Cut the gelatin into fine strips. Add it with the saccharin to the mixture and stir for 10 minutes. Put it into a jar and leave it to set. The keeping properties of this marmalade are not very good, and if it is desired to store it for any length of time a small quantity of sulfurous acid—40 parts per million—preferably in form of potassium metabisulfite, should be added.

Starchless Breakfast Foods

Soybean Middlings	Honey or Malt Syrup

The soybean middlings are moistened with the syrup, toasted and then put into packages.

Bran (With All the Starch Removed)	90 parts	Soybean (Fine Granulation)	10 parts

Sugarless Chocolate Bars

Bitter Chocolate	20 oz	Vanillin	Enough to flavor
Skimmed Milk Powder	50 oz	Saccharin	Enough to sweeten
Fatless Soybean Flour	30 oz		

Mapleine Syrup, Sugarless

Powdered Gum Karaya	1 oz	Mapleine Extract	9 cc
Water	32 oz	Caramel Color	Enough to suit
Saccharin	3 gr	Salt	To taste

Whip the gum up with the cold water to a thick liquid. Then add flavor, sweetener and color.

Lemon Pie Filling, Sugarless

Powdered Gum Karaya	1 oz	Lemons (Juice and Skin)	2
Water	12 oz	Saccharin	2 gr
Butter	¾ oz		
Egg Yolk	1 oz		

Add water gradually to the gum and then bring to a boil. Remove from fire. Add remaining liquid ingredients in the order given. Pour into pan lined with pie crust which has been baked and cover with meringue.

Meringue, Sugarless

Egg White	1	Lemon Extract	¼ tsp
Saccharin	¼ gr	Salt	To taste

To partly beaten egg white, add the other ingredients in order given and beat until stiff. Spread over surface and bake in moderately hot oven.

Custard Filling

Egg	1	Vanilla	1 tsp
Water	½ cup	Saccharin	1 gr
Heavy Cream	½ cup	Nutmeg and Salt	To taste

Beat egg lightly; add salt, cream, water, saccharin and vanilla. Strain into pie pan lined with pie crust. Grate nutmeg over the top and bake in moderately hot oven until filling sets.

Pumpkin Filling

Eggs	2		Cinnamon	1	tsp
Heavy Cream	2	tbsp	Allspice	½	tsp
Mashed Pumpkin	1	cup	Ginger	½	tsp
Water	1	tbsp	Cloves	½	tsp
Saccharin	1	gr	Salt	½	tsp

Beat egg lightly; add other ingredients in order given. Strain into pie pan lined with pie crust. Bake in a moderately hot oven 30 minutes.

White or Cream Sauce

Butter	3	tbsp	Water	½ cup
Gum Karaya	2	tbsp	Salt and Pepper	To taste
Heavy Cream	½	cup		

Heat butter and mix with water and cream. When mixture is heated, gradually sift in the gum karaya, beating or stirring until the gum is dissolved, and season. Serve hot. If thin sauce is desired use only 1 tablespoonful of gum.

Health Dry Cereals

Formula No. 1

Carob Flour (St. John's Bread)	20 oz	Cracked Wheat	20 oz
Soybean Middlings	60 oz	Salt	To taste

Cook with water or milk and serve with cream.

Formula No. 2

Whole Wheat Cereal	80 oz	Agar-Agar	5 oz
Flaxseed Meal	15 oz		

Formula No. 3

Corn Meal	80 oz	Flaked Dates	20 oz

Formula No. 4

Whole Wheat Cereal	40 oz	Whole Rice Bran	5 oz
Whole Rice Cereal	40 oz	Flaxseed	5 oz
Whole Wheat Bran	10 oz		

Formula No. 5

Whole Wheat Cereal	95 oz	Rice Polishings	5 oz

Formula No. 6

Whole Wheat Cereal	85 oz	Honey	To suit
Soybean Middlings	15 oz		

Treat the above with honey and then dry it.

Formula No. 7

Whole Wheat Cereal	85 oz	Spinach Powder	5 oz
Carrot Powder	10 oz		

Coffee Substitutes and Health Drinks

Formula No. 1

Roasted Whole Wheat	75 oz	Flaxseed	5 oz
Roasted Whole Rye	20 oz		

Formula No. 2

Roasted Barley	90 oz	Dry Figs	5 oz
Roasted Bran	5 oz		

Formula No. 3

Roasted Whole Wheat	75 oz	Roasted Soybean Middlings	25 oz

Flour, Meals, Etc.

Alfalfa flour is rich in alkaline minerals and invaluable for acidosis.

Lentil, lima bean, and pea flour are very nutritious.

Soybean flour, with or without a high percentage of fat, is used to make bread, biscuits, noodles, mayonnaise, candy, etc.

Noodles

Kelp noodles are made with whole wheat and kelp. Spinach noodles are made with Semolina flour and spinach. Vegetable whole wheat noodles are made with whole wheat flour and a combination of carrot, spinach, and potato powders.

Tea Substitute

Alfalfa tea is rich in alkaline organic minerals and is sometimes combined with ground peppermint.

Maté tea is a South American nature drink.

Kelp tea is a mixture of seaweed rich in organic iodine.

Vegetable Gelatin

Irish moss is a starch-free, powdered product for preparing puddings, jellies, and for thickening hot drinks, soups and sauces.

Agar-agar is a product of the sea, and lends itself to the preparation of many attractive fruit agar jellies, salads and desserts.

Agar Jelly

Agar-Agar	3 g	Sugar	80 g
Fruit Juice	1 cup	Water	2 cups

Mix the agar with the sugar. Now boil the water. Dissolve the sugar in it, and boil for 1 minute. Shut off the gas, add the fruit juice and enough lemon juice to acidify the dessert.

Vegetable Seasoning

Sea Salt	80 oz	Dried Alfalfa	2 oz
Powdered Kelp	20 oz	Tri-calcium Phosphate	1 oz

Nuts

Nuts are very nutritious and, unlike cereals, they are rich in fat and proteins and very poor in carbohydrates or starch. Nuts are heat- and energy-producing foods. They are also tissue builders because of their protein content. Unless special care and attention are taken, nuts become rancid in a few months. They should not be stored in warm or damp places. If they must remain in a container a long time, it is best to transfer them from one fresh container to another.

The following data on nuts, although limited, show clearly how they are classified and graded for commercial use.

Almonds

Almonds are the seeds of a tree, similar to the peach tree, indigenous to the Mediterranean countries. Before the fruit is ripe, it is impossible to tell if the tree will produce a sweet or a bitter almond, however the same tree will always produce the same kind of nut.

The principal commercial almonds are:

Jordan almonds Malaga almonds Valencia almonds Palma almonds Girgenti almonds	Shipped with large seeds
Bari or Sicilian almonds	Small seeds; cheap in price and of good quality
Berber almonds Jaffi almonds	From North Africa; small; bitter; frequently broken; not clean

Domestic and imported almonds come in three grades: (1) Paper Shell, (2) Soft Shell, (3) Hard Shell.

In the United States almonds are cultivated in California, where they are grown by budding into seedlings of either the bitter or sweet variety. They are also imported in large quantities from France, Spain, Italy, and the Levant.

Grades of Unshelled Almonds

Nonpareil IXL Ne Plus Peerless Drake Seedling Languedoc Hard Shell	First grade should crack 95% or better with good kernels.

Grades of Shelled Almonds

Nonpareil Jumbo Nonpareil Medium Nonpareil Small	All unbroken meats; free from shell and foreign matter
Nonpareil Sheller Run	Not more than 25% broken meats; ungraded for size; free from shell and foreign matter
IVL Medium	All unbroken meats; free from shell and foreign matter
IXL Sheller Run	Not more than 25% broken meats
Jordanettes	All unbroken meats; free from shell and foreign matter
Drake Medium	All unbroken meats; free from shell and foreign matter
Drake Sheller Run	Not more than 25% broken meats; free from shell and foreign matter
Seedling Sheller Run	Seedlings of the Drake, Texas, Languedoc, or kindred variety; not more than 25% broken meats; free from shell and foreign matter

Brazil Nuts (Cream Nuts, Para Nuts)

These nuts come from a large tree which grows in Brazil, French Guiana, and British Guiana.

They grow inside of a hard, round shell, about 5 inches in

diameter, which contains from 15 to 25 compactly arranged nuts. The shells are very hard to break. Harvesting begins early in January. If nine out of every ten nuts cracked are good, it is considered a "good crack."

Babassu

This is a variety of the Brazil nut which is not edible. It is so rich in oil that it will burn freely, if lighted.

Grades

Large Washed, Large Medium, Medium

Butternuts

These brown-husked, rugged nuts come from a large tree grown in the United States. They contain oil and have a delicious flavor.

Coconut (Cocoanut)

This is the fruit of a tropical palm which reaches a height of 50 feet or more and which is found in nearly all tropical regions. On an average, each tree yields about 80 units each year. The kernel of the cocoanut (copra) contains about 70% fat, from which the cocoanut oil, or cocoanut butter, is expressed. The whole nuts and various desiccated products are exported.

Dried Cocoanut

Dried coconut is imported from Jamaica, Panama, Trinidad and Porto Rico.

Grades

Thread
- 1 Regular, about 3 inches in length
- 2 Extra large, about 4 inches in length

Both white; about 1/18 inch in diameter

Shred	Shorter than Thread
Macaroon	Pulverized; moisture varying from 2 to 6%

Cola Nuts

These are the seeds of the cola tree, which is cultivated in West Africa, West India, and South America. The seeds contain 1.5% caffeine and theobromine. They are used in chocolate and cola beverages.

Chestnut

There are five species of the true chestnut, three of which are American. In Italy and Spain they form an important part of the diet. Marrons glacés are candied chestnuts, popular both in Europe and America.

Hickory

This is the fruit of a tree native to America. Hickory nuts are highly prized in the United States, and were a common food among the Indians. Large quantities of the nuts are shipped to Europe.

Filberts (Hazelnuts)

Filberts are imported from Italy, Spain, and Turkey.

Grades of Shelled Filberts

Italian (Naples)	Two varieties, Long Naples, Round Naples
Spanish (Barcelona)	Round in shape, of fairly uniform size
Turkish (Levant)	Mostly round in shape and size

Grades of Unshelled Filberts

Long Naples	Long in shape
Round Naples	Round in shape

Barcelona	Round in shape
Unselected	Required to crack 85% sound

Both varieties are known to the trade as filberts, as they are practically alike in appearance when the husks are removed. The husk of the filbert is much longer than that of the hazelnut.

Peanut

The peanut, a native of Brazil, is now cultivated in Africa, India, China, and in the United States in Virginia, North Carolina, Georgia, Texas, and Alabama. The peanuts of the Spanish variety have a small and almost round kernel. They contain more oil than the Virginia peanuts. The kernel of the latter is large and oblong.

Grades of Unshelled Peanuts

Hand-picked Virginia Jumbo	Contain not more than 11% of imperfect or immature nuts
Hand-picked Virginia Fancy	Not more than 15% of imperfect or immature nuts
Hand-picked Virginia Extra	Not more than 10% of imperfect or immature nuts
Hand-picked White Spanish	Not more than 10% of imperfect or immature nuts
Hand-picked Red Spanish	Not more than 15% of imperfect or immature nuts
Hand-picked Runner	Not more than 10% of imperfect or immature nuts
Recleaned Runner	Not more than 20% of imperfect or immature nuts

Grades of Shelled Peanuts

Extra Large Virginia Shelled	Not more than 528 nuts per pound; not more than 5% of imperfect or unsound nuts or foreign matter

No. 1 Virginia Shelled	Well selected; not more than 7% of imperfect or unsound nuts or foreign matter
No. 2 Virginia Shelled	Broken and shriveled nuts; not more than 10% of imperfect or unsound nuts or foreign matter
No. 3 Virginia Shelled	Oil stock
No. 1 Spanish Shelled	Not more than 5% of imperfect or unsound nuts or foreign matter
No. 2 Spanish Shelled	Not more than 7% of imperfect or unsound nuts or foreign matter
No. 3 Spanish Shelled	Oil stock
No. 1 Runner Shelled	Not more than 5% of imperfect or unsound nuts or foreign matter
No. 2 Runner Shelled	Not more than 7% of imperfect or unsound nuts or foreign matter
No. 3 Runner Shelled	Oil stock

Grades of Blanched Peanuts

Whole Virginia Jumbo Blanched	Made from Extra Large Virginia Shelled; 80% whole nuts
Split Virginia Jumbo Blanched	Made from Extra Large Virginia Shelled; very small portion whole nuts, nearly all splits
Virginia Blanched	Made from Virginia grades that have been roasted before shelling, shelled, and blanched; largely splits, ungraded for size

No. 1 Virginia Blanched	Made from No. 1 Virginia Shelled; largely splits, ungraded for size
No. 2 Virginia Blanched	The small particles that are screened out in making No. 1 Virginia Blanched
Spanish Blanched	Splits
No. 1 Runner Blanched	Made from No. 1 Runner Shelled; largely splits, ungraded for size
No. 2 Runner Blanched	The small particles that are screened out in making No. 1 Runner Blanched

Chinese Peanuts

Grades Unshelled

11–13 count per ounce

Grades Shelled

28–30 count per ounce
30–32 count per ounce
38–40 count per ounce

Pecan

This is the fruit of a North American, nut-bearing tree belonging in the same genus as the hickory. There are shelled and unshelled pecans on the market. Paper shell pecans are shipped from Georgia, Texas, and Oklahoma.

Georgia Paper Shell

Grades Shelled

Mammoth Halves	Firm, perfect halves, counting about 250 to the pound
Jumbo Halves	Large, well-shaped halves, counting about 450 to the pound

Selected Halves	Well-formed halves, counting about 650 to the pound
Selected Pieces	Large meaty pieces, free from all foreign matter
Granules	Chopped into fine pieces

Oklahoma and Texas Pecans

These are thin shell, but not paper shell. They are generally not graded, but sold as Orchard Run. The nut is of medium size and full meated.

Pignolia (Pine Nut)

These nuts are the seeds of pine trees. The kernels, about the size of puffed rice, are very rich in food value and form an important article of diet for large numbers of people, especially Indians and Mexicans in the Southwest of the United States and in Mexico, where the nuts grow plentifully. Pignolia nuts also grow in southern Europe and some are imported into the United States.

Pili Nuts

This is a very valuable nut growing wild in the forests of the Philippine Islands. It is likely to find its way into our markets when its value becomes known. The nut of the pili is triangular and pointed at each end. It has a kernel pleasant to eat, raw or roasted.

Pistachio Nut (Pistache)

The pistache, or green almond, is a small, thin-shelled nut, which is delicious to eat when roasted and salted. While the pistache is grown to a limited extent in California, the main supply comes from Syria, other Mediterranean countries, and the desert countries of Asia and Africa.

Grades Cracked

Extra Fancy	Large size; all split; full meats; new crop; finest flavor
Fancy	Medium size; 95% full meats
Regular	Smaller size; 92% full meats

Grades Shelled

No. 1 Cilician Extra Fancy	Uniform large size; mostly green; extra picked; clean; sweet; free from shells, particles and foreign matter
No. 1 Cilician Fancy	Uniform size; mostly green; picked; clean; sweet; free from shells, particles and foreign matter

Walnut

The name walnut, is generally restricted to the European or English species and to the American black walnut.

In Southern California, the Placentia is regarded as the best all-around variety, as it is well adapted to the soil and climate and bears nuts of a very fine quality. The first quality Placentia nuts are regarded as standard, and most of the nuts sold as budded are of this variety. However, any of the leading nuts which equal No. 1 Placentias in quality may be sold as budded.

Fancy Varieties and Grades

Varieties	Grades
Eureka Franquette	Grades: Jumbo, Extra Fancy, Fancy
Mayette Payne Concord	Grades: Jumbo, Fancy

Standard Grades and Varieties

No. 1 Soft Shell	A walnut of the Santa Barbara soft shell variety of a size that will not pass through a 1-inch square opening; must crack from 90–100% good
Baby Soft Shell	Same as No. 1, but smaller nuts which pass through a 1-inch square opening; crack 90–100% good
Large Budded	A nut of the Placentia-Perfection variety or any other budded or grafted nut of the same general characteristics and of a size that will not pass through a 1⅛-inch square opening; crack 90–100% good
Medium Budded	Same as Large Budded, but will pass through a 1⅛-inch square opening; crack 90–100% good
Near Grade	A substandard grade of walnut which cracks anywhere from 80–90% good, but otherwise equals, in size and appearance, the standard grade
Off Grade	A substandard grade cracking 75–90% good; but, besides being below the standard in cracking, they are below the standard in size and appearance
Culls	Those walnuts that, either on account of cracking below 80% good or on account of defects in appearance, are not suitable for distribution to the public as unshelled walnuts and are, or at least should be, converted into by-products (shelled walnuts)

Grades Shelled

Light Halves	All unbroken halves; light in color
Light Ambers	Averaging 50% halves and 50% pieces; light amber color

Standard Ambers Halves and pieces mixed; dark amber color

BREAD MAKING

The Milling of Wheat

Before wheat is milled into flour, it is carefully screened, scoured, washed and dried. The cleaned wheat is then moistened under controlled temperature and moisture conditions. Wheats are also frequently blended to obtain a flour of high quality.

The wheat is then passed through a series of break rollers which are grooved. The break rollers, which revolve at different speeds, break the wheat into pieces coarse enough to permit easy removal of bran and germ. This cracked wheat is called middlings. By means of an air-blowing arrangement, the light bran particles are removed and the purified middlings are then passed through a series of reduction rollers, which cause unremoved germ and fibrous material to flake out against the smooth-faced rolls.

The middlings, which have been ground and reground, are separated into various streams of flour by sieving and bolting. The bran particles which do not pass through the sieve are termed shorts, and go into making feedstuff. We see that a perfectly manufactured white flour depends upon the complete removal of the bran, germ, and coarser particles.

In modern milling practice, 100 pounds of cleaned wheat usually yield 70 to 76 pounds of middlings, to be ground into flour of various grades, and about 24 to 30 pounds of feed. The term per cent separation, or per cent extraction, is the percentage of flour actually secured after all of the bran and feed have been removed. If 75 pounds of middlings, free from bran and germ, were milled into 75 pounds of flour, this would be considered as 100% separation or extraction. If 45 pounds were obtained, it would be a 60% separation flour. Any desired percentage separation may be obtained. Short separation flours represent a lower percentage separation or more highly refined flour than flours of longer separation. The grade of flour depends upon the perfect elimination of fiber, germ and other impurities of the whole wheat kernel.

Flour

Flour readily takes up moisture or loses it. Too high a moisture content may result in a musty flour. New crop flour is sometimes high in moisture content. Such flours should be stored until normal in moisture.

Protein in flour is very important, since that which is water-soluble nourishes the yeast in the dough during fermentation, while the water-insoluble protein, gluten, gives bread its skeleton structure or foundation.

Gluten is composed of gliadin and glutenin. Gliadin is of a sticky nature, imparting adhesiveness to the gluten, while glutenin imparts tenacity and strength. The combination of these two dissimilar proteins together with water is known as gluten.

The quantity as well as the quality of the gluten is important from the baker's viewpoint. Hard spring flour is slightly higher in gluten than grades of winter wheat flour. An average patent flour contains 10.8 to 12.5% of gluten. Shorter separation flours contain less gluten but of superior quality than higher per cent extraction flour milled from the same wheat.

Starch represents about 70% of the average bread flour. The small amount of maltose and sucrose present in the flour aids materially during the fermentation period. Soluble starch, cellulose, and dextrin may also be present, but in very small amounts. Damaged flour, resulting from intense heat and moisture, may also develop soluble starch and sugars to such an extent as to render it unsound. The normal fat content of patent flour is seldom over 1.5%. The clear grades of flour contain more fat than the shorter separation of patent grades. A high percentage of fat in flour may influence the development of rancidity while in storage.

Enzymes are present in flour. An enzyme is a chemical compound of animal or vegetable origin which causes chemical transformations to take place without itself being a part of those changes. There are many enzymes occurring in nature, each having a particular function to perform. For example, the enzyme called zymase will cause cane sugar to break down into alcohol and carbon dioxide gas. It is also the action of this same enzyme which is present in the yeast that causes the bread dough to rise. Diastase is an enzyme which causes starches to break down into maltose and dextrose.

The important enzymes present in wheat flour are diastase and protease. Diastase liquefies the starch granules in the dough, and then converts the liquefied starch into maltose, dextrose, and dextrins. Some believe that there are two types of diastase present in flour, a starch-liquefying enzyme and a starch-saccharifying enzyme. Diastase is necessary in the making of bread. It supplies greater gassing power to the dough, due to the conversion of starch into sugars, which in turn ferments into carbon dioxide and alcohol.

Protease, or the proteolytic enzyme in flour, has the function of softening the gluten in the dough batch, thereby imparting to it greater stretching power, strength, elasticity, and oven spring.

The strength of a flour and the ability of it to produce a large yield of loaves depend a great deal upon the quality of its gluten. A flour which will hold together during fermentation and which is not too sensitive to sudden changes of temperatures is called a stable flour. A flour which will produce a good loaf even if taken just a little before or after its fermentation period, is said to have a good fermentation tolerance.

The amount of water required to make a dough with flour, ideal for baking bread, and a dough which does not become sticky, or too stiff, is termed as percentage absorption of the flour. The average bread-making flour will take about 60 pounds of water for each 100 pounds of flour. A dough which will slacken noticeably during fermentation should be made a little stiffer at the start, or less water should be used.

The following table briefly shows the sources of important enzymes required in bread making.

Enzyme	*Source*	*Function*
Diastase	Flour	
	Malt syrup	
	Malt flour	
	Malted wheat flour	Converts starch into malt sugar
	Honey	Liquefies starch
Protease	Flour	
	Malt syrup	
	Malt flour	
	Malted wheat flour	Softens gluten
	Soybean flour	Makes gluten elastic
	Carob flour	Makes protein soluble
Invertase	Yeast	Converts cane sugar into invert sugar
Maltase	Yeast	Converts malt sugar into dextrose
Zymase	Yeast	Converts invert sugar and dextrose into carbon dioxide and alcohol; the alcohol disappears during baking and the carbon dioxide raises the dough

STANDARD FORMULAS FOR BREAD MAKING
(Straight Doughs)

	White Bread	*Raisin Bread*	*Rye Bread*	*Graham Bread*	*Whole Wheat Bread*	*French Bread*
	(Quantities in Pounds Unless Otherwise Stated)					
Flour	100	100	—	65	12	100
Water	60	60	60	58	66	58
Yeast	1¾	4	1½	2	1½	1¼
Yeast food	¼	—	—	¼	¼	6 oz
Salt	1¾	1¾	—	2	1½	1¾
Sugar	1	3-4	—	2	1½	—
Malt syrup	1½	1½	10 oz	1¼	1	2
Milk	4	4	—	—	2	—
Shortening	2	3-5	¾	2	2	2½

STANDARD FORMULAS FOR BREAD MAKING (*Cont.*)

	White Bread	*Raisin Bread*	*Rye Bread*	*Graham Bread*	*Whole Wheat Bread*	*French Bread*
	(Quantities in Pounds Unless Otherwise Stated)					
Raisins	—	35–50	—	—	—	—
Ist Clear	—	—	66	—	—	—
White rye flour	—	—	8	—	—	—
Dark rye flour	—	—	26	—	—	—
Graham flour	—	—	—	35	—	—
Whole wheat flour	—	—	—	—	100	—
Caraway seed	—	—	4 oz	—	—	—
Ground caraway seed	—	—	4 oz	—	—	—
Temperature	78° F	80° F	78° F	78° F	79° F	79° F
1st Turn	1¼ hr	¾ hr	2½ hr	1¼ hr	1½ hr	2¼ hr
2nd Turn	½ hr	¼ hr	¾ hr	—	¾ hr	¾ hr
To bench	¼ hr	10 min	¼ hr	½ hr	½ hr	¼ hr

Soy Bean Flour Bread

Formula No. 1

Soy Flour	65 lb	Yeast	15 lb
Wheat Flour	260 lb	Shortening	15 lb
Sugar	10 lb	Water	About 210 lb
Salt	5 lb		

Mix 3 minutes and ferment at 90° F.

First punch	45 min	Bake	30 min
To bench	15 min	Temperature of Oven	445° F
Proof	45 min		

Formula No. 2

Whole Soy Flour	25 lb	Shortening	2 lb
Whole Wheat Flour	25 lb	Yeast	2 lb
Clear	50 lb	Sugar	1½ lb
Dry Milk	3 lb	Dry Malt	1½ lb
Salt	1¾ lb	Water	About 10 gal

The straight dough method is employed. A rather wide range in the quantity of water to be used is permitted. This is done to allow for the particular water absorption of the whole wheat flour and the clear that may be used by the baker. A straight dough is made, but the whole soy flour is soaked for half an hour with a portion of the water before the dough is made.

STANDARD FORMULAS FOR BREAD MAKING
(Sponge Doughs)

Sponge	*White Bread*	*White Bread*	*Rye Bread*
Flour	60 lb	60 lb	Rye 30 lb Clear 30
Water	36	36	34
Yeast	1¼	1¼	1¼
Yeast food	¼	¼	¼
Malt syrup	¾	¾	¼
Shortening	1	1	
Temperature	76° F	76° F	76° F
Time	4 hr	4 hr	4 hr
Dough			
Flour	40 lb	40 lb	Clear 40 lb
Water	24	26	24
Salt	2	2	1¾
Sugar	1	3	
Malt syrup	½	½	1¼
Milk (Sweet condensed)	5 (Powder)	2–3	
Shortening	1½	1½	1
Temperature	79° F	79° F	79° F
To bench	15 min	15 min	20–30 min

There are two different systems of bread making from the standpoint of mixing and handling the doughs. They are:

A Straight Dough Method
B Sponge Dough Method

Straight Dough Method

In the straight dough method all of the ingredients are mixed together at one time, and the dough is fermented and conditioned uniformly to the proper extent. This method is simple, and produces a loaf the same way each time, having desirable characteristics such as compact grain, pleasing flavor, improved eating and keeping qualities.

Sponge Dough Method

In this method, only a part of the flour, water, yeast, yeast food, and sometimes a little of the malt syrup and shortening are mixed together at first. This dough is allowed to rise and mature, usually to a greater degree than a straight dough. After having matured to the proper extent, it is mixed with the balance of the ingredients. The final dough is then again transferred to the troughs and allowed to rise. It is conditioned for a short time and then made up into loaves in the regular way.

There are some advantages to the sponge method. A sponge dough requires less yeast than a straight dough, and can stand a longer time when ripe, or conditioned to the proper extent, than a straight dough. A loaf of greater volume, soft, and white crumb is produced by the sponge method. Cheaper grades of flour, or stronger flour, can be used in sponges to better advantage. The dough can be better regulated, especially when new or young flour is used. If the sponge comes up too quickly or too slowly, you can hasten or hold back the dough by regulating the temperature of the water and increasing or reducing the amount of salt.

The advantage of straight doughs over sponge doughs lies in the saving of time and labor, as all the ingredients are mixed at one time. The length of fermentation can be controlled at will, according to the amount of yeast, salt, sugar, and yeast foods used. A straight dough needs more mixing than a sponge dough, or the gluten must be worked more, for elasticity. A sweeter loaf can be made by the straight dough method, and a fine, rich flavor can be obtained with extra additions such as sugar, malt syrup, honey, milk, etc. If a long-time sponge dough method is employed, the flavor of the loaf is apt to be slightly sour and is therefore much less desirable than the flavor of a loaf made by the straight dough method.

Method of Bread Making

Straight Dough Process	*Sponge Dough Process*
Preparation of ingredients	Preparation of part of ingredients
Dough is mixed	Sponge is mixed
Dough placed in trough and fermented	Sponge placed in trough and fermented
Dough transferred to bench	Sponge placed in mixer, broken up, and mixed with other ingredients, and finally placed back in trough
Dividing and scaling	Allowed to rise again
Rounding of loaf	Dough sent to bench, dividing and scaling
Additional proofing	Rounding of loaf
Molding	Additional proofing
Panning	Molding
Proofing in pan	Panning
Baking	Proofing in pan
Cooling	Baking
	Cooling

Mixing the Yeast

The yeast to be used should first be made into a uniform mixture with water at room temperature. The salt, yeast foods, or malt syrup should be added as a solution to the dough, after the addition of the yeast.

Matured Dough

When the dough has been properly matured in the trough. it is transferred to the bench or to the dividing machine. Here the dough is cut into individual pieces of proper weight. Under normal conditions, for white pan bread, 1 pound of bread will be produced from 17½ to 18 ounces of dough. There are many types of dividing machines. A divider cutting off one piece of dough is called a one-pocket divider. In dividing the large mass of dough, either by hand or machine, the gas in the dough is further expelled. Before each piece can be molded the dough must be given a chance to develop sufficient pliability for molding. Furthermore, the sides and edges of the pieces are open, permitting gas formed by the continued action of the yeast to escape. Right after dividing the dough, each piece is rounded or balled up so as to put a skin around it. When the individual pieces of dough are rounded out, they should be allowed to rest, free from all drafts, for about 15 minutes. This additional proofing will noticeably increase the volume, and make the bread softer, and in this condition the bread will mold better. This short intermediate period of proofing is sometimes called short-proof, or overhead, or first proof.

Molding the Loaves

After the short-proof process, each piece of dough is molded into a dough loaf. This can be done by a machine. The dough loaves, after being molded, are placed in the pans which have been lightly greased on the bottom and sides. Before being baked, it is necessary for the dough loaves to rise once more. For ordinary pan bread, the loaves should be allowed to become at least double in size during proofing. The pan-proofing cabinet should be well insulated so that a uniform temperature of 95 to 98° F and a relative humidity of 80 to 85% are maintained. At the proofing temperature the yeast functions vigorously, causing the dough to rise rapidly, and the high humidity prevents the dough loaves from crusting over. Proper pan-proof period may range from 30 to 60 minutes. This final proofing is important, and should be accurately controlled. A healthy dough, containing liberal quantities of malt, yeast, salt, and milk will stand up best with all the abuse it receives during the proofing process and will produce a good quality loaf of bread. During the first few minutes of baking in the oven, the yeast functions more vigorously than at any previous time. When the dough temperature reaches 140 to 150° F, fermentation ceases. There is a final rising of the dough which is secured because of the rapid production of more gas, due to the activity of the yeast. Another reason is the fact that a considerable amount of carbon dioxide gas produced in the dough prior to its entrance into the oven remains dissolved in the mixture of the dough and, at the increased temperature of the dough in the oven, this dissolved carbon dioxide is driven off and aids materially in producing the desired oven spring. The alcohol produced during the fermentation evaporates in the oven.

The exact time of baking and the oven temperature depend upon the size and type of bread desired. Oven temperatures range from 375 to 450° F. For ordinary one-pound loaves of pan bread ½ hour bake at approximately 425° F is about right.

Date Nut Loaf Bread

Bread Flour	3½ lb	Salt	1¼ oz
Wheat Flour	1½ lb	Chopped Dates	4½ lb
Baking Powder	⅛ lb	Chopped Walnuts	1 lb
Baking Soda	1¾ oz	Hot Water	4½ pt
Brown Sugar	2¼ lb	Eggs	1 lb
Shortening	1½ lb		

Flour the dates before chopping and then soak the chopped dates and walnuts for 5 minutes in hot water. Now add the baking soda. While thoroughly creaming the sugar, shortening, and salt, add the eggs. Now add the warm date-walnut mixture to the cream mass and stir. The flour and baking powder, having been sifted and mixed, should now be mixed thoroughly with the creamed mass.

Bake at 380° F for 45 minutes, and remove the loaf from the pans while warm.

Prune Bread

Flour	50 lb	Milk	¾ lb
Water	28 lb	Malt Syrup	½ lb
Yeast	1½ lb	Shortening	1½ lb
Sugar	2 lb	Chopped Prunes	15 lb
Salt	1 lb		

Allow the prunes to soak overnight in a small amount of water. Then add the sugar and shortening. After mixing the dough in the usual way, the entire batch is mixed and worked over.

The dough is fermented at 80 to 85° F.

The first punch is taken 1 hour and 30 minutes later; second punch, 45 minutes later; to the bench, 15 minutes later.

Allow to proof 15 minutes before panning.

Give medium proof and then bake in medium oven.

Bran Bread

Clear Flour	5	lb	Salt	2	oz
Whole Wheat Flour	1	lb	Shortening	8	oz
Bran	1	lb	Baking Soda	3	oz
Buttermilk	8	lb	Cream of Tartar	1¼	oz
Brown Sugar	6	lb	Chopped Nuts	2	lb
Eggs	1½	lb			

Dissolve the baking soda in the buttermilk. Now cream together the shortening, sugar and salt. Then add the eggs, buttermilk and nuts. Now add flour, bran and cream of tartar, and mix until the batter rises a little.

Bake in paper-lined tins at 350° F for 45 minutes.

Aids in Bread Making

Salt

Salt improves the taste and flavor of bread, helps to control the activity of the yeast during fermentation, and assists in preventing the development of objectionable bacterial action or wild type of fermentation. Salt has a binding or strengthening effect on gluten, and so adds strength to any flour.

Yeast

Yeast is a microscopic, one-celled plant belonging to the group known botanically as fungi. It ordinarily multiplies by a process known as budding, and under suitable conditions causes fermentation. Each yeast plant is rounded or oval and measures about 1/3600 inch in diameter. Each cell is surrounded by a thin membrane of cellulose, and the interior of each cell is made up of fine granular protoplasm.

The chief enzymes of yeast are invertase, maltase, zymase, and protease.

Sugar

The following sugar products are used in bread making:

Brown sugar	Cane sugar
Molasses	Honey
Maple sugar	Corn syrup
Malt sugar	Lactose (From milk)
Corn sugar	

Sugar supplies the energy, gas production, sweetness, and crust color in bread.

Shortening

Shortenings are edible fats or oils of vegetable or animal origin used in baked goods. They produce a soft velvety crumb, termed short. Those which are fluid at ordinary temperatures are called oils, and those which are solid are called fats.

Shortenings, when analyzed, are found to be mixtures containing some of the following substances in various proportions:

1 Stearin, a hard, natural fat of animal origin. Vegetable stearin is made by saturating vegetable oils with hydrogen gas, and is known as hydrogenated fat.

2 Palmitin, a fat secured from both animal and vegetable sources.

3 Olein, an oil secured from animal and vegetable sources.

4 Linolein, an oil present in cottonseed oil.

Animal Fats

Lard is an excellent shortening. It is composed of a mixture of stearin, palmitin, and olein. There are many types of lard, depending on from which part of the hog the fat is secured. Leaf lard is the highest grade, obtained from the fat surrounding the kidneys, and rendered at a high temperature.

Beef fats are separated into a hard fat known as oleo-stearin, and soft fats or oils known as oleo. The oleo oils are used in making oleomargarine, while the oleo-stearin is used in the preparation of compound shortenings.

Butter is the fat separated from milk or cream by churning. It is used mainly in cakes and pastries.

Vegetable Oils

Vegetable oils such as corn, cottonseed, and coconut are first refined and deodorized if they are to be used in bread making. While these oils can be used in bread making, they are more apt to become rancid than the solid fats.

Hydrogenated Vegetable Shortening

When hydrogen gas is circulated through vegetable oils at high temperatures and under certain other controlled conditions, a gradual change takes place, resulting in a white solid fat. What happens is that two principal constituents, olein and linolein, react with the hydrogen and are converted into a solid fat, stearin. The longer the hydrogen is passed through the oil, and the more the oil takes up of it, the harder and whiter will the resulting fat become.

Compound Shortening

As the name signifies, this group is made by blending vegetable oils with hard fats in proper proportion to obtain a product comparable to the consistency of lard. Cottonseed oil, corn oil, or other vegetable oils blended with vegetable or animal stearin form a compound shortening which is white, colorless, and of plastic consistency.

The animal stearin used in the manufacture of compounds is usually oleo-stearin, or the hard fat secured from tallow.

Oleomargarine

This product is usually made of refined oleo oil, churned in with some pure butter, neutral lard, and milk. Sometimes vegetable oils are also used. Nut margarine contains cocoanut oil. Its shortening power is good.

Shortening used in bread improves its eating qualities, appearance, keeping qualities, and food value. Always keep shortening containers tightly covered, and store in a cool, dark, dry place.

Milk Products

In bread making the following milk products are used:

Liquid whole milk
Evaporated whole milk
Sweetened condensed whole milk
Dried whole milk
Dried ½ skimmed milk
Dried skimmed milk
Skimmed evaporated milk
Skimmed condensed sweetened milk

Soy Products

Soy Flour

Soy flour is a food ingredient used in many food formulas, especially in the baking industry. It fulfils a definite requirement and is now strongly entrenched in the industry. It aids in the dispersion of fat and moisture. Soy flour, averaging 20 to 22% soy oil, has a definite shortening value as well as the effect of lecithin which is a well-known, natural, emulsifying agent.

Soy flour can be used in the bakery. Manufacturers of the flour recommend the following proportions based on the weight of the flour:

For bread, rolls, etc.—	2½ to 5%
For dark bread, sweet doughs and pastries—	5 to 15%
For special soy bread loaf—	up to 25%
For cakes—	10 to 15%

Slight formula changes can be made by reducing the amount of eggs and milk since soy flour is rich both in fat and protein. Since the soy is alkaline and may check fermentation, and additional quantity of yeast and a longer fermentation period are found desirable.

Soy flour may be added, sifted with the white flour, or it may be whipped in with part of the water like dry milk is dissolved, or soy and milk powder can be dissolved in part of the water before making the dough. Where large amounts of shortening are used, the soy may be creamed in.

APPROXIMATE COMPOSITION OF SOY FLOURS

	Per Cent High Fat Soy	*Per Cent Medium Fat Soy*	*Per Cent Low Fat Soy*	*Per Cent Very Low Fat Soy*
Water	5–6	5–6	5–6	5–6
Protein (N x 6.25)	41–42	46–47	49–50	52
Fat	20–22	15	5.5–6.5	1
Minerals	4.5	4	5.5	6
Lecithin	1.5–2.5	—	—	—
Absorption of water per 100 parts	120	132	165	172

Soy Products and Uses

A defatted high protein soy flour is recommended as an added ingredient for bread, cakes, rolls, cookies, cake mixes, pancake, waffle and muffin mixes, macaroni, spaghetti, noodles, dehydrated soups, ice cream sherbets, chocolate coating, candy and other foods.

A high-fat type soy flour is recommended as an added ingredient for sweet goods, doughnuts, pie dough, sweet dough mixes, biscuits, cookies, candy, soups, ices and many other food products.

A coarse defatted soy flour is a valuable ingredient in the

manufacture of meat loaves, sausage, franks, chili, dehydrated soups, sauces, gravies and other foods.

Small or large defatted soy grits present a new ingredient for candy makers, biscuit manufacturers, bakers, meat packers and other food processors.

Puffed soy grits, small or large, toasted light or dark, are being used as topping for biscuits, candy and bread. Soy puffs are sometimes blended with chopped nuts. They are bland in flavor, contain over ⅘% protein and are practically fat free.

White Bread with Soy Flour
(Straight Dough)

White Flour (Patent)	25	lb	Shortening	6½	oz
Milk Powder	12	oz	Water (Variable)	16	lb
High Fat Soy Flour	12	oz	Yeast	8	oz
Sugar	1	lb	Yeast Food (Optional)	1	oz
Salt	8	oz			

Mix to a medium stiff dough. The temperature of the dough when mixed should be approximately 79 to 80° F. The approximate fermentation first rise is 1 hour and the second rise is ½ hour.

To bench in 20 minutes. Round and proof for 15 minutes. Give maximum proof in pan (about 55 minutes). Bake at 425° F with medium top heat for 30 to 45 minutes depending upon size of loaf.

Soy flour may be suspended in water with the milk solids or added along with the milk solids on top of the flour. Shortening is adjusted for 2%. More may be used if desired. Allow for the usual average fermentation. Give ¾ to full proof in pans.

Soy Sweet Dough
(Rich)

Whole Eggs (Or Part Yolks)	2¾	lb	Salt (Variable)	5½	oz
			High Fat Soy Flour	2	lb
Flavor Blend (Variable)	1	oz	Shortening	3½	lb
			Sugar	3¾	lb

Cream the above ingredients together until light.

Water (Variable)	9 lb	Yeast	1½ lb

Dissolve yeast in part of water. Add along with balance of water and stir into the above.

Bread Flour	17 lb

Add and mix to a smooth, medium-soft dough (until dough leaves the side of the mixing bowl).

Temperature of dough when mixed should be 80° F.

Ferment for approximately 2 hours. Adjust fermentation to allow for time needed on make-up table. Give ¾ to full proof with light steam and bake at 365° F with medium bottom heat.

Part of shortening may be replaced with butter. Various flavor combinations may be used, such as cardamom (or coriander), cinnamon (or mace) and vanilla (or lemon) etc.

Soy Quick Yeast Raised Doughnuts

Whole Eggs or Yolks	1 lb	Salt (Variable)	4 oz
Shortening	2¾ lb	Flavor Blend (Variable)	¾ oz
High Fat Soy Flour	1½ lb	Sugar	1½ lb

Cream together until light.

Water (Variable)	8½ lb	Yeast	1 lb

Dissolve yeast in part of water. Add along with balance of water and stir in.

Bread Flour	15 lb	Baking Powder (Variable)	5 oz

Blend together by sifting. Add and mix only long enough to produce a smooth soft dough.

Temperature of dough when mixed should be 80° F.

Roll out and cut at once. Give ¾ proof. Do not overproof and fry at 370° F. Dip in glaze while hot.

Part of shortening may be replaced with butter. Various flavor combinations may be used such as cardamom (or coriander), cinnamon and vanilla coriander, cinnamon (or mace) and vanilla (or lemon) etc. If desired, regular yeast-raised doughnuts may be made from either the rich basic or lean soy-sweet doughs.

Soy Yellow Cake

Water	3 lb	Emulsifier Shortening	2¼ lb
Sweetose	3 lb		

Weigh the above ingredients into mixing bowl.

Sugar	4¼ lb	Baking Powder	
Cake Flour	5 lb	(Variable)	5¾ oz
High Fat Soy Flour	1 lb	Soda (Optional)	¼ oz
Salt (Variable)	3 oz		

Blend together by sifting. Add to above mix until smooth and then mix for about 5 minutes on low speed:

Whole Eggs	2½ lb	Flavor Blend	1 oz
Water (Variable)	1½ lb		

Mix together. Add in 2 or 3 equal parts mixing smooth after each addition and then mix for about 5 minutes on low speed.

Scrape down bowl properly to insure a smooth batter. Temperature of finished batter should be 70–75° F. Bake deep layers at about 385° F.

Soy White Cake

Water	3 lb	Emulsifier	
Sweetose	3 lb	Shortening	2 lb 9½ oz

Weigh above ingredients into mixing bowl.

Sugar	5 lb	Baking Powder	
Cake Flour	5 lb	(Variable)	5¾ oz
High Fat Soy Flour	1 lb	Cream of Tartar	
Salt (Variable)	3 oz	(Optional)	½ oz

Blend together by sifting. Add to above. Mix together until smooth and then mix for about 5 minutes on low speed.

Egg Whites	3¾ lb	Flavor Blend	1 oz
Water (Variable)	1¼ lb		

Mix together. Add in 2 or 3 equal parts, mixing smooth after each addition and then mix for about 5 minutes on low speed.

Scrape down bowl properly to insure a smooth batter. Temperature of finished batter should be 70 to 75° F. Bake deep layers at about 385° F.

Soy Devil's Food Cake

Water	4 lb	Emusifier Shortening	2¼ lb
Sweetose	4 lb		

Weigh above ingredients into mixing bowl.

Sugar	5 lb	Salt (Variable)	3½ oz
Cake Flour	5 lb	Soda (Variable)	1¼ oz
High Fat Soy Flour	1½ lb	Baking Powder	
Lightly Dutched Cocoa	1½ lb	(Variable)	5 oz

Blend well together by sifting. Add to above. Mix together until smooth and then mix for about 5 minutes on low speed.

Whole Eggs	3¼ lb	Vanilla (Variable)	1½ oz
Water (Variable)	2¾ lb		

Mix together. Add in 2 or 3 equal parts, mixing smooth after each addition and then mix for about 5 minutes on low speed.

Scrape down bowl properly to insure a smooth batter. Temperature of finished batter should be 70 to 75° F. Bake deep layers at about 385° F. This formula is adjusted for a lightly dutched cocoa. If a natural cocoa is used, increase the soda and reduce the baking powder.

Soy Chocolate Cake

Water	3 lb	Emulsifier Shortening	2 lb
Sweetose	4½ lb		

Weigh into mixing bowl.

Sugar	5 lb	Soda (Variable)	1¼ oz
Cake Flour	5 lb	Baking Powder	
High Fat Soy Flour	1½ lb	(Variable)	4½ oz
Salt (Variable)	3½ oz		

Blend well together by sifting. Add to above. Mix together until smooth and then mix for about 5 minutes on low speed.

Whole Eggs	3 lb	Vanilla (Variable)	1½ oz
Water (Variable)	4 lb		

Mix together. Add in 2 or 3 equal parts, mixing smooth after each addition.

Melted Bitter Chocolate	2½ lb

Add and mix in thoroughly. Then mix for about 5 minutes on low speed.

Scrape down bowl properly to insure a smooth batter. Temperature of finished batter should be 70 to 75° F. Bake deep layers at about 385° F.

Soy Fudge Cream Icing

Water (Variable)	4 lb	6X Sugar	20 lb
Sweetose	1½ lb	High Fat Soy Flour	2 lb
Emulsifier Shortening	3¾ lb	Salt (Variable)	1½ oz

Blend sugar, soy flour and salt together by sifting. Add to other ingredients and cream light on low or medium speed.

Water	1 lb	Vanilla (Variable)	1½ oz
Gelatin	1½ oz		

Soak gelatin in cold water. Then heat to 140° F to dissolve. Add to above and cream to desired lightness.

For best results the temperature of the fats and the finished icing should be 70 to 75° F. Flavor and color may be added as desired. Regulate consistency of icing by varying amount of the water used.

Soy Butterscotch Icing

Brown Sugar (No. 9 or No. 10)	8 lb	Water	1½ lb
Butter	2 lb	Sweetose	2¼ lb

Cook brown sugar and butter, stirring constantly to develop flavor. Add water gradually and cook to 242° F. Remove from heat and stir the sweetose in well.

6X Sugar	20 lb	High Fat Soy Flour	6 lb
Hydrogenated Shortening	3 lb	Salt (Variable)	1 oz
Hard Fat	2 lb	Hot Water	5 lb

Sift sugar, soy flour and salt. Add to fat and water, and cream light. Then add above hot syrup slowly while mixing at low speed.

Hot Water (Variable)	3 lb

Add gradually while mixing and continue to mix until smooth.

Temperature of icing when applied to cake should be 100 to 110° F. Icing may be reheated in a hot water or steam bath. Regulate consistency of icing by varying amount of water used. For a soy butterscotch nut icing, add about 5 pounds of coarse-chopped nutmeats to above batch.

Soy Chocolate Fudge Icing

6X Sugar	25 lb	Salt (Variable)	1½ oz
High Fat Soy Flour	3 lb	Vanilla (Variable)	1½ oz

Sift sugar, soy flour and salt together, and place in mixing bowl.

Water (Variable)	5½ lb	Sweetose	2¼ lb

Bring to a boil. Add to above and mix smooth.

Hydrogenated		Butter	1½ lb
Shortening	3 lb	Bitter Chocolate	4¾ lb

Melt together. Add to above and mix on low speed until icing cools slightly and takes on a fudgy appearance.

Temperature of icing when applied to cakes should be 100 to 110° F. Icing may be reheated in a hot water or steam bath. Regulate consistency of icing by varying amount of water used. For a soy chocolate nut fudge icing add about 3 pounds coarse-chopped nutmeats to above batch.

Baking Powder

The addition of baking powder and water to a dough or batter, when heated liberates carbon dioxide gas. Another way of producing carbon dioxide is by the addition of yeast. The expansion of the dough, due to the liberation of the gas, produces baked products that are light, fluffy and crisp. By the use of baking powder, this expansion is brought about quickly, whereas considerable time is required when yeast is used.

According to the definition and standard adopted by the United States Department of Agriculture, a baking powder is the leavening agent produced by the mixing of the acid reacting substance and sodium bicarbonate, with or without starch or flour. It yields not less than 12% of available carbon dioxide.

The acid reacting substances in baking powder can be any one of the following:

(1) Tartaric acid or its acid salts
(2) Acid salts of phosphoric acid
(3) Compounds of aluminum
(4) Any combination in substantial proportions of the foregoing.

Although the substances named in this definition and standard for baking powder are those ordinarily used in commercial baking powders, manufacturers are not prohibited from using other harmless substances that may be found suitable for some specific purpose. If any substance that is not included among those named in the definition is used in a baking powder, this fact should be plainly stated upon the label so that the purchaser will be fully informed.

There are five or perhaps six types of baking powder on the market today.

(1) Cream of tartar baking powder
(2) Mono-sodium phosphate baking powder
(3) Mono-calcium phosphate baking powder or phosphate powder
(4) Phosphate alum baking powder
(5) Straight alum baking powder
(6) Calcium baking powder

As with other food products, it is desirable to have baking powders as free from impurities as possible. They must be within reasonable government regulation with regard to lead, arsenic, zinc and tin.

Chemical Composition

The approximate chemical composition of baking powders may be indicated by the following theoretical formulas for their manufacture. The examples shown below are calculated to give a theoretical total carbon dioxide strength of 14% for household baking powders, and 17% for commercial powder. The formulas given below are based on the chemically pure substances without the presence of moisture.

In these formulas it is assumed that 100 pounds of the acid-reacting substances will neutralize respectively the following weights of bicarbonate of soda:

Acid-Reacting Substance	*Pounds of Sodium Bicarbonate Neutralized*
Cream of Tartar	44
Tartaric Acid	116
Mono-calcium Phosphate	80
Sodium Aluminum Sulfate	104
Sodium Acid Pyrophosphate	75

Although values are theoretical, they will serve as a basis for commercial manufacture.

Cream of Tartar and Tartaric Acid
Baking Powder

Sodium Bicarbonate	26.73
Corn Starch	22.40
Tartaric Acid	5.97
Potassium Bitartrate (Cream of Tartar)	44.90
Total	100

A baking powder which is quite generally used in the household, is a baking powder made up of cream of tartar (the acid salt mentioned in the first baking powder formula) and tartaric acid. These two substances are combined with 26.73 parts of baking soda and 22.4 parts of corn starch to make up a baking powder of the same strength. The quantity of tartaric acid can be varied in a baking powder of this kind. As this substance is varied, it is also necessary to vary the cream of tartar and corn starch content, for only a definite acidity is necessary to break up the baking soda, regardless of the type of acid used. Of course, since these amounts vary with the strength of acid used, it is necessary to use more or less and so vary the quantity of starch.

Calcium Acid Phosphate
Baking Powder

Sodium Bicarbonate	26.73
Corn Starch	39.84
Mono-calcium Phosphate	33.43
Total	100

The mono-calcium phosphate baking powder is the most common of the phosphate baking powders, and is often sold under the name of imitation cream of tartar. When sold in this way it is toned down with corn starch so that it will have the strength of cream of tartar.

In making up a combination of this salt with baking soda, 26.73 parts of baking soda, 33.43 parts of mono-calcium phosphate and 39.84 parts of corn starch, are used. The chemical reactions of the above baking powder are complicated, varying under different conditions. The chemical equation herewith presented probably represents what takes place under ordinary conditions.

$$3CaH_4(PO_4)_2 + 8NaHCO_3 \rightarrow Ca_3(PO_4)_2 + 4Na_2HPO_4 + 8CO_2 + 8H_2O$$

Phosphate Alum Baking Powder or
Combination Baking Powder
(S.A.S. Phosphate)

Sodium Bicarbonate	26.73	26.73
Corn Starch	40.07	42.97
Mono-calcium Phosphate	13.28	4.62
Sodium Aluminum Sulfate (Calcined)	19.92	25.68
Total	100	100

A combination baking powder gives a chemical reaction more complicated than a straight powder. The three types of baking powder mentioned above probably constitute more than 90% of the baking powder produced in this country. The combination powders represent a large proportion of the total production. When a combination baking powder reacts, the mono-calcium phosphate is activated first and, since this reaction takes place in the cold, there is an immediate leavening action. The sodium aluminum sulfate reaction requires heat and, therefore, will not act upon the baking soda until the dough is put into the oven.

Cream of Tartar Baking Powder

Sodium Bicarbonate	26.73
Potassium Bitartrate	59.86
Corn Starch	13.41
Total	100

Mono-sodium Phosphate Baking Powder

Sodium Bicarbonate	26.73
Anhydrous Mono-sodium Phosphate	38.20
Corn Starch	35.07
Total	100

Straight Alum Baking Powder

Sodium Bicarbonate	26.73
Sodium Aluminum Sulfate	25.68
Corn Starch	47.59
Total	100

The examples shown above are calculated to give a theoretical yield of carbon dioxide ranging from 12 to about 18% from the reaction between sodium bicarbonate with an acid or acid salt. The corn starch is added as a filler, to modify the amount of carbon dioxide produced to the same strength in all of the above examples, and to retard spoiling due to the presence of moisture in the powders and in the air.

Calcium Baking Powder

Recently a baking powder has been introduced which is rather complicated in its makeup, but which seems to be very efficient in its action. It is made up of certain sodium phosphates, mono-calcium phosphate and calcium lactate. A definite amount of starch is added to this combination as a filler.

There are three definite reactions with this baking powder and, as each reaction takes place, some carbon dioxide gas is given off. The powder reacts somewhat in the cold, to a greater degree at a slightly elevated temperature, and completely in the heat of the oven. It is believed that about 25% of the gas is liberated during the first two reactions and 75% in the final reaction.

One advantage of this baking powder is its water-absorbing power when it is used in the cake mixture. A greater quantity of moisture must be added to liberate all the carbon dioxide from the powder. This may be due to the liberation of free lactic acid, with the formation of calcium phosphate. This reaction is probably reversible with a final reaction in which sodium lactate is produced along with calcium phosphate. There are probably a number of phosphates formed in addition to the lactate. Theoretically, at least, it can be said that this is a triply reacting baking powder since one reaction occurs in the cold, another at a slightly elevated temperature and a third in the heat of the oven.

The lactic acid, if formed and permitted to remain in the free state, would certainly have an opportunity to act on the gluten of the flour, causing the gluten to take on a greater quantity of moisture. It is probably this reaction which is responsible for the increased quantity of moisture required when this type of powder is used.

Baking Powder
Formula No. 1

Bicarbonate of Soda	35 lb	Acid Calcium Phosphate	12 lb
Corn Starch	24 lb	Albumen	2 lb
Sodium Aluminum Sulfate	29 lb		

Baking Powder
Formula No. 2

Bicarbonate of Soda	102 lb	Acid Calcium Phosphate	45 lb
Corn Starch	86 lb	Sodium Aluminum Sulfate	72 lb
Acid Sodium Phosphate	40 lb	Albumen	½ lb

Baking Powder
Formula No. 3

Sodium Bicarbonate	28 lb	Acid Calcium Phosphate	12 lb
Corn Starch	41 lb	Theoretical Gas: 14½%	
Sodium Aluminum Sulfate	19 lb		

Baking Powder
Formula No. 4

Granulated Sodium Bicarbonate	28 lb	Acid Calcium Phosphate	35 lb
Redried Corn Starch	37 lb	Theoretical Gas: 14½%	

Cream of Tartar Baking Powder

Bicarbonate of Soda	27 lb	Corn Starch	14 lb
Cream of Tartar	54 lb		

Baking Powder—Tartar Substitute

		Or	
Acid Sodium Phosphate	40½ lb	Acid Sodium Phosphate	27 lb
Sodium Aluminum Sulfate	45 lb	Sodium Aluminum Sulfate	30 lb
Corn Starch	64½ lb	Corn Starch	43 lb

Alum Baking Powder

Bicarbonate of Soda	28 lb	Acid Sodium Phosphate	20 lb
Sodium Aluminum Sulfate	19 lb	Corn Starch	33 lb

Baking Powder

Bicarbonate of Soda	35 lb	Calcium Acid Phosphate	13 lb
Sodium Aluminum Sulfate	25 lb	Corn Starch	27 lb

Alum Baking Powder

Soda Alum	28 lb	Corn Starch	43 lb
Bicarbonate of Soda	29 lb	Theoretical Gas: 15.4%	

Phosphate Baking Powder

Granular Calcium Acid Phosphate	56 lb	Dried Corn Starch	18½ lb
Granular Bicarbonate of Soda	25½ lb	Theoretical Gas: 13¼%	

Cream of Tartar and Tartaric Acid Baking Powder

Cream of Tartar	50 lb	Corn Starch	20½ lb
Tartaric Acid	3 lb	Theoretical Gas: 13¾%	
Bicarbonate of Soda	26½ lb		

The cream of tartar and the soda should be in the granular form. The starch super-dried (otherwise use more starch).

Phosphate and Alum Baking Powder

Soda Alum	22 lb	Corn Starch	40 lb
Calcium Acid Phosphate	11 lb	Theoretical Gas: 14¼%	
Bicarbonate of Soda	27 lb		

Chemicals Used in Baking Powders

Alkalies

Bicarbonate of soda is in general use because it is easily neutralized, gives off a large amount of gas, is cheap and is commercially pure.

Besides this, carbonate of magnesium ($MgCO_3$), is used in small proportions. It produces about the same amount of gas per pound as sodium bicarbonate, but requires twice the amount of acid to neutralize it. It is extremely light, but since its bulk is about 7 times as great as sodium bicarbonate, it increases the bulk of the baking powder. For the same reason, baking powder made with magnesia keeps better because the other ingredients are not as closely packed. Also, an excess of magnesia does not cause the product to turn yellow, as does a soda excess.

The Neutral Substances or Fillers

Flour, corn starch, rice flour are most commonly used as fillers.

Wheat flour often contains larvae. Corn starch made by a chemical process is the best. Egg albumen is not of any real economic value, but is used as a filler to slow up the action of the powder. Starch and albumen thicken up the water so that a good body of foam is made, whereas any baking powder that contains only a small amount of starch and no albumen will not foam up in this way.

Acids and Acid Salts

Tartrates { cream of tartar
tartaric acid
Phosphates
Alums

Cream of tartar is most commonly used. Its neutralizing strength is only 44; that is, 100 parts of cream of tartar neutralize 44 parts of sodium bicarbonate.

When cream of tartar baking powder is used, the oven must be very hot, so that the biscuit can be baked immediately after mixing, or failure is certain. The same thing is true of tartaric acid, but it is not true to the same extent with any other acid ingredient in common use.

Tartaric acid has 2½ times the neutralizing strength of cream of tartar. It is an extremely active acid, and can be used with safety only in small proportions, unless it is used in granular form and with a large proportion of starch filler. It is a valuable ingredient, however, when used in proportion of 2 or 3%, because it begins the leavening process the instant it is wet. Leavening is, therefore, more continuous and uniform.

The neutralizing strength of phosphates is about the same as that of cream of tartar. The granular form is believed to possess better keeping qualities. Acid phosphate of soda, sodium pyro-phosphate, is widely used, although the cost is much higher than that of calcium phosphate. A baking powder which contains no other acid salt except phosphates does not have good keeping qualities, consequently, the best way to use it is with alum, so that its properties are preserved until they are drawn out by the proper application of moisture and heat.

An alum is the double salt of aluminum and some other base. Only three of the alums, however, have been in use for baking powder: potash, ammonia and soda alum. Potash alum leaves a bitter taste. Ammonia alum is very good, but there is a prejudice against the use of ammonia. Soda alum has no odor in baking, no unpleasant taste, produces the largest amount of gas and leaves the smallest amount of residue. It is the cheapest of all acid ingredients, has the best keeping quality and preserves all the other ingredients. It is most commonly used with phosphates. It takes up all the moisture, keeping the baking powder very dry.

Chemical action in baking powder occurs when the baking powder comes in contact with moisture and heat. Soda should be used in the granular form in pure phosphate and pure cream of tartar baking powders. The other ingredients, except the starch, can also be used to advantage in this form, because it has been discovered that the ingredients in granular form keep much better than those in powdered form.

CHAPTER VII

PIE FILLINGS AND PIES

Pie Fillings

The manufacture of pie fillings is very similar to that of preserves, the essential difference being that starches, instead of pectin powder, are used for thickening purposes. In making pie fillings insufficient cooking is just as bad as over-cooking. In the first case, the product will be too liquid and run out of the pie, while in the second case, the product will be too firm or rubbery and not pleasant to eat.

Apple Pie Filling

Evaporated Apples	50 lb	Mixed Spices	9 oz
Cane Sugar	50 lb	Sodium Benzoate	4 oz
Corn Syrup	150 lb	Salt	8 oz
Corn Starch	12 lb	Water	10 gal
Tapioca Starch	4 lb		

Soak fruit in water overnight. The following morning soften the fruit with live steam and then put through the meat grinder.

Add ten gallons of water to the kettle and warm up just enough to take the chill out of the water. Now draw off three gallons for the corn starch and one gallon for the tapioca in separate mixing receptacles.

To the water left in the kettle, add the steamed evaporated apples. Start the steam. Then add cane sugar. Stir around to dissolve it and add the corn syrup and salt. Cook the contents of the kettle until the temperature reaches 214° F. At this point add the tapioca starch which has been made into a paste. The batch should not be cooked too vigorously, and should be stirred until it becomes transparent. Now practically shut off the steam and add slowly the corn starch, previously made into paste with the water. Cook up until heavy and the corn starch becomes transparent.

Now add the mixed spices and a solution of the sodium benzoate. Transfer the pie filling to the cooling table, and when cooled down to about 130° F, pour into pails.

Blueberry or Huckleberry Pie Filling

Formula No. 1

Fruit	100 lb	Sodium Benzoate	4 oz
Cane Sugar	75 lb	Citric Acid	9 oz
Corn Syrup	75 lb	Powdered Pectin No. 80	1½ lb

Mix the pectin powder with about 5 pounds of cane sugar.

Add 6 gallons of water to the kettle. Bring to a boil and add the pectin mixture slowly with mixing. When completely dissolved, add the cane sugar, corn syrup, solution of sodium benzoate, and then the fruit. Bring the contents to a full boil, and when it sheets off the paddle, shut off steam. Now add the solution of citric acid, and mix in. Transfer the filling to the cooling table, and when cool run into pails.

Formula No. 2

Fruit	100 lb	Tapioca Starch	2 lb
Cane Sugar	75 lb	Sodium Benzoate	6 oz
Corn Syrup	75 lb	Citric Acid	9 oz
Corn Starch	6 lb	Water	10 gal

Add 10 gallons of water to kettle and warm up to remove the chill. Remove 8 gallons to a receptacle and mix the 8 pounds of starches in it.

To the 2 gallons left in the kettle, add the sugar and the fruit, and bring to boiling. Add the corn syrup. Cook to 214° F and then add the starch paste slowly, with stirring, until the starches become transparent. Cook it all up to the proper heaviness. Shut off the steam, and then add the solution of sodium benzoate. Stir around. Add the solution of citric acid and stir. Now transfer the filling to the cooling table, paddle around on the table to mix properly, and when cooled down run into pails.

Cherry Pie Filling

Formula No. 1

Fruit (Montmorency or Morello Cherries)	60	lb	Shortening	3	lb
			Salt	8	oz
Cane Sugar	50	lb	Sodium Benzoate	3	oz
Corn Syrup	45	lb	Dry Cherry Color	¼	oz
Corn Starch	8	lb	Water	6	gal
Tapioca Starch	1	lb	Citric Acid	4	oz

Place 6 gallons of water in a kettle and warm up to remove the chill. Remove 1½ gallons and ½ gallon into receptacles, for the mixing of the corn and tapioca starch.

To the balance of water in the kettle add the sugar. Turn on steam. Then add the corn syrup, and while it is cooking add the shortening, salt, and the fruit. Cook up to 214° F. Now add the tapioca starch paste and cook to transparency. Add the corn starch paste and cook to proper transparency and heaviness. Shut off steam and add a solution of sodium benzoate, then citric acid. Paddle the contents around. Transfer to cooling table, and when cooled off pour into pails.

Formula No. 2

Sour Cherries	150 lb	Citric Acid	8 oz
Cane Sugar	130 lb	Water	12 gal
Corn Starch	10 lb	Shortening	5 lb
Tapioca Starch	6 lb		

Add 12 gallons of water to kettle and warm up to remove the chill. Now draw off 3 gallons for the corn starch and 2 gallons for the tapioca, and in separate receptacles, make into starch pastes.

To the water left in the kettle add the cane sugar, and while cooking add the shortening and then the fruit. Boil the contents up to 214° F. Then add the tapioca starch paste and cook until it becomes transparent. Add the corn starch paste. Cook up to the proper heaviness. Shut off steam and add a solution of citric acid. Now transfer to the cooling table and then into pails.

Fig Filling

Formula No. 1

Fig Paste	50 lb	Sodium Benzoate	
Dried Figs	50 lb	(Optional)	4 oz
Cane Sugar	100 lb	Water	10 gal
Salt	8 oz		

Soak the figs and fig paste overnight in water. In the morning put the figs through the meat grinder.

Bring the 10 gallons of water in the kettle to a boil and add the cane sugar and then the fig paste. Dissolve the sodium benzoate and salt in a little water and add to kettle. Bring the contents of the kettle up to 224° F or to proper heaviness.

Formula No. 2

Fig Paste	50 lb	Salt	8 oz
Dried Figs	50 lb	Sodium Benzoate	
Cane Sugar	100 lb	(Optional)	5 oz
Corn Syrup	130 lb	Water	10 gal

Make in same way as recommended above for Formula No. 1.

Formula No. 3

Dried Figs		Sodium Benzoate	5 oz
Dried Apples	15 lb	Salt	8 oz
Cane Sugar	100 lb	Water	15 gal
Corn Syrup	125 lb		

Soak apples and figs overnight in separate containers or soften the apples with live steam. Put both through the meat grinder.

To the water in the kettle add the cane sugar. Bring to boiling and dissolve the sugar. Add the fig and apple paste, then the corn syrup. Dissolve the salt and sodium benzoate in a little water and add to the kettle. Boil the contents of the kettle to 224° F or to the proper heaviness.

Lemon Pie Filling

Formula No. 1

Powdered Lemon Juice	¾	lb	Citric Acid	1	lb
Cane Sugar	110	lb	Salt	2	lb
Corn Syrup	165	lb	Lemon Oil (Dissolved in Alcohol)	5	oz
Shortening	10	lb	Sodium Benzoate	9	oz
Corn Starch	32	lb	Egg Color Solution	5	oz
Tapioca	10	lb	Water	36	gal
Powdered Egg Yolk	3	lb			

Measure the 36 gallons of water into the kettle. Remove chill. Into separate mixing receptacles, draw off 6 gallons for corn starch, 3 gallons for tapioca starch, ½ gallon for egg powder.

Start cooking the balance of the water in the kettle and add the cane sugar, corn syrup, shortening, sodium benzoate and salt, and boil to 213 to 214° F.

Whip up the tapioca starch with the 3 gallons of water into a paste and add the paste to the kettle, stirring around with the paddle until the starch gels or becomes transparent. This should take place in 1 or 2 minutes.

Now whip up the corn starch with the 6 gallons of water. Add the color to it, and then add this to the kettle. Cook the batch until the starch gels or becomes transparent. Avoid too long cooking or the resulting pie filler will be too heavy. Paddle the mixture thoroughly after the corn starch has been added. Shut off steam.

Whip up the egg powder until free from lumps with the ½ gallon of water drawn before. Add to kettle and quickly mix it in thoroughly. Now draw off about 3 gallons from the kettle and whip up with the lemon juice powder and the citric acid dissolved in a little water. Then put it back into the kettle.

Transfer the batch onto the cooling table, and while paddling add the lemon flavor. When cooled down to about 130° F fill the pails.

Egg Color Solution

Tartrazine	9 oz	Water	2½ gal
Orange I	1 oz		

Formula No. 2

Powdered Lemon Juice	4 oz	Citric Acid	2 oz
Cane Sugar	25 lb	Salt	2 oz
Corn Syrup	20 lb	Lemon Extract	2 oz
Shortening	2 lb	Sodium Benzoate	2 oz
Corn Starch	9 lb	Egg Color Solution	½ oz
Tapioca	1 lb	Water	52 lb
Powdered Egg Yolk	1 lb		

Add the following ingredients to kettle:

Water	37 lb	Shortening	2 lb
Cane Sugar	25 lb	Salt	2 oz
Corn Syrup	20 lb		

Cook them to 214° F.

Water	15 lb	Corn Starch	9 lb
Tapioca Starch	1 lb		

Whip up into paste and add with paddling to the kettle. Cook until the starches gel and drop in sheets from the paddle. Shut off the steam.

Whip up with ½ gallon of water the egg powder, lemon powder, citric acid, color and sodium benzoate, and then add to the kettle. Stir around with the paddle and then transfer to the cooling table. Add flavor to pie filler while on cooling table.

Formula No. 3

Water	6 lb	Lemon Rind	2½ oz
Sugar	4½ lb	Salt	½ oz

Bring the above to a good boil.

Cold Water	2 lb	Corn Starch	1 lb

Suspend the starch in the water. Add this to the above boiling mixture and cook clear.

Egg Yolks	1 lb

Beat the yolks slightly. Then stir into the boiled mixture.

Lemon Juice	1 lb	Sugar	1 lb

Stir in the lemon juice. Add the sugar last. Mix smooth. Then immediately fill baked pastry shells.

Formula No. 4

Water	3 qt	Sugar	5 lb
Lemon Gratings	2 oz	Agar-Agar	½ oz
Salt	1 oz		

Cook to boil and continue boiling for 3 minutes.

Corn Starch	14 oz	Water	1 qt

Add to the above mixture and cook until clear. Remove from fire.

Eggs (Whipped Smooth)	8	Lemon Juice	18 oz

Thoroughly incorporate eggs into the mixture and then add the lemon juice.

Formula No. 5

Water	6 lb	Salt	½ oz
Sugar	4 lb	Lemon Gratings	2½ oz
Corn Syrup	8 oz	Agar-Agar	½ oz

Cook the above ingredients to boil and continue boiling 3 minutes.

Corn Starch	14 oz	Powdered Egg Yolk	7 oz
Water	2½ lb		

Add these to the boiling mixture and cook until clear. Remove from fire.

Lemon Juice	1⅛ lb	Sugar	1 lb

Add the lemon juice and then the sugar last.

Mince Meat Pie Filling

Water	8	gal	Seedless Raisins	100	lb
Evaporated Apples	50	lb	Spices	1½	lb
Brown Sugar	40	lb	Sodium Benzoate	9	oz
Corn Syrup	100	lb	Salt	1½	oz
Beef Suet	15	lb	Wine	2	gal
Meat	60	lb	Alcohol	2	pt

Soak the apples and raisins overnight. To the water in the kettle add the cane sugar, corn syrup, salt and sodium benzoate, and bring to a boil. Add the apples and raisins, which have been put through the grinder. Boil the contents to 214° F.

Cook meat until soft, then put through the meat grinder. Transfer to kettle above and cook to proper consistency. Shut off the steam, and now add spices, wine, and alcohol and mix with agitator.

The beef suet after being put through the grinder can be incorporated, hot or cold, into the batch. When the entire batch is in the cooling pan, or cooling down in the kettle, add the suet and continue agitating; or the beef suet can be added with the meat to the hot batch and thoroughly mixed in.

Orange-Pineapple Cake Filling

Formula No. 1

Crushed Pineapple	48	(No. 10 tins)
Cane Sugar	350	lb
Corn Syrup	40	lb
Agar-Agar	2½	lb
Sodium Benzoate	12	oz
Salt	2½	lb
Orange Color*	16	oz
Orange Oil	8	oz
Water	8	gal

Soak the agar-agar in 8 gallons of water overnight.

Add the pineapple juice to the kettle. Bring to a boil. Now add the sugar, corn syrup, salt, a solution of sodium benzoate and the color. While it is still boiling add the fruit and cook up to 216° F. Now add the agar-agar solution, putting it through a strainer first, and continue boiling to 221° F. Shut off the steam, transfer to cooling table, and when temperature is around 140° F fill pails.

*Orange Color Solution

Orange I	4 oz	Water	1 gal

Formula No. 2

Crushed Pineapple	24 (No. 10 tins)
Ground Orange Peel	90 lb
Cane Sugar	400 lb
Liquid Pectin (See Jelly Chapter)	15 gal (Or 5 gal Douglas Pectin)
Citric Acid	8 oz
Water	15 gal

Add 15 gallons of water to kettle. Bring to boil, then add the sugar, pineapple and ground orange peel. Cook to 221° F and add the pectin juice. Cook until temperature is 223° F or until the batch sheets off the paddle. Shut off steam. Add a solution of the citric acid and then transfer to cooking table. A little orange color and sodium benzoate solution may be added, if desired.

Peach and Apricot Pie Filling

Evaporated Peaches or Apricots	75 lb	Corn Syrup	75 lb
Cane Sugar	75 lb	Sodium Benzoate	5 oz

Soak peaches or apricots overnight, cook up fruit, sugars and sodium benzoate to around 224° F.

Raspberry Pie Filling

Raspberry Fruit	25 lb	Red Color	12 oz
Raspberry Pulp	25 lb	Powdered Pectin No. 80	3½ lb
Cane Sugar	10 lb	Water	15 gal
Corn Syrup	300 lb		

Mix pectin with the 10 pounds of cane sugar.

Boil the 15 gallons of water and then add the pectin powder. When it is dissolved, add the corn syrup, fruit and color, and cook to 221° F, or to a good sheet test.

Run hot filling 30-pound pails, having 5 ounces of acid solution in them. True fruit flavor can be added if desired.

Color Solution

Amaranth	4 oz	Water	1 gal

Acid Solution

Tartaric Acid	4 lb	Water	1 gal

Raspberry-Peach Filling

This filling is made with the addition of 30 pounds of evaporated peaches to the above formula.

The peaches should be soaked overnight, and in the morning put through the grinder.

Butterscotch Pie Filling

Brown Sugar	6 lb	Vanilla Extract	1 oz
Water	5 lb	Corn Starch	1¼ lb
Butter	8 oz		

Dissolve the corn starch in a small part of the water. Place the other ingredients in a kettle and bring to a boil. Slowly add the suspended starch and thicken. Allow to cool before using. For best results cover with a greased paper. Before using stir thoroughly to obtain a smooth plastic filling. A little yellow color may be added to improve the appearance.

Butterscotch Pie Filler Powder

Formula No. 1

Light Brown Sugar	22 oz	Corn Starch	5½ oz
Skimmed Milk Powder	3½ oz	Salt	¼ oz
Powdered Egg Yolk	1 oz	Water	4½ pt
Roasted Powdered Skimmed Milk	3½ oz		

Formula No. 2

Light Brown Sugar	23½ oz	Corn Starch	5 oz
Skimmed Milk Powder	6 oz	Salt	¼ oz
Roasted Powdered Skimmed Milk	2½ oz	Water	4½ pt

Butterscotch pie filling can be made by bringing to a boil 2½ pints of water, then stirring in the above ingredients into the balance of 2 pints of water to a smooth mixture, adding this to the boiling water, and cooking until thick and glossy, 185 to 190° F. For better flavor, add 2 ounces of butter. Add vanilla or maple flavor.

Lemon Pie Filler Powder

Granulated Sugar	28 oz	Citric Acid	⅒ oz
Powdered Whole Eggs	1 oz	Salt	¼ oz
Powdered Lemon Juice	2½ oz	Water	4½ pt
Tapioca Starch	6½ oz		

The lemon pie filler may be modified considerably to reduce the cost. Less lemon juice can be used and lemon flavor substituted, etc.

The above when cooked properly, in the manner suggested for butterscotch pie powder filling, results in a good product. The addition of 2 ounces of butter will make a richer filling.

Chocolate Pie Filling

Formula No. 1

Granulated Sugar	20 oz	Corn Starch	5½ oz
Dried Whole Milk	5 oz	Salt	¼ oz
Powdered Egg Yolk	1¼ oz	Water or Milk	5 pt
Cocoa Powder	4 oz		

Formula No. 2

Granulated Sugar	20 oz	Cocoa Powder	3½ oz
Powdered Skimmed Milk	3½ oz	Salt	¼ oz
Powdered Sugar	2½ oz	Water or Milk	5 pt

Bring 3 pints of water to a boil. Now mix into a smooth paste the ingredients with the balance of 2 pints of water or milk. Then add this to the boiling water. Cook until thick and glossy at 185 to 190° F. Add 2 ounces of butter for better quality.

Formula No. 3

Sugar	6 lb	Cocoa	12 oz
Water	5 lb	Butter	4 oz
Vanilla Extract	1 oz	Corn Starch	1 lb

Dissolve the corn starch in a small part of the water. Place the other ingredients in a kettle and bring to a boil. Add the suspended corn starch and continue boiling until gelatinized. Cover with a greased paper and allow to cool before using. Stir thoroughly to get a smooth filling.

Gelatin Icing Powder

Gelatin	10 lb	Gum Arabic	2 lb
XXXX Sugar	10 lb		

One ounce of this powder added to ½ pint hot water, then stirred well, can be used instead of egg whites. It can be colored and flavored.

Imitation Egg Powder Compound

Powdered Egg White	2 lb	Gum Karaya	3 lb
Powdered Egg Yolk	1 lb	Potato Flour	4 lb

Lemon Powder for Pie Fillings

Corn Starch	75 lb	Powdered Sugar	20 lb
Citric Acid	5 lb	Lemon Oil	6 oz

Vanilla Powder

Powdered Mexican Vanilla Bean	50 lb	Granulated Sugar	50 lb
		Oil Bitter Almond	½ oz

Pie Filling Powder Base for All Flavors

Corn Starch	45 lb	Salt	1 lb
Tapioca Starch	15 lb	True Fruit Flavor	To Suit
Citric Acid	1 lb	Imitation Fruit Flavor	To Suit
Gum Karaya	½ lb	Certified Food Color	To Suit

This recipe, when properly mixed with sugar, will make the filling for pies. Cherry, strawberry, and raspberry flavors may be used. For lemon, a small percentage of powdered egg yolk will improve the taste of the product.

One pound of the above formula is dispersed in 1 quart of warm water. Then add this to a hot solution of 3 pounds of sugar in 3 quarts of water. Boil, and remove from fire as soon as it thickens.

Pies

A really good pie crust should contain at least 65 to 80% shortening, based on the weight of the flour. Good quality can be obtained with proper manipulation of the flour and shortening before the water is added. Poor pie crust will not bring good business.

Flaky Pie Crust

Flour	3 lb	Salt	1¼ oz
Shortening	2¼ lb	Water (Variable)	12 oz

Mix all the shortening with all of the flour to obtain an irregular mixture with little lumps of fat throughout. Add cold water and mix very little. For a short mealy type of crust, mix all the shortening with half the flour. Add balance of flour and mix it in, to break up the creamed mass. Add cold water and mix just enough to incorporate it. For a short flaky type of crust, mix half the shortening with all the flour until you obtain distribution of the fat. Add balance of the shortening and mix in lightly, so that it is left in little lumps throughout the dough. Next, add cold water and mix lightly.

Pastry Pie Dough

Pastry Flour	8 lb	Salted Butter	1 lb
Shortening	4 lb		

Mix the above ingredients together well.

Cold Water	3½ lb	Sugar	2½ oz
Salt	4 oz		

Dissolve the salt and sugar in half of the water.

Mix shortening with flour so that the distribution of fat will form an irregular flaky mixture. Then rapidly add the cold water and mix very little so that the flakes will be left throughout the dough. Place in ice box at 40 to 42° F for 8 to 10 hours before using.

This pastry pie dough is ideal for pie tops, turnovers, cheese straws and tart shells.

Pie Crust

Cake Flour	25 lb	Water	6 lb
Shortening	18 lb	Skimmed Milk Powder	¼ lb
Salt	½ lb		

Mix the skimmed milk powder with the flour. Then mix all the shortening with the flour. Add the cold water slowly to the batch and mix very little.

Rolled Pie Crust

Flour	5 lb	Shortening	2¼ lb
Shortening	1¼ lb	Ice Water (Variable)	2 lb
Salt	2½ oz		

Mix the flour, salt and 1¼ pounds shortening together well. Mix the 2¼ pounds shortening in, in large lumps. Mix the ice water in lightly. After the water has been incorporated, roll the dough out to a thickness of approximately ¼ inch and fold a third of the dough over half the remaining two-thirds. Then fold the final third of the dough over the top to make the third layer. Roll out again to a thickness of ¼ inch. Fold to give three layers again and roll out to a thickness of approximately ½ inch.

Then cut the dough into pieces which, when rolled out thin, will be the approximate size for the top crust of a pie. These pieces can be put on pans in the refrigerator and used as needed.

Cold Pack Fruits Used in Pies (Sugar Added)

The term Cold Pack refers to all fruits prepared by mixing the fruit with cane sugar and freezing to a solid mass. The proportion varies from 2 to 5 parts of fruit to 1 part sugar.

All Cold Pack fruits and berries must be thawed and thoroughly drained. The following formulas are for 4 plus 1 pack. The water in these formulas is used in addition to the drained juice. Water and sugar may be varied to suit individual requirements.

In all formulas, drain fruit and cook drained juice, water, sugar and agar-agar to boil for 3 minutes. Add thickener. Cook to reboil. Then pour over drained fruit and allow to cool.

Basic Rule for Fruit Pie Filling

For each quart of liquid and 2 pounds sugar, use ¼ ounce agar-agar and ½ ounce corn starch. Agar and starch must be cooked to a rolling boil and boiled continuously for 3 minutes.

Apple Pie
(Cold Pack)

Apples	30 lb	Agar-Agar	¾ lb
Water	3½ qt	Cinnamon	¾ oz
Sugar	5 lb	Nutmeg	¼ oz

Thicken the above ingredients with the following:

Corn Starch	6 oz	Water	1 pt

Apple Meringue Pie

Applesauce		Salt	½ oz
Water	1 lb	Cinnamon	¼ oz
Sugar	1½ lb		

Bring the above ingredients to a boil.

Water	8 oz	Egg Yolks	8 oz
Corn Starch	4 oz		

Mix the corn starch and the water and stir in the egg yolks. Gradually stir into the boiling mixture. Cook until smooth.

Melted Shortening	8 oz	Flavor	To taste

Fold in the shortening and flavor. Pour while hot into baked pie shells. Cool and top with a meringue. Sprinkle lightly with cinnamon and brown at 400° F.

Apple Pie

Sugar	2 lb	Nutmeg	1/8 oz
Corn Starch	2 oz	Salt	1/8 oz
Cinnamon	1/8 oz		

Mix the above filling ingredients together thoroughly. Sprinkle the filling over the bottom of the pie crust. Fill up with sliced apples. Cover with more sugar and dot with small pieces of shortening. Add top crust and bake at 425 to 450° F.

Apricot Pie Filling
(From Dried Fruit)

Dried Apricots	25 lb	Sugar	20 lb
Water	5 gal	Salt	3 oz

Soak above ingredients for 2 to 3 hours.

Water	1½ gal	Corn Starch	4 oz
Sugar	5 lb	Agar-Agar	2 oz

Just before cooking apricots, bring the above to a boil for 3 minutes.

Then cook apricots slowly for 10 to 15 minutes. Add the cooked stabilizer solution and remove from stove.

This procedure may be followed for peaches, figs, prunes, etc.

Apricot Pie
(From Canned Fruit)

Water	2¾ qt	Salt	1 oz
Sugar	9 lb	Agar-Agar	1 oz

Cook to boil and continue boiling for 3 minutes. Add the following, and cook until clear.

Starch	9 oz	Tapioca Flour	9 oz
Water	1 qt		

Add and mix three cans of solid pack pre-heated apricots and juice.

Apricot Cream Pie

Sugar	1 lb	Flour	1 oz
Milk	2 lb	Shortening	2 oz
Corn Starch	1 oz	Salt	¼ oz
Cooked or Canned Apricots	1 lb	Egg Yolks	3 oz

Make a paste of the flour and starch in part of the milk. Bring balance of milk, sugar, salt and shortening to a boil. Add paste and cook until thick. Add beaten yolks and remove from fire. Mix in apricots which have been drained of their juice. Let filling cool, and pour into baked pie shells. Top with meringue or whipped cream.

Blueberry Pie
(Cold Pack)

Blueberries	30 lb	Sugar	10 lb
Water	5 qt	Agar-Agar	1 oz
Salt	1½ oz		

Thicken the above with the following:

Tapioca Flour	12 oz	Corn Starch	12 oz

Cherry Pie

(Cold Pack)

Sour Cherries	30	lb	Corn Starch	8	oz
Juice (No Water)	6	qt	Water	1	pt
Agar-Agar	1	oz	Sugar	3½	lb

Cook the cherries, agar-agar and juice, and boil for 3 minutes. Thicken with the corn starch and water. Add the sugar and cook to boil.

Cherry Pie

(From Canned Fruit)

Drain 2 No. 10 cans of cherries.

Agar-Agar	½	oz	Corn Starch	6	oz
Drained Juice	2	qt	Sugar	5	lb
Juice or Water	1	pt			

Cook the agar-agar and the drained juice to boiling and continue boiling for 3 minutes. Add the juice or water and the corn starch, and cook until clear. Then, while it is still cooking, stir the sugar in until it starts to boil. Remove from fire and pour over the drained cherries.

Cherry Blossom Pie

Liquid Whole Milk	3¾	lb	Salt	½	oz
Sugar	1¾	lb			

Bring the above to a boil.

Liquid Whole Milk	1¼	lb	Egg Yolks	10	oz
Corn Starch	6	oz			

Mix the corn starch in the milk and stir in the egg yolks. Add to the above boiling mixture and cook smooth.

Shortening	8	oz

Remove from fire and stir in the shortening.

Egg Whites	12	oz	Granulated Sugar	7	oz

Whip the egg whites and sugar light and fold into the cooled filling. Line the bottom of a baked pie shell with cooked cherries. Cover with the above cream filling and top with meringue. Brown in oven at 400° F.

Chiffon Pie Base

Sugar	4½ lb	Corn Starch	½ oz
Water	1 pt	Agar-Agar	1 oz

Cook to a rolling boil and continue boiling for 3 minutes.

Egg Whites	1 qt

Stiffly beat the egg whites. Slowly pour hot syrup into beaten egg whites and continue whipping to a peak. Add any type of fruit or berries. Place in baked shell and top with whipped cream.

Chocolate Cream Pie

Sugar	2½ lb	Milk Powder	8 oz
Salt	¼ oz	Water	3 lb

Bring the above ingredients to a boil.

Water	1 lb	Egg Yolks	10 oz
Corn Starch	5 oz		

Mix the starch in the water and add the egg yolks. Stir into the above solution. Cook until smooth.

Melted Chocolate	8 oz	Vanilla Extract	To taste
Shortening	2 oz		

Stir in the melted chocolate, shortening and flavor. Pour the hot filling into baked pie shells. Cool and top with whipped cream.

Chocolate Malted Milk Pie

Milk	2 lb	Salt	⅛ oz
Sugar	1 lb	Malted Milk Powder	5 oz
Egg Yolks	5 oz	Melted Chocolate	4 oz
Corn Starch	2¾ oz		

Mix the corn starch, egg yolks and salt together, and thin down with some milk. Bring balance of the milk, sugar and malted milk powder to a boil, and stir in the starch mixture rapidly. Lower heat to prevent lumping and cook until thick. Stir in the melted chocolate and then cool. Pour into baked pie shells. Top with whipped cream.

Coconut Cream Pie

Liquid Whole Milk	6 lb	Salt	1 oz
Granulated Sugar	3 lb		

Bring to a boil.

Liquid Whole Milk	2 lb	Shortening	8 oz
Corn Starch	10 oz	Flavor	To taste
Egg Yolks	1¼ lb		

Mix the corn starch in the milk and stir in the egg yolks. Gradually stir into the boiling mix. Cook until smooth.

Remove from fire and fold in the shortening and flavor. Line the bottoms of baked pie shells with shredded coconut, fill with the above filling and cool. When cool, top with a light meringue, sprinkle with shredded coconut and brown in oven at 400° F.

Lemon Chiffon Pie

1st Mix

Sugar	4 lb	Water	2½ qt
Agar-Agar	1 oz	Grated Lemon Rind	2 oz
Salt	1 oz		

Cook to a rolling boil and continue boiling for 3 minutes.

Corn Starch	12 oz	Water	1 pt

Add to the above mixture and cook until clear.

Remove from fire and whip in 8 whole eggs thoroughly. Then add 1 quart lemon juice.

2nd Mix

Sugar	4 lb	Egg Whites	1 qt
Water	1 qt		

Cook the sugar and water to 234° F. Beat the egg whites stiffly. Slowly pour hot syrup into beaten eggs. Continue whipping until stiff.

Fold first mix into second mix with wire whip. Fill baked shells and top with whipped cream.

Lemon Meringue Pie

Water	3¾ lb	Salt	½ oz
Sugar	3¼ lb	Lemon Gratings	3 oz

Bring to a boil.

Water	1¼ lb	Lemon Juice	1¼ lb
Corn Starch	8 oz	Melted Shortening	5 oz
Egg Yolks	10 oz		

Mix the starch in the water and stir in the egg yolks. Add to the above solution and cook until clear.

Stir in the lemon juice and melted shortening.

Pour the filling while hot into baked pie shells. Cool and top with meringue. Brown at 400° F.

Orange Pecan Pie

Water	2 lb	Corn Starch	5 oz
Sugar	2 lb	Salt	¼ oz
Egg Yolks	8 oz	Melted Shortening	2 oz
Orange Juice	8 oz	Toasted Pecans	6 oz
Orange Gratings	2 oz		

Mix the corn starch, salt and egg thoroughly, and thin down with some of the water. Bring balance of the water, sugar, orange juice and gratings to a boil, and stir in the corn starch mixture rapidly. Lower heat to prevent lumping and cook until thick. Stir the melted shortening into the thickened mass before it is cool. Pour the filling into baked pie shells and sprinkle pecans. Top with a meringue.

Peach Pie
(Cold Pack)

Peaches	30 lb	Salt	1½ oz
Water	1½ qt	Agar-Agar	¾ oz
Sugar	6 lb		

Thicken the above ingredients with the following:

Water	1 pt	Corn Starch	14 oz

Peach Pie

Peaches	1	No. 10 Can
Water	1	qt
Sugar	6	lb
Salt	1½	oz
Agar-Agar	¼	lb
Color	To suit	

Cook to boil and continue boiling for 3 minutes.

Corn Starch	11	oz
Water	1	pt
Peaches	2	No. 10 Cans

Add the corn starch and water, and cook until clear. Remove from fire and add the peaches. The peaches may be cut into small pieces so they will cook into a jam and enhance the flavor of the pie. One can of apricots to four cans of peaches may be added to improve flavor.

Pineapple Pie

Crushed Pineapple	7 lb	Salt	½ oz
Sugar	2 lb	Water	1 lb

Bring to a boil.

Water	1 lb	Egg Yolks	4 oz
Corn Starch	5 oz		

Mix the corn starch in the water and add the egg yolks. Stir into the above boiling mixture. Cook until clear. Pour while hot into baked pie shells. Cool and top with whipped cream.

Pineapple Pie

(Cold Pack)

Pineapple	30 lb	Salt	2 oz
Water	3 qt	Orange Flavor	To suit
Sugar	10 lb	Agar-Agar	1 oz

Thicken the above ingredients with the following:

Water	1 qt	Corn Starch	1¼ lb

Pineapple Chiffon Pie

1st Mix

Pineapple	1 No. 10 Can
Water	1 pt
Sugar	1 lb
Salt	1 oz
Agar-Agar	1 oz
Grated Orange or Emulsion	1 oz

Cook to rolling boil and continue boiling for 3 minutes.

Starch	8 oz	Water	1 pt

Add to the above mixture and cook until clear.

2nd Mix

Sugar	4 lb	Egg Whites	1 qt
Water	1 qt		

Cook the sugar and water to 234° F. Beat the egg whites until stiff. Slowly pour hot syrup into beaten egg whites and continue whipping until stiff.

Fold first mix into second mix with wire whip. Fill baked shells and top with whipped cream.

Pumpkin Pie

Formula No. 1

Brown Sugar	2¼ lb	Corn Starch	1 oz
Salt	2 oz	Spice Mix*	¾ oz

Mix the above ingredients together.

Whole Eggs	1½ lb	Liquid Whole Milk	4½ lb
Canned Pumpkin	4½ lb	Melted Shortening	6 oz

Add the eggs and whip in. Stir the pumpkin in thoroughly. Add the whole milk and mix to a smooth liquid. Stir the

* See page 418.

melted shortening in. Pour the filling into unbaked pie shells and bake at 400° F.

*Spice Mix

Cinnamon	1 oz	Ginger	½ oz
Allspice	½ oz	Mace	½ oz

Mix ingredients together.

Pumpkin Pie

Formula No. 2

Pumpkin	1 No. 10 Can
Sugar	3½ lb
Salt	2 oz
Spices	1½ oz
Tapioca Flour	4 oz
Agar-Agar	½ oz
Milk (Variable)	4 qt
Eggs	10

Place pumpkin in mixing bowl. Mix all the dry ingredients and thoroughly incorporate with pumpkin, using paddle. When it is smooth, add slightly whipped eggs, and mix thoroughly. Add 3 to 4 quarts milk. Allow mix to stand for 3 to 4 hours before baking. In small batches, eggs can be whipped and added last, after batch has stood for 2 to 4 hours. Bake at 450° F for 30 to 45 minutes, depending on size of pie.

Raisin Pie

(From Dried Raisins)

Raisins	25 lb	Salt	2 oz
Water	2 gal	Agar-Agar	½ oz
Sugar	12 lb		

Soak the raisins in the water for 4 to 6 hours or overnight. Add the sugar, salt and agar-agar.

Water	1 qt	Lemon Peel	2 oz
Corn Starch	1 lb	Cinnamon	½ oz
Flavor			

Bring the above ingredients to a boil and continue boiling for 3 minutes.

Raspberry Pie
(Cold Pack)

Raspberries	30 lb	Water	1½ qt
Sugar	6 lb	Agar-Agar	1 oz
Salt	1½ oz		

Thicken the above with the following:

Corn Starch	12 oz	Water	1 qt

Rhubarb Pie
(Cold Pack)

Rhubarb	30 lb	Sugar	8 lb
Water	1½ qt	Agar-Agar	¾ oz
Salt	1 oz		

Thicken the above with the following:

Water	1 pt	Tapioca Flour	8 oz
Corn Starch	8 oz		

Strawberry Pie
(Cold Pack)

Strawberries	30 lb	Salt	1½ oz
Water	1½ qt	Agar-Agar	1 oz
Sugar	6 lb		

Thicken the above with the following:

Water	1 pt	Corn Starch	1 lb

Strawberry Pie

Water	2 gal	Agar-Agar	1 oz
Sugar	12½ lb	Corn Starch	2 oz
Salt	2½ oz	Color	To suit

Cook to rolling boil and boil 3 minutes.

Water	2 qt	Hard Frozen Strawberries	40 lb
Corn Starch	2 lb		

Add the corn starch and water to the above and cook until clear. Remove from fire and immediately add the strawberries.

The cold frozen berries added immediately to the hot glacé reduce the temperature of the glacé. The filling should be used as soon as possible. In this way the berries will retain most of their natural goodness. Allowing the filling to stand over a long period causes berries to bleed and shrink excessively.

CHAPTER VIII

COOKIES, DOUGHNUTS AND CAKES

Cookies

The dictionary definition states that cookies are small sweet cakes, sometimes spiced. This definition embraces a wide variety of bakery products such as crackers, biscuits, common cookies, soft drop cake cookies, flour-confections, cup cakes and other small cakes. The following facts concerning the development and use of formulas apply to all of these cookie products.

When the formulas for all cookie products are included in a single group, they seem to specify almost all conceivable proportions of ingredients. In fact, by mixing and molding materials by hand, great variations in the proportions of flours, sugars, fats, eggs, milk and flavor-carrying ingredients can be effected, and the final doughs or batters can be baked into products having an attractive appearance and pleasant eating-qualities.

However, when the proportions of ingredients are kept within certain narrow limits, they produce definite characteristics. When the structure of the cookies is formed chiefly by eggs, e.g., sponge drops and ladyfingers, they may be called sponge cookies. Cookies which contain a large proportion of fat, e.g., pound cake, may be termed short-breads. Variety cookies, e.g., vanilla wafers and sugar cookies, have small amounts of fat and eggs. They are somewhat analogous to batter cakes. Moreover, machines have been built which do not only duplicate hand operations but also make it possible to produce economically certain new characteristics.

Each type of cookie is mixed in a manner similar to the way in which various types of cakes are mixed. Sponge cookies, among which are ladyfingers, almond macaroons and meringue kisses, are made by whipping eggs and then folding in the dry ingredients. Short-breads are mixed by creaming fat, sugar and eggs and then adding the liquid and dry materials. Short-breads can be mixed by incorporating the fat and flour and then adding the eggs and sugar. Variety cookies are made by creaming the sugar, shortening, eggs, and adding the liquids and flour. This is similar to the way short-breads are made. The methods are analogous to the methods used in making baking powder batter cakes.

Cookie doughs are going through two stages of development. First, doughs that formerly were made up by hand-rolling or cutting or by hand-bagging, are being run on machines built to imitate hand work. Second, these doughs are being modified somewhat for better running or to improve eating-qualities.

Use of Sugar

In order to develop formulas for crunchy cookies, large proportions of sugar should be used. When more sugar than flour is used, it is advisable to pay particular attention to the flour. Be sure to use a flour strong enough to carry the sugar. If a weak flour is used, the cookies fall and do not have the proper thickness or shape.

When making crisp cookies, very little milk or water should be incorporated in the batters in order to secure proper spread. Instead of using extra liquids, spread should be obtained in other ways. Coarse granules of sugar produce more spread than fine particles of powdered sugar. If part of the sugar is added with the flour, it causes more spread and more of a macaroon top effect than if all of the sugar is creamed with the shortening. Dissolving the sugar in the liquids makes for a uniform spread and smooth top.

In order to produce a macaroon top on cookies, use a stiff dough containing high proportions of sugar and leavening agents. Part of the sugar, preferably coarse granulated, should be added just before finishing the mixing. Also, if desired, put extra granulated sugar on top of the cookies before baking. Use plenty of steam in the oven.

Macaroon-top cookies are very attractive when the tops of the raised peaks have a rich golden-brown color. This appealing quality is produced in cookies made with formulas which specify the proper proportion of Nulomoline.

The spread of cookies is altered by incorporating the sugars used in different ways, or by balancing one type of sugar with another. If a smoother crust is desired than that secured with coarse granulated sugar in the dough, finer sugars should be used. Fine granulated and powdered sugars impart a smoother appearance.

When a smooth crust and a richer, more appetizing bloom are desired, a larger part of the total sugar should be Nulomoline. Its maximum levulose content not only imparts a flashier color without forced baking, but also produces greater spread and spring with a smooth, instead of a cracked, crust.

Use about 1 pound of Nulomoline for every 10 pounds of sugar in doughs for crisp cookies, in order to secure better spread. The Nulomoline also imparts, without forced baking, a rich, attractive, non-fading bloom. It properly distributes and holds very small percentages of moisture in crisp cookies and, thereby, eliminates checking and reduces losses due to breakage.

Two other qualities secured with Nulomoline are in contrast to those obtained with granulated sugar. Nulomoline attracts and holds moisture instead of throwing it off. Consequently, the presence of Nulomoline in the cookies aids in retaining and evenly distributing moisture which otherwise would escape from the cookies upon their removal from the oven. As a result, losses from checking are materially reduced and there is less strain and contraction of the cookies while cooling.

Moreover, while cookies which contain very high proportions of sugar, tend to grain, become crumbly and lose their snap, the levulose content of Nulomoline stops graining, and cookies made with it stay fresh longer.

Intensifying Flavors with Sugar

Primarily, sugar acts as a sweetening agent. In addition, it is valuable in bringing out or intensifying other flavors. This is particularly true with mild flavors such as banana.

Experiments conducted in the Nulomoline Testing Bakery have shown that, although a given amount of Nulco Meloban produces a mild banana flavor, it becomes much more pronounced when extra sugar is used. In fact, when Nulco Meloban is incorporated in sweet cookies, it is strong enough to carry through such ingredients as cocoa and produces very pleasing rich combinations.

Keeping Coconut Macaroons Soft and Chewy

Although it may seem that there are many different kinds of coconut macaroons, actually, almost all are alike in several respects, differing mostly in appearance and eating-qualities.

The simplest macaroons are modified confections. They are compact, flattened balls of sweetened coconut, slightly toasted on the top and bottom. This is the soft, chewy dough ball type of macaroons, usually made by specialty concerns and biscuit bakers. They are distributed in such places as the bakery and candy department of retail grocery and five-and-ten-cent stores.

In contrast to the compact dough ball type of macaroons are those of the meringue type, which are considerably lighter and fluffier. These are not simply toasted on the surface but are thoroughly baked all the way through. These chewy macaroons are not quite as soft as the dough ball variety, but are firmer because of the extra baking and set. This type is made mostly by retail bakers and by soft cake and biscuit bakers. In between the dough ball and meringue types are many varieties, but practically all are modifications of these two. The similarities of all varieties are apparent when they are considered from the point of view of their structure, rather than of their appearance.

All coconut macaroons consist of particles of coconut, surrounded by syrup which holds the coconut together. This is all there is to the dough ball type. However, if part of the sugar is dissolved in egg whites, instead of solely in water, the meringue type is produced. The egg whites trap air during mixing and then, during baking, swell and coagulate, or set. This explains why the meringue type requires thorough baking, whereas only toasting is needed for the dough ball type, which contains no ingredients which set, even if heated to high temperatures.

All other varieties of coconut macaroons are simply modifications of the two distinct types just mentioned. Sometimes starches or flour are mixed into either one or the other type; occasionally, leavening agents are also added. The quantity of eggs used in the meringue type varies considerably from only 5 pounds in a 100-pound batch to as much as 20 pounds or more. Small amounts of eggs, flour and baking powder usually are used. Of course, fresh or frozen egg whites may be replaced by equivalent amounts of dried whites (egg albumen).

One variety is made by incorporating a small proportion of fat into a modified meringue type containing flour and baking powder. This produces an almond macaroon-top appearance, smooth and glossy with shallow cracks.

Balancing the Syrup

As pointed out in a preceding paragraph, the syrup in a macaroon serves to surround and hold the coconut particles together. Moreover, it functions to sweeten the whole macaroon and to impregnate the coconut so as to soften and sweeten it. It can be readily understood why the syrup controls the quality of the finished macaroon. For retained freshness and chewiness, the syrup must not dry out, crystallize or harden, but should hold moisture and remain soft. For these properties, moisture retention and crystal control, the levulose contained in Nulomoline is most advantageous.

Laboratory and shop tests show that macaroons which contain from 5 to 20 pounds of Nulomoline in each 100-pound batch have long-lasting, soft, chewy eating-qualities. The Nulomoline also imparts a rich, golden crust color and makes the macaroons more attractive.

Selecting and Handling the Flour

When developing formulas for fruit or other specialty cookies, a flour should be selected which will carry the fruit combination and produce the cookie shape desired. The formula should not be limited by a flour which necessitates formula-juggling in order to get an acceptable shape. If, from the many types of flours available, satisfactory appearance is not obtained on the first mixing, the mixing time can be lengthened or shortened. Increased mixing usually tightens the dough, allowing the flour to carry the fruit and preventing the cookies from spreading too much during baking. Added absorption increases the spread; a tighter dough retards spread.

Revising Standard Formulas

For improved eating-qualities, some of the standard formulas can be revised as follows:

(1) Select a cereal other than wheat flour, e.g., oatmeal. (2) Use butter for at least part of the shortening. (3) Use Grandma's Old-Fashioned Molasses, a delicious, flavorful sweetener, as part of the total sugary materials. (4) Use natural flavor-carrying materials, such as ground raisins, figs, chopped nuts and Nulco Meloban. They add extra taste appeal and are low in cost.

Combinations of these materials produce unusually good eating-qualities, especially if high percentages of sugars are included to impart crunchiness. Doughs for crisp, crunchy cookies should be stiff to secure a rough, home-like finish when run on a wire-cut cookie depositing machine. When the oatmeal is in fine splinters, the dough feeds through the die and cuts off without dragging on the wire. If the oatmeal is of large flakes, they can be made finer by mixing them by themselves until broken up into small particles.

Handling Soft and Stiff Doughs

The consistency of almost all doughs must be kept within very narrow limits for the sheeter, bar press, short-bread and all other machines except the wire-cut. Practically all doughs developed specifically for other machines, with little modification, can be used on a wire-cut machine, but many wire-cut cookies cannot be made on any of the other machines.

Both soft doughs, such as those for vanilla wafers and macaroons, and stiff doughs, such as those for rough-topped sugar or short-bread cookies, can be run on a wire-cut machine. For this reason, many special cookies can be developed for production with a wire-cut machine.

Making Many Varieties on the Wire-Cut Machine

Although formulas for molasses-honey, hard cake cookies made on a sheeter are somewhat similar to those for honey jumbles, the doughs for honey jumbles are softer. Both contain high proportions of Nulomoline, sugary materials and low proportions of shortening. Honey jumbles are usually run on a wire-cut machine, are not baked hard and are kept soft by using Nulomoline for a large part of the sugary materials. The wire-cut machine is used also for a wide variety of cookies, containing great variations in sugar, shortening and flavor-carrying ingredients, and for flour-confections such as almond and coconut macaroons. Moreover, the wire-cut machine and depositor are used for handling soft doughs or batters, such as those for vanilla wafers and soft drop cake cookies.

Competitive conditions force many bakers to emphasize one or more of three phases of production. Band oven installations for cookie production are driving selling prices down. This tends to cause quality to be sacrificed for low production costs on many items. Without such equipment, bakers should not attempt to make similar varieties. Rather, they should make better cookies, by paying particular attention to all workmanship that enhances appearance, or by using unusual ingredients to improve eating-qualities.

Some materials, not normally used in cookies which are made on other machines, are particularly suitable for use with a wire-cut machine.

Soft Drop Cake Cookies

Between soft layer cakes and common cookies is a group of products which are often called drop cakes, or soft cookies.

These are not exactly soft cakes or dry cookies but have a few qualities of both. Soft cookies are particularly suited for warm, humid sections of the country where it is difficult to keep cookies crisp, but they are popular everywhere, if prevented from drying out and becoming crumbly.

Soft cookies should take color quickly in the oven without forced baking that dries out too much moisture. Also, they should retain their moisture after baking to keep them in fresh eating condition. For these purposes the use of Nulomoline is beneficial.

From 1½ to 3 pounds of Nulomoline to every 10 pounds of sugar quickly produce a rich, ruddy bloom that prevents excessive drying in the oven and gives cookies an appetizing appearance. It attracts and holds moisture, and keeps soft cookies in good condition for a long time.

Soft layer cakes are baked in pans which have sides that hold the soft doughs in shape, both during baking and until most of the eggs in the dough sets, and are able to continue to hold the shape and lightness of the moist finished product.

Cookies are baked in pans which have no sides. The formed pieces of dough are stiff enough in consistency to retain their shape during baking and are dry enough after baking to hold their shape without having a high egg content.

The doughs for soft drop-cookies are softer than for common cookies but stiffer than for layer cakes. The total egg content plus milk or water usually equals less than three-quarters, but more than one-half, of the weight of the flour. Cookies are stiff enough to hold their shape before and during baking without being dropped on pans with sides, but are so soft and moist before and after baking that they require considerable gas to retain lightness and large size.

The doughs or batters for drop cakes, or soft cookies, should be built up carefully during creaming and mixing until light, fluffy and strong. The doughs or batters should be mixed carefully, just as those for cakes made according to the usual creamed sugar-shortening-eggs method. The creamed portion and the final batter should not be curdled or broken down, as they are for so many ordinary cookie doughs.

When mixing is done carefully, the final grain and texture are uniform and not too many of the cell walls are broken down. When batches are curdled, the cookies are more crumbly. The dough or batters are usually dropped on pans greased with shortening and dusted with flour or on pans with a depression for each cookie.

Information about these soft cookies or drop cakes is useful in making many profitable items that are quite popular. For example, several varieties of soft cookies or drop cakes can be used for making five-cent varieties. Make gold, chocolate or banana round drop cakes, about 3 to 4 inches in diameter and about ½ inch thick. Top these big, soft cookies with marshmallow and enrobe the marshmallow in a quick-setting coconut fat icing, or sprinkle with coconut, decorettes, etc. As an alternative, two cookies can be sandwiched with marshmallow or imitation whipped cream. Another group of varieties may be made by baking the dough into finger forms, about 4½ inches long, 1½ inches wide and ½ inch thick, and then topping or sandwiching.

Vanilla Wafers

Vanilla wafers constitute a group of cookies which have certain unique and well-defined characteristics. They have a typical, light, fluffy, tender, pinhole grain, which is in contrast to the flat grain of hard sweets and common cookies. Wafers are made from carefully creamed, soft batters which usually contain a considerable quantity of eggs and which are deposited in a wire-cut machine. These soft batters for wafers are in contrast to the curdled stiff doughs which contain little or no eggs and which are used in making other cookies. Vanilla wafers are somewhat similar to soft drop cake cookies, except that the wafers are baked out until quite dry, whereas the soft drop cake cookies are quite moist after baking. The formulas for vanilla wafers usually specify about 35 pounds of shortening, preferably part butter, 60 pounds of sugar, 5 pounds of Nulomoline, 10 pounds of eggs and sufficient leavening for every 100 pounds of flour.

The distinctly unique grain and texture and tender eating-qualities of vanilla wafers have been popular for a long time. Other cookies having similar grain and texture but different flavor and color, e.g., molasses, chocolate and lemon, can also be made.

Changes in flavor and color stimulate sales. The chocolate-molasses base specified in this formula also improves the appearance and eating-qualities of base-cake cookies for marshmallow topped varieties.

Tender Mellow Base-Cakes

Vanilla wafers, as made on a wire-cut machine or depositor, are often modified and used as base-cake cookies for marshmallow topped varieties. These modified vanilla wafers for base-cakes usually contain less shortening, little or no eggs, but more Nulomoline. In base-cakes made on a wire-cut machine, use about 10 pounds of Nulomoline to every 100 pounds of a combination of sugar and Nulomoline for summer or humid weather; for winter and dry climates, use about 30 pounds of Nulomoline.

Base-cakes for cookies to be topped with marshmallow are also made on the rotary machine and on the sheeter. The rotary machine reduces production costs a great deal, especially when used along with a band oven. Base-cakes with a higher proportion of shortening may be made with a rotary machine, rather than with a wire-cut or sheeter, without increasing total cost. Moreover, the pieces made on a rotary machine are more uniform in shape. Uniformly shaped base-cake cookies are also made on the sheeter.

Producing a Porous Grain

In making base-cakes on a rotary machine or sheeter, the formula is adjusted to specify a considerable proportion of the total leavening agents as ammonia; this produces a quick spring without too much spread. It also makes the bottoms and insides more porous, so that the cakes quickly absorb moisture from the marshmallow, becoming mellow, soft and tender.

Securing Softness

Base-cakes made on a rotary machine or sheeter should specify more Nulomoline than used in short-bread cookies or in common cookies, so that the baked pieces will rapidly become mellow before and after being topped with marshmallow. After baking, base-cakes should be stored in a humid room with a relative humidity of about 65% for about 2 or 3 days until they become somewhat softened, before topping with marshmallow. The large amount of Nulomoline, 10 to 30% of the total sugar, in rotary and sheeter machine base-cakes is the same as that specified for wire-cut base-cakes.

Holding Moisture in Soft Cookies

Nulomoline is a valuable ingredient in all soft cookies because it attracts and holds moisture, in this way checking drying. For an explanation of what happens during the mixing and baking of soft cookies, reference should be made to the preceding comments concerning chewy cookies. Soft cookies differ from chewy cookies in that more water, which should be retained during baking and merchandising, is used in making soft cookies. Nulomoline aids in keeping them moist.

Using Sugar to Develop Crisp, Macaroon-Top Cookies

Crisp cookies are not cookies which have been dried out during or after baking, but rather, they are cookies which have crunchy, enjoyable eating-qualities imparted by the ingredients used. Some crisp cookie formulas are limited in application because they specify less sugar than flour, in proportions insufficient to impart crunchiness. These low sugar content formulas were developed without regard to the fact that true cookies are sweet cakes; or they were developed in an effort to use soft flours, which do not require large proportions of enriching ingredients for tenderness, but which cannot carry enough sugar for crispness.

Classifying Cookies by Machines and Proportions of Ingredients

Although the term cookies covers a wide variety of bakery products, all of these may be classified into at least three distinctly different groups, according to: (1) the machine used in making them, (2) the proportions of ingredients used, and (3) the finished characteristics.

The particular machine used limits, to some extent, the varieties of cookies which can be turned out. The rotary machine limits production to cookies of the short-bread type, the bar press to bar cookies, the sheeter to crackers, biscuits, hard cookies and similar products made from dry doughs which can be sheeted. Although each of these machines produces a large variety of products in each distinct class, none of the machines will produce to advantage the products which the others are particularly suited to turn out. The wire-cut machine will handle almost any kind of dough and make almost all varieties of cookies, except biscuits, crackers and embossed pieces. However, the wire-cut machine is particularly suited to batter doughs for vanilla wafers and to short, stiff doughs for snaps and similar varieties.

When cookie products are classified according to the proportions of ingredients used, they fall into certain distinctly different groups, and most of the varieties in each class are made to the best advantage on a certain machine. For example, the term cracker usually refers to all lean dry products which are made, almost without exception, on a sheeter.

Crackers

Crackers, essentially, are mixtures of flour and water with little or no sugar and eggs, and with more or less shortening. When flour is mixed with enough water to form a stiff dough, 100 pounds of flour plus 30 pounds of water, and then rolled into thin sheets and baked, water or matzoh crackers are produced. When the dough is fermented for about 18 hours, with about 4 ounces of yeast and ½ pound of Xpando or malt, seasoned with about 1 pound of salt, caused to spring and color with about ½ pound of soda and tenderized and flaked with about 10 pounds of shortening, soda crackers are produced. For cheese crackers about 30 pounds of cheese are incorporated into a soda cracker dough. When a soda cracker dough is given a shorter fermentation period and extra fat is rolled in, cream crackers result. When the basic dough is very stiff, not fermented, and no leavening is used, and about ½ pound of salt and 5 pounds of fat are worked in, pilot crackers result. When the basic dough is softer, including about 1½ pounds of soda and acid leavening, about 15 pounds of shortening and about 5 pounds of sugar, lunch-acid crackers are the product. By incorporating a little more or less sugar, shortening and milk, other varieties of acid crackers are made. When acid crackers are sprayed with coconut fat, butter crackers are produced. Butter thins are made with about 100 pounds of flour, 15 pounds of shortening, 3 pounds of sugar and about 10 pounds of Nulomoline for rich crust color and prevention of checking and breaking.

When about one-quarter of the white flour in a plain dough is replaced with whole wheat flour, the flavor improved with about 40 pounds of a combination of molasses, Nulomoline and sugar, and the dough shortened with about 15 pounds of fat and aerated with about 1 pound of soda and ½ pound of ammonia, graham crackers are produced.

The Sheeter Machine

The sheeter machine is used not only for crackers, but also for other varieties of cookie products made from dry doughs which can be rolled out and then cut and embossed, e.g., hard or semi-hard sweets, hard cookies, embossed items, and hard cakes. These varieties contain considerably more sugar and shortening than most crackers. The lean varieties, such as hard sweets which contain only about 10 pounds of shortening, 2½ pounds of Nulomoline and 30 pounds of sugar to about 100 pounds of flour, and especially those made from doughs which receive and hold an impression, are often called biscuits. The richer varieties, such as the typical, large, American hard types which contain about 15 pounds of shortening, 5 pounds of Nulomoline and about 50 pounds of sugar to 100 pounds of flour, are classified as common cookies. Those with extra sugar are called snaps. The hard cake cookies, or honey molasses types, are made on a sheeter by using large proportions (75 pounds) of sugary materials, i.e., sugar, Nulomoline, molasses or honey, and low proportions (5 pounds) of shortening.

Fig and Molasses-Coconut Bars

Some sheeter machines have a bar press attachment which is used in making varieties such as molasses-coconut bars and fig bars. Molasses-coconut bars contain about 15 pounds of shortening, 30 pounds of Nulomoline and about 50 pounds of other sugary materials, including molasses, for each 100 pounds of flour. Bar cookies are also made on some wire-cut machines which can handle stiff doughs. The jackets for fig bars usually contain about 15 pounds of shortening, 10 pounds of Nulomoline and about 40 pounds of sugar. Bar press machines, separate from the sheeter, for fig bars are available.

The Rotary Short-Bread Cookie Machine

When high proportions of shortening, from 30 to 50 pounds, and from 35 to 65 pounds of sugar for each 100 pounds of flour are incorporated in doughs, they become too tender to roll out and sheet on a machine. These short-bread doughs must be pushed through a wire-cut machine die or, preferably, handled on rotary machines which are designed especially for handling short-bread varieties.

De Luxe Graham Crackers
(Rich Brown Crust and Crumb, Good Spring, Flaky, Crispy and Fine Flavor; Basic Formula for Sheeter

* Sugar	15	lb
Grandma's Old-Fashioned Molasses	15	lb

Nulomoline	5	lb
Salt	1¼	lb
Water	20	lb
** Flour—Clear	80	lb
Whole Wheat	20	lb
Soda	1	lb
Ammonium Carbonate	½	lb
Shortening	15	lb

Heat the sugar, Grandma's Molasses, Nulomoline, salt and water to about 180° F; have the temperature of this syrup such that the heat will warm the final dough to 120 to 130° F.

Meanwhile, separately blend the flours and leavening agents.

Combine the hot syrup with the flour blend and the shortening. Mix the dough smooth and run it while warm.

Molasses Brittle-ettes

(Ginger Snaps with Good Crack, Medium Molasses Content, for Sheeter and Wire-Cut Machine or Hand-Made)

	Small Batch		Large Batch	
Fine Granulated Sugar	4	lb	40	lb
Shortening	2	lb	20	lb
Milk (Solid Basis)	½	lb	5	lb
Salt	1½	oz	1	lb
Ginger	2½	oz	1½	lb
Nulomoline	½	oz	5	lb

* Other proportions of Grandma's Molasses, sugar and Nulomoline can be used: (1) a greater total, up to 55 pounds, or (2) less molasses and more sugar, to make a total of 35 pounds. Increased amounts of Grandma's Molasses overcome the satiating taste of dark flours and produce a richer brown crust color.

** Other proportions of white and dark flours can be used, such as, 5 bran, 5 cracked wheat, 20 whole wheat and 70 clear.

	Small Batch	Large Batch
Grandma's Molasses	4½ lb	45 lb
Water	¾ lb	8 lb
Soda	2½ oz	1½ lb
Flour	10 lb	100 lb

Cream the sugar, shortening, dry milk, salt and ginger.

Add the Grandma's Old-Fashioned Molasses, Nulomoline and water.

Separately blend the soda and flour. Add this blend to the creamed mixture and mix smooth.

The basic formula may be used either for sheeter, wire-cut or hand-rolled and cut cookies. Change the consistency of the dough to fit the procedure desired by using different amounts of water.

Golden Fig Bar Jacket
(Basic Dough, to be Filled with Fig Jam, for Bar Press Machine.)

Powdered Sugar	40 lb	Whole Eggs	12 lb
Salt	½ lb	Soda	½ lb
Shortening (Or Part Butter)	20 lb	Ammonium Carbonate	¼ lb
Nulomoline	10 lb	Milk (Liquid Basis)	18 lb
Flavor	As desired	Flour	100 lb

Cream the sugar, salt, fat, Nulomoline and flavor. Gradually add the eggs.

Dissolve the soda and the ammonium carbonate in the milk and add it along with the flour. Mix smooth.

For a leaner dough, replace part or all of the eggs with milk; replace part or all of the milk with water; reduce the shortening.

General Fig Bar Filling
(Basic Formula)

Figs (Ground)	500 lb
Water	Sufficient to soften the figs
Fine Granulated Sugar	200 lb
Corn Syrup (Glucose)	100 lb
Nulomoline	200 lb

Salt	2–3 lb
Flavor	As desired

Soak the figs with sufficient water long enough to soften them. Grind to suitable fineness. Mix the ground figs, other ingredients and enough water for proper consistency until all are uniformly distributed.

For other fig pastes vary the proportions of sugar, corn syrup and Nulomoline, one to another, but keep the total of these in about the proportion to the ground figs as specified in the above formula.

Tasty Chewy Cookies

(Molasses Coconut Bars, Rich Crust, Crisp, Chewy Eating-Qualities, for Hand Rolled or Bar Machine.)

	Small Batch	Large Batch
Sugar	3½ lb	35 lb
Shortening (Preferably Part Butter)	2 lb	20 lb
Salt	1½ oz	1 lb
Grandma's Molasses	2½ lb	25 lb
Nulomoline	1½ lb	15 lb
Coconut (Macaroon)	4 lb	40 lb
Flavor (Mild-Vanilla, Lemon or Spice)	As desired	As desired
Water	1¼ lb	12 lb
Ammonium Carbonate	¼ oz	⅛ lb
Flour	10 lb	100 lb
Soda	1¼ oz	¾ lb

Cream the sugar, shortening and salt. Add the Grandma's Molasses, Nulomoline, coconut and flavor, and mix until distributed.

Meanwhile, dissolve the ammonium carbonate in the water and separately blend the soda and flour.

Add the ammonium carbonate solution and the flour blend to the creamed ingredients while mixing slowly, and mix the dough until smooth.

The consistency of this dough is suitable for the bar machine. For rolling out and cutting by hand or bagging by hand, change the water content to secure suitable consistency.

Economy Ginger Snaps

(With Crumbs and Low Shortening Content, High Production, Low Cost Cookies Made on a Rotary Short-bread Machine.)

Powdered Sugar	10	lb	Crumbs	15	lb
Salt	1	lb	Ginger	1¼	lb
Nulomoline	5	lb	Shortening	17½	lb
Grandma's Old-Fashioned Molasses	50	lb	Soda	2¼	lb
			Flour	100	lb

Bring the sugar, salt, Nulomoline and Grandma's Molasses to a boil.

Meanwhile, place the crumbs and ginger in a mixer and mix until the crumbs are broken into very fine particles. Pour the hot molasses solution onto the crumbs, add the shortening and mix smooth.

Separately blend the soda and flour and add the blend to the crumb mixture. Mix the dough smooth.

Allow the dough to rest, at least an hour or more, until cool and until it tightens sufficiently to become dry enough to run on the machine without sticking.

Honey Varieties

(Basic Formulas for Peppernuts, "Honey" Gingerbread Men, Santa Claus and Other Cut Out Christmas Cookies, Cobblestone "Honey" Cakes, etc.)

Variety Peppernuts

(A basic Pfeffernuss Formula with Which to Make Modifications to Suit Individual Preferences, Using a Stock Dough, For Wire-Cut Machine and Hand Production.)

	Small Batch	Large Batch
Whole Eggs	1 lb	10 lb
Ammonium Carbonate	¾ oz	½ lb
Soda	¾ oz	½ lb
Salt	¾ oz	½ lb
Cinnamon	1¼ oz	¾ lb
Allspice	¾ oz	½ lb

	Small Batch	Large Batch
Cloves	3/8 oz	1/4 lb
Mace	3/8 oz	1/4 lb
Cardamom	3/8 oz	1/4 lb
Pepper (Optional)	1/4 oz	1/8 lb
Anise	As desired	As desired
Stock Dough (See Next Formula)	20 lb	200 lb
Shortening (Preferably Part Butter)	1 lb	10 lb
Nuts (Chopped Walnuts, Cashews, etc.)	1 1/2 lb	15 lb
Fruit (Mixed Glaced; Ground)	1 lb	10 lb

Stir together the eggs, ammonium carbonate, soda, and salt. Add to the egg mixture the cinnamon, allspice, cloves, mace, cardamom, pepper and anise, and mix until smooth. Add this spice-egg mixture in small amounts to the stock dough and mix well after each addition. Add the shortening and mix in.

Just before finishing the mixing, add the nuts and fruit, and mix the dough smooth.

For hand work, roll into cylinders about 3/4 inch in diameter, cut about 1/2 inch long, round to form balls and place on pans.

For wire-cut machine production, make the dough softer by adding extra eggs or milk until suitable for depositing.

Increase or decrease the leavening agents; more ammonium carbonate produces higher puff; more soda makes a wider spread.

Change the flavor by using different combinations and quantities of spices.

The proportion of nuts and ground fruits may be changed or the total increased, decreased or omitted.

Make the peppernuts more or less tender by increasing, decreasing or omitting the shortening.

Bake at about 380° F.

To keep the peppernut cookies soft, they should be coated as soon as possible after baking, with the snow-bank peppernut coating and rolled in powdered sugar.

Be sure the package labels comply with the food laws. Some laws require that when the word, honey, is used on labels, honey must be used solely or in preponderant proportions in the product.

Stock Dough
(For Use in Peppernuts, Christmas Cookies, Cobblestone "Honey" Cakes, etc.)

	Small Batch	Large Batch
* Nulomoline (Honey, Grandma's Molasses)	10 lb	100 lb
Flour	10 lb	100 lb

Heat the Nulomoline barely to a boil or preferably to about 180° F. Add the flour to it and mix into a dough.

Use this dough immediately or, preferably, after aging for a few days, weeks or months, for peppernuts, Christmas cut-out cookies and cobblestone honey cakes.

Snow White Cup Cakes
(Close Grain, Fair Volume, Chewy)

	Small Batch	Large Batch
Fine Granulated Sugar	5 lb	50 lb
Dry Milk	1¼ lb	12 lb
Flour	2½ lb	25 lb
Shortening	2½ lb	25 lb
Nulomoline	1½ lb	15 lb
Flavor	As desired	As desired
Soda	1½ oz	1 lb
Water	5 lb	50 lb
Flour	7½ lb	75 lb
Egg White	3½ lb	35 lb
Fine Granulated Sugar	3½ lb	35 lb
Salt	3¼ oz	2 lb
Cream of Tartar (Or Other Acid Leavener)	3½ oz	2¼ lb

Cream the first portion of sugar, the powdered milk, flour, shortening, Nulomoline and flavor.

Dissolve the soda in the water and, then, while continuing

* The Nulomoline may be combined with, or replaced by, all or part of Grandma's Old-Fashioned Molasses or honey, about ¼, ½ or ¾, depending upon individual preferences.

the creaming, carefully add about one-third of the dissolved soda. Add the remainder along with the second portion of flour.

Meanwhile, whip the egg whites, sugar, salt and cream of tartar until fairly stiff. Carefully add the whipped eggs to the creamed mixture. Mix the batter until it is smooth.

Old Fashioned Health Muffins
(Rich, Flavorfully Sweet, Large Size, Made with Whole Wheat Flour.)

	Small Batch	Large Batch
Shortening (Preferably Part Butter)	1½ lb	15 lb
Flour (Clear)	1½ lb	15 lb
Nulomoline	1 lb	10 lb
Salt	2½ oz	1½ lb
Eggs (Whole)	1¼ lb	12 lb
Grandma's Molasses	2 lb	20 lb
Milk (Liquid Basis)	7 lb	70 lb
Soda	¾ oz	6½ oz
Flour (Clear)	2 lb	20 lb
* Flour (Whole Wheat)	6½ lb	65 lb
Baking Powder	5¼ oz	3¼ lb
Raisins, Currants, Figs, Blueberries, etc. (Optional)	1–2 lb	10–20 lb

Cream the shortening, first portion of clear flour, Nulomoline and salt. While continuing the creaming, add the eggs. Gradually add the Grandma's Molasses.

Meanwhile, dissolve the soda in the milk and blend the second portion of clear flour, whole wheat flour and baking powder.

Add the dissolved soda along with the blended ingredients to the creamed mass. Mix the batter until it is smooth.

* Other flours may be used to replace all or part of the whole wheat and clear flours, such as patent white flour, bran and corn meal.

Buttermilk Cookies

Sugar	3½	lb	Sour Milk or		
Malt Syrup	½	lb	Buttermilk	2	qt
Shortening	1	lb	Flour	7	lb
Eggs	15		Salt	1	oz
Soda	3	oz	Lemon Flavor		

Drop on greased and dusted pans, bake in hot oven.

Sugar Cookie Mix

XXXX Sugar	3	lb	Whole Milk Powder	5	oz
Winter Wheat Flour	6	lb	Salt	1½	oz
Cottonseed Oil	6	oz	Egg Color	Trace	
Bicarbonate of Soda	1	oz	Dandy Cake Flavor	Trace	
Granulated Egg Yolk	1	oz			

Spicy Cookie Mix

Brown Sugar	3	lb	Salt	1½	oz
Winter Wheat Flour	6	lb	Cinnamon	1	oz
Cottonseed Oil	6	oz	Allspice	¼	oz
Bicarbonate of Soda	1	oz	Ginger	¼	oz
Granulated Egg Yolk	1	oz	Nutmeg	¼	oz
Buttermilk Powder	5	oz			

Mix 5 pounds of cookie mix with 1 quart of water and make dough. Roll out and cut.

Two pounds of cookie mix makes 3½ to 4 dozen cookies.

This is a preparation that will produce quickly and at a low cost the most delicious cookies.

Ginger Cookies

Sugar	2	lb	Soda	2	oz
Malt Syrup	1	lb	Ginger	1	oz
Shortening	1	lb	Salt	½	oz
Molasses	1	qt	Flour	6	lb

Cut with 3-inch cutter and dampen with water. Bake in medium oven.

Ginger Snaps

Sugar	4	lb	Water	1	pt
Malt Syrup	1	lb	Soda	2½	oz
Shortening	30	oz	Ammonium Carbonate	½	oz
Molasses	2	qt	Flour	12½	oz

Cut with 2½-inch cutter and bake in medium oven. These snaps should be washed with an egg wash before baking or should be baked in light steam.

Honey Cakes

Sugar	½	lb	Ammonium Carbonate	1	oz
Shortening	½	lb	Soda	Pinch	
Honey	1	qt	Salt	½	oz
Malt Syrup	1	lb	Flour	To make medium dough	
Molasses	1	pt			

Let stand in cool place after mixing, to age for an hour or so. Bake in medium oven.

Oatmeal Cookies

Sugar	6	lb	Ground Raisins	2	lb
Malt Syrup	2	lb	Flour	8	lb
Shortening	4	lb	Rolled Oats	4	lb
Eggs	14		Salt	1	oz
Milk	1½	pt	Cinnamon	1	oz
Soda	3¼	oz			

Cut with 2½ to 3-inch cutter. Wash with egg wash and bake in medium oven.

Butterscotch Cookies

Brown Sugar	2½	lb	Vanilla	¼	oz
Shortening	1	lb	Pastry Flour	3	lb
Salt	1	oz	Soda	¼	oz
Beaten Eggs	1	lb			

Cream the shortening and the dry ingredients. Add the beaten eggs and vanilla, creaming well. Sift the pastry flour and soda together and add. Mold into long piece about 1½ inches in diameter. Place in refrigerator for several hours. Then cut about ½ inch thick. Bake at 350° F.

Date Nut Cookies

Sugar	2 lb	Bicarbonate of Soda	1 oz
Shortening	1 lb	Whole Eggs	4
Butter	8 oz	Honey	8 oz
Salt	1 oz		

Cream the shortening, butter and dry ingredients together until light. Continue creaming and add the eggs in small portions. Add the honey.

Chopped Pecans	8 oz	Pastry Flour	4 lb
Chopped Dates	1¼ lb	Baking Powder	1½ oz

Add the pecans and dates to the above mixture. Sift the dry ingredients together and add.

Scale dough in 12-ounce pieces and cut twelve cookies out of each piece. Flatten on lightly greased pan, wash with 50% egg and milk solution, and bake at 400° F.

Oatmeal Cookies

Light Brown Sugar	3 lb	Cinnamon	¾ oz
Shortening	1½ lb	Frozen Egg Yolk	8 oz
Salt	2 oz	Water (Variable)	1 lb
Soda	¾ oz		

Cream the above ingredients, except the water, light. Add the water. Then add the following:

Pastry Flour	3 lb	Currants	1¼ lb
Rolled Oats	2 lb		

Roll out, thin and cut. Wash with egg wash and bake at 370° F.

Chocolate Nougat Cookies

Shortening	12	oz	Ginger	¼	oz
Sugar	1	lb	Allspice	¼	oz
Salt	½	oz	Natural Cocoa	10	oz
Soda	1	oz	Whole Eggs	4	
Vanilla Extract	½	oz	Honey	1	qt
Maple Extract	½	oz			

Add the eggs gradually to all the ingredients, except the honey, creaming well after each addition. Then gradually add the honey and cream well.

Molasses Cookies

Sugar	4½	lb	Bicarbonate of Soda	3	oz
Shortening	2¼	lb	Cinnamon	1	oz
Salt	1½	oz	Ginger	½	oz
Milk Powder	4	oz	Cloves	½	oz
Vanilla Extract	¾	oz			

Cream the above ingredients together well.

Light Molasses	1½	qt	Pastry Flour	12	lb
Water	3	lb			

Mix the molasses into the above mixture well. Sift and add the flour. Roll out on cookie cloth to ¼-inch thickness and cut with scalloped cutter. Place on lightly greased pans and wash with a light egg wash. Bake at 350° F.

Fancy Butter Wafers

Sugar	3	lb	Lemon	½	oz
Shortening	2½	lb	Cold Egg Whites	1	lb
Butter	2	lb	Pastry Flour	6	lb
Vanilla Extract	1	oz			

Cream the above ingredients, except the egg whites and flour, until light. Then gradually add the egg whites. Sift the flour and add it last.

Bag out with medium size star tube on ungreased pans into various shapes, such as rings, small round drops, scrolls, etc.

Decorate with small pieces of candied cherry, pineapple, citron, etc. Bake at 350° F.

Plantation Tarts

Sugar	4	lb	Whole Eggs	2	lb
Cinnamon	½	oz	Corn Syrup	3	lb
Cloves	¼	oz	Molasses	1	lb
Salt	1	oz	Liquid Whole Milk	1	lb
Shortening	8	oz	Ground Cracker Crumbs	1½	lb

Stir the sugar, cinnamon, cloves, salt and shortening together. Add the eggs and cream light. Add the corn syrup and molasses, and mix thoroughly. Add the milk and crumbs. Pour into unbaked tart shells and bake at 400° F.

Pecan Tarts

Corn Syrup	8	lb	Corn Starch	12	oz
Whole Eggs	1½	lb	Butter	1	lb
Sugar	10	oz	Salt	½	oz

Stir all together and pour into unbaked shells with pecans in bottom of shell. Bake in medium oven. During the baking, the pecans come to the top of the pie and form an appetizing topping. A rich filling topped with pecans gives zest to these unusual tarts.

Doughnuts

Good doughnuts can be made only by the proper manipulation of the dough. The use of good ingredients, proper choice and proper care of frying fat and the correct frying temperature are imperative.

Proper dough consistency is most important in producing cake doughnuts with the desired expansion. The whole character of the finished doughnut, its grain, texture, volume and absorption are affected by the expansion of the doughnut during the frying process. Even with a well balanced formula, poor quality doughnuts may be produced if the dough consistency is not adjusted to permit correct expansion as the doughnuts are fried.

Effects of Stiff Dough

When dough consistency is too stiff your doughnuts will have:

(1) A rough, broken surface on one side
(2) Excessive absorption
(3) Thick crust
(4) Cracks developing as they are fried
(5) Poor expansion
(6) Improper break

Effects of Slack Doughs

Slack doughs produce doughnuts with:

(1) A distorted appearance
(2) Large blow holes
(3) Excessive absorption
(4) Poor expansion
(5) Poor grain and texture
(6) Improper break

The importance of determining the exact amount of liquid to incorporate in your doughnut mixes cannot be too strongly emphasized. The small amount of experimental work that may be required, in some cases, to adjust the dough consistency to your own conditions will more than pay for itself, by improving the quality of your doughnuts.

Effect of Proper Dough Consistency

When dough consistency is right your doughnuts will:

(1) Expand properly to produce good symmetry
(2) Be uniform in appearance
(3) Have low absorption
(4) Have thin crust
(5) Have good grain and texture
(6) Break perfectly during frying

Obtaining Proper Dough Consistency

Controlling dough consistency to produce the best quality doughnuts depends upon:

(1) Determining the right amount of liquid to incorporate in the doughnut mix

(2) Mixing the dough for the correct length of time

Dough-mixing time also affects dough consistency. Doughnut doughs made from balanced formulas adjusted for the exact amount of liquid may give poor doughnuts if the dough is not properly mixed. The exact mixing time for any doughnut dough will depend upon:

(1) The richness of the dough

(2) The type of flour in the formula

(3) The type of doughnut machine used or whether it is hand rolled and cut

Determining Proper Mixing Time

Rich doughnuts require more mixing than lean doughnuts. Some flours in a doughnut dough require more mixing than others.

Determining the correct length of time to mix your doughnut doughs requires careful study of the effect of mixing time on the character of the doughnut. Mixing time for doughnut doughs will vary from 30 seconds to 3 or 4 minutes. You can easily adjust the mixing time of your doughs, to produce the best possible doughnuts, by means of a few trial batches.

Dough Handling Pointers

Doughnut doughs should be mixed according to the dough temperature. Cold doughs require more mixing than warm doughs. A dough temperature of about 75° F is normal for most cake doughnut mixes. If you are making machine doughnuts allow the dough to stand for 15 to 20 minutes before it is put through the mechanical dropper. This practice permits the dough to open up somewhat. Some leavening action, while the dough is standing, helps prevent the formation of objectionable tight center cores. Doughnuts made by rolling and cutting out the dough should be given a chance to rest on the wire screen for a period before they are dropped into the frying fat. This resting period, which may vary from 15 minutes to 1 hour, helps produce doughnuts with the desired expansion grain and texture. Keep your frying fat in the best possible condition. A few important pointers are listed below.

(1) Avoid heating frying fat to high temperature (above 400° F).

(2) Regulate the heating of the frying fat to eliminate hot spots which cause breakdown.

(3) Regularly strain out the burnt food particles which accumulate during frying.

(4) Don't hold the frying fat at high temperatures for long periods when there is no food in the kettle.

(5) Fry in the smallest quantity of fat that is practical and in this way insure rapid turnover of fresh fat in the kettle.

(6) Clean frying kettles thoroughly after each day's or night's frying.

(7) Avoid contamination of the fat with alkali and metals which may hasten its breakdown.

Doughnuts

Formula No. 1

Granulated Sugar	2	lb	Egg Yolks	8	oz
Shortening	6	oz	Whole Eggs	8	oz
Salt	1½	oz	Liquid Milk (Variable)	4	lb
Pastry Flour	7¼	lb	Flavor	To taste	
Baking Powder	4	oz			

Mix together at medium speed, varying the mixing time from 30 seconds to 4 minutes depending upon the ingredients, shop conditions, etc.

Use the formula as set up for machine-made doughnuts. If

you cut down on the milk in the mix, you will get a dough which can be rolled and cut by hand.

Fry in shortening at 375 to 390° F.

Formula No. 2

XXXX Powdered Sugar	3 lb	Bread Flour	4 lb
		Pastry Flour	4 lb
Shortening	8 oz	Baking Powder	6 oz
Salt	2 oz	Egg Yolks	2½ lb
Skim Milk Powder	2 oz	Liquid Milk (Variable)	4 lb
Soda	½ oz	Flavor	To taste

Mix together at medium speed, varying the mixing time from 30 seconds to 4 minutes, depending upon the ingredients, shop conditions, etc.

Use the formula as set up for machine-made doughnuts. If you cut down on the milk in the mix, you will get a dough which can be rolled and cut by hand.

Fry in shortening at 375 to 390° F.

Formula No. 3

Sugar	2 lb	Flour	8 lb
Shortening	8 oz	Eggs	1 lb
Salt	1 oz	Buttermilk	5 lb
Soda	1 oz	Flavor	To taste
Baking Powder	2 oz		

Mix together at medium speed, varying the mixing time from 30 seconds to 4 minutes depending upon the ingredients, shop conditions, etc.

Fry in shortening at 375 to 390° F.

Chocolate Doughnuts
Plunger Machine

Sugar	6 lb	Salt	3 oz
Shortening	¾ lb	Vanilla	2 oz
Cocoa	1¼ lb		

Cream the above ingredients together.

Eggs	3	lb	Cake Flour	8	lb
Milk	10	lb	Baking Powder	½	lb
Bread Flour	5	lb			

Add the eggs, then the milk. Sift the dry ingredients together and add. Fry at 385° F.

Yeast Raised Doughnuts
Formula No. 1

Sugar	1	lb	Bread Flour	4½	lb
Shortening	1	lb	Pastry Flour	2½	lb
Salt	3	oz	Flavor	To taste	
Soft Flour	1	lb	Water (Variable)	4	lb
Cooked Mashed			Yeast	12	oz
Potatoes	2	lb			

Mix together at medium speed, gradually adding the water in which the yeast has been dissolved. Continue mixing for about 3 minutes to a smooth dough.

Take the dough from the mixer at 80° F. Let the dough raise for ½ hour, take to the bench, make up and proof. Fry in shortening at 360 to 370° F.

Formula No. 2

Brown Sugar	10	oz	Bread Flour	2½	lb
Salt	2	oz	Pastry Flour	3	lb
Cooked Mashed			Whole Eggs	10	oz
Potatoes	2	lb	Liquid Milk (Variable)	4½	lb
Shortening	1	lb	Yeast	12	oz
Whole Wheat Flour	2	lb	Flavor	To taste	

Mix together at medium speed, gradually adding the water in which the yeast has been dissolved. Continue mixing for about 3 minutes to a smooth dough.

Take the dough from the mixer at 80° F. Let the dough raise for ½ hour, take to the bench, make up and proof. Fry in shortening at 360 to 370° F.

Formula No. 3

Sugar	1¾ lb	Pastry Flour	7 lb
Shortening	3 lb	Baking Powder	5 oz
Salt	4 oz	Eggs	2 lb
Mace	1 oz	Liquid Skim Milk (Variable)	9 lb
Nutmeg	1 oz		
Bread Flour	8 lb	Yeast	1¼ lb

Mix together at medium speed, gradually adding the milk in which the yeast has been dissolved. Continue mixing about 3 minutes to a smooth dough.

Take the dough from the mixer at 80° F. Let the dough raise for 20 minutes, take to the bench, make up and proof. Fry in shortening at 360 to 370° F.

Glazed Potato Doughnuts
(Yeast)

Sugar	1 lb	Cooked Mashed Potatoes	2 lb
Shortening	1 lb		
Salt	3 oz	Mace	¼ oz
Soft Flour	1 lb	Lemon Flavor	To taste
		Vanilla Flavor	To taste

Mix together for 10 minutes.

Water (Variable)	4 lb	Yeast	12 oz

Dissolve the yeast in part of the water and mix well. Stir in the rest of the water alternately with the flour.

Hard Flour	4½ lb	Soft Flour	2½ lb

Sift together and add alternately with the water and mix for about 3 minutes.

Take the dough from the mixer at 80° F. Let the dough raise for ½ hour, take to the bench, make up, proof and fry in shortening at 360 to 370° F. Scale the doughnuts about 16 ounces per dozen.

Glaze the doughnuts by dipping them into the glaze as they come from the frying kettle. Take the doughnuts from the glaze and allow them to drain on wire screens until the glaze sets up. Doughnuts can be covered with chopped nuts, coconut or granulated sugar if desired.

Extra Rich Yeast Raised Doughnuts

Sugar	1¾	lb	Pastry Flour	7	lb
Shortening	3	lb	Baking Powder	5	oz
Salt	4	oz	Whole Eggs	2	lb
Mace	1	oz	Liquid Skim Milk	9	lb
Nutmeg	1	oz	Yeast	1¼	lb
Bread Flour	8	lb			

Place all the dry ingredients, including the shortening, in the mixing bowl. Then, at medium speed, add the eggs and the milk in which the yeast has been dissolved. Continue mixing to a smooth dough. Under most conditions, this dough can be mixed in from 2 to 3 minutes by this method.

Take the dough from the mixer at 78 to 85° F. Let rest for 20 minutes and take to the bench. After cutting, give about 20 minutes proof.

Glaze the doughnuts by dipping them into a honey glaze as they come from the frying kettle. Take the doughnuts from the glaze and allow them to drain on wire screens until the glaze sets or hardens. Doughnuts may be covered with chopped nuts, coconut or granulated sugar if desired.

Doughnut Cocoa Glaze

XXXX Sugar	8	lb	Hot Water	2½ lb
Cocoa	1	lb	Vanilla	To taste
Powdered Skim Milk	4	oz		

Sieve the dry ingredients together. Add the hot water and vanilla. Use the glaze warm. Dip the hot doughnuts in the glaze immediately after they come from the frying kettle. Allow to drain on screens until glaze sets.

Doughnut Dusting Sugar

XXXXXX Powdered Sugar	8 lb	Shortening	8 oz

Cream together for 10 to 15 minutes.

Powdered Skim Milk	8 oz	Corn Starch	1 lb

Add and mix for several minutes. Run mix through a very fine sieve.

Doughnut Glaze

Formula No. 1

Water	2½ lb	Gelatin *	1 oz
Corn Syrup	8 oz		

Dissolve gelatin in the water. Add corn syrup and bring to boil.

XXXX Powdered Sugar	10 lb	Flavor	To taste

Pour above hot syrup over sugar and mix smooth. Stir in the flavor.

Dip yeast-raised doughnuts into the glaze as they come from the frying kettle. This glaze works best around 90 to 100° F. The glaze will set up while draining on the wire.

Formula No. 2

(1) Granulated Sugar	1 lb	Water	1 qt
Agar	3 oz	Vanilla	If desired

Mix sugar and agar, and boil for 3 minutes in 1 quart of water.

(2) Mix smooth 10 pounds of powdered sugar with 1 quart of hot water.

Pour hot (1) over (2) and mix thoroughly.

Doughnut Honey Glaze

XXXX Powdered Sugar	10 lb	Hot Water	1 lb
Honey	6 oz		

Mix to a paste.

* The amount of gelatin may be varied depending upon its strength.

Gelatin *	3 oz	Hot Water	1 lb

Dissolve, add and mix smooth. Bring to a temperature of 90 to 100° F for use.

Dip yeast-raised doughnuts into the glaze, as they come from the frying kettle. Take the doughnuts from the glaze and allow them to drain on wire screens until the glaze sets up. Doughnuts may be covered with chopped nuts, coconut or sugar.

Doughnut Icing Glaze

Water	4 lb	Corn Starch	½ oz
Tapioca Flour	1 oz	Agar-Agar	½ oz

Bring to boil for 3 minutes. Place in cake machine with the following:

Vanilla Extract	½ oz	Butter	2 oz
Salt	¼ oz	Corn Syrup	1 lb

Then add 10 pounds of icing sugar and mix well.
Use while warm.

Doughnut Icing

Water	1½ lb	Corn Syrup	8 oz

Bring the above to boil.

Gelatin *	½ oz	Water	8 oz

Dissolve gelatin in hot water and add to the above syrup.

XXXX Powdered Sugar	12 lb

Add the above syrup slowly to the XXXX sugar and beat until smooth. Bring to temperature of 90 to 100° F for use.

Ice the doughnuts by dropping them into the icing made from the formula given above as soon as they come from the frying

* The amount of gelatin may be varied depending upon its strength.

kettle. After complete immersion remove and allow to drain on a wire screen.

Doughnut Orange Icing

Ground Oranges	2 lb	XXXX Powdered Sugar	8 lb

Grind oranges and mix to desired consistency with powdered sugar.

PREPARED FLOURS

Doughnut Flour

The following directions should be followed carefully to obtain a quality product:

a Sieve the prepared doughnut flour into a mixing bowl.

b Make a hole in the center of the flour and pour cold water into it.

c Fold flour into the water, mix well but avoid over-mixing. A cake mixing machine can be used.

d Doughnuts should be fried soon after mixing.

e The temperature of the dough should be between 70 to 75° F, to get maximum expansion and minimum grease absorption.

f In the summer, add cold water, in the winter, warm water, to the flour. In this way the temperature of the dough can be controlled.

g The temperature of the dough is important because the dough is governed by the temperature at which it hits the grease. The best frying results are obtained at 390 to 400° F.

h Under certain conditions the preparation of flour and water may have to vary slightly. Only by experimentation can this be determined. For bench work use approximately 2 pounds cold water and 5 pounds prepared doughnut flour.

Doughnut Flour

Formula No. 1

Soft Wheat Patent	88 lb	Shortening	5 lb
Hard Wheat Patent	10 lb	Soda	13 oz
Granulated Sugar	32 lb	Sodium Acid Pyrophosphate Salt	1 lb
Powdered Skim Milk	5 lb	Ground Mace	2 oz
Powdered Egg Yolk	5 lb		

The above formula is an example around which to construct a product to suit the individual's taste. The flavoring material may consist of solid spices or their corresponding oils.

Formula No. 2

Sugar	20	lb	Vanillin	To suit	
Shortening	3½	lb	Powdered Egg	2	lb
Mace Oil	⅓	oz	Skim Milk	5	lb
Cinnamon Oil	10	drops	Cake Flour	61	lb
Salt	1	lb	Baking Powder	3	lb

Use fine granulated sugar. Cream the shortening with the sugar and add the spice oils. Beat egg with sugar and add some creamed shortening. Then add this to the creamed sugar. Sift the cake flour, milk powder and baking powder, and add to creamed sugar. Put through a mixer and sifting machine.

Formula No. 3

Sugar	29	lb	Powdered Egg	4	lb
Shortening	3½	lb	Skim Milk	4	lb
Mace Oil	⅓	oz	Cake Flour	57	lb
Salt	1	lb	Baking Powder	3½	lb
Vanillin	To suit				

Mix the same as suggested above for formula No. 2.

To make doughnuts use 16 to 17 pounds of doughnut flour with 3¼ quarts of water. Baking temperature is around 390° F.

Powdered Doughnut Sugar

Pulverized Dextrose	75 lb	Shortening (Around 96° F m. p.)	5 lb
Cake Flour	10 lb	Vanillin	1 oz
XXXX Sugar	9 lb		

Melt the shortening and cream with some of the dextrose sugar. Gradually work in the balance of the dextrose and, when thoroughly incorporated, add the balance of the ingredients. Put the entire mix through a mixer and sifting machine.

Pancake Flour

A good pancake flour not only adds to the cereal consumption of many households but also offers a possibility of marketing certain grades of flour to better advantage. What should a good pancake flour be like? A thousand types and tastes can be produced by combining different flours and other ingredients in various combinations and it is impossible to say which is the best. It is all a matter of taste. However, no one likes a soggy, tough or rubbery pancake. The ideal pancake should be light, fluffy and tender and, in compounding a prepared flour that is to yield such pancakes, a number of facts must be observed.

A satisfactory pancake flour should contain:

Wheat, soft and hard
Corn
Buckwheat
Rice
Rye
Soy
Powdered skim milk or powdered buttermilk
Corn sugar or cane sugar
Salt
Leavening: bicarbonate of soda and acid phosphate

These ingredients may be combined to produce:

A wheat pancake flour
A corn pancake flour
A buckwheat pancake flour
A mixed pancake flour of no particular type

Pancakes, made from wheat flour only, have a tendency to be rubbery. To avoid this feature, clears and longer extractions should be used. Hard wheat flour, although not as important in pancake flours as it is in the most prepared doughnut mixes, helps to improve the texture and volume of the pancake. Whole wheat or small amounts of fine middlings are very helpful additions to soft wheat flours.

A wheat pancake flour can also be improved by the addition of corn or rice flour. Both of these flours, especially rice flour, make for a short pancake. Where improvement of the physical structure of the pancake is desired without impairing the original taste, rice should be used rather than corn. Corn flour has a strong characteristic taste and as its content approaches 50% of the total, the taste of corn begins to predominate.

Corn pancake flours should contain some wheat to counteract excessive mealiness and bitterness. The addition of wheat improves the gas retention of the batter and the texture of the corn pancake. There is no need of adding rice flour to a typical corn pancake flour.

Buckwheat flour also has a strong flavor. It produces a rather syrupy batter and a short but heavy cake. Wheat may be added in quantities up to 30 to 50%, to improve the cakes in volume and tenderness. With increasing percentage of wheat, better gas retention is obtained, resulting in better volume. Rye and soy flours may be regarded as flavoring ingredients, but only up to 10% of rye or soy flour in the flour mix can safely be recommended. Particular precaution is necessary when rye flour is used. Rye has the tendency to produce a soggy pancake with heavy crust. Adequate amounts of corn flour, at least 50% corn per 10% of rye, should be included in the flour mixture to insure proper baking quality. However, a small amount of rye is sufficient to impart its aroma to the pancake.

Powdered milk lends a richer flavor to the pancake. Dry skim milk is a more common ingredient than dry buttermilk. The latter produces a cheesy flavor and is preferably used in wheat pancake flours. Between 2 and 5% of powdered milk, based on the flour used, should be sufficient in most formulas. The large proportion of milk sugar in dry milk is helpful in producing the much desired golden-brown outside of the pancake.

The brown color may also be obtained by adding a small percentage of sugar. Corn sugar has very much less sweetening power than cane sugar, but a better browning effect. Sweetness in a pancake, however, is of comparatively little importance as pancakes are commonly eaten with some kind of syrup.

Since the golden-brown crust can also be achieved by the addition of milk, addition of sugar to the mix may not be necessary. Between 2 and 5% of cane sugar, or up to about 10% of corn sugar, may be used. However, as little as 1% of corn sugar can be sufficient to produce the desired coloring effect.

About 2%, or just a trifle more, of salt is adequate in most cases. Many people prefer about 2½%, but it seems wise to stay somewhat on the low side, since extra seasoning can easily be added by the housewife.

Tenderness and volume of the pancake will largely depend on types and mixture of flours used, but also on the quantity and type of leavening added to the mix.

Mono-calcium phosphate, which, in combination with bicarbonate of soda, represents the most common leavening in self-rising flours, lends itself admirably to pancake baking. Most of the gas evolution should take place in the cold batter, giving it a foamy, fluffy structure before it goes on the griddle. Mono-calcium phosphate, a relatively fast-acting baking acid, serves this purpose very well.

Mono-calcium phosphate is usually applied at a ratio of 100 parts to 80 parts of sodium bicarbonate. In a pancake flour, a ratio of 100 parts of phosphate to 82 to 85 parts of soda (according to the acid strength of the phosphate), may often show an advantage over the common ratio of 100/80. This slight shift towards the alkaline side makes for a more tender pancake.

The amount of leavening added to the pancake flour will depend on the types of flour used. Buckwheat and corn flour require more leavening for best results than plain wheat formulas.

Generally, a leavening mixture based on 1.5 to 2% of soda will be found sufficient in a wheat pancake flour. Normally, 2 to 2.5% should give better results in buckwheat and corn cakes.

After considering any local preference for this or that type of pancake, not forgetting the fact that flour mixtures containing 50% or more wheat are subject to a government tax, the producer can easily decide just what kind of pancake flour should be turned out for local consumption.

FORMULAS FOR PANCAKE FLOURS

Mixed Type

Soft Wheat Clear	35	lb	Cane Sugar	3	lb
Buckwheat Flour	35	lb	Salt	2¼	lb
Corn Flour	15	lb	Bicarbonate of Soda	2	lb
Rice Flour	15	lb	Mono-calcium		
Powdered Skim Milk	3	lb	Phosphate	2⅖	lb

Buckwheat Type

Buckwheat Flour	60	lb	Salt	2¼	lb
Corn Flour	15	lb	Bicarbonate of Soda	2¼	lb
Soft Wheat Clear	25	lb	Mono-calcium		
Powdered Skim Milk	3	lb	Phosphate	2¾	lb

Corn Type

Corn Flour	65	lb	Salt	2¼	lb
Soft Wheat Clear	35	lb	Bicarbonate of Soda	2½	lb
Powdered Skim Milk	3	lb	Mono-calcium		
Corn Sugar	5	lb	Phosphate	3	lb

Containing Soy Flour

Hard Wheat Patent	70	lb	Cane Sugar	5	lb
Corn Flour	20	lb	Salt	2¼	lb
Soy Flour	10	lb	Bicarbonate of Soda	2	lb
Powdered Skim Milk or			Mono-calcium		
Buttermilk	5	lb	Phosphate	2¾	lb

Containing Rye Flour

Corn Flour	60	lb	Salt	2¼	lb
Soft Wheat	30	lb	Bicarbonate of Soda	2½	lb
Rye	10	lb	Mono-calcium		
Corn Sugar	6	lb	Phosphate	3	lb

Mixing a pancake flour in the mill offers hardly any more difficulties than compounding a self-rising biscuit flour. Where a small mixer is used, it is convenient to compound first a basic mix, consisting of all ingredients, but containing only a small part of the flour. This mix should be large enough to supply as many batches as are made within a day. In this manner part of the scaling job and some possibilities of errors are avoided; also a more uniform distribution of the leavening is obtained.

Pancake flours do not usually keep during storage as well as self-rising biscuit flours. The types of flours mostly used in pancake mixes are more subject to insect pests, rancidity and mustiness. Utmost care and cleanliness in the milling and compounding are, therefore, of great importance.

Prepared Waffle Flour

Prepared waffle flours are, in some localities, used a great deal for waffle baking. Corn and wheat flour mixtures generally yield good waffles, shorter extractions of wheat being more suitable for waffle than for pancake baking. Buckwheat formulas yield heavy and rubbery waffles and the percentage of buckwheat in the flour mixture must be cut down if a satisfactory waffle is to be obtained.

Formula No. 1

Flour (General Purpose Type)	100	lb
Soda (Sodium Bicarbonate)	1½	lb
Mono-calcium Phosphate	1³⁄₁₆	lb
Sodium Acid Pyrophosphate	1⅛	lb
Flour Salt	2	lb
Powdered Whole Egg	3	lb
Sugar	12½	lb
Shortening (Plastic)	20	lb
Dry Milk Solids	12½	lb or more

Mix flour and soda for 2 or 3 minutes before adding remainder of dry ingredients. Continue mixing to make total time of 20 minutes.

Add shortening. If a horizontal ribbon type mixer is used, continue mixing for 45 minutes. If trowel type is used, continue mixing for 8 to 12 minutes, depending upon consistency of shortening and mixing-room temperature. If the horizontal ribbon type mixer is strung with cutting wires (as sometimes used for pie crust mixes), continue mixing for 25 to 30 minutes.

Do not continue mixing beyond the period necessary to get uniform, homogeneous distribution of shortening.

Mix 1 pound of waffle flour with 10 fluid ounces of water. Grease the iron and bake until golden-brown.

Formula No. 2

Powdered Sugar	3 lb	Winter Wheat Flour	8 lb
Cottonseed Oil	1 lb	Baking Powder	6 oz
Whole Egg Powder	12 oz	Salt	2 oz

Incorporate the oil into the flour, follow with the egg powder, sugar and the salt. Add the baking powder last. The oil and flour are best prepared in a mixing machine and, when dry enough, transferred to a powder mixing machine, provided with a sifter on top (Day Mixer).

Waffle Desserts

Chocolate Cream Waffle

Divide a plain waffle in quarter or half sections. Spread whipped cream over one section, then cover with other section. Pour chocolate syrup over it. Top with whipped cream.

Waffle A La Mode

Place a portion of ice cream on half of a hot plain waffle.

Fruit Waffle A La Mode

Follow waffle a la mode recipe. Cover ice cream and waffle with crushed fresh fruit or crushed canned pineapple.

Plain Waffles

Flour	1½ lb	Milk	1½ lb
Baking Powder	1 oz	Melted Shortening	4 oz
Salt	¼ oz	Sugar	1 oz
Eggs	9 oz		

Mix and sift dry ingredients. Add milk and beaten egg yolks and stir well. Fold in the stiffly beaten egg whites. Add melted shortening.

Quantity of milk depends somewhat on size of eggs and kind of flour.

Self Rising Flours

Pancake Flour

Mono-Calcium Phosphate	7 lb	Kiln-dry White Corn Flour	100 lb
Bicarbonate of Soda	3¼ lb	Medium Grade Wheat Flour	86¼ lb
Fine Salt	3 lb		

Pancake Flour (Blend Type)

*Flour blend	100 lb	Flour Salt	2 to 2½ lb
Soda (Sodium Bicarbonate)	2 lb	Dextrose (Corn Sugar)	3½ lb
Mono-Calcium Phosphate	2½ lb	Dry Milk Solids	3½ lb

Mix flour and soda for 2 or 3 minutes before adding remainder of ingredients. Continue mixing to make total time of 20 minutes. Horizontal ribbon type mixer is generally used.

Buckwheat Flour

Mono-Calcium Phosphate	7 lb	Buckwheat Flour	130 lb
Bicarbonate of Soda	3¼ lb	Low-Grade Wheat Flour	56¾ lb
Fine Salt	3 lb		

Biscuit Flour

Mono-Calcium Phosphate	7 lb	Fine Salt	3 lb
Bicarbonate of Soda	3¼ lb	Winter Wheat Flour	186¾ lb

* Flour may be a wheat flour blend, a buckwheat blend, a cornmeal blend or a wheat, corn, rye and rice mixture.

Griddle Cake Mixture

Sugar	5½ lb	Phosphate Baking Cream	4½ lb
Salt	2 lb	Malted Milk	4 oz
Vegetable Shortening	10 lb	Pastry Flour	90 lb
Milk Powder	6½ lb	Rice Flour	5 lb
Soda	2½ lb	Corn Flour	5 lb

Mix ingredients together dry. When ready to use, add 1 to 1½ pounds of water slowly to each pound of mixture. Cook quickly on hot griddle.

Because of the controlled leavening action, this batter can be held over for hours or used the following day.

Cakes and Prepared Cake Mixes

The ingredients used in cake formulas contribute certain properties to the finished product. The ingredients may be tougheners, tenderizers, moisteners, driers and flavorers. The function of flour in cake is that of a structure builder. It is a toughener and drier. Sugars contribute sweetness, flavor, eating-qualities and tenderness. Dry sugar is a drier, whereas liquid sugars which contain some moisture might be considered as moisteners. Shortening produces eating-and keeping-qualities, also tenderness. A shortening with good emulsifying properties also helps in carrying liquid in the mix. Shortening is not a drier. It is a tenderizer, and must be considered as a moistener. Milk solids add richness and structure to the cake, and contribute to both the toughness and tenderness of the product. Egg whites are structure builders in angel food cake, meringue and egg white marshmallow. Egg yolks and whole eggs are composed of moisture and fat. The proteins of the egg yolks are structure builders or tougheners. The fat of the egg yolk must be considered as a tenderizer. Egg yolks contain 50% moisture and must be considered as moisteners. Nevertheless, when partially replacing whole eggs, egg yolks have a drying effect. Since they contain 50% moisture and the whole eggs contain approximately 74% moisture, the egg yolks absorb some of the moisture from the whole eggs.

Leavening results in the production of volume or lightness. It may almost be considered a tenderizer. All of the above ingredients combine to bring about the flavor of the cake. Flavors and sugars are most effective for flavor production. Other ingredients have a slight, or no, effect on flavor at all.

Cakes of good eating- and keeping-qualities can be made by adhering to the following recommendations, which call for balanced formulas and have a high sugar and liquid ratio.

a The weight of the sugar should exceed the weight of the flour.

b The weight of the fat should not exceed the weight of the eggs.

c The weight of the liquids (milk and eggs) should equal, or exceed, the weight of the sugar.

d One ounce of baking powder should be used for each pound of flour.

With these rules, cakes are balanced to produce good eating- and good keeping-qualities.

Almond Loaf Cake

Sugar	3 lb	Salt	1½ oz
Shortening	2 lb	Vanilla Extract	½ oz
Almond Paste	4 oz	Almond Extract	¼ oz

Cream the above ingredients for 5 minutes at low speed.

Liquid Milk	3 lb	Sugar	1 lb

Dissolve the sugar in the milk and add to the above.

Cake Flour	3¾ lb	Powdered Sugar	12 oz
Baking Powder	1 oz		

Sift together, add and mix for 5 minutes.

Whole Eggs	2½ lb

Add and mix for 5 minutes.

Sprinkle chopped almonds or candied fruit on the cakes. Bake at 360° F.

Angel Food Cake

XX Sugar	1¼ lb	Cream of Tartar	½ oz

Combine together by sifting once.

Salt	½ oz	Egg Whites	2 lb

Beat with egg whip on 3rd speed until eggs show beater marks (approximately 3 minutes). Then add above mix slowly. Beat until mix stands to a wet peak (approximately 3 minutes).

Vanilla Extract	¾ oz	Almond Extract	¼ oz

Fold into the above mixture.

Cake Flour	12 oz	XX Sugar	1 lb

Sift 4 times and fold into above mixture.

Bake at 300 to 325° F approximately 30 minutes depending on size of cake.

Chocolate Angel Food Cake

Egg Whites	2 lb	Cream of Tartar	¼ oz
XX Sugar	1 lb	Vanilla Extract	½ oz
Salt	½ oz		

Beat with egg whip on third speed until eggs show beater marks (approximately 3 minutes). Combine, by sifting once, the sugar, salt and cream of tartar. Add slowly to the egg whites and beat until mix stands to a wet peak (approximately 3 minutes). Add the vanilla extract.

Cake Flour	9 oz	XX Sugar	1 lb
Cocoa	2½ to 3 oz		

Sift 4 times and fold in.

Bake at 325° F.

Golden Angel Food Cake

Egg Whites	2 lb	Salt	½ oz
XX Sugar	1¼ lb	Cream of Tartar	½ oz

Beat the egg whites until beater marks appear. Sift sugar, salt and cream of tartar together, and add to egg whites.

Cake Flour	12 oz	Egg Yolks	10 oz
XX Sugar	1 lb		

Sift cake flour and sugar 4 times, and fold in. Smooth out by hand. Do not beat. Fold in egg yolks.

Bake at 325° F.

Flour and egg yolks may be incorporated by machine if desired. Care must be taken not to overmix.

Angel Food Macaroon Cup Cakes

XX Sugar	1¼ lb	Cream of Tartar	½ oz
Salt	½ oz		

Combine together by sifting once.

Vanilla Extract	¾ oz	Almond Extract	¼ oz

Beat with egg whip on third speed until eggs show beater marks (approximately 3 minutes). Then add above mix slowly. Beat until mix stands to a wet peak (approximately 3 minutes). Add the extracts.

Cake Flour	14 oz	XX Sugar	1 lb

Sift 4 times. Fold into above mixture.

Macaroon Coconut	12 oz	Water	2 oz

Combine these and fold into mixture.

Bake at 325° F.

Apple Sauce Cake

Brown Sugar	1 lb	Cinnamon, Allspice, Mace and Cloves	¼ oz
Shortening and Butter	1½ lb		
Salt	⅛ oz		

Cream the above ingredients.

Molasses	12 oz	Milk	1½ pt
Malt Syrup	1½ lb		

Add to the above mixture.

Bread Flour	2¼ lb	Baking Powder	2½ oz
Cake Flour	2 lb		

Sift and add.

Ground Apples	2	lb	Flour	8	oz

Add and mix in well.

Currants	1½	lb	Candied Grapefruit	4	oz
Walnuts	4	oz	Lemon and Orange Peel	2	lb
Chopped Cherries	1	lb	Seeded Raisins	2	lb

Add.
Bake in loaf cake tin at 350° F.

Banana Cake

Sugar	3¼	lb	Bananas	1½	lb
Shortening	1¾	lb	Salt	1¼	oz

Cream for 5 minutes at low speed.

Liquid Milk	1½	lb	Sugar	12	oz

Dissolve sugar in milk and add to above.

Cake Flour	3	lb	Baking Powder	1½	oz

Sift together, add, and mix for 5 minutes.

Whole Eggs	2	lb

Add and mix in for 5 minutes.
Bake at 360° F.

Butter Sponge Cake

Sugar	6	lb	Egg Yolks	8	oz
Whole Eggs	2½	lb	Salt	½	oz

Beat warm for approximately 15 minutes at fast speed.

Cake Flour	6	lb	Baking Powder	2½	oz

Sift 3 times. Then fold into the above.

Melted Butter	2	lb	Hot Liquid Milk	4	lb

Combine these and fold in while hot at approximately 180° F.
Bake at 375 to 400° F, at once.

Note: Heat sugar in the oven on a piece of paper. When

sugar starts to brown around the edges, add to eggs and beat. Then warm to approximately 110° F.

Cheese Cake

Cottage Cheese	5 lb	Corn Starch	4 oz
Cake Flour	8 oz	Salt	1 oz
		Juice and rind of one lemon	

Cream the above together.

Whole Eggs	8 oz

Add to above mixture.

Milk	1 qt	Shortening	4 oz

Bring to a boil.

Sugar	3½ lb	Water	1 pt

Boil to 240° F.

Egg Whites	1 qt

Beat and add the boiled syrup.

As soon as the boiled sugar has been added to the egg whites, mix the boiled milk in the cheese mixture. Then combine the cheese mixture and the boiled meringue and get in the oven as soon as possible. Bake at 375° F.

French Cheese Cake

Dry Cottage Cheese	5½ lb	Bread Flour	12 oz
Sugar	2 lb	Salt	¼ oz

Mix together thoroughly.

Egg Yolks	1 qt

Add slowly to the above mixture.

Milk	1 qt

Add in 3 parts until mixture is very smooth.

Egg Whites	1 qt	Cream of Tartar	½ oz

Whip fairly stiff.
Sift together.

Cake Flour	2 lb	Baking Powder	½ oz
Powdered Sugar	12 oz	Salt	¼ oz

Fold the above into whites carefully.

Shortening	6 oz	Butter	4 oz

Add and mix as mealy type pie dough.

Milk (Variable)	¾ pt

Add and mix smooth.

Roll out dough for various size pans the same as for pie dough. Fill the pans with cheese mix and bake in oven at 350° F for about 35 minutes depending on size of pan. Leave damper open for 15 minutes, close, and finish baking until the desired color is obtained.

Cheese Cake Filling

Cottage Cheese	5 lb	Dry Milk Solids	2 oz
Butter	½ lb	Lemon Flavor	To taste
General Purpose Flour	½ lb	Egg Whites	1½ lb
Whole Eggs	1½ lb	Sugar	1½ lb
Salt	¼ oz	Water	About 1 lb

Put cheese through coarse sieve. Rub cheese, butter and flour together. Add egg yolks, salt, dry milk solids and lemon flavor. If sugared yolks are used instead of whole eggs, cut weight of eggs to 1 pound.

Beat egg whites and sugar to stiff meringue, and add to mixture. Quantity of eggs used will determine quantity of water needed to give filling the desired consistency. Stir well before pouring into cheese cake pan. Sprinkle with cinnamon.

If desired, a layer of spiced and sugared dry bread crumbs can be substituted for the layer of basic sweet dough in bottom of baking pan.

Bake at medium heat, 300 to 350° F. If oven is too hot, the cheese filling will crack. Cracking may be prevented by drawing cake from oven when partially baked and piercing skin on cake surface to let out the excess moisture.

Cherry Loaf

Sugar (Fine Granulation)	1½ lb	Salt	⅛ oz
Shortening	1 lb	Cherry Flavor	¼ oz

Cream the above ingredients.

Unbeaten Egg Whites	1 lb

Add to the above mixture.

Milk	1 lb

Add.

Maraschino Cherries (Cut Up)	1 lb	Walnuts	¼ lb

Add.

Cake Flour	2½ oz	Baking Powder	1½ oz

Sift and add.
Bake in loaf tins at 375° F.

Chocolate Malted Milk Cake

Sugar	3¼ lb	Salt	¾ oz
Shortening	1⅝ lb	Vanilla Extract	½ oz

Cream for 5 minutes at low speed.

Liquid Milk	3 lb	Sugar	8 oz

Dissolve sugar in milk and add to above mixture.

Cake Flour	3 lb	Malted Milk Powder	8 oz
Cocoa	5 oz	Baking Powder	2 oz

Sift together, add and mix for 5 minutes.

Whole Eggs	1¾ lb

Add and mix in for 5 minutes.
Bake at 360° F.

High Ratio Chocolate Malted Milk Cake

Cake Flour	5 lb	Baking Soda	¾ oz
Skim Milk Powder	1½ lb	Powdered Egg Yolk	12 oz
Malted Milk Powder	1½ lb	Vanilla Extract	½ oz
Granulated Sugar	7 lb	Dark Cocoa Powder	2 lb
Shortening	2½ lb	Salt	3 oz
Baking Powder	3½ oz		

Cream sugar and shortening, then add to the rest of the ingredients, which have been sifted together, and mix thoroughly. This mixture will require about 10 to 12 pounds of water to make a smooth creamy dough.

Bake at 375° F.

Chocolate Marshmallow Roll
(Two Sheets)

Sugar	1½ lb	Egg Yolks	8 oz
Whole Eggs	8 oz		

Beat the above ingredients until light.

Liquid Milk	1 lb

Add slowly

Cocoa	3 oz	Salt	½ oz
Baking Powder	1 oz	Cake Flour	1½ lb

Sift 4 times and fold into above mix.
Bake at 400° F.

When cool, cover with a thick layer of marshmallow and roll up. Cover with chocolate icing.

Chocolate Peppermint Roll

Egg Whites	1 lb	Sugar	14 oz

Beat egg whites until they point but not until dry. Fold in sugar.

Egg Yolks	1	lb	Sugar	14	oz
Water	4	oz			

Beat egg yolks and water until light. Add sugar and beat until fluffy. Fold in above egg white and sugar mix.

Cake Flour	7	oz	Baking Powder	½	oz
Cocoa	8	oz	Salt	¼	oz

Sift 3 times and fold in.

Vanilla Extract	¾	oz

Add to the mixture.

Bake at 400° F.

Formula makes two sheets. When baking is complete, remove paper and roll up. When the roll is cold, unroll and fill with peppermint filling.

Peppermint Filling

Whipping Cream	1	lb	Peppermint Extract	¼	oz
Sugar	4	oz			

Whip the whipping cream. Fold in the sugar and peppermint extract.

Milk Chocolate Cake Powder
(Rich Mix)

Cake Flour	5	lb	Cocoa Powder Natural	12	oz
Granulated Sugar	6	lb	Salt	2	oz
Powdered Whole Milk	12	oz	Baking Soda	½	oz
Shortening	2½	lb	Nulomoline	1	lb
Powdered Whole Egg	3	oz	Baking Powder	3	oz
Powdered Egg Yolk	1½	oz	Vanilla Flavor	As desired	

All ingredients should be sifted. Cream shortening, sugar and Nulomoline. Add slowly while creaming, powdered milk, eggs, cocoa, baking soda, powder, salt and flour. Flavor with vanilla. This mixture will require about 4 pounds of water.

Bake in layer cake forms at about 375° F.

Milk Chocolate Cake

Sugar	3¾ lb	Salt	1¼ oz
Shortening	1⅞ lb	Soda	¼ oz

Cream for 5 minutes at low speed.

Egg Yolks	11 oz	Melted Unsweetened Chocolate	1 lb

Add to above.

Liquid Milk	1⅞ lb	Vanilla Extract	½ oz

Add gradually.

Cake Flour	3 lb	Baking Powder	2 oz

Sift together, add, and mix until smooth.

Liquid Milk 1⅞ lb

Add and mix for 5 minutes.

Egg Whites 11 oz

Add gradually and mix for 5 minutes.

Bake at 360° F.

Milk Chocolate Cake

Cake Flour	4¾ lb	Shortening	2 lb 13 oz
Cocoa	5 oz		

Mix together for approximately 6 minutes at low speed.

Sugar	7 lb	Vanilla	1 oz
Baking Powder	5 oz	Skim Milk Powder	9 oz
Salt	2½ oz	Water	2 lb

Add and cream for approximately 4 minutes at low speed.

Egg Whites	3¾ lb

Add and cream for approximately 4 minutes at low speed.

Water	2 lb 8½ oz

Add, mixing for approximately 4 minutes at low speed.

Scale 13 ounces in 8-inch, deep, round, layer cake tins and bake at 360° F.

For loaves and cups, add 1 to 1½ ounces of flour per pound of batter.

Date Loaf and Layer Cake

Boiling Water	3 pt	Chopped Dates	2 lb

Stir together and cool to room temperature.

Brown Sugar	3⅛ lb	Soda	½ oz
Shortening	1½ lb	Salt	1 oz

Cream for 5 minutes at low speed. Then add to date mixture.

Cake Flour	2½ lb	Baking Powder	⅝ oz

Add and mix for 5 minutes.

Whole Eggs	1¼ lb

Add and mix for 5 minutes.

Chopped Walnuts	1 lb

Mix in.

Bake at 360° F.

Devil's Food Layers

Cake Flour	5 lb	Gum Tragacanth	2 oz
Baking Powder	3 oz	Skim Milk	1 lb
Powdered Whole Milk	1 lb	Cocoa Powder	1 lb
Granulated Sugar	5 lb	Baking Soda	2 oz
Shortening	2 lb	Salt	2 oz
Egg Albumen	4 oz		

Sift together all but the sugar and shortening, which should be creamed and then added to the sifted mixture. This mixture will require about 7 pounds of water.

Bake in oven heated to about 400° F.

Devil's Food Mix

Brown Sugar	15	lb	Ground Cinnamon	4	oz
Bicarbonate of Soda	6	lb	Vanillin	3	oz
Cream of Tartar	4	lb	Powdered Buttermilk	7½	lb
Winter Wheat Flour	30	lb	Chocolate Color	4½	oz
Dutch Cocoa	15	lb	Amaranth	1½	oz

Recipe

Granulated Sugar	2½	lb	Eggs	1	pt
Shortening	1	lb			

Cream the above.

Devil's Food Mix	2	lb	Salt	1	oz

Add to the above mix and mix thoroughly, then add:

Flour	2¼ lb

Thin down with water (2 to 2½ pints) to desired consistency. Bake in temperature of 350 to 375° F.

Devil's Food Cake
(Powdered Mix)

Cake Flour	5	lb	Shortening	2	lb
Powdered Whole Milk	8	oz	Egg Albumen	2½	oz
Granulated Sugar	2½	lb	Baking Powder	3½	oz
Brown Sugar	4½	lb	Salt	1⅗	oz
Cocoa Powder	1	lb	Vanillin	As desired	

All ingredients should be sifted. Cream shortening and sugar. Then add, while creaming, milk, cocoa, albumen, baking powder and the balance.

This mixture will require about 8 pounds of water. Bake at 360 to 375° F.

Devil's Food
Formula No. 1

Sugar (Powdered XXXX)	1¾ lb	Sour Milk	3¼ pt
		Milk	¾ pt
Malt Syrup	4 oz	Soda	¾ oz
Shortening	1 lb	Flour	2 lb
Eggs	6	Cocoa	5 oz

Sift the flour and cocoa together. This can be baked in layers or sheets and iced with chocolate or marshmallow icing.

Devil's Food
(Layers or Sheets)
Formula No. 2

Sugar	10 lb	Soda	2 oz
Shortening	4 lb	Cocoa	1½ lb
Salt	2 oz		

Cream together.

Eggs	6 lb	Vanilla Flavor	To suit

Add to the above mixture.

Water	2 lb	Invert Syrup	¾ lb
Sweetened Condensed Milk	4 lb		

Add mixed together.

Flour	9 lb	Baking Powder	2 oz

Sift together and add.

Water	2 lb

Add.
Bake at 350 to 375° F.

Devil's Food
(Layers, Cups and Sheets)
Formula No. 3

Sugar (Fine Granulation)	9	lb	Salt	2	oz
Shortening	3½	lb	Vanilla	2	oz

Cream together.

Eggs	4	lb

Add slowly.

Milk	4	lb

Add.

Flour	10	lb	Baking Powder	5½	oz

Add mixed together.

Cocoa	2	lb	Soda	¼	lb
Sugar	2	lb			

Sift and make a paste of these and add to above mixture.

Cold Water	4	lb

Add.
Bake at 350 to 375° F.

High-Ratio Devil's Food Fudge Cake
Formula No. 1

Cake Flour	4	lb	Baking Powder	3	oz
Skimmed Milk Powder	12	oz	Baking Soda	1	oz
Sugar	7	lb	Shortening	2½	lb
Dark Cocoa Powder	1	lb	Salt	3	oz
Powdered Egg Yolk	1¾	lb	Vanilla Extract	To taste	
Powdered Egg Albumen	1	oz	Cinnamon	½	oz

Carefully mix and sift the above ingredients. The amount of water needed for the proper dough consistency is about 8 pounds. Mix until smooth. This formula may be used for layer cake, cup cakes, loaf cakes, etc.

Bake in oven at about 375° F.

Formula No. 2

Flour	2 lb	Shortening	1⅜ lb

Mix from 3 to 5 minutes. Scrape down at least once during this stage.

Sugar	3½ lb	Baking Powder	1½ oz
Cocoa	8 oz	Cinnamon	¼ oz
Salt	1½ oz	Liquid Skim Milk	1¼ lb
Soda	½ oz		

Add to the blended mass and mix for 3 to 5 minutes, scraping down at least once.

Whole Eggs	1⅞ lb	Vanilla	To taste
Liquid Skim Milk	1⅞ lb		

Scale off eggs, milk and flavor together, and add approximately half of it to the bowl. Mix until smooth. Then add the balance of the liquid ingredients and continue mixing for a total of 3 to 5 minutes in this stage, scraping down again to insure a smooth batter.

This quantity of batter should be mixed in a small bench type mixing machine.

Mixing Speed: For the whole mixing operation, use low speed if a three-speed machine is used, or second lowest speed on a four-speed machine. Use paddle for making this cake.

Devil's Food Cake

Formula No. 1

Cake Flour	2 lb	Cocoa	12½ oz
Shortening	2¾ lb		

Mix together at low speed for approximately 6 minutes.

Sugar	8⅛ lb	Cinnamon	1 oz
Salt	2½ oz	Milk Powder	12½ oz
Baking Powder	2½ oz	Water	2½ lb
Soda	1¾ oz		

Add and mix in at low speed for approximately 4 minutes.

Egg Whites	1⅞ lb	Whole Eggs	1⅞ lb

Add in 2 parts and mix in low speed for approximately 4 minutes.

Water	3 lb

Add and mix in low speed for approximately 8 minutes.

If a finer grain and texture are desired, use 2½ ounces soda only instead of baking powder and soda combination.

Caution:

Scrape bowl in each stage.

In second stage mix slowly to prevent lumps.

When using bench type machines, cut the time in each stage in half.

Quantity of sugar is based on total weight of flour.

Formula No. 2

Sugar	3 lb	Soda	1½ oz
Shortening	1¾ lb	Vanilla Extract	¾ oz
Salt	1 oz		

Cream for 5 minutes at low speed.

Melted Unsweetened Chocolate	10 oz

Add to above.

Buttermilk	4½ lb	Sugar	12 oz

Dissolve sugar in milk and add to above.

Cake Flour	3 lb	Baking Powder	1¼ oz

Sift together, add and mix for 5 minutes.

Whole Eggs	1⅞ lb

Add and mix for 5 minutes.
Bake at 360° F.

Formula No. 3

(A)	XX Sugar	6¾ lb	Shortening	3¾ lb
(B)	Whole Eggs	1¼ lb	Vanilla Extract	¾ oz
	Egg Yolks	12 oz		

Cream thoroughly until light.

Beat together slightly. Add to (A) gradually and continue creaming for 3 to 4 minutes.

(C)	Cake Flour	4½ lb	Salt	2½ oz
	Cocoa	1 lb	Soda	1½ oz
	Cinnamon	¾ oz		

Blend and sift together. Then add to (A) and (B) alternately with half of (D) and mix for 2 to 3 minutes.

(D)	Buttermilk	3¾ lb	Cold Water	1 lb

Add remaining milk and mix until smooth. Add water last and beat for 2 to 3 minutes.

Scaling for 8 inch layer: 12 to 14 ounces. Oven temperature: 350 to 360° F. Variations: Sheets, cakes and cup cakes. Use chocolate fudge icing.

Devil's Food Layer Formula

Flour	100 lb	Shortening	55 lb

Mix for 3 minutes at low speed.*

Sugar	140 lb	Soda	1³⁄₁₀ lb
Cocoa	20 lb	Baking Powder	5 lb
Salt	4 lb	Liquid Skim Milk	50 lb

Add to the blended mass and mix for 3 minutes at low speed.

Whole Eggs	65 lb	Flavor	To taste
Liquid Skim Milk	90 lb		

* Considering low speed as first speed, use first speed on a three-speed machine and second speed on a four-speed machine.

Add eggs and milk together carefully and mix for 3 minutes at low speed.

Keep the batter smooth in all stages by scraping down the bowl properly.

English Nut Cup Cakes

Sugar	2½ lb	Salt	1 oz
Shortening	1⅛ lb	Vanilla Extract	½ oz

Cream for 5 minutes at low speed.

Liquid Milk	2 lb	Sugar	1 lb

Dissolve sugar in milk and add.

Cake Flour	2¾ lb	Baking Powder	1 oz

Sift together, add and mix for 5 minutes.

Whole Eggs	2¼ lb

Add and mix in for 5 minutes.

Chopped Walnuts	1 lb

Fold in.

Bake at 360° F.

These cup cakes should be baked in paper cups.

Egg Cup Cake

Shortening	10 oz	Baking Powder	1⅗ oz
Sugar	30 oz	Salt	½ oz
Flour	30 oz	Vanillin	Enough to suit
Powdered Egg Yolk	5 oz		
Powdered Skimmed Milk	5 oz		

Bake in moderate oven (375° F).

Gingerbread

Flour	20	oz	Ginger	5	g
Sugar	4	oz	Nutmeg	1	g
Molasses	10	oz	Cloves	½	g
Shortening	4	oz	Baking Powder	4	g
Whole Yolk	1	oz	Baking Soda	5	g
Salt	2	g	Boiling Water	7	oz
Cinnamon	5	g			

Combine shortening, salt and spices. Cream in sugar until light. Add eggs and beat thoroughly. Add molasses. Add to creamed mixture the blended flour, soda and baking powder. Mix well, then add boiling water, beat until smooth. Bake at 350° F for 50 to 60 minutes.

Gingerbread Mix

Sugar	12	lb	Butter	4	lb
Salt	4	oz	Totally Hydrogenated Vegetable Shortening	12	lb
Concentrated Butter Flavor	⅖	oz	Soda	2½	lb
Cinnamon	8	oz	Cake Flour	64	lb
Ginger	8	oz			

Mix all ingredients together, except the cake flour, until a smooth blend is produced. Then add the cake flour and mix until it is incorporated. This mixture may be packaged in 1-or 2-pound packages for household or institutional use.

Gingerbread Made From Prepared Mix

Prepared Gingerbread Mix	2	lb	Molasses	1½	lb
Whole Eggs	6	oz	Water	12	oz

Place the whole eggs, molasses and gingerbread mix in the machine bowl, and stir at low speed to wet. Beat 1 minute at high speed. Add about 3 ounces of water and mix until smooth. Continue to add water in small amounts and each time mix smooth again until all the water is used. Then beat ½ minute at high speed.

Weigh 7 ounces into buttered papered-bottom loaf tin 4 by 4 by 2½ inches. Bake at 325° F with medium top and low bottom oven temperature. Bake until just fairly done.

Fruit Salad Cake

Cake Flour	5 lb	Shortening	2⅝ lb

Mix together for approximately 6 minutes at low speed.

Sugar	6½ lb	Baking Powder	5 oz
Salt	2½ oz	Skim Milk Powder	7 oz

Add and cream for approximately 4 minutes at low speed.

Egg Whites	3½ lb

Add and cream for approximately 4 minutes at low speed.

Water	1 lb

Add, mixing in for approximately 4 minutes at low speed.

Fruit Mix *	2½ lb

Add, mixing in for approximately 4 minutes at low speed.

Scale 13½ ounces into 8-inch, deep, round, layer cake tins and bake at 360° F. For loaves and cups, add 1 to 1½ ounces of flour per pound of batter.

* Fruit Mix

Peel of 1 Orange		Peaches	2 lb
Maraschino Cherries	4 oz	Pineapple	2 lb

Grind the above together.

Add juice of:

Oranges	2	Lemon	1

FRUIT CAKES

	Golden Fruit Cake		*White Fruit Cake*		*Spice Fruit Cake*	
Flour	5	lb	5	lb	5	lb
Sugar	3½	lb	3½	lb	—	
Brown Sugar	—		—		5	lb
Molasses	2	lb	—		1	lb
Honey	1	lb	1	lb	—	
Butter	12	oz	1	lb	4	oz
Shortening	3	lb	3	lb	3	lb
Milk	4	oz	8	oz	3	lb
Caramel Color	—		—		3	oz
Whole Eggs	3	lb	4	lb	3	lb
Salt	3	oz	2	oz	4	oz
Cinnamon	1½	oz	—		1	oz
Mace	½	oz	—		½	oz
Cloves	1	oz	—		½	oz
Vanilla Extract	½	oz	¼	oz	¼	oz
Almond Flavor	—		1	oz	¼	oz
Lemon Flavor	1	oz	—		—	
Rum Flavor	1	oz	—		—	
Baking Powder	5	oz	5	oz	5	oz
Fruit Mixture:						
Seedless Raisins	6	lb	7	lb	6	lb
Sliced Citron	3	lb	3	lb	1	lb
Sliced Lemon Peel	¼	lb	—		—	
Sliced Orange Peel	½	lb	3	lb	—	
Walnut, Pecan Pieces	3	lb	—		1	lb
Cherry Pieces	2	lb	3	lb	2	lb
Chopped Dates	2	lb	2	lb	2	lb
Water	2	lb	—		—	
Sugar Syrup	1	lb	—		—	
Almonds	—		2	lb	—	

Cream shortening, add sugar, syrups, beat well, then add egg, milk, etc. Sift flour with baking powder, salt and spices, and add to creamed sugar, etc. Sift part of the flour over the fruit, add to batter and add the nuts last. Place in loaf pans lined with greased paper. Cover tightly.

Bake with light steam for 1½ to 3 hours at 325 to 350° F.

Remove covers at end of first hour and finish baking.

Cover fruit mixture with damp cloth overnight.

Light Fruit Cake

Formula No. 1

Sugar	2 lb	Salt	1 oz
Shortening	2 lb	Flavor	½ oz

Cream together.

Eggs	2 lb

Cream into the above mixture.

Flour	2 lb	Baking Powder	¼ oz

Add and blend well.

Almonds	1½ lb	Glace Cherries	2 lb
White Raisins	2 lb	Glace Pineapple	1½ lb
Citron	2 lb		

Add to the mixture.
Scale into loaf tins and bake at slow heat.

Formula No. 2

Bleached Raisins	4½ lb	Citron	12 oz
Currants	3 lb	Candied Pineapple	2½ lb
Candied Cherries	2½ lb	Brandy	1 lb

Prepare fruit and soak overnight in brandy.

Cake Flour	3¼ lb	Shortening	2¾ lb

Cream until light at low speed.

Sugar	3¼ lb	Vanilla Extract	½ oz
Whole Eggs	1¾ lb	Salt	1¼ oz
Egg Whites	1 lb		

Beat until light. Add to above mixture and cream for 5 minutes at medium speed.

Corn Syrup	12 oz

Add.

Pecans	1¾ lb	Blanched Almonds	1½ lb

Fold in.
Now fold in prepared fruit. Bake at 350° F.

Dark Fruit Cake

Raisins	3¼ lb	Pineapple	1½ lb
Currants	3 lb	Candied Cherries	2¼ lb
Citron	1¾ lb	Cinnamon	1 oz
Lemon Peel	6 oz	Allspice	½ oz
Orange Peel	6 oz	Brandy	1½ lb
Figs	1 lb	Sugar	4⅛ lb
Dates	1 lb	Shortening	1¾ lb

Prepare fruit and soak overnight in brandy.

Salt	1¼ oz	Soda	½ oz

Cream together until light.

Whole Eggs	1½ lb

Add gradually.

Molasses	12 oz	Honey	6 oz

Add to above mixture.

Liquid Milk	12 oz

Stir in.

Cake Flour	4¾ lb

Fold in.

Blanched Almonds	1¾ lb

Add to the above.
Now fold in prepared fruit.
Bake at 350° F.

Fruit Food Bars

Dried Ground Prunes	60 lb	Powdered Citric Acid	3 oz
Dried Ground Raisins	60 lb	Salt	6 oz
Dried Ground Apricots	12 lb	Dried Ground Orange Peel	3 oz
Powdered Sugar	25 lb		
Corn Syrup	3 lb		

Mix sugar, acid and salt, and combine with other ingredients by passing through a food grinder several times. The dried fruits can be softened first in a kettle if difficult to get proper mixing. Spread on slab or oiled paper. When hard, cut to size then roll in sugar or ground cocoanut, or coat with chocolate. Variations of the above formula can be made by using dried ground figs, dates or fondant.

Fruit Bars

Cake Crumbs	6 lb	Water	2 qt
Molasses	1½ qt		

Soak until crumbs are soft and then cream up.

Sugar	4 lb	Salt	1 oz
Malt Syrup	2 lb	Ginger	½ oz
Shortening	3 lb	Cinnamon	1 oz
Soda	6 oz		

Mix these with the above.

Raisins or Currants	5 lb	Flour	11 lb
Coconut	2 lb		

Add to above mixture.

Scale off in 12-ounce pieces and roll out in long strips. Lay four strips on pan, flatten and bake in medium oven. These bars may be iced with water icing after baking, or washed with egg before baking, and left plain.

Cut into 1-inch slices when cool. The fruit may be omitted and many sorts of cookies made out of this mixture.

Fudge Layer Cake

Cake Flour	5 lb	Cocoa	1¼ lb
Shortening	2⅛ lb		

Mix together for approximately 6 minutes at low speed.

Sugar	7½ lb	Salt	3 oz
Baking Powder	3 oz	Skim Milk Powder	14¾ oz
Soda	1 oz	Water	3¾ lb

Add and cream for approximately 4 minutes at low speed.

Egg Whites	1¾ lb	Whole Eggs	1¾ lb

Mix together, add and cream for approximately 4 minutes in low speed.

Water	3 lb 1¼ oz	Flavor	1 oz

Add, mixing in for approximately 4 minutes in low speed.

Scale 12½ ounces into 8-inch deep, round, layer cake tins and bake at 360° F.

For loaves and cups, add 1 ounce of flour per pound of batter.

High-Ratio Golden Cake

Cake Flour	5 lb	Powdered Egg Yolk	12 oz
Skimmed Milk Powder	14 oz	Baking Powder	4 oz
Sugar	7 lb	Salt	3 oz
Shortening	2 lb	Vanilla Extract	3 oz

Carefully mix these ingredients into a uniform powder. Add about 8 pounds of water and mix thoroughly until a smooth consistency is formed.

Bake in oven at 375° F.

This formula may be used for strawberry shortcake, jelly rolls, cup cakes, layer cakes, etc.

Jelly Roll

Sugar	2¼ lb	Flour	3 lb
Malt Syrup	4 oz	Baking Powder	1 oz
Shortening	¼ lb	Ammonium Carbonate	½ oz
Egg Yolk	1½ pt	Salt	½ oz
Milk	1½ pt		

This makes a jelly roll or marshmallow roll of good keeping-qualities. If a chocolate roll is desired, add a thick paste made of 8 ounces of cocoa and water.

High-Ratio Platinum White Cake

Cake Flour	5 lb	Gum Tragacanth	10 oz
Skimmed Milk Powder	12 oz	Baking Powder	5 oz
Granulated Sugar	7 lb	Cream of Tartar	1 oz
Shortening	2 lb	Vanilla Extract	3 oz
Powdered Egg Albumen	4 oz	Salt	3 oz

Carefully mix and sift all the ingredients to a uniform powder. Add about 7 pounds of water and mix until smooth and creamy. This formula may be used for layer cakes, cup cakes, ring cake, sheet cake and loaf cakes.

Baking temperature depends upon the form of cake to be baked.

White Layer Cake

Formula No. 1

Cake Flour	5 lb	Shortening	2¾ lb

Mix together for approximately 6 minutes at low speed.

Sugar	7 lb	Skim Milk Powder	10 oz
Salt	2½ oz	Water	2⅜ lb
Baking Powder	4¾ oz		

Add and cream for approximately 4 minutes at low speed.

Egg Whites	3¾ lb

Add and cream for approximately 4 minutes at low speed.

Water	2 lb	Vanilla Extract	1½ oz

Add, mixing in for approximately 4 minutes at low speed.

Scale 13½ ounces into 8-inch deep, round, layer cake tins and bake at 360° F. For loaves and cups, add 1 ounce of flour per pound of batter.

Formula No. 2

Sugar	4 lb	Salt	3 oz
Shortening	2¼ lb		

Cream for 5 minutes at low speed.

Liquid Milk	5 lb	Vanilla Extract	½ oz
Sugar	1¾ lb		

Dissolve sugar in milk and add to above.

Cake Flour	5 lb	Baking Powder	4 oz

Sift together and mix for 5 minutes.

Egg Whites	3 lb

Add and mix in for 5 minutes.

Bake at 360° F.

For cup cakes and loaf cakes, add 1 ounce of flour to each pound of batter and mix about 3 minutes longer.

White Layer Formula

Flour	100 lb	Shortening	50 lb

Mix for 3 minutes at low speed.*

Sugar	125 lb	Baking Powder	6 lb
Salt	4 lb	Liquid Skim Milk	40 lb

Add to the blended mass and mix for 3 minutes at low speed.

Egg Whites	65 lb	Flavor	To taste
Liquid Skim Milk	50 lb		

* Considering low speed as first speed, use first speed on a three-speed machine or second speed on a four-speed machine.

Add eggs and milk together carefully and mix for 3 minutes at low speed.

Keep the batter smooth in all stages by scraping down the bowl properly.

Lemon Cup Cakes

Cake Flour	5	lb	Powdered Egg Yolk	12	oz
Powdered Whole Milk	1	lb	Powdered Egg Albumen	½	oz
Granulated Sugar	6	lb	Powdered Lemon Juice	7	oz
Baking Powder	5	oz	Salt	1½	oz
Shortening	2	lb			

Cream sugar and shortening. Sift all other ingredients and mix well together. Then work in about 5 pounds of water and mix until smooth and creamy. Fill into baking cups and bake in an oven at 375 to 400° F for about 20 minutes. Use a lemon fondant icing.

High-Ratio Honey Macaroon Cake

Flour	2½	lb	Shortening	1⅜	lb

Mix from 3 to 5 minutes. Scrape down at least once during this stage.

Light Brown Sugar	1⅞	lb	Baking Powder	2½	oz
Granulated Sugar	1¼	lb	Honey	6	oz
Salt	1½	oz	Liquid Skim Milk	1	lb

Add to the blended mass and mix from 3 to 5 minutes, scraping down at least once.

Whole Eggs	1½	lb	Vanilla	To taste
Liquid Skim Milk	1⅝	lb		

Scale off the eggs, milk and flavor together, and add approximately half of the mixture to the bowl. Mix until smooth, scrape down and mix until smooth again. Then add balance of liquid ingredients and continue mixing for a total of 3 to 5 minutes in this stage, scraping down again to insure a smooth batter.

This quantity of batter should be mixed in a small bench type mixing machine.

Baking Temperature: For a 7-inch layer scaled 10 ounces bake at 375° F. Using this as a basis, you can determine what temperature to use on other sized cakes.

Loaves, Rings and Cups: This makes an excellent mix for these varieties.

Mixing Speed: For the whole mixing operation, use low speed if a three-speed machine is used or second lowest speed on a four-speed machine. Use the paddle for making this cake.

Mocha Layer Cake

Sugar	3 lb	Salt	1 oz
Shortening	2 lb		

Cream for 5 minutes at low speed.

Liquid Milk	2¼ lb	Sugar	12 oz

Dissolve sugar in milk and add to the above.

Cake Flour	3 lb	Baking Powder	1½ oz
Cocoa	1½ oz		

Sift together, add and mix for 5 minutes.

Whole Eggs	2

Add and mix for 5 minutes.

Ground Walnuts	12 oz

Fold in.

Bake at 360° F.

Orange Cocoanut Layer Cake

Cake Flour	5 lb	Shortening	2⅜ lb

Mix together for approximately 6 minutes at low speed.

Sugar	6½ lb	Salt	2½ oz
Baking Powder	4 oz	Macaroon Cocoanut	1 lb
Skim Milk Powder	8 oz	Water	2¼ lb

Add and cream for approximately 4 minutes at low speed.

Whole Eggs			2⅝	lb

Add and cream for approximately 4 minutes at low speed.

Orange Juice	6 oz	Grating of 2 Oranges and		
Lemon Juice	2 oz	1 Lemon		
		Water	1⅝	lb

Add, mixing in for approximately 4 minutes at low speed.

Scale 12 ounces into 8-inch deep, round, layer cake pans. Bake at 360° F.

Plantation Golden Layer

Cake Flour	3¾ lb	Shortening	2³⁄₁₆	lb

Mix together for approximately 6 minutes in low speed.

Cake Flour	1¼ lb	Brown Sugar	7	lb
Skim Milk Powder	8¼ oz	Water (To Dissolve		
Baking Powder	3¾ oz	Brown Sugar)	2½	lb

Add and cream for approximately 4 minutes at low speed.

Egg Yolks	3⅛ lb	Water	2½	lb

Mix together, add mixing in for approximately 4 minutes at low speed.

Scale 13 ounces into 8-inch deep, round, layer cake pans. Bake at 360° F.

This mix may be used for loaf or cup cakes, no change being necessary.

High-Ratio Spice Cake

Cake Flour	5	lb	Powdered Egg		
Skimmed Milk Powder	10	oz	Albumen	1	oz
Granulated Sugar	3½	lb	Baking Powder	5	oz
Brown Sugar	3	lb	Salt	2	oz
Powdered Molasses	1	lb	Cinnamon	50	g
Shortening	3	lb	Cardamom	20	g
Powdered Egg Yolk	1	lb	Cloves	10	g
			Nutmeg	10	g
			Ginger	10	g

Cream shortening and sugar. Then add to the other ingredients, which have been sifted all together. This mixture will require about 7 pounds of water.

Bake in oven at about 375° F.

Old Fashioned Spice Cake

Cake Flour	5 lb	Ground Raisins	10 oz
Shortening	2½ lb		

Mix together for approximately 6 minutes at low speed.

Brown Sugar	3 lb	Spice Blend *	3½ oz
Sugar	3¼ lb	Skim Milk Powder	10 oz
Salt	2½ oz	Water	2⅛ lb
Baking Powder	4 oz		

Add and cream for approximately 4 minutes at low speed.

Whole Eggs	3 lb

Add and cream for approximately 4 minutes at low speed.

Molasses	12 oz	Water	2⅜ lb

Add, mixing in for approximately 4 minutes at low speed.

Scale 12½ ounces into 8-inch deep, round, layer cake tins and bake at 360° F. This mix may be used as above for cup cakes. If loaf or round top cakes are desired, add 1 ounce of flour for each pound of batter required.

* Spice Blend

Cinnamon	6 oz	Cloves	1 oz
Nutmeg	2 oz	Allspice	1 oz
Mace	2 oz	Ginger	½ oz

Sift together several times.

Cake Spice Mixture

Allspice	⅜ lb	Mace	¾ lb
Nutmeg	⅜ lb	Cloves	¾ lb
Ginger	⅜ lb	Cinnamon	2¾ lb

Cake Spices

For Fruit and Spice Cakes

Ground Cinnamon	1 lb	Ground Cloves	4 oz
Ground Ginger	¼ oz	Ground Mace	1 oz

Apple Pie Spice

Ground Mace	½ lb	Cayenne Pepper	1¼ lb
Ground Cinnamon	1½ lb		

(If mace is high, use nutmeg.)

Mince Meat Spice

Cinnamon	67½ lb	Cloves	22½ lb
Allspice	22½ lb	Mace	14½ lb

Spice Mix

Cinnamon	1 lb	Cloves	4 oz
Allspice	8 oz	Ginger	4 oz
Mace	8 oz	Cake Flour	12 oz
Nutmeg	4 oz		

Mix well and store in an air-tight container.

Spice Blend

Cinnamon	4 oz	Allspice	¼ oz
Nutmeg	½ oz	Mace	¼ oz
Ginger	¼ oz	Ground Cardamom	¼ oz
Cloves	¼ oz		

Sift several times.

Sponge Cake Mix

XXXX Sugar	6 lb	Winter Wheat Flour	6 lb
Granulated Egg Yolk	1½ lb	Salt	1½ oz
Skim Milk Powder	1⅞ lb	Lemon Oil	15 drops
Baking Powder	7 oz	Egg Color	Trace
Albumen	8 oz		

Chocolate Sponge Cake Mix

Same as Sponge Cake Mix but use:

Cocoa	¾ lb	Flour	5¼ lb

Add nothing but water. This will produce any form of sponge cake, sponge layers, sheet layers, loaf cakes, lunch cakes, jelly rolls, etc.

Take 12 ounces of lukewarm water and mix with 2 pounds of sponge cake mix and make dough. Then put in machine and beat for 4 minutes. Scale into pans. Bake in a moderate oven.

Deluxe Sponge Cake

Sugar	3 lb	Cake Flour	10 oz
Whole Eggs	2 lb	Salt	1 oz

Beat until light.

Liquid Milk	1½ lb	Vanilla Extract	½ oz
Cake Flour	1 lb		

Whip together and add to above.

Cake Flour	10 oz	Baking Powder	1¼ oz

Sift together and fold in.
Bake at 380° F.

Sponge Jelly Roll
(Cold Rolled)

Sugar	2⅜ lb	Salt	½ oz
Whole Eggs	1⅜ lb	Vanilla Flavor	½ oz
Egg Yolks	10 oz		

Beat until light.

Hot Liquid Milk	1 lb

Fold in.

Cake Flour	2 lb	Baking Powder	½ oz

Sift together 3 times and fold in carefully.
Bake at 400° F.

American Pound Cake

Sugar	4½ lb	Cake Flour	1¼ lb
Shortening	3¾ lb	Vanilla Extract	¾ oz
Salt	1½ oz	Lemon Extract	¼ oz

Cream for 5 minutes at low speed.

Liquid Milk	3 lb	Sugar	2 lb

Dissolve sugar in milk and add to above.

Cake Flour	3¾ lb

Sift and add.

Whole Eggs	3¾ lb

Add and mix for 5 minutes.
Bake at 350° F.

Pound Cake

Sugar	4 lb	Milk Powder	4 oz
Butter or Shortening	2⅛ lb	Flour (Short Patent)	1 lb

Cream together.

Eggs	2½ lb	Flavor	To suit
Salt	½ oz	Invert Syrup	2 oz

Add to the above mixture.

Milk	1 lb

Add.

Flour	3 lb	Baking Powder	½ oz

Blend together flour and baking powder. Add.
Then add:

Milk	½ lb

For fruit pound cake, half of the short patent flour should be bread flour. Add about 4 to 6 pounds of fruit. The fruit should be dusted with about ½ pound bread flour. Bake in wood-lined pound cake tins at 300 to 325° F.

White Pound Cake

Cake Flour	5	lb	Gum Tragacanth	2	oz
Powdered Whole Milk	1	lb	Vanilla Extract	1	oz
Granulated Sugar	5	lb	Salt	1	oz
Shortening	2½	lb	Baking Powder	3	oz
Powdered Egg Albumen	4	oz			

Cream shortening and sugar, then add to the other ingredients, which have been sifted all together. Mix carefully. This mixture will require about 5 pounds of water.

Bake in a slow oven at 300 to 325° F.

Yellow Pound Cake

Cake Flour	5	lb	Powdered Egg Yolk	1½	lb
Powdered Whole Milk	10	oz	Powdered Mace	3	g
Granulated Sugar	6	lb	Salt	2½	oz
Shortening	3	lb			

Cream shortening and sugar, then add to the other ingredients, which have been sifted all together. Mix thoroughly. This mixture will require about 4 pounds of water.

Bake 1 hour in a moderate oven at about 350° F.

Pound Cake Mix

Mix 1 pound of cake flour with ½ pint of cold water. Make dough, put in molds and bake at 400° F.

Winter Wheat Flour	3	lb	Salt	1½	oz
Granulated Egg Yolk	½	lb	Cottonseed Oil	5	lb
Baking Powder	¼	lb	Vanol (Concentrated Vanilla Flavor)	¼	oz
Milk Powder	10	oz	Lemon Oil	¼	oz
Sugar	2⅛	lb			

Prepared Powdered Cake Mixes

Spicy Cake Mix

Flour	8 lb	Ground Nutmeg	½ lb
Ground Cinnamon	1½ lb	Ground Cardamom	½ lb
Cream of Tartar	2 lb	Ground Ginger	½ lb
Bicarbonate of Soda	3 lb	Ground Cloves	½ lb
Ground Mace	½ lb		

Recipe

Shortening	¾ lb	Granulated Sugar	3¾ lb

Rub up.

Eggs	5 lb	Salt	1 oz
Spicy Cake Mix	1 lb		

Add to above.

Sour Milk or Buttermilk	3 qt	Ground Raisins	2 lb
Cake Flour	7 lb		

Add gradually.

Bake in slow oven. This will make layer, sheet and loaf cake, also cup cake and muffins. A 2-layer, 7-inch cake with icing will cost about ten cents to make. This is a trade winner.

Pound Cake Flavor

Cornmeal	10 lb	Peanut Oil	1 pt
Granulated Egg Yolk	3 lb	Mace Oil	1 oz
Rice Flour	6 lb		

Pound Cake Flavor

Mace	¼ oz	Tapioca Flour	5 lb
Nutmeg	¼ oz	Lemon Oil	¼ oz
Apple Pie Flavor	2 oz	Orange Oil	¼ oz
Sugar	5 lb	Vanillin	¼ oz

Yellow Mix

Sugar	6 lb	Baking Powder	3¼ oz
Malt Syrup	1½ lb	Ammonium Carbonate	1 oz
Shortening	3 lb	Flour	11 lb
Milk	3⅓ qt	Salt	1½ oz
Egg Yolk	1 qt		

This is an excellent formula for layers, sheet cakes, and loaf or wine cakes. Scale from it. By the addition of a little hard wheat flour and a little more baking powder, jumbles, drop cakes and a wide variety of small cakes can be run with a pastry bag. The variety is unlimited.

Yellow Layer

Flour	100 lb	Shortening	50 lb

Mix for 3 minutes at low speed.*

Sugar	125 lb	Baking Powder	6 lb
Salt	4 lb	Liquid Skim Milk	40 lb

Add to the blended mass and mix for 3 minutes at low speed.

Whole Eggs	50 lb	Flavor	To taste
Liquid Skim Milk	60 lb		

Add eggs and milk together carefully and mix for 3 minutes at low speed.

Keep the batter smooth in all stages by scraping the bowl down properly.

Yellow Layer Cake

Formula No. 1

Cake Flour	5 lb	Shortening	2½ lb

Mix together for approximately 6 minutes at low speed.

* Considering low speed as first speed, use first speed on a three-speed machine and second speed on a four-speed machine.

Sugar	7 lb	Skim Milk Powder	10 oz
Salt	2½ oz	Water	2¼ lb
Baking Powder	4 oz		

Add and cream for approximately 4 minutes at low speed.

Whole Eggs	2⅞ lb

Add and cream for approximately 4 minutes at low speed.

Water	2⅛ lb	Vanilla	To taste

Add, mixing in for approximately 4 minutes at low speed.

Scale 12 ounces into 8-inch deep, round, layer cake tins. Bake at 360° F.

For loaves and cups, add 1 to 1½ ounces of flour per pound of batter. Should a finer grain and texture be desired, reduce baking powder content to 3¾ ounces.

Formula No. 2

Cake Flour	5 lb	Powdered Egg Yolk	1 lb
Baking Powder	5 oz	Salt	2 oz
Powdered Whole Milk	1 lb	Vanilla Extract	1 oz
Sugar	6½ lb	Corn Starch	8 oz
Shortening	2 lb		

Sift together flour, baking powder and milk. Cream the sugar with the shortening and add the egg. Carefully mix all of the ingredients. This mixture will require about 5 pounds of water.

Bake in oven at 420° F for about 15 to 20 minutes.

Formula No. 3

Sugar	3 lb	Salt	2½ oz
Shortening	1½ lb		

Cream for 5 minutes at low speed.

Liquid Milk	4¾ lb	Vanilla Extract	1 oz
Sugar	3½ lb		

Dissolve sugar in milk thoroughly and add to the above mixture.

Cake Flour	5 lb	Baking Powder	3 oz

Sift together. Add and mix for 5 minutes.

Eggs Yolks	1¼ lb	Whole Eggs	1¼ lb

Add and mix in for 5 minutes.
Bake at 360° F.

Yellow American Pound Fruit Cake

Sugar	4 lb	Salt	1 oz
Shortening	2 lb	Vanilla Extract	½ oz
Cake Flour	1 lb	Lemon Extract	¼ oz

Cream for 5 minutes at low speed.

Liquid Milk 1¼ lb

Add gradually.

Cake Flour 3 lb

Sift, add and mix for 5 minutes.

Whole Eggs 2 lb

Add and mix for 5 minutes.

Mixed Cherries, Peel, Raisins, etc. 3 lb

Fold in.
Bake at 350 to 360° F.

Yellow Pound Cake

Formula No. 1

Cake Flour	5 lb	Shortening	3⅜ lb

Mix together for approximately 4 minutes at low speed.

Sugar	6 lb	Skim Milk Powder	3¾ oz
Salt	2½ oz	Pound Cake Flavor	1½ oz
Baking Powder	¼ oz	Water	2 lb 4¼ oz

Add and cream for approximately 4 minutes at medium speed.

Whole Eggs	3 lb 7 oz

Reduce the speed to low, add slowly and mix for approximately 6 minutes.

Scale into wooden-lined pound cake pans. Bake at 375° F.

Remove cakes from the pan after they come from the oven.

Formula No. 2

Flour	100 lb	Shortening	70 lb

Mix for 3 minutes at medium speed.*

Sugar	120 lb	Liquid Skim Milk	50 lb
Salt	4 lb		

Add to the blended mass and mix for 6 minutes at medium speed.

Whole Eggs	70 lb	Flavor	To taste

Add eggs and mix carefully for 5 minutes at low speed.

Keep the batter smooth in all stages by scraping the bowl down properly.

White Pound Cake

Formula No. 1

Flour	100 lb	Shortening	65 lb

Mix for 3 minutes at medium speed.*

Sugar	125 lb	Baking Powder	3 lb
Salt	4 lb	Liquid Skim Milk	50 lb

* Considering low speed as first speed, for medium speed use second speed on a three-speed machine or third speed on a four-speed machine. This applies to the first two stages. In the third stage, use first speed on a three-speed machine or second speed on a four-speed machine.

Mix for 6 minutes at medium speed.

Egg Whites	65 lb	Flavor	To taste
Liquid Skim Milk	30 lb		

Add milk and eggs together in two parts, and mix for 5 minutes at low speed.

Keep the batter smooth in all stages by scraping the bowl down properly.

Formula No. 2

Cake Flour	13 oz	Skim Milk Powder	14 oz
Shortening	3 lb	Salt	1½ oz
Sugar	5¼ lb		

Cream for 8 minutes at low speed.

Egg Whites	2⅞ lb

Add slowly, creaming for 8 minutes at medium speed.

Water	3¼ lb	Pound Cake Flavor *	1½ oz

Add, mix in for approximately 4 minutes at low speed.

Cake Flour	4 3/16 lb

Sift, add and mix for approximately 4 minutes at low speed.

Scale into wooden-lined pound cake pans and bake for 45 minutes at 300° F. Then increase heat to 350° F and bake until done.

White American Pound Fruit Cake

Sugar	4 lb	Cake Flour	1 lb
Shortening	3¼ lb	Lemon Flavor	¼ oz
Salt	1¼ oz	Orange Flavor	¼ oz

Cream for 5 minutes at low speed.

* Pound Cake Flavor

Vanilla Extract	8½ oz	Lemon Extract	½ oz
Orange Extract	1 oz		

Liquid Milk	1¼ lb

Add gradually.

Cake Flour	3 lb	Cream of Tartar	¼ oz

Sift together, add and mix for 5 minutes.

Whole Eggs	2 lb

Add.

Mixed Cherries, Peel, Raisins, etc.	3 lb

Fold in.
Bake at 350 to 360° F.

Sweet Dough Formula
(Sweet Yeast Goods)

Sugar	14 oz	Shortening	1 lb
Malt	2 oz	Salt	1 oz
Mace	To taste		

Cream together to desired lightness.

Whole Eggs or Egg Yolks	12 oz

Add and mix in.

Liquid Skim Milk		Yeast *	8 oz
(Variable)	2$\frac{1}{16}$ lb	Flavor	To taste

Dissolve yeast and flavor in milk, and add.

Bread Flour	3 lb	Cake Flour	1½ lb

Add and beat smooth.
Take dough from mixer at 80° F.
Give three-quarters to a full rise and then fold.

* You will be able to adjust the fermentation time and the amount of yeast to meet your own conditions.

Pan Coat Grease

Shortening 103° F	4 oz	Flour	1 oz

Mix the above ingredients to a smooth paste. Apply on pans.

Cake Icings and Fillings

Improving Water Icings

Water icings are made by mixing sugars with water. These icings are in general use and illustrate the importance of using different classes of sugars and methods of manipulation. Consistency, i.e., coherence and firmness or the ability to be spread and to stay put, and good results are not just a matter of proportions of ingredients. They are determined, also, by the properties of the materials used and the way in which they are put together.

In the preparation of icings, the water dissolves some of the sugar crystals to form a syrup which surrounds and holds together the other particles. When powdered sugar is mixed with cold water, it forms plain water icing which is short, semi-transparent and has little gloss. Because cane and beet sugars are natural driers, they force the water to evaporate quickly and, therefore, plain water icings rapidly become dry, hard, brittle, cracked and spotted. This also occurs in all other icings unless proper precautions are taken to prevent their occurrence. In order to produce body, opaqueness, gloss and other desirable qualities in a water icing, it is only necessary to make a few changes in the basic formula and process.

Powdered cane and beet sugars, ground for icing purposes, usually contain small percentages of starch which gelatinize when heated with water. More sugar dissolves in a given quantity of hot water than in a like amount of cold water. These facts may be used to change the characteristics of plain water icings and other varieties. Using hot water, or heating the plain water icing, causes the starch to swell and in this way more sugar dissolves, increasing the opaqueness and gloss.

When greater opaqueness and more gloss are desired, a small percentage of fat may be added, to whiten the icing during creaming and to make it shiny when used. For more body and quicker set, replace part or all of the water with egg white, or dissolve, in each pound of water used, from ½ to 1½ ounces of egg albumen or gelatin, the egg white and gelatin thicken the syrup and set during cooling and drying. In order to overcome brittleness, cracking and peeling, and to retain softness and freshness, use about 1½ ounces of Nulomoline to every pound of icing. This controls sugar crystallization and retains moisture.

The Advantages of Fondant

When finer crystals, more smoothness and higher gloss than can be obtained with the most finely powdered sugar are desired, a fondant is made. Non-grittiness is produced by forcing the sugar to form very tiny crystals while the fondant is being made. The ratio of sugar to water in finished water icings and in fondants is practically the same; 7 parts sugar to 1 part water. However, in the icing, the remaining particles are comparatively large, whereas in making the fondant, all the sugar is first dissolved, and the crystallization controlled to form extremely fine granules.

Fondants are made by controlling the heat, first to dissolve an excess of sugar, and then to cause the extra sugar to form fine crystals. Enough water is used to dissolve all the sugar (about one to three) and the extra water is boiled off by cooking the mixture to 236 to 244° F. The cooked syrup for fondant is then cooled to about 90 to 110° F, after which it is creamed, beaten or agitated. During creaming, part of the sugar dissolved in the syrup crystallizes, the rest being retained. If only cane or beet sugars are used and the syrup is creamed at the wrong temperature, coarse grains are formed and the fondant is of poor quality. However, if Nulomoline is used in making the syrup, it checks the syrup from graining during cooling and before creaming, and causes the sugar to form extremely fine crystals while the syrup is being creamed to make fondant. Fondants are kept tender and soft by including about 1 pound of Nulomoline in each ten-pound batch of syrup.

The beating or creaming may be done by hand, in a cake machine or, preferably, on a special fondant beater. The proper production of fondant requires special equipment and skilled operators. It is advisable that bakers buy their fondant from supply houses which usually offer an excellent product at a reasonable price. It is usually more economical and satisfactory to buy fondant than make it under average bake-shop conditions.

Fondant can be used to replace equivalent proportions of sugar and water in almost all ordinary icings for improvement in quality. Moreover, fondants should be used as a base in making many varieties of high-quality icings. With the addition of fats, eggs, gelatin, flavors, colors, etc., many varieties of fat, light and fluffy icings can be developed.

Handling Sugar and Water in Cold-and-Hot-Process Icings

The size of the sugar crystals influences the speed at which they dissolve in water, but does not influence the amount that can be dissolved in a given quantity of water. One pound of water dissolves 2 pounds of granulated or powdered sugar at 63° F; larger amounts of sugar dissolve when the temperature rises. For example, if a pound of water is heated with 3 pounds of sugar at 147° F, the sugar dissolves completely. At 189° F, a pound of water will dissolve 4 pounds of sugar. For example, a syrup at 189° F can hold in solution twice as much sugar as a syrup at 63° F. However, the additional sugar will not remain in the syrup but will crystallize out, until at room temperature, about 65° F, the ratio between sugar and water will reach 2:1. As the water evaporates from the icing syrup, the dissolved sugar crystallizes out, forming larger and coarser grains. Therefore, Nulomoline is used in icings to check drying and to cause the sugars to form fine crystals.

Syrup and Sugar in Icings

When icings are freshly made, they usually contain much more water than remains after the iced cookies are put on display or packed in caddies or packages. The icings on cookies contain only about 12% or less moisture, just sufficient to make an icing of stiff fondant consistency. This water holds about twice as much sugar in solution (24%); the rest of the sugar is in crystal form (64%).

However, if an icing is made solely with 12 pounds of water and 18 pounds of sugar, it is too stiff for icing made either by hand or machine. For softer consistency, more water is needed, and icings are often made with up to 22 pounds, or more, of water.

Icings with 22 pounds of water have 44 pounds of sugar in solution and only 34 pounds in crystal form. The extra 20 pounds of sugar, which are not dissolved when less water is used dissolve in the extra 10 pounds of water. Afterwards, the extra sugar crystallizes out as the water evaporates from the icings during surface-drying. Some of the extra 20 pounds of sugar forms new crystals, but much of it adds onto and enlarges the crystals comprising the original 34 pounds. These enlarged crystals make icings hard and gritty and produce dull appearance. Moreover, these large sugar particles grow even larger in size when icings continue to dry more than is desirable.

To secure smoothness and gloss, excess drying must be checked, and sugar crystallization must be controlled. These are the functions for which Nulomoline is especially advantageous. The maximum levulose content of Nulomoline attracts and holds moisture; it retains the proper water content in finished icings and keeps them pliable and fresh. Also, the levulose has a remarkable influence in increasing the total amount of sugars which remain in solution and in retarding the crystallization of other sugars. Nulomoline, therefore, produces icings with finer crystals, more gloss and much smoother eating-qualities.

The same precautions should be taken with all other kinds of icings. About 10 pounds of Nulomoline should be added to each hundred-pound batch of icing to retain softness and freshness.

Controlling the Moisture Content of Icings

The time required for an icing to set or become surface-dried, so that it can be handled and wrapped, depends considerably upon its moisture content. Icings with less water set more quickly; therefore, it is good practice to use the least amount of water possible. Flat icings, e.g., plain water icings and fondant, should contain about 12% water; those with large amounts of fat, e.g., butter creams, should contain less, and those with water-absorbing materials, e.g., cocoa, more. Fluffy icings, lightened with gelatin or eggs, should contain a total of 16 to 22% water.

Changing the Consistency of Icings

If flat icings do not handle easily, it is usually better to change their consistency by adding a simple syrup, made by dissolving 2 pounds of sugar in 1 pound of water, or by keeping the icings warm, rather than by adding water. When icings are warmed, the syrup and fats become more fluid, allowing the sugar crystals to slide by each other more easily. Of course, fluffy icings lightened by shortening cannot be heated without losing volume, but practically all others can be warmed up to as high as 110° F, or kept warm to make spreading easier.

For dipping icings, any flat icings, e.g., water icing or fondant, can be used. Add sufficient simple syrup, made of 2 pounds of sugar dissolved in 1 pound of water, to make the icings thin enough for dipping purposes. Warming the icing to about 100° F also thins it. Warmed icings need less or no extra syrup for thinning and, consequently, set more quickly and require less time for surface-drying.

Chocolate Icings (Fudge and Others)

Chocolate fudge is one of the most popular cake icings. For exceptional quality, it should be made according to a candy maker's procedure and with the same ingredients that he uses. The fat in block chocolate, which is hard at temperatures below about 85° F, must be kept warm or softened while it is being incorporated in icings, so that hard lumps do not form. Also, the chocolate should be made plastic so that it will be workable for spreading on a cake and remain soft afterwards to prevent hardening and cracking on the cake. The final color of chocolate icings depends considerably upon the temperature of the batches at the time they are used. The color of chocolate icings darkens when batches contain extra fat and are warm. The starch and vegetable fibres, which naturally make up a large percentage of chocolate, slowly soak up and swell when water is present, particularly if heated. This swelling produces body and smoothness.

The nature of chocolate brings up a few problems, which should be considered when selecting the other ingredients and the procedure to be used in making a chocolate fudge icing. A fudge icing should have characteristic candy-like richness and eating-qualities. Much of the richness can be secured from butter, which is soft at ordinary temperatures. If butter is mixed with the firm fat naturally contained in chocolate (unsweetened chocolate is about half cocoa butter which is hard and brittle), the icing will be soft while being used.

In addition to chocolate and butter, two other ingredients are needed to complete the icing—liquids and sugars. As the liquid ingredient, milk is preferable, as it imparts quick setting and non-sticky properties, particularly when cooked. The hazards of cooking milk are lessened by cooking it with butter, to prevent scorching and to cut down foaming. When chocolate is heated in a cooking kettle, there is danger of its becoming scorched; and even slight burning destroys flavor and produces dark specks. This hazard can be eliminated by chipping the chocolate into small pieces and placing it in a mixing bowl by itself. It can then be melted by pouring over it, hot syrup or hot fat which has been heated separately.

Sugars which produce the most desirable results should be selected. Granulated sugars should be chosen for economy, when the sugar is to be dissolved. A very finely powdered (6X or 8X) icing sugar should be used when the sugar is not to be dissolved. Nulomoline should be added to control grain in the cooked syrups when they are to be poured on the chocolate. It also retains moisture and prevents the final icing from drying, spotting and peeling, keeping it soft, glossy and fresh.

Developing Flavor During Cooking

Icings, perhaps better than any other products made by the baker, illustrate the fact that some pleasing flavors come from skillfully blended materials. The tempting goodness of caramel, butterscotch and cream fudge defies extraction by the flavor chemists. These appealing combinations are developed by the baker during actual icing production.

Caramel flavor is developed by making two changes in a cold-process icing consisting of powdered sugar and water. First, the water is replaced by milk and part of the powdered sugar by brown sugar. Second, the sugar and milk are cooked in order to develop the universally appealing flavor of caramel. When butter is used instead of milk, the inimitable flavor of butterscotch is secured. By cooking together the three ingredients, sugar, milk and butter, the rich flavor of cream fudge is brought out. By adding chocolate, chocolate fudge is produced. Chocolate and caramel combined form milk chocolate.

Sweetened condensed, evaporated and dry milks have a milk-caramel flavor which is acquired while these materials are being made. They can be used to advantage in caramel and fudge icings. Tricol, a flavor-processed sweetener for balanced flavor, color and sweetness, has an exquisite, uniform, sugar-caramel flavor, and is an exceptionally good product to use in caramel and butterscotch icings.

Eliminating Stickiness in Icings

Some substances, when they are dissolved in, or moistened by, water, become sticky, whereas others do not. Stickiness in icings may be overcome by tying up the water with materials which will make non-sticky prepared syrup bases. These non-sticky prepared bases, when mixed with dry materials such as powdered sugar, produce icings which can be wrapped almost immediately.

Bakers are familiar with the fact that when starches are gelatinized, they take up large amounts of water and form stiff gels; in the same way, agar-agar, Japanese gelatin, sets up water. When these starch and agar-agar bases are used instead of water, they induce the growth of micro-organisms with consequent spoilage of icings and cakes. However, spoilage in prepared bases can be prevented by including in them the maximum content of sugar. Sugar in high concentration acts as a preservative. Each pound of water used in a prepared base should have dissolved in it at least 2 pounds of a combination of Nulomoline and sugar. This large amount of sugar and Nulomoline, while acting as a preservative, also prevents starch from forming too-stiff gels. In this way, the high sugar-and-Nulomoline content also allows the inclusion of sufficient starch to offset excess sweetness, without causing too much stiffness that would make the incorporation of the prepared base in the icing difficult. Well balanced icings contain properly prepared syrup bases and the necessary amounts of Nulomoline to keep them soft and pliable and to prevent them from drying, cracking and peeling.

Added Bottled Flavors

The almost unlimited number of bottled flavors which are available offers every baker the possibility of achieving individuality through resourceful blending and care in incorporation. Vanilla and lemon flavors are good but are often thoughtlessly dashed into too many batches, resulting in the frequent complaint that all bakery goods taste the same.

Bottled flavors are highly concentrated and should be handled with extreme care. The exact amount needed to produce elusive pleasantness should be determined, carefully measured and thoroughly distributed to avoid over-flavored parts of the icings.

Inaccuracies caused by using concentrated flavors are lessened by diluting the flavors so that the least amount taken for any one batch is an ounce or more. The dilutions can often be made with water, but the flavors are less likely to spoil if diluted in a special emulsion. An emulsion is also useful in combining two or more concentrated flavors into one diluted flavor for special icings or cakes.

Making Light-Fluffy Icings

The principles, which have been discussed about the functions of ingredients, influence of heat and use of flavors, apply to light-fluffy icings, as well as they do to flat icings.

Perhaps the most important difference between light-fluffy icings, and flat icings, is in the size of the finished product. Light icings make cakes look larger. The increase in the bulk alters the consistency and makes the icing operation easy.

When butter and creaming fats, such as hydrogenated vegetable shortenings, are creamed with powdered sugar, they take on air. In these butter creams, whole eggs, egg yolks or egg whites are sometimes included for the smoothness and body which they add. Moreover, yolks contain considerable lecithin, which makes water mix more readily with fat; by creaming yolks with butter or shortening, an emulsion is formed which produces smoother, more permanent icings. When egg whites are used, they are creamed along with the other ingredients, just as whole eggs and yolks, or whipped separately with syrup to make marshmallow which, in turn, is added to the creamed portion.

Egg whites are used without fats to make light icings. For cold-process, fluffy icings, either dried egg white (egg albumen) which has been dissolved, or whole egg white is simply beaten with icing sugar; for Royal icings, about 6 pounds of icing sugar are added for each pound of egg white. However, for light-fluffy boiled icings a hot process is used; a syrup is poured onto egg white or dissolved egg albumen, while it is being beaten. This syrup is made by dissolving the sugar in an excess amount of water. The extra water is then boiled off by cooking to a definite temperature. The crystallization of the sugar is controlled with Nulomoline, in order to form a smooth, fine grain in the finished icing.

Under boiled icings are classified many different varieties made by heating the ingredients used in their production; but the term boiled icing is often used only in reference to one particular group, the light-fluffy icings which are similar to the "seven minute" or "Lady Baltimore" icings made by housewives, when they whip egg whites and sugar while warming them in a double boiler. Although these icings are popular in the home, they are seldom used by bakers because they frequently grain and become sandy. However, this problem can be overcome by bakers if they take the characteristics of light-fluffy boiled icings into consideration.

In one respect, light-fluffy boiled icings are a cross between fondant icing and marshmallow. Fondant consists of very fine crystals of sugar surrounded and held together by syrup (12% water and 88% sugars). In fondant, there is not enough water to dissolve all the sugar, but in marshmallow, there is considerably more (28%), holding in solution all the sugars (68%). Marshmallow, then, is syrup which is aerated with about 2% of egg white solids or gelatin.

The water content of boiled icings (20%) is about halfway between that of fondant and marshmallow. From this point of view, a light-fluffy boiled icing is somewhat like marshmallow into which icing sugar has been stirred. However, the granules of icing sugar are much coarser than those in fondant or in well-balanced, properly made, light-fluffy boiled icings.

A light-fluffy boiled icing is an aerated syrup in which sugar crystals are suspended. This is what is produced when a housewife beats warmed sugar and egg whites. The 20% water contained in the 24% egg white first dissolves about 40% sugar at 63° F; gradually, as the beating is continued and the temperature rises, more and more sugar is dissolved. About 60% sugar is dissolved at 147° F. However, at about 140° F, the egg whites start to coagulate (become cooked), and the icing becomes stiffer and acquires a pleasing taste. The sugar, which is dissolved at high temperatures, recrystallizes as the icing cools and these crystals remain suspended in the icing. The crystals must be kept extremely small in order to make smooth, glossy, boiled icings.

If the baker is equipped with a mixing bowl that can be heated, boiled icings can be made in the same way that "seven minute" icings are made at home. These must be made in small batches and used quickly to prevent the sugar, which recrystallizes during cooling, from forming large grains, in this way producing sandiness.

A more convenient way to make boiled icings is to heat the sugar separately from the eggs. By making a hot syrup, considerable sugar can first be dissolved and then formed into extremely fine crystals while the icing cools. Moreover, the heat cooks the eggs, assuring stiffness and good taste.

Since water is required to make the hot syrup, a correspondingly smaller amount of eggs must be used. For example, if the egg whites are reduced to 15%, 7% water is used for making the hot syrup. This is enough to dissolve 53% sugar by cooking the syrup to 245° F. This syrup, plus the egg whites, totals 75% of the ingredients, which means that 25% sugar should be dissolved in the egg whites by stirring the two together slowly for several minutes.

When making boiled icings, about 1 pound of Nulomoline should be added to each ten-pound batch to control sugar crystallization. This prevents hot syrups from graining before they are poured into beaten egg whites and produces extremely fine crystals during the cooling of the finished icings. When making boiled icings for wrapped cakes, some substance which will prevent stickiness should be incorporated, e.g., starch, agar-agar and gelatin, extra egg-white solids using reconstituted dry albumen instead of fresh whites, or a water-absorbing gum just before finishing the whipping.

Although icings made with egg albumen are more tender, gelatin is often used because it sets when cool and does not have the odor often associated with egg albumen. Both egg albumen and gelatin are prepared for use in a similar manner; usually 1 weight of egg albumen or gelatin is dissolved in 2 weights of water. The egg albumen is dissolved in lukewarm water. Hot water causes coagulation which results in insoluble particles. Gelatin is soaked in cold water and then warmed until fluid. Heating at over 140° F for long periods is injurious. The advantages of each are procured by using combinations of egg albumen and gelatin in icings. Commercial whip or meringue powders occasionally are such combinations.

Fats, egg and gelatin are all good materials with which to incorporate air into icing. If used with proper combinations of other ingredients, they produce superior light-fluffy icings.

Making Marshmallow

Marshmallow is often called light-fluffy icing, but there is an important difference between the two. Icings contain sugar crystals, whereas the sugar in marshmallow should be completely dissolved. In other words, marshmallow is syrup which has been aerated; the air which is beaten into the syrup is held by egg whites, egg albumen or gelatin. That this is a fitting definition is not so apparent when freshly made marshmallow is seen before it is beaten. A study of the marshmallow syrup batch before beating tells what will happen to the marshmallow afterward.

Marshmallow should stay, as long as possible, the way it is right after being freshly made and used. Therefore, the syrup must be such that it will not change on standing. It should be: non-drying, non-graining and non-fermentable.

If only cane or beet sugars are used in making marshmallow and sufficient water is used to keep these sugars dissolved, the marshmallow is fermentable and is likely to spoil. For safety, it is necessary to have not over 32% total water in marshmallow, and for fillings between layer cakes for non-soaking, not over 25%. These figures show that marshmallow made with only cane or beet sugar is bound to become sandy, as a part of these sugars will crystallize and form coarse grains.

However, freedom from graining and fermentation is easily secured by having part of sugar in the form of Nulomoline. It prevents the sugar used from crystallizing. Drying, which would cause graining and toughness, is retarded, and the marshmallow is kept tender. Use from 1 to 5 pounds of Nulomoline in each 10 pounds of hot- or cold-process marshmallow.

Making Fillings

For cake filling, use icings as they are or, if desired, with flat and butter cream icings, add a little extra water or extra prepared marshmallow before spreading them. Extra light-fluffy cream fillings, somewhat similar to genuine whipped cream, are modified butter cream icings which contain a high proportion of fats. In these, it is preferable to use a combination of creaming shortening, butter and coconut fat for maximum stability and elimination of greasy taste. Twenty to 25% of the total batch weight in imitation whipped creams is fat in contrast to the 15% in butter creams. These imitation whipped creams are used to fill layer cakes, sponge cups and fingers, to spot on sponge popovers and tarts, and to sandwich soft cookies.

Although icings are sometimes used as fillings, they are usually intended to become surface-dry. Consequently, icings contain only enough water to form a syrup which holds the fine sugar crystals together. In contrast, fillings may be sticky and, therefore, may contain much more moisture.

In fact, sufficient water may be used in fillings to dissolve all of the sugar and form a syrup. However, syrups alone are too soft and liquid for spreading in between layer cakes. But when the syrups contain one or more materials which cause them to stiffen, they hold the layers apart and do not allow excessive soakage. This describes the type of fillings which are discussed here. They are thickened syrups.

Inasmuch as fillings contain considerable moisture, they are apt to spoil unless precautions are taken to prevent the growth of yeasts, molds or bacteria. Fortunately, sugars are excellent preservatives, if they are present in sufficiently high concentrations—more than 67% and up to 80% for maximum safety from spoilage. Therefore, fillings should contain slightly more than 2 pounds of a combination of Nulomoline and sugar for each pound of water. However, 1 pound of water will not keep more than 2 pounds of cane or beet sugar in solution. The extra sugar required for preservation tends to crystallize out of the solution. When part of the sugar is Nulomoline, crystallization is checked and a greater amount is kept dissolved in a given quantity of water.

When making solutions of maximum concentration, a combination of about 3 pounds of Nulomoline for each one of granulated sugar should be used. About 1 pound of water to dissolve each 3 pounds of sugar are then added and the solution is cooked at from 220° F for a concentration of 67% to 230° F for a concentration of 80%. The temperature of cooking depends upon the final concentration desired.

Low concentration syrups (below 50%) are easily thickened, but stiffening those of high concentration is a difficult problem. To solve it, numerous experiments with many thickeners were made in our testing bakery. Some of the facts concerning the results obtained are presented in the following paragraphs.

Pectin is perhaps the most effective and economical ingredient with which to stiffen high density syrups but is used only in making tart marmalades, jams and jellies. For jellies without tartness, agar-agar is valuable. Although gums (karaya, tragacanth, locust bean) are expensive and only form soft pastes, they are advantageous sometimes when used along with other thickeners. Milk and eggs custardize, or thicken, when cooked and are very useful, except when their fat content prevents the fillings from being whipped light and fluffy. The proteins in flour coagulate and form objectionable lumps. Although small proportions of starches form very stiff pastes in solutions of low concentration, a greater amount forms only soft pastes in syrups of high concentration. Gelatin not only sets syrups but also aerates them when beaten.

Competent authorities recognize that the properties of thickeners are complex and can be understood fully only by working with them. This becomes evident especially when starch and gelatin are used together; their properties supplement one another and results not obtainable otherwise can be secured with such a combination.

Icing Paste or Cream Base

XXXX Sugar	60 lb	Glycerin	8 lb
Corn Syrup	30 lb	Shortening (110° F)	2 lb

The above ingredients are all melted down by means of heat. Stir thoroughly with agitator until thoroughly cold. The addition of color and fruit flavor will then yield the desired fruit-flavored icing cream. Where a chocolate icing is desired, add cocoa powder.

Vanilla Icing

XXXX Powdered Sugar	30 lb	Hot Water	4 pt
Vanilla Bean Extract	3 oz	Egg White	14 oz
Corn Syrup	1 lb		

Prepare in same manner as for icing pastes.

Chocolate Icing

XXXX Powdered Sugar	24 lb	Invert Sugar Syrup	2 lb
Coconut Butter (m.p. 86° F)	1 lb	Bitter Chocolate	4 lb
		Hot Water	2 pt

Prepare in same manner as for icing pastes, but melt the chocolate and coconut butter first.

Chocolate Fudge Icing
Formula No. 1

Granulated Sugar	8 lb	Salt	¼ oz
Water	2½ pt		

Heat to 230° F.

Invert Sugar Syrup	2 lb	Bitter Chocolate	4 lb

Shut off steam and then add to the above.

Stir until smooth. Then transfer to machine bowl and, at low speed, sift in 3 pounds of XXXX powdered sugar. Keep mixing, and when mixture starts to stiffen add 1½ pounds of shortening, ½ pound of butter and ½ ounce of vanilla flavor. Mix well.

Formula No. 2

Bitter Chocolate	5 lb	Corn Oil	8 oz
Shortening	4 lb		

Warm above mixture to a uniform paste.

Water	2 qt	Corn Syrup	3 lb
Agar-Agar	3 oz	Cane Sugar	1 lb

Mix the sugar with the agar-agar and add slowly to the boiling water. Then add the corn syrup. Boil for 5 minutes on a slow flame. Transfer to mixing bowl. Add the bitter chocolate, shortening and corn oil, and mix smooth in machine. Now add 18 pounds of powdered sugar. Beat again until smooth. Flavor and color if desired.

Butterscotch Icing

Formula No. 1

Brown Sugar	8¼ lb	Corn Syrup	6 oz
Butter	1⅞ lb	Water	7 lb

Scale all together into a copper kettle and boil to 245° F.

Pour out into a shallow pan and cool. When cold, put into a cake machine bowl and grain off like fondant.

Formula No. 2

Brown Sugar	8 lb	Water	4 pt
Butter	4 lb	Salt	2 oz

Boil to 245° F.

XXXX Powdered Sugar	25 lb	Liquid Skimmed Milk	2 pt
Shortening	5 lb		

Cream up light.

To this creamed mass add the hot butterscotch syrup and whip up to the desired consistency.

Banana Icing

Cake Flour	10 lb	Salt	12 oz
Shortening	20 lb		

Whip the above together.

Water	15 pt

Add to above and mix well.

Icing Sugar	100 lb	Banana Paste	20 lb

Add and continue whipping until icing is light.

Butterscotch Paste Icing

Butter	5 lb	Skimmed Milk Powder	4 lb
Brown Sugar	10 lb		

Make a solution of the milk with 6 pints of water.

Scorch the butter in a kettle, then stir in the brown sugar and heat to 260° F. To this hot sugar syrup, add the liquid milk and cook up to 230° F, stirring briskly.

Butter Cream Icing

Powdered Sugar	10 lb	Shortening	5 lb
Powdered Milk	2 lb	Eggs	1 pt
Salt	3 oz	Vanilla Flavor	To suit

Mix up smooth in machine and then add 10 pounds of boiled meringue icing.

Caramel Boiled Icing

Brown Sugar	6 lb	Water	2 lb

Boil to 245° F.

Egg Whites	2 lb	Salt	1 oz
Sugar	1 lb		

Start egg whites beating when syrup reaches 200° F. Beat half-way up. Add sugar and salt, and beat until firm. Add hot syrup slowly, continuing beating.

Powdered Sugar	2 lb

Add to above and beat until firm.

Caramel Icing

Invert Syrup	1½ lb	Sweetened Condensed	
Butter	1½ lb	Milk	4½ lb

Cook on slow fire to medium brown color, stirring constantly.

Add and mix to incorporate:

Stock Icing	20 lb

Add simple syrup and thin to desired consistency. Use warm (110 to 120° F).

To make milk chocolate icing, add melted chocolate or cocoa.

Caramel Sauce

White Sugar	5 lb	Butter	3/10 lb
Water	5 lb	Flour	1/10 lb
Cream	1 lb		

Brown the white sugar. Add water and boil until it forms a soft ball in cold water. Stir quickly. Let come just to a boil. Turn heat off and add butter.

Praline Butter Cream Icing

(A)	Powdered Icing Sugar	5 lb	Salted Butter	8 oz
	Shortening	12 oz	Salt	⅓ oz

Soften these together.

(B)	Egg Whites	6 oz	Vanilla Extract	½ oz
	Cold Water	4 oz	Citric Acid	¼ oz

(C) Roasted Finely Chopped Almonds 10 to 12 oz

Cream (A) until light and gradually add (B), the egg whites first. Whip for 2 to 3 minutes. Then add the water, vanilla and citric acid, and continue mixing until fluffy. Last, add the roasted chopped almonds.

Chocolate Icing

Fondant	10½ lb	Vanilla Extract	1 oz
Water	8 oz	Evaporated Milk or Simple Syrup	To thin
Chocolate Liquor	2½ lb		
Butter	1½ lb		

Place fondant and water in machine bowl and mix smooth. Melt chocolate in a double boiler. Then add to the fondant and mix smooth. Allow to cool. After the fondant and chocolate mixture has cooled to room temperature, add the butter and flavor. Cream light. Use the evaporated milk or simple syrup to thin to the proper spreading consistency.

If a light-fluffy icing is desired, be sure to let the fondant and chocolate mixture cool to around room temperature before adding the butter. But if a flat dark icing is preferred, add the butter while the mixture is still warm so that it will melt.

An entirely different result will be obtained if the fondant and water are warmed to 100° F before adding the chocolate and butter. This gives a smooth, free-flowing, high-gloss icing for cup cakes, éclairs, etc.

Still another variation can be made by adding 2 pounds of homemade boiled icing.

Chocolate Fudge Icing

Powdered Icing Sugar	5¼ lb	Egg Whites	1¹⁄₁₆ lb
Prepared Syrup	1¼ lb	Shortening (Melted Soft)	½ lb
Water	1 lb	Sweet Butter (Melted Soft)	4 oz
Salt	¾ oz		
Vanilla Extract	¾ oz		
Melted Bitter Chocolate	2¹⁄₁₆ lb		

Put syrup and water on fire. Bring to a boil. Place the powdered sugar in a mixing bowl and start mixing. Pour the hot syrup gradually onto sugar. Mix thoroughly until well incorporated. Add the salt, vanilla and chocolate, then the egg whites, and beat until smooth. Finally, add the melted shortening and butter, and mix thoroughly.

This icing should be kept in a warm water-bath while using in order to keep in soft enough condition to spread.

Note: If a softer icing is desired, an equal part of syrup and water can be added.

Basic Fondant

Basic fondant is a smooth cream fondant preparation which is used in nearly every department of the candy factory and bakery. Basic fondant is used in cast cream centers of all types, crystallized creams, fudge cast caramels, nougat coconut work icings and grained or semi-chewing candies.

The following formula for basic fondant can be produced in large or small batches on any type of equipment and it will produce a smooth, creamy-textured fondant that may be stored for long periods without prematurely drying.

Sugar	80 lb	Standardized Invert Sugar	10 lb
Corn Syrup	20 lb	Water (To Dissolve the Sugar)	

Heat all together stirring the batch occasionally until it boils. Then wash down all grains of sugar that adhere to the inside of the kettle and boil the batch rapidly to 238 to 248° F. Cool the resulting syrup to 110 to 120° F.

For cast cream centers, boil the basic fondant to 238 to 244° F.

For fudge nougat caramels and coconut work, boil the basic fondant to 244 to 248° F.

Fondant Base Icing or Frosting

Sugar	80 lb	Water	20 lb
Corn Syrup	8 lb	Cream of Tartar	2 oz

Cook the above to 242° F. Cool to about 110° F, then stir in cake mixer until it becomes creamy and white. Put this base away, and when needed, slightly warm it, but do not heat the fondant over 95° F. If fondant is too heavy, thin down with simple syrup. This icing can be made up with chocolate, fruits and nuts. Always apply the icing warm.

Honey Macaroon Icing

Shortening	1 lb	Skim Milk Powder	3 oz
Salt	½ oz		

Cream the above ingredients up light.

Water	8 oz	Honey	5 oz
Vanilla	To taste	Brown Sugar	1 lb

Mix together. Add to above and mix in.

4X Powdered Sugar	1 lb

Add and whip light.

Top the iced cake with toasted macaroon cocoanut.

Maraschino Cherry Fudge Icing

Confectioners XXXX Sugar	5 lb	Maraschino Cherry Juice	10 oz
Diced Maraschino Cherries	1 lb	Fondant	6 lb
		Shortening	12 oz
		Butter	12 oz

Place sugar, diced cherries and cherry juice in machine bowl. Mix to a smooth paste, using low speed. Add fondant, shortening and butter to the mixture. Mix thoroughly until icing is smooth and of the desired consistency.

These directions are for a three-speed machine. On a four-speed machine, use second speed.

Suggestions:

A five-gallon pail of maraschino cherries should yield approximately 30 pounds of drained fruit and 12 pounds of juice. The shortening and butter should be at 70° F. Icings made from hard butter or shortening will not be smooth and will contain small lumps of hard fat.

If the stated weight of cherry juice is not available, make up the shortage with water.

French Pastry Cream (Fluff Cream)

(A)	Milk Fluid	4 lb	Condensed Maple Syrup (2½ Fold)	1 lb
	Sugar	8 oz		
(B)	Flour (Sifted)	5 oz	Salt	¼ oz
	Egg Yolk	1 lb		

Beat these together and add to (A) while beating.

(C)	Egg White	1 lb
	Sugar	8 oz

Beat together until stiff.

After (B) has been added to (A) and the mixture cooks thick, stir in (C) before removing from the fire. Use egg whip and stir briskly.

Lemon Icing

XXXX Powdered Sugar	30 lb	Corn Syrup	1 lb
Lemon Juice	½ lb	Hot Water	4 pt

Add the above ingredients to the hot water and mix together in the machine.

Orange Icing

XXXX Powdered Sugar	30 lb	Corn Syrup	1 lb
Orange Juice	1 lb	Hot Water	4 pt

Prepare in same manner as lemon icing.

Orange Sunshine Icing

Shortening	8 oz	Salt	½ oz
Butter	8 oz	Cake Flour	2 oz

Cream until light, using a wire whip or four wing beater.

Egg Yolks	3 oz	Lemon Juice	4 oz
Orange Juice	10 oz		

Add, gradually mix and continue creaming. Add gratings from three oranges and two lemons.

Powdered Sugar	7½ lb

Sift, add and cream until light.

Special Icing

Powdered Sugar *	9⅜ lb	Salt	1 oz
Shortening	2½ lb	Flavor	To taste
Special Stock (See Below)	3⅛ lb	Cake Flour	1½ lb

* Corn sugar may be used to the extent of 25% replacement of powdered sugar.

Cream one-third of the sugar (3 pounds 2 ounces) with the shortening until light. Add the special stock alternately with the balance of the sugar, salt and flavoring. Then add the cake flour. Whip until light.

Special Stock

Sugar	2½ lb	Corn Starch	8 oz
Water	6¼ lb		

Bring sugar to boil with 5 pounds of water. Dissolve the corn starch in 1 pound 4 ounces of water. Add to the sugar mixture and cook until clear. Use cold.

Variations

This icing will carry fruit. For a chocolate icing use 1 pound of cocoa or chocolate and reduce the powdered sugar by a corresponding amount. Add the cocoa or chocolate when creaming the sugar and shortening.

Old Fashioned Spice Cake Icing

Powdered Sugar	5 lb	Salt	1 oz
Shortening	1¼ lb	Spice Blend (See Spice Cake Formula)	¾ oz
Skim Milk Powder	4 oz		
Pastry Flour	4 oz		

Place in mixing bowl and cream at high speed for approximately 8 minutes.

Ground Raisins	4 oz	Chopped Pecans	4 oz

Add and mix together for 2 minutes at high speed.

Non-Sticky Chocolate Icing

Water	2 pt	Bitter Chocolate	1½ lb
Agar-Agar	1 oz	Shortening	2 lb
Granulated Sugar	3½ lb	Icing Sugar	10 lb
Corn or Invert Syrup	2 lb	Vanilla Flavor	To suit

Soak the agar-agar overnight in the water. Add the granulated sugar to the agar solution and heat until dissolved. Shut off the steam and add the bitter chocolate, shortening, and stir until dissolved. Now add the corn syrup, icing sugar and flavor, and mix until smooth.

This icing is applied at 100 to 110° F. Warm to this temperature when needed. If the icing becomes too heavy after being used several times, dilute with sugar syrup.

Non-Sticky and Bun Icing

(A)	Keltose	2½ oz to 5 oz
	Sugar	2 lb
	Warm Water	10 to 12 qt
(B)	Salt	½ oz
	4X Icing Sugar	100 lb
	Cake Flour	3 lb
	Vanilla Extract	2½ oz

Dry mix Keltose and sugar in (A) and add it slowly to the water in mixing machine. Mix until Keltose is completely dissolved.

Dry mix (B) and add to (A) in mixing machine. Mix thoroughly until smooth. Add vanilla.

Note: If lustrous finish is desired, 5 to 10 pounds of shortening may be added in the mixer after (B) is mixed with (A).

Stock Icing (A)
(For all Unwrapped Goods)

Corn Syrup	14 lb	Agar-Agar (or Similar Product)	2 oz
Water	15 lb	Salt (As Desired)	2 oz
Corn Starch	2 oz		

Bring to boil and boil for 3 minutes.

Confectionery Sugar	100 lb

Place in machine at low speed.

Add hot syrup to sugar at low speed and blend thoroughly. Add flavor and speed up machine for a few turns until icing is smooth and glossy.

Use this icing the same as fondant. Use warm at about 110° F.

If icing dries too much or is too brittle, increase agar slightly.

Use simple syrup to thin icings.

Stock Icing (B)
(For all Wrapped Goods—Cellophane)

Corn Syrup	4¾ lb	Agar-Agar (or Similar Product)	7/10 oz
Water	15 lb	Salt (As Desired)	2 oz
Corn Starch	7/10 oz		

Bring to boil and boil for 3 minutes.

Confectionery Sugar	100 lb

Place on machine at low speed.

Add flavor and speed up machine for a few turns until icing is smooth and glossy.

Use this icing the same as fondant. Use warm at about 110° F.

Do not use this icing for unwrapped goods. It will become too brittle unless wrapped.

White Cream Icing

(A)	Water	3 lb	Corn Starch	1 oz
	Agar-Agar	1 oz	Sugar	4 oz

Cook to boil for 3 minutes.

(B) 4X Sugar	35 lb	Powdered Milk	1½ lb
Shortening	6 lb	Salt	2 oz

Cream.
Then add slowly:

Water	4 lb

Then add hot syrup (A) slowly.
Then add last, and mix well:

Nougat Fluff	5 lb	Flavor	To suit

White Fudge Icing

Liquid Skimmed Milk	3½ lb	Butter	1½ lb
		Vegetable Shortening	1½ lb

Heat to 160° F.

XXXX Powdered Sugar	18 lb	Salt	¼ oz

Add the above to these.
Mix until smooth, then add 2 ounces of vanilla extract.

Caramel Icing Base

Water	7 pt	Butter	5 lb
Brown Sugar	20 lb		

Heat the above to 320° F. Shut off steam.

Granulated Sugar	16 lb	Water	10 pt
Powdered Skimmed Milk	4 lb		

Mix the sugar and milk together and then make into paste with the water.
Whip into the above mix until smooth.
Where a darker icing is desired, add caramel color.

Simple Boiled Icing

Egg Whites	2½ lb	Salt	¼ oz
Sugar	1 lb	Vanilla Flavor	To suit

Beat until stiff.

Sugar	6½ lb	Water	2 pt
Corn or Invert Sugar	½ lb		

Cook to 236 to 240° F.

Add the hot syrup to the beaten egg white slowly, with beating, until stiff. Chopped nuts or fruits or any desired flavor may be introduced.

Vanilla Cream Fudge Icing

Powdered Icing Sugar	9¼ lb	Citric Acid	½ oz
Prepared Syrup	⅝ lb	Shortening (Melted Soft)	½ lb
Water	½ lb	Sweet Butter (Melted Soft)	6 oz
Salt	1 oz		
Vanilla Extract	¾ oz		
Egg Whites	9 oz		

Put syrup and water on fire. Bring to a boil. Place two-thirds of the powdered sugar in a mixing bowl. Start mixing and gradually pour the hot syrup on the sugar. Mix thoroughly until well incorporated. Add the salt, vanilla and then the egg whites and citric acid, and beat until smooth. Finally, add the remaining sugar, then the soft shortening and butter. Mix thoroughly and smoothly.

Note: If a softer icing is desired, an equal part of syrup and water may be added.

Rum, almond extract and other flavors can be added to this icing for variety. Chopped roasted almonds and walnut meat or chopped glazed fruits may also be added if desired.

Almond Paste Filling
(Can be Spread)

Almond Paste	1 lb	Sugar	1 lb

Cream the above together.

Whole Eggs	4 oz

Add to the above and mix in.

Cake Crumbs	2 lb	Salt	½ oz
Liquid Skim Milk (Variable)	12 oz		

Add and mix together thoroughly.

Almond Filling
(Can be Spread)

Ground Almonds or Almond Paste	1 lb	Sugar	1 lb
		Egg Yolks	2 oz
Shortening	1 lb	Pastry Flour	4 oz

Mix together thoroughly.

Applesauce Pecan Filling
(Can be Spread)

Chopped Pecans	1½ lb	Brown Sugar	1 lb
Cake Crumbs	2 lb	Applesauce	1 lb

Mix together to a smooth-spreading paste.

Basic Cream Filling

Formula No. 1

Water	3 qt	Powdered Milk	1 lb
Sugar	2½ lb	Salt	1 oz

Cook to boil.

Corn Starch	13 oz	Water	1 qt

Add and cook until clear. Remove from fire.

Eggs	8

Beat into the above mixture.

Have 1 pint finished cold water meringue ready and whip in with hand whip.

This formula is suitable for Boston or walnut creams. For cocoanut creams, add 8 ounces cocoanut.

Formula No. 2

Water	2½ gal	Agar-Agar	6/10 oz
Sugar	7½ lb	Salt	3 oz

Bring to boil for 3 minutes.

Water	2 qt	Corn Starch	3 lb

Add and cook until clear.

Water	7 pt	Meringue Powder	20 oz
Sugar	12 lb		

Whip to stiff meringue.
Add cooked batch to meringue and blend to a smooth cream.

Date Nut Filling

(Can be Spread)

Pitted Dates	2½ lb	Water	1½ lb
Brown Sugar	8 oz		

Bring to a boil and continue boiling for 5 minutes.

Chopped Pecans	1 lb

Add and stir in.

Fig Filling

(Can be Spread)

Dried Figs	10 lb

Soak in water until soft.

Sugar	10 lb	Water	3 lb

Boil to 240° F and add the figs.

Filbert Filling
(Can be Spread)

Roasted Ground Filberts	1 lb	Cake Crumbs	3 lb
Sugar	2 lb	Liquid Skim Milk (Variable)	1 lb
Cinnamon	½ oz		
Whole Eggs	6 oz		

Mix together to a smooth-spreading paste.

Fruit Filling
(Can be Spread)

Formula No. 1

Cake Crumbs	2 lb	Chopped Cherries	8 oz
Raisins	1 lb	Cinnamon	1 oz
Chopped Nuts	8 oz		

Mix together thoroughly.

Liquid Skim Milk (Variable)	1 lb

Add and mix to proper consistency.

Formula No. 2

Ground Dates	10 lb	Sweetened Apple Sauce	4 lb
Fig Jam	10 lb	Chopped Nut Meats	3 lb

Mix the above ingredients.

Fruit Nut Filling
(Can be Spread)

Sugar	3 lb	Salt	1 oz
Shortening	1 lb	Almond Paste	12 oz

Cream together thoroughly.

Whole Eggs	1½ lb

Add slowly and cream in.

Ground Filberts	2½ lb	Chopped Cherries	1 lb
Shaved Almonds	8 oz	Crushed Pineapple	1 lb
Chopped Black Walnuts	1 lb	Cake Crumbs	4 lb

Mix together thoroughly and stir in.

Water	Enough

Add enough water to bring filling to the proper consistency.

Honey Fruit Filling (Cream Type)

Brown Sugar	1½ lb	Honey	1 lb
Shortening	1½ lb	Salt	1 oz

Cream together thoroughly.

Chopped Fruits	1½ lb

Add and mix in.

Flour	8 oz

Add and mix smooth.

Nut Filling (Can be Sprinkled)

Sugar	2 lb	Ground Nuts	4 oz
Cinnamon	4 oz		

Mix together thoroughly.

Orange Butter Cake Filling

Sugar	5 lb	Egg Yolk	1⅞ lb
Corn Syrup	1¾ lb	Powdered Orange Juice	10 oz
Corn Starch	1¼ oz		
Salt	½ oz	Powdered Lemon Juice	2½ oz
Butter	2½ lb		
Water	2½ lb	Cold Water	10 oz

Place sugar, corn syrup, corn starch, salt, butter and water in a copper kettle. When this is warm enough to melt the butter, add slightly beaten egg yolk. Cook to 228° F stirring to prevent burning. When this temperature is reached add the mixture of orange and lemon powder dissolved in the second portion of water. Recook to 212° F, stirring to prevent burning.

Cool before using. Keep covered and cool.

Orange Butter Filling
(Can be Spread)

Water	1½ pt	Lemon Juice	1½ oz
Sugar	12 oz	Butter	2 oz
Ground Orange	8 oz	Shortening	1 oz

Bring to a good boil.

Water	½ pt	Egg Yolks	6 oz
Corn Starch	4 oz		

Mix together thoroughly and add to the boiling mixture, stirring constantly. Cook until thick; cool, and use.

Orange Filling
(Can be Sprinkled)

Ground Oranges	8 oz	Currants	8 oz
Sugar	5 lb		

Mix together thoroughly.

Orange Filling

Powdered Sugar	3 lb	Salt	½ oz
Butter	8 oz	Milk Powder	4 oz
Shortening	1 lb		

Cream until light.

Water	8 oz	Oranges (Rinds and Juice)	2

Stir together, add and cream lightly.

Stabilizers For Dry Fruit Fillers

To be used with wet fruit (by pouring over or combining) for holding shape and prevention of leakage.

Formula No. 1

Granulated Cane Sugar	40 oz	Corn Starch	60 oz
Citric Acid	1 oz	Color and Flavor	If desired
Salt	2 oz		

Formula No. 2

Granulated Cane Sugar	30 oz	Tapioca Starch	15 oz
Citric Acid	1 oz	Corn Starch	50 oz
Salt	2 oz	Color and Flavor	If desired

The above stabilizers can be used as glaze for fruit tarts and fresh-fruit open pies. Use enough fruit juice out of a quart to mix up with 6¼ ounces of Formula 1 or 2. To the balance of the (quart) juice, in a kettle, add 6 ounces sugar. Boil to 185° F. Then stir in the cold mixture. Cook until thick and clear.

Shut off the heat. Add 7 ounces sugar, ½ ounce lemon juice and stir. Brush over fruit tarts while still hot, or pour over berry pies, etc.

Cake and Nut Toppings

Almond Brittle Topping

Sugar	2 lb	Corn Syrup	8 oz
Water	1 lb		

Boil at 275° F.

Chopped Almonds	12 oz

Add and boil to 300° F.

Pour into bun pans in thin sheets. Cool and break up into fine pieces with a rolling pin. Apply to flat pieces of coffee cake dough.

Butter Topping

4X Powdered Sugar	1¼ lb	Shortening	8 oz
Butter	6 oz	Cake Flour	4 oz

Cream together thoroughly.

Whole Eggs	4 oz	Vanilla Extract	To taste
Liquid Skim Milk	6 oz		

Mix together and stir into the above.

Apply with a bowl knife to the coffee cake before putting it in the oven.

Crumb Topping

Shortening	8 oz	Ground Crushed Macaroons	3 oz
4X Sugar	10 oz	Cake Flour	1⅛ lb
Sugar Yolk	3 oz		
Vanilla Extract	¾ oz		

Cream shortening and sugar until light. Add sugar yolks and vanilla. Cream thoroughly. Then add the flour and mix in to break up the creamed mass. Gradually add the crushed macaroons. Rub through coarse sieve for a better, more uniform, finished topping.

DO NOT use this topping too heavily as it will prevent the cake from rising to its full value.

Crumb Topping

Almond Paste	45 lb	Gum Acacia	5 lb
Granulated Sugar	40 lb	Coconut Butter (m.p. around 98° F)	3 lb
Skimmed Milk Powder	7 lb		

The above mixture is mixed very thoroughly, then placed in a revolving heated compartment until the batch becomes toasted. It is then broken up and sifted to proper size.

Glazed Nut Topping

Pectin Stock Syrup	4 lb	Vanillin	To suit
Butter and Shortening	2 lb	Ground Sliced Nuts	1 lb
Salt	½ oz		

The above mixture is creamed lightly and spread on coffee cake dough. It should be applied to dough before proofing. It makes a nice luster topping.

Honey Nut Topping

Butter	1 lb	Honey	2 lb
Shortening	1 lb	Ground Nuts	3¼ lb
Sugar	1 lb	Water	1¼ lb

Bring to a boil and apply, while warm, to the coffee cakes.

Toasted Butter Topping

(A) Shortening	2 lb	Butter	2 lb

Heat slowly to a golden brown tint over a slow fire.

(B) Cake Flour	6 lb	Brown Sugar	10 lb
Cinnamon	3 lb		

Mix and sift together.

Now gradually work in (A) with (B), with stirring, until the mixture turns to soft crumbs.

Pecan Coconut Nut Topping

Fine Granulated Sugar	6 lb	Chopped Pecans	2 lb
Macaroon Coconut	3 lb	Salt	½ oz

Mix well.

Water	1 qt	Shortening	1 lb

Bring to a boil. Add to the above and mix to a soft paste consistency.

Now spread the paste evenly and not too thickly on toasting pans, or toast in baking oven, until layer develops a deep golden tint. The layer should be turned occasionally to get uniform heating.

Almond Brittle Topping

Granulated Sugar	6 lb	Corn Syrup	2 lb
Water	3 pt		

Cook to 275° F.

Add 1 pound of chopped almonds and cook to 310° F. Pour

onto pans, not too thick, and when cooled off, break up into fine almond brittle, and sift.

Sesame Seed and Peanut Topping

Granulated Sugar	3 lb	Water	2 pt
Corn Syrup	3 lb		

Cook to 240° F.

Add ½ pound of sesame seeds and ¾ pound of chopped peanuts, and cook to 300° F. Pour on pans, not too thick, let cool, break up fine and sift according to size pieces required.

Peanut Butter Topping
(May also be used as a filling)

Shortening	2 lb	Peanut Butter	3 lb
Honey	2 lb	Salt	2 oz

Cream together thoroughly.

Brown Sugar	12 lb	Pastry Flour	4 lb

Mix together, add and incorporate thoroughly with the above mixture.

Pecan or Cocoanut Topping

Brown Sugar	2 lb	Shortening	4 oz
Butter	4 oz		

Rub together.

Finely Chopped Pecans	1½ lb

Mix in.

Note: Macaroon cocoanut may be substituted for the pecans. Wash the tops of the dough pieces with egg wash and apply before baking.

Streusel Topping

Sugar	1¼ lb	Shortening	1 lb
Salt	¼ oz	Egg Yolks	6 oz
Mace	⅛ oz	Lemon Flavor	To taste

Cream together thoroughly.

Pastry Flour	1½ lb

Add, and rub by hand to a streusel.

Butterscotch Fudge

Corn Syrup	1 lb	Skimmed Milk	20 lb
Brown Sugar	40 lb	Butter	5 lb

Cook above to 240° F and let cool down to 150° F.

Burnt Sugar Coloring	3 oz	Butter	5 lb
Skimmed Milk	10 lb	Salt	2 oz
Lemon Powder	1 oz	XXXX Sugar	80 lb

Add to above mixture.
When ready to use, warm to 110° F.

Maple Fudge

Concentrated Maple Syrup (2½ Fold)	4 lb	Hard Coconut Butter (92° F)	4 lb
Sugar	16 lb	Cream (40%)	3 qt
Corn Syrup	8 lb	Salt	3 oz

Fondant Batch

Regular Stock Fondant	20 lb	Nougat Cream	4 oz

Note: For the *old-fashioned* or grainy variety, add 6 pounds of powdered 4X sugar to the creaming batch.

Mix all the ingredients of the syrup batch together in the kettle, with sufficient added water to complete the solution of the sugar, and cook to 245° F. Then melt in the fondant and nougat cream adding the powdered sugar last. Finish with chopped walnuts or pecans as desired and pour out on a slab.

White Fudge

Stock Icing (A or B)	20 lb	Butter or Part Shortening	3½ lb

Blend together. Use warm at 110 to 120° F. Add fruits and nuts as desired.

Chocolate Fudge

Stock Icing (A)	12 lb	Melted Chocolate	2 lb
Salt	¼ oz	Butter	8 oz
Hot Water	8 oz		

Blend together. Use warm at 110 to 120° F.

Cocoa may be used instead of chocolate as follows:

Cocoa	12 oz	Butter	1¾ lb
Hot Water	12 oz		

Chocolate Fudge (For Wrapping)

Water	4½ lb	Agar-Agar	1 oz
Sugar	15 lb	Corn Starch	1 oz

Boil to 228° F. Remove from stove.

Corn Syrup	6 lb	Butter and Shortening	4½ lb
Chocolate Liquor	6 lb		

Add and stir in. Place on machine and add:

Confectionery Sugar	20 lb	Water	2 lb

Use warm at 110 to 120° F.

For a lighter body, add a portion of nougat fluff to the warm icing.

Chocolate Fudge Base

Bitter Chocolate	30 lb	Glycerin	1 lb
Cocoanut Shortening	25 lb	Corn Syrup	25 lb
Cocoa Powder	5 lb	Vanillin Flavor	To suit

Melt slowly chocolate, shortening, cocoa and glycerin. Heat to 150° F then add the corn syrup.

Chocolate Fudge Sauce

Cocoa Powder	5 lb	Fluid Whole Milk	20 lb
Unsweetened Chocolate	4 lb	Water	30 lb
Cane Sugar	25 lb	Vanillin	To flavor (about 2 oz for 100 lb)
Light Brown Sugar	10 lb		
Salted Butter	10 lb		

Combine the cocoa powder, cane sugar and light brown sugar by mixing them thoroughly while dry. Heat chocolate and butter in a double boiler. Heat the water and milk to about 180° F. Then add the dry mix by stirring constantly until it is completely dissolved. Add the melted chocolate and butter gradually, continuing to stir until a homogeneous blend of ingredients is obtained by bringing the entire mix to a boil. Cool and flavor with vanillin.

To obtain a very smooth product, it is recommended to run the entire mix through a colloid mill or similar apparatus.

Cake Shines and Glazes

Apricot Gloss

Formula No. 1

Apricot Paste	40 lb	Agar-Agar	4 lb
Apricot Pulp	80 lb	Water	15 gal
Cane Sugar	250 lb	Citric Acid	4 oz

Bring the water to a boil. Then add 200 pounds of cane sugar and the fruit with the dissolved citric acid. Cook to 223° F. Transfer to cooling table and work in the solution of agar-agar mixture until uniformly mixed.

Formula No. 2

Agar-Agar	2 lb

Soak overnight in water to cover agar-agar.

Sugar	100 lb
Water	1 qt
Apricot Pulp	62½ lb (10 Cans of 6 lb 4 oz)

Bring to a boil in a 25-gallon copper kettle.

Add the above agar-agar, and again bring to a boil under good stirring. When thoroughly mixed, shut off steam immediately. Strain under a roller strainer into 30-pound pails.

Let the agar-agar soak overnight in 10 gallons of water. In the morning transfer to kettle, bring to a boil and add the 50 pounds of sugar. Cook to 223° F, and then mix the batch with the apricot cook on the cooling table.

If the above gloss is too tough add less agar-agar, if too soft, add more agar-agar.

Agar-agar: Malayan name for a seaweed, known also as Ceylon moss and Jaffna moss. Agar-agar is employed by bacteriologists for the cultivation of bacteria. In the food industry it serves to gelatinize food products.

Agar-agar in powder form is added sometimes to pudding powders (1%) to make the pudding smooth.

Apricot Shine

Dry Apricots (Soaked for 24 Hours)	3 lb	Granulated Sugar	5 lb
		Water	

Cook apricots with sugar and enough water to cover. Force through a sieve. When cool, place in crock jar and keep in icebox.

When using, boil for 3 to 4 minutes and apply very hot with a brush.

Cake Shine

Water	4 pt	Gelatin	4 oz
Sugar	5 lb		

Bring water to a boil and dissolve the gelatin in it. Immediately add the sugar. When dissolved shut steam off. Apply this shine to cakes while hot.

Fruit Glaze Stocks for Frozen Berries

All frozen berries, except strawberries, should be added frozen hard to a cold glaze instead of a warm glaze. Only strawberries are added to a warm glaze. The glaze should be made up in advance as a stock and thoroughly cooled before the frozen berries are added. Use the following formula for blackberries, loganberries, raspberries, youngberries and blueberries.

Water	6 qt	Agar-Agar	1 oz
Sugar	15 lb	Corn Starch	2 oz
Salt	2 oz		

Cook to rolling boil and continue boiling for 3 minutes.

Water	2 qt	Tapioca Flour	1 lb
Corn Starch	1¼ lb		

Add and cook until clear. Remove from fire. Let stand until thoroughly cold. Then incorporate:

Hard Frozen Berries	30 lb

Let filling stand to warm up to room temperature before using.

Honey Glaze

4X Powdered Sugar	10 lb	Hot Water	1 lb
Honey	6 oz		

Mix to a paste.

Gelatin	3 oz	Hot Water	1 lb

Dissolve, add and mix smooth.

Glazes and Washes for Bakery Goods

Glucose	2 lb	Water	1 lb

Bring to a good boil and apply to pieces as they come from the oven.

Pectin Glaze for Fruit Tarts

No. 100 Pectin Powder	2 oz	Water or Fruit Juice	1¼ qt
Sugar	4½ lb	50% Phosphoric Acid	1¼ oz

Mix pectin with 1 pound of sugar. Bring liquids to boil and add remaining sugar. Shut off steam and stir the sugar until dissolved. Then add the pectin-sugar mixture, stirring it around. Allow liquid to cool, then add the acid. This will give a high gloss to fruit tarts.

Pectin Glaze or Wash for Danish Pastry, Sweet Goods, etc.

No. 100 Powdered Pectin	1 lb	Corn Syrup	50 lb
Powdered Cane Sugar	40 lb	Water	4 gal

Mix 1 pound of pectin with 5 pounds of sugar in a dry pan. Bring water to a boil and add the pectin-sugar mixture slowly, with stirring. Then add balance of sugar and corn syrup, and cook to 220° F.

This syrup is known as stock syrup and is thinned down by adding 1 pint of water to each gallon, for glazing on pastry, etc. Apply this glaze on pastry as soon as taken from oven. With the addition of 1 ounce of 50% solution of citric acid to 10 pounds of stock syrup, it makes a glaze for any kind of fruit, upon fruit tarts, fruit cakes, etc. After adding acid solution, pour quickly on cakes and spread with a knife before the formed jelly becomes too thick. Add acid to small portions of stock solution at one time.

Stock Fruit Glaze for Fruit Pie Filling

Sugar	18 lb	Corn Starch	2 oz
Water	6 qt	Salt	2 oz
Agar-Agar	1⅗ oz		

Cook to boil for 3 minutes.

Water	1 qt	Tapioca Flour	1¾ lb

Add to the above and cook clear.

This glaze may be made up and held for several days. It can be used cold with any type of fruit for covered or open pies and tarts.

For fresh strawberries, it works best when used fairly hot.

It is handy fo. quick preparation of canned fruits. It may be added to fruit and juice of a No. 10 can in proportion to suit individual taste and requirements.

Glaze for Sweet Rolls, Coffee Cakes, etc.

Water	2 lb	Sugar (Brown or Granulated)	2 lb
Corn Syrup	2 lb		
Agar-Agar	3/5 oz		

Bring to boil for 3 minutes.

For adding shine to rolls, etc., apply to rolls or cakes as they come from the oven. Replace water with fruit juice if desired.

This is a good glaze for baked pie shells for open fruit pies to prevent soggy crusts.

Stock Fruit Glaze for Open Pies

Liquid (Water and Fruit Juice)	3 qt	Salt	1½ oz
		Corn Starch	1 oz
Sugar	7½ lb	Agar-Agar	1 oz

Bring to boil and boil for 3 minutes.

Water	1 qt	Fruit	3 No. 10 Cans
Corn Starch	14 oz		

Add and cook clear.

This glaze is especially designed for use with any type of fruit or berry for open pies, tarts and cakes. When canned or cold pack fruit is used, the fruit juice may be used as part of the liquid. This glaze is fine for fresh peach special pies topped with whipped cream.

Glaze for Fresh Berry and Open Face Pies

(A)	Water	22 pt	Sieved Fruit (Apricot or Peach)	10 lb
	Sugar	18 lb		
	Agar	2 oz	Salt	3 oz

Mix the sugar with the agar, and then boil the entire batch for 3 minutes. Then add (B).

(B)	Water	1 qt	Color and Flavor	If desired
	Corn Starch	30 oz		

Mix together.

The above, when cooked, makes up a clear filling and will seal the fruit when spread over it.

BUTTER CREAMS

Butter Cream
(Cooked)

Sugar	2½ lb	Water	Sufficient to dissolve
Corn Syrup	2½ lb		

Cook the above to 240° F. Then add:

Condensed Maple Syrup (2½ Fold)	9 oz

Separately beat to a stiff fluff:

Egg White	5 oz	Butter	1 lb
Shortening	2¼ lb		

Add the cooked syrup slowly to the beaten whites. Cream. Then add the creamed shortening and butter.

Butter Cream Filling
(Cream Type)

4X Powdered Sugar	2 lb	Butter	6 oz
Shortening	6 oz		

Cream together thoroughly.

Whole Eggs	8 oz

Add and cream in.

Bread Flour	14 oz

Add and mix in.

Butter Cream Filling

Shortening	17 lb	Salt	3 oz
4X Sugar	20 lb	Water	6 lb
Powdered Milk	3 lb		

Cream the dry ingredients, thoroughly. Add the water slowly.

Water	6	lb	Agar-Agar	1⅗	oz
Sugar	8	oz	Corn Starch	1⅗	oz

Boil 3 minutes, add slowly to the above mixture. Beat at high speed for 5 minutes.

Nougat Fluff	12 lb

Add and beat until light.

Flavor to suit.

This recipe is very good for icings, fillings, cream rolls, éclairs, cakes, etc.

Butter Cream
(For Decorating)

Stock Icing (A)	12	lb	Butter and Shortening	2¼	lb
Powdered Milk	7	oz	Salt	½	oz

Butter Cream
(For Wrapping)

Stock Icing (B)	20	lb	Butter and Shortening	5 lb

Cream until light.

Butter Cream Filling

4X Sugar	10	lb	Butter	10	lb
Powdered Milk	3	lb	Vanilla Extract	1½	oz

Cream at medium speed.

Nougat Fluff	5 lb	Cold Water	4 lb
Whole Eggs	10		

Add eggs to the creamed mixture and cream light. Add nougat fluff and, at high speed, add water gradually.

Nougat Fluff

Corn Syrup (Warm)	5 lb	Powdered Egg Whites	2 lb

Cream at low speed.

Warm Water	6 lb

Add. Whip at high speed to peak.

Sugar	25 lb	Corn Syrup	25 lb
Water	8 lb		

Add slowly. Whip at high speed to sharp peak.

Boil to 240° F.

Cool to 160° F with 25 pounds cold corn syrup.

Use this fluff to blend with any type of icing including butter cream.

Nougat fluff may be made up and kept on hand in large quantities. It will keep, covered, for several weeks. It is always handy in making candy-like icings and butter cream.

Basic Fondant Cream

Sugar	75 lb	Nulomoline Syrup	10 lb
Corn Syrup	15 lb		

This fondant can be stored without drying out. Add 5 gallons of water and the sugars to the kettle, and boil rapidly to 238 to 248° F. Cool to 110 to 125° F and beat into fondant.

Basic Cream

(A)	Water	7 pt	Honey	½ pt
	Sugar	12 lb	Lemon Emulsion	¼ oz
	Meringue Powder	20 oz	Vanilla	3 oz

Beat the above to a stiff meringue.

(B)	Water	3 gal	Salt	4 oz
	Sugar	7 lb	Corn Starch	3 lb
	Agar-Agar	2 oz		

Mix the sugar with the agar. Draw off 1 gallon of water and make a paste with the starch. Boil the 2 gallons of water. Add the sugar-agar mixture. When clear, add the starch paste solution, and cook for 3 minutes. To the stiff meringue above (A) add this hot mixture and beat up until cool enough to handle.

This mix may be used with the addition of color and flavor to make cream pies. It may also be mixed with acid-tasting fruits.

Meringues

Boiled Meringue Icing

Egg Whites	2 pt	Water	2 pt
Meringue Powder	8 oz	Color and Flavor	

Beat to a mazette.

Powdered Agar-Agar	2¾ oz	Cream of Tartar	1 oz
Sugar	12 lb	Water	2 qt
Corn Syrup	3 lb		

Mix the sugar with the agar. Boil water. Add cream of tartar. Then add the sugar-agar mixture slowly, then the corn syrup and cook at least 10 minutes. Add this slowly to the above mazette and beat for 10 minutes in high speed.

Light Meringue Icing

Salt	¼ oz	Egg Whites	10 lb
Vanilla Flavor	To suit		

Beat until stiff.

Sugar	32 lb	Water	10 pt
Corn Syrup	3 lb		

Boil to 240° F.

Add the hot syrup to the beaten egg whites and beat to a good consistency. Chopped nuts or fruit juices may be added.

Heavy Meringue Icing

Egg Whites	10 lb

Beat until stiff.

Sugar	45 lb	Salt	½ oz
Corn Syrup	12 lb	Vanilla Flavor	To suit
Water	12 pt		

Boil to 240° F.

Add the hot syrup to the beaten egg whites and beat to a good consistency. Flavor, nut meats, fruits or juices may be added.

Meringue
(Non-Weeping and Cracking)

Water	13 oz	Keltose	¼ oz
Corn Starch	¾ oz	Sugar	24 oz

Bring water to a boil. Mix the above ingredients and add to the boiled water with stirring. Break up lumps. Then add in a fine stream to the meringue below whipped to a peak, and continue whipping at third speed.

Egg White Liquid	22 oz	Salt	⅕ oz
Sugar	2 oz	Cream of Tartar	⅕ oz

Boiled Meringue

(1) Water	1 qt	Meringue Powder	6–8 oz
Sugar	3 lb		

Whip to a stiff meringue.

(2) Water	1½ pt	Agar	1 oz
Sugar	2 lb		

Mix sugar and agar, and boil for 3 minutes.

When the meringue (1) is thoroughly whipped, reduce the machine to second speed, and then add the hot agar mixture (2) slowly so that it is entirely absorbed as poured in.

Meringue Powder
Formula No. 1

Powdered Egg Albumen	50 lb	Fine Granulated Sugar	28 lb
Powdered Skimmed Milk	22 lb	Powdered Gelatin	2 oz
		Vanillin	3 oz

Formula No. 2

Powdered Egg Albumen	65 lb	Powdered Citric Acid	5 oz
Tapioca Flour	20 lb	Vanillin	3 oz
Fine Granulated Sugar	10 lb		

Formula No. 3

Powdered Egg Albumen	35 lb	Gum Karaya	5 lb
Fine Granulated Sugar	30 lb	Vanillin	3 oz
Corn Sugar	30 lb		

Eight ounces of any of the above meringue powders are sifted into 1 quart of water. Beat well. Then gradually add 3½ pounds of sugar, and whip until stiff.

Egg White Meringue

Egg Whites	2 qt	Agar-Agar	2 oz
Granulated Sugar	6 lb	Water	3 qt

Sprinkle pinch of salt over egg whites and beat until dry.

Mix the sugar and the agar. Add this mixture to the boiling water and boil for 3 minutes.

Add this hot mixture slowly to the beaten egg whites, beating slowly all the time. Do not add the hot mixture faster than the egg white takes it up.

Fresh fruit such as prunes, apricots, peaches may be added to the above meringue for a fruit whip pie. Baking is unnecessary. Fruit juice may be used instead of water in cooking the agar.

Egg White Topping

Egg Whites	3 qt	Sugar	1 lb

Whip together.

Water	4½ qt	Corn Starch	2 oz
Sugar	10 lb	Agar-Agar	2 oz

Cook to boil and boil for 3 minutes.

For whipping egg whites, use medium speed throughout. Start egg whites whipping and, at the same time, start cooking sugar, water and agar-agar. Watch the time closely from the instant boiling starts. This is important. Under-boiled topping is watery.

Whip egg whites until they start to pull from the sides of the bowl. Do not whip until dry. Pour the stabilizer syrup in a very fine stream into the egg whites while they are boiling hot. Continue the whipping at the same speed until all the syrup has been added. Be careful to pour syrup very slowly to prevent it from going to the bottom of the bowl and cooking the egg whites, causing an uneven, lumpy meringue.

Continue whipping after all the syrup has been added for a few minutes to expel some of the heat and to develop a soft peak. Add flavor as desired. The meringue is now ready for the pies and should be used immediately. Place pies on pans and brown in hot oven at 450 to 500° F.

Marshmallow

Marshmallow Topping or Filling

Sugar	5 lb	Agar-Agar	3/10 oz
Corn Syrup	2 lb	Corn Starch	3/10 oz
Water	2 lb		

Bring to boil and continue boiling for 3 minutes.

Water	1 lb	Powdered Egg White	4 oz

Place in bowl on machine. Whip on low speed, then on fast speed to a firm peak.

Add hot syrup slowly in a fine stream while whipping and continue whipping to soft peak or as desired.

For lighter marshmallow, increase water and egg white slightly. This marshmallow can be used for blending with stock icing, butter cream, or any type of icing requiring a smooth light body.

This is an all-purpose marshmallow for rolled cakes, horns, slices, layer cakes, snow balls, and for icing and decorating. Fruits or preserves may be added for variety icings. It will not become tough, stringy or rubbery. It keeps short and tender and may be made up several hours before using.

Marshmallow Icing

Granulated Sugar	10 lb	Cream of Tartar	¾ oz
Corn Syrup	2 lb	Water	1½ pt

Boil to 240° F.

Egg Whites	2 lb

Whip up lightly.
Then add hot syrup to beaten egg whites, and beat up light.

XXXX Powdered Sugar	3 lb

Stir into the above mixture and mix until smooth.

Billowy Marshmallow

Gelatin	1 lb	Standardized Invert Sugar	10 lb
Cold Water	8 lb	Flavor and Color	As desired
Egg White	15 lb		
Icing Sugar	65 lb		

Soak the gelatin in the cold water. Heat the soaked gelatin carefully until fluid (about 140° F). Add the fluid gelatin to the egg whites, 35 pounds of sugar and the standardized invert sugar. Whip this mixture until it is fairly stiff. Just before finishing the whipping, add the remaining 30 pounds of sugar along with the flavor and color. Beat until it is of the consistency of marshmallow.

For banana flavor, use about 4 ounces of ripe banana powder containing dry milk solids for each ten pound batch. Incorporate it in any way convenient, preferably first blended with about an equal weight of sugar.

Marshmallow Fluff
(Stock)

Corn Syrup	3½ lb	Water	14 oz
Sugar	3½ lb		

Boil to 230° F. Then remove from fire.

Corn Syrup	2½ lb	Invert Sugar Syrup	1 lb

Add to the above in a clean beating bowl.

Water	13 oz	Salt	¼ oz
Egg Albumen	4½ oz	Cream of Tartar	¼ oz
Vanilla Extract	½ oz		

Dissolve. Add to above. Then beat until stiff on medium speed. Fill containers for storage.

This fluff spreads easily. It never gets tough or rubbery.

Marshmallow Topping

Gelatin	40 oz	Syrup (See Formula Below)	31 lb
Cold Water	2 lb		
Flavor and Color	As desired		

Soak the gelatin in the cold water. Then heat until fluid but not over 140° F. Add the fluid gelatin to the special syrup below and beat to marshmallow consistency. Just before the beating is finished incorporate the flavoring and color.

Syrup

The syrup must be varied to take care of summer and winter conditions so two types of formula are given here.

	Summer	Winter
Water	36 lb	33 lb
Sugar	90 lb	63 lb
Invert Sugar	90 lb	120 lb
Corn Syrup	60 lb	60 lb

Dissolve the sugar in the water by heating. Hot water will be sufficient for the winter formula but the increase of sugar in the summer formula will require heating to 150° F. Add the invert sugar and corn syrup to the warm dissolved sugar. Stir until thoroughly mixed. This syrup may be made in any quantity desired and used as needed.

Cut Marshmallow
(To Be Coated)

Sugar	54 lb	High-Grade Gelatin	3¼ lb
Corn Syrup	46 lb	Water	24 lb
Standardized Invert Sugar	12 lb		

Dissolve the gelatin in the water. Then add the sugar and standardized invert sugar, and stir until they are dissolved. When this is done the temperature of the batch should register approximately 175° F. Shut off the heat. Then add the corn syrup and stir through the batch. Beat until one American gallon measure filled with this marshmallow weighs 3¾ to 4 pounds net. Add color and flavor as wanted.

Spread into trays that have been lined with ordinary wrapping paper. Leave the trays in a room of normal temperature until the marshmallow has set. This will take from 5 to 10 hours. When firm enough, the marshmallow can be dumped out of the trays, the paper moistened and removed, and the marshmallow cut into pieces of the desired size. After cutting, allow the pieces to dry sufficiently. Then coat with chocolate.

Creole Marshmallow

Refined Sugar	60 lb	Water	25 lb
High-Grade Molasses	15 lb	Roman Punch or Vanilla Flavor	As desired
Corn Syrup	30 lb		
Gelatin	2½ lb		

Mix the gelatin with 10 pounds of cold water and set aside for at least ½ hour. Place the remaining 15 pounds of water in a clean kettle. Add the sugar and heat until dissolved.

Turn off the heat. Add the molasses and corn syrup. Then add the soaked gelatin. Mix until the gelatin is completely dissolved.

Place the batch in a marshmallow beater and beat until quite light.

Add the flavor, mixing well. Then cast into clean starch which has been tempered to approximately 95 to 100° F. Allow to remain at normal room temperature for approximately 12 hours in order to set.

After removing from the starch, dust with a powder made up of 10 pounds of dry corn starch and 90 pounds of powdered sugar.

The marshmallow should remain in the dusting material for approximately 3 hours before packing the product.

Tenderlite Marshmallow

This is made with gelatin for filling and topping layer cakes, enrobing cup-cakes and for snowballs to be topped with coconut, etc. (Semi-cold process).

Gelatin	3½ oz	Invert Sugar Syrup	2½ lb
Cold Water	6½ oz	Corn Syrup	1½ lb
Hot Water	1¼ lb	Flavor and Color	As desired
Sugar	4 lb		

Soak the gelatin in the cold water. Heat the soaked gelatin carefully until fluid (about 140° F).

In the meantime, prepare a syrup by dissolving the sugar in the hot water (heat to 190° F) and adding, to the dissolved sugar, the invert sugar syrup and corn syrup. Stir the syrup until the invert sugar syrup and corn syrup are dissolved and then combine these with the fluid gelatin.

Beat the whole mass until it is of the consistency of marshmallow. Add the flavor and color to the marshmallow just before finishing the beating.

For molasses flavor, use 2½ pounds of open-kettle molasses (highest quality) for each 10 pounds of the total batch, and reduce the invert sugar by ½ pound, the sugar by 1½ pounds, and the water by ½ pound.

For banana flavor, use about 4 ounces of ripe banana powder containing milk solids for each ten-pound batch. Incorporate it in any convenient way preferably first blended with about an equal weight of sugar.

Whip Powder
For Marshmallow Type Filling

Egg Albumen Powder	7½ lb	Fine Sugar	50 lb
Gelatin	15 lb	Keltose	3¾ lb

Blend the above ingredients. Use 7½ pounds in each one-hundred-pound batch of marshmallow.

Whip Powder (Formula Above)	7½ lb	Water (120–140° F)	30 lb
		Sugar	62½ lb

Dissolve the whip powder in the hot water. Add the sugar and heat until light and fluffy.

From 5 to 10 pounds of sugar may be replaced with water in each one-hundred-pound batch, depending on the body desired.

Gelatin and Egg White Marshmallow Powder

Sugar	3 lb	Vanillin or Other Flavor	To suit
Gelatin Powder	2½ oz		
Powdered Egg Albumen	½ oz		

Beat the above ingredients up with 1 quart of warm water, using second or third speed. If too stiff, dilute with more water. Color may be added while beating. Marshmallow may be used in filling cream rolls, layer cakes and tarts. If desired, corn syrup up to 1 pound may be added and beating continued at third speed.

Marshmallow Topping
Formula No. 1

(A)	Egg Albumen Powder	4¾ oz	Water	4¾ oz
(B)	Keltose	¼ oz	Cream of Tartar	⅜ oz
	Water (180° F)	6½ oz	Salt	⅛ oz
(C)	Sugar Syrup (140° F)	9 lb		

Dissolve the egg albumen in the first portion of the water. While mixing the second portion of water very fast, slowly sift in the Keltose and mix until free from lumps.

Combine the egg and Keltose solutions and stir in the salt and cream of tartar.

Start whipping hot syrup, and slowly add the egg albumen solution. Whip until stiff.

Formula No. 2

(A)	Corn Syrup	5 lb	Vanilla Extract	1 oz
	Powdered Egg Whites	3 oz		

Weigh into mixing bowl.

(B)	Corn Syrup	1 lb	Gum Arabic	1⅝ oz
	Sugar	2½ lb	Water (To Dissolve Gum)	13 oz
	Salt	½ oz		

Weigh into kettle and boil to 242° F.

Bring (B) to 242° F and allow to stand while (A) is beating for 10 minutes. Add (B), while hot, very slowly with beater running. After all is in, beat for an additional 10 minutes.

Marshmallow Topping or Filling

This is a marshmallow which will not become tough or rubbery on aging. It remains short and tender and has an excellent texture. It is well stabilized against syrup separation. It remains smooth and does not have to be applied immediately after being made.

Sugar	20 lb	Water	8 lb
Corn Syrup	8 lb		

Boil to 230 to 232° F. Remove from heat.

Keltose	3 oz	Liquid Egg Whites	5 lb
4X Sugar	2 lb		

Sift Keltose and sugar together. Add to hot syrup and stir to dissolve. Whip egg whites to firm peak.

Slowly pour the hot syrup into the beaten egg whites and continue whipping to a sharp peak.

Marshmallow Filling
(Non-Soak)

Egg Albumen Powder	2½ lb	Sugar	62 lb
Warm Water	30 lb	Keltose	6 oz

Blend the egg whites with approximately 10 pounds of the sugar and dissolve the blend in the warm water (120 to 140° F from hot water tap). Separately blend the Keltose with the rest of the sugar, and add the blend to the dissolved egg whites. Stir this mixture at low speed for 3 or 4 minutes and then whip until it is of marshmallow consistency.

This marshmallow is very tender but has a firm body. It is well stabilized against soakage and is suitable for filling layer cakes, cream horns, etc. If a lighter body is desired, 5 pounds of sugar may be replaced by 5 pounds of water in each one-hundred-pound batch.

Marshmallow Topping
(Made in Upright Vertical Beater)

Corn Syrup	200 lb	Powdered Albumen	2 lb
Water	3 lb	Vanillin Flavor	To suit

Place 5 pounds of cold corn syrup in the bowl of the mixer. To this add the powdered egg white. Cream at low speed for about 1 minute until the egg white is dissolved. Then add the water. Whip at high speed, until a good body is produced.

Now, slowly add the corn syrup (which has been cooked to 240° F) and whip at high speed to the desired peak. Add vanillin last. Part of the corn syrup may be replaced by invert or cane sugar syrup.

Marshmallow Topping
(Made in Horizontal Type Beater)

Corn Syrup	200 lb	Powdered Albumen	3 lb
Water	5 lb	Vanillin Flavor	To suit

Cook corn syrup to 240° F and transfer to horizontal type beater. Before starting beater, add the powdered albumen directly to the hot syrup, along the back side of the beater. Cover the machine and heat for ½ to ¾ minute. Then add very slowly the above amount of cold water. Beat to the right peak. Part of the corn syrup may be replaced by other syrup.

Marshmallow

Granulated Sugar	12 lb	Powdered Egg Albumen	4 oz
Water	3 qt	Gelatin	4 oz
Powdered Tartaric Acid	¼ oz	Vanilla Extract	¼ oz

Cook the water and sugar to 235° F. Now add the acid and cook to 240° F. This will invert the sugar, keep the marshmallow soft and prevent crystallization. Dissolve the gelatin in 1 quart of hot water. Whip the albumen to a stiff meringue with 8 ounces of water. Pour the boiled syrup into the meringue and start whipping again. Slowly pour in the gelatin and whip until light. Add vanilla flavor last.

Marshmallow Filler

Formula No. 1

Sugar	100 lb		
or			
Sugar	50 lb	and Heavy Glucose	50 lb
		Tartaric Acid	2 oz

Boil in a kettle, with the precautions given under fondant formula.

Albumen	3 lb	XXXX Sugar	15 lb
Water	6 qt		

Beat up and add to the boiling syrup, letting the stirrer go at high speed.

Powdered Gelatin	1 lb	Corn Starch	3 lb

Mix well and add to the above. Add some vanilla, or any other desired flavor, and continue to work machine until the desired consistency is reached.

To make marshmallow paste leave out the filler, i.e., the 3 pounds of corn starch.

Formula No. 2

Cane Sugar	100 lb	Water	25 lb
Corn Syrup	25 lb	Invert Syrup	30 lb

Boil to 240° F.

Egg White	20 lb	Cream of Tartar	¼ oz

Beat until stiff and then beat into the above hot mixture.

Then dissolve 1 pound of gelatin in 5 pints of hot water. Add to the above sugar-egg white mass, and beat until stiff. Add vanilla just before the end of the beating.

Formula No. 3

Cane Sugar	100 lb	Water	5 gal
Corn Syrup	12 lb	Invert Syrup	20 lb

Boil to 240° F.

Egg White	25 lb	Cream of Tartar	¼ oz

Beat until stiff and then beat into the above hot mixture.

Then dissolve 20 ounces of gelatin in 5 pints of hot water. Add to the above sugar-egg white mass, and beat until stiff. Add vanilla flavor just before the end of the beating.

Marshmallow Stock for Cream Fluff Filling

Sugar	10 lb	Water	3 pt
Corn Syrup	2 lb	Cream of Tartar	¼ oz

Boil to 250° F.

Egg White (Do not beat too stiff)	2 lb	Vanilla Flavor	¾ oz

Beat for 2 minutes at high speed. Then add above hot syrup quickly to the beaten egg white and beat stiff.

Then add 2 ounces of gelatin previously dissolved in 6 ounces of hot water and beat 2 to 4 minutes. Remove from bowl and allow to cool. This is called stock.

Marshmallow Cream Fluff Filling
(To Prepare)

XXXX Powdered Sugar	8 lb	Butter	2 lb
Shortening	6 lb	Powdered Skimmed Milk	2 lb

Cream until light at medium speed.

Whole Eggs	20 oz	Vanilla Flavor	2 oz

Add slowly to the above and cream in well.

Stock Marshmallow	8 lb

Add to the above and cream in well.

Cold Water	4 pt

Add very slowly at medium speed and then continue to cream at medium speed for 2 to 3 minutes.

Marshmallow Cream
(Made with Soy Albumin)

Gelatin and albumin have always been used in the manufacture of marshmallow. Soy albumin, which is a bland white powder made from isolated soy proteins, is a new interesting product and produces foam structures of good volume and stability. Like gelatin, it is capable of producing air cells. A good combination to use is 20% soy albumin and 80% gelatin. Soy albumin materially retards the moisture of the marshmallow from getting into the biscuit. It is also claimed that it will prevent syneresis and will retard shrinkage and toughness in the marshmallow upon aging.

The following marshmallow formulas are recommended for biscuits.

	Coated Biscuits	Uncoated Biscuits
Syrup	42 lb	42 lb
Gelatin	11 oz	11 oz
Warm Water	5 lb	6½ lb
Soy Whip (Soy Albumin)	3 oz	3 oz

Place the syrup in the bowl of the beater at room temperature. Add soy whip. Mix it at low speed until it is completely dissolved. This requires about 2 minutes. Now add solution of gelatin in warm water and whip at high speed to the desired peak and body. If the marshmallow is too loose and fluffy, reduce the water content. If it is too short in volume and heavy in body, increase the water. The syrup may be all corn syrup or a combination of corn, invert and sugar.

Short Nougat

Sugar	50 lb	Salt	8 oz
Corn Syrup	50 lb	Flavoring	As desired
Full-Fat Soy Flour	7 to 12 lb		
Egg Albumin or Substitute	2 lb		

Place 5 pounds of corn syrup in nougat beater with the egg, while the beater is in operation. Boil the sugar and the remaining corn syrup to the desired degree, usually about 270° F, and pour this slowly into the egg mixture in the beater. Beat to the desired lightness.

Adjust the beater to slow speed and add the soy flour and salt. Mix well. Flavor and finish as usual.

Low-fat soy flour can be used if desired.

This formula is suggested to increase the nutritive balance of the candy through introduction of soy flour. It should be regarded as experimental only and should be thoroughly tested under proposed operating conditions before its manufacture is undertaken.

CHAPTER IX

FLAVORS, SYRUPS AND BEVERAGES

Flavoring Extracts

These extracts consist of small percentages of true extracts, oils or essences in a solution of alcohol. The chemical duplicates of natural flavoring extracts are usually called imitations. The aromatic characteristics of a great many spices, nuts, herbs, fruits and even flowers are sold, the best known ones being lemon, orange, cloves, cherry, cinnamon, anise, peppermint, pineapple, strawberry, vanilla, etc.

Many natural essences are obtained by extracting the essential aromatic oil from the blossoms, fruits, fruit rinds, either by expression, absorption, distillation or maceration. The first method, that of expression, is applied to citrus rinds, e.g., lemon, lime or orange. The second method, absorption, is generally applied by steeping in alcohol, e.g., vanilla beans. The third method, distillation, is accomplished by distilling, e.g., peppermint, in water. By the use of the fourth method, maceration, the raw material is simply allowed to soak in water, or in a combination of water and alcohol, long enough for soluble substances to be extracted. There are some artificial flavors that compare quite favorably with the natural substance, e.g., vanillin and wintergreen.

Terpeneless flavors are those which have had terpenes or hydrocarbon components removed from the essential oils. They are stronger, more easily soluble, longer lasting and, therefore, economical. They are, however, very costly and at times lack some of the richness of the original oil itself.

All flavoring extracts should be transparent, free from any

cloud, and should be able to stand on the shelves without becoming rancid or cloudy.

Allspice Extract
(60% Alcohol)

Pimento Oil	1 pt	Water	¼ pt
Alcohol	2 pt		

Anise Extract
(72% Alcohol)

Anise Oil	3 pt	Water	½ pt
Alcohol	9 pt		

Bergamot Extract
(58% Alcohol)

Bergamot Oil	5 pt	Water	1½ pt
Alcohol	8½ pt		

Birch Beer Extract
(63% Alcohol)

Methyl Salicylate	106 oz	Water	6 gal
Oil of Sassafras	18 oz	Red Color Solution	5 oz
Ethyl Alcohol	12 gal		

Bitter Almond Oil
(62% Alcohol)

Bitter Almond Oil	2 pt	Water	1 pt
Alcohol	5 pt		

Cardamom Extract
(68% Alcohol)

Cardamom Oil	2½ pt	Water	2½ pt
Alcohol	7½ pt		

Cassia Extract
(57% Alcohol)

Cassia Oil	3⅓ pt	Water	2½ pt
Alcohol	7½ pt		

Caraway Extract
(78% Alcohol)

Caraway Oil	1⅕ pt	Water	1½ pt
Alcohol	8½ pt		

Celery Extract
(75% Alcohol)

Celery Oil	1½ pt	Water	4 pt
Alcohol	12 pt		

Cinnamon Extract
(57% Alcohol)

Cinnamon Oil	3⅓ pt	Water	2½ pt
Alcohol	7½ pt		

Coffee Extract
(Made by Maceration)

Ground Freshly Roasted Coffee	2½ pt	Water	5½ pt
Alcohol	4½ pt	Vanillin	½ g

Macerate or allow to soak for 8 days. Then strain liquid. Maceration is the process of soaking ingredients in a liquid until they have been thoroughly penetrated and the soluble portions are softened and dissolved.

Fenugreek Extract
(70% Alcohol)

Fenugreek Seeds	8 lb	Water	10 pt
Alcohol	22 pt		

Macerate for 1 month; filter.

Lemon Extract
(95% Alcohol)

Lemon Oil	6½ oz	Alcohol	7½ pt

Lime Extract
(95% Alcohol)

Lime Oil	6½ oz	Alcohol	7½ pt

Mace Extract
(Nutmeg Extract)
(83% Alcohol)

Mace Oil	8 oz	Water	½ pt
Alcohol	7 pt		

Maple Extract
(Like Mapleine)

Coffee Extract	20 gal	Coumarin	1 oz
Fenugreek Extract	108 oz	Alcohol	4½ gal
Lovage Oil (1:10)	5 oz	Hot Water	8 gal
Vanillin	13 oz	Caramel	10 gal

Maple Extract

Alcohol	20 oz	Peru Balsam	⅕ oz
Caramel	20 oz	Coffee	16⅘ oz
Vanillin	2⁷⁄₁₀ oz	Hickory Bark	5½ oz
Lovage Extract	1³⁄₁₀ oz	Water	20 oz
Fenugreek Extract	15½ oz		

Maple Extract
(For Compounding Flavors)

Powdered Fenugreek Seeds	1 lb	Ground Lovage Root	¼ lb
		Alcohol	½ gal

Let stand a few days. Strain.

Pulverized Coffee	2 lb	Boiling Water	1 gal

Boil for 10 minutes. Cool and strain.
Dissolve:

Vanillin	2 oz	Celery Oil	1 dr
Rue Oil	1 dr	Glycerin	1 pt

Ground Fenugreek Seeds	3¼ lb	
Alcohol	7 pt	(extract: 108 oz)
Water	3 pt	(extract: 54 oz)

Mix all these ingredients together with 2 pints of heavy syrup or glucose. Add enough water to make 2 gallons.

Maple Flavor

Fenugreek Extract	2 lb	Propylene Glycol	5 lb
Caramel Color	¼ lb	Water	10 lb

Mix the above ingredients thoroughly.

Mocha Flavor

Coffee Extract	7½ lb	Propylene Glycol	10 lb
Caramel Color	8 lb	Water	22 pt

Mix the above ingredients thoroughly.

Onion Extract

This extract is made from well-peeled onions which must be finely cut. In order that the extract should not be too mucous in character, the onions are first well expressed, the juice being discarded. To every pound of the residue add 2 pounds of alcohol. Let stand 3 days and then express. Treat the residue once again with 2 pounds of 50% alcohol. Then express after 3 days and unite the liquids. To each pound of extract add ¼ dram true volatile mustard oil.

Orange Extract
(95% Alcohol)

Orange Oil	5 oz	Alcohol	95 oz

Peppermint Extract
(92% Alcohol)

Peppermint Oil	8 oz	Alcohol	7 pt

Spearmint Extract
(82% Alcohol)

Spearmint Oil	8 oz	Alcohol	7 pt

Tonka Extract
(50% Alcohol)

Cut Tonka Beans	50 lb	Water	25 gal
Alcohol	25 gal		

Macerate for 1 month.

Pure Vanilla Extract
(39% Alcohol)
Formula No. 1

Cut Mexican Beans	10 lb	Alcohol	30 gal
Cut Bourbon Beans	50 lb	Water	30 gal

Place in a bag.

Pure Vanilla Extract
(48% Alcohol)
(By Maceration)
Formula No. 2

Cut Vanilla Beans	1 lb	Water	3 pt
Alcohol	5 pt		

Let stand for 72 hours.

Pure Vanilla Extract
(50% Alcohol)
Formula No. 3

Vanilla Beans (Cut Very Fine)	14 oz	Distilled Water	½ gal
Cologne Spirit (190 Proof)	½ gal	Cane Sugar	8 oz

Vanilla and Tonka Extract

Cut Vanilla Beans	4 oz	Alcohol	3 pt
Cut Tonka Beans	2 oz	Water	1½ pt
Glycerin	8 oz		

Macerate for 8 days; filter.

Pure Vanilla Flavor

Oleoresin Vanilla	4 oz	Water	To make 1 gal
Alcohol	2 pt		

Dissolve flavoring ingredients completely in the alcohol before adding any water. Filter clear after 2 or 3 days.

Imitation Vanilla Flavor
(Concentrated)

Vanillin	90 oz	Glycerin	20 pt
Coumarin	20 oz	Caramel Color	24 pt
Alcohol	25 pt		

One pint of above added to 2 gallons of warm water will make a good imitation vanilla flavor.

Imitation Vanilla Flavor
(Non-Alcoholic)
Formula No. 1

Vanillin	1 oz	Water	8 pt
Coumarin	⅛ oz	Caramel Color	To suit, add last
Glycerin	10 oz		

Imitation Vanilla Flavor
(Non-Alcoholic)
Formula No. 2

Vanillin	1½ oz	Water	5 pt
Coumarin	½ oz	Caramel Color	To suit
Glycerin	42 oz		

Imitation Vanilla Flavor
(Non-Alcoholic)
Formula No. 3

Ethyl Vanillin	1 oz	Water	3 pt
Coumarin	¼ oz	Caramel Color	To suit
Glycerin	72 oz		

To insure solution of vanillin in above formulas, heat glycerin to temperature not exceeding 140° F. Then add vanillin and coumarin. Stir. The water must not be too cold when added or it may precipitate out some crystals of vanillin.

Vanilla Flavor

10X Concentrated Pure Vanilla	1 lb	Methyl Coumarin	¼ oz
Vanillin	½ lb	Propylene Glycol	7 lb
Coumarin	¼ lb	Caramel Color	½ lb
		Water	30 pt

Add the flavorings to propylene glycol. Mix in the water and caramel color.

Essences

Absinthe

Wormwood Oil	30 oz	Neroli Oil	1 oz
Anise Oil	40 oz	Alcohol (70%)	1660 oz
Fennel Oil	12 oz		

Boonekamp
(May Bitters, Wormwood Bitters)

Calamus Oil	1 oz	Anise Oil	1 dr
Orange Oil	1 oz	Cloves	2 dr
Wormwood Oil	1 dr	Cinnamon Oil	2 dr

Macerate with pumice.

Alcohol	22 oz	Water	22 oz

Brandy

Prune Oil	2 oz	Or	
Butyric Ether	1 dr	Grape Oil	5 oz
Cognac Oil	4 dr	Acetic Ether	4 oz
Wine Ether	1 dr	Tincture of Allspice	1 oz
Alcohol	4 oz	Tincture of Gall	3 oz
		Alcohol	100 oz

Benedictine

Sweet Orange Oil	100 oz	Celery Oil	2 oz
Lemon Oil	85 oz	Cinnamon Oil	4 oz
Angelica Oil	9 oz	Rose Oil	2 oz
Mace Oil	4 oz	Neroli Oil	1 oz
Bitter Almond Oil	4 oz	Alcohol	4009 oz

Caraway

Caraway Oil	4 oz	Alcohol	5 pt 6 oz
Anise Oil	¼ oz	Water for	1 gal
Glycerin	4 oz		

Caraway Cordial

Lemon Oil	¼ oz	Carvol	14¾ oz
Bitter Orange Oil	¼ oz	Alcohol	2 gal
Fennel Oil	¾ oz		

Double Caraway Cordial

Lemon Oil	½ oz	Carvol	11 oz
Fennel Oil	1 oz	Alcohol	2 gal
Star Anise Oil	3½ oz		

Citral

Citral	2 dr	Dilute Alcohol	1 pt

Cognac

Formula No. 1

Amyl Alcohol	10 oz	Raisin Extract	10 oz
Oenanthic Ether	10 oz	Alcohol	100 oz
Alcohol	100 oz	Or	
Or		Cognac Oil	15 oz
Pelargonic Ether	1 oz	Ethyl Nitrate	30 oz
Alcohol	20 oz	Thyme Oil	10 oz
Or		Ethyl Acetate	30 oz
Cognac Oil	1 oz	Vanillin	3 oz
Ethyl Acetate	10 oz	Alcohol	1760 oz

Oil Cognac
Formula No. 2

Tincture of Prunes	480 g	Oil Cognac	28 g
Ethyl Butyrate	21 g	Oenanthic Ether	42 g

Oil of Green Cognac
Formula No. 3

Sebacic Ether	5 g	Cognac Oil	3 g
Pelargonic Ether	2 g	Oenanthic Ether	90 g

Cognac
Formula No. 4

Cognac Ether	650 g	Ethyl Acetate	165 g
Rum Ether	650 g	Oenanthic Ether	5 g
Sweetened Saltpeter		Sugar Color	335 g
Spirit	165 g	Alcohol (90%)	4000 g

Chartreuse

Melissa Oil	6 oz	Or	
Angelica Oil	30 oz	Lemon Oil	50 oz
Clove Oil	6 oz	Sweet Orange Oil	40 oz
Peppermint Oil	40 oz	Neroli Oil	10 oz
Hyssop Oil	6 oz	Angelica Oil	15 oz
Nutmeg Oil	6 oz	Fir Oil	8 oz
Cinnamon Oil	6 oz	Cinnamon Oil	10 oz
Alcohol	2000 oz	Wormwood Oil	5 oz
		Mace Oil	3 oz
		Coriander Oil	5 oz
		Ethyl Butyrate	5 oz
		Rose Oil	1 oz
		Violet Oil	1 oz
		Alcohol	3060 oz

Fruit

Lemon Oil	20 oz	Neroli Oil	1 oz
Orange Oil	30 oz	Benzaldehyde	10 oz
Vanillin	2 oz	Alcohol	1000 oz
Strawberry Essence	10 oz		

Ginger Ale

Ginger Essence	1 pt	Rose Essence	½ oz
Lemon Essence	1 oz	Tincture of Cinnamon	1 dr
Ginger Oil	1 oz	Pineapple Essence	½ dr
Vanilla Extract	1 oz	Capsicum Essence	2 dr

Curacao
(Bitter Orange Peel)

Sweet Orange Oil	4 dr	Alcohol	8 oz
Lemon Oil	2 dr	Water	8 oz
Rose Oil	4 min		

Add 2 ounces bitter orange peel extract to improve.

Curacao Oil

Benzaldehyde	15 g	Petitgrain Oil	75 g
Oil Cassia	30 g	Orange Oil	650 g
Geraniol Extra	30 g	Lemon Oil	150 g
Linalyl Acetate	50 g		

Gin

Juniper Oil	1 oz	Caraway Oil	¼ oz
Nutmeg Oil	1 dr	Alcohol	112 oz
Caraway Oil	6 min	Or	
Fusel Oil	10 min	Juniper Oil	10 oz
Alcohol	16 oz	Coriander Oil	1 oz
Or		Caraway Oil	1 oz
Juniper Oil	8 oz	Bitter Almond Oil	1 oz
Glycerin	8 oz	Alcohol	115 oz
Orange Oil	¼ oz		

Cordial Gin

Juniper Oil	5 oz	Alcohol	1 gal
Bitter Almond Oil	2 drops		

Holland Gin
Formula No. 1

Juniper Oil	5 oz	Acetic Ether	5 drops
Caraway Oil	2 drops	Alcohol	1 gal

Holland Gin Oil
Formula No. 2

Lemon Oil	3 g	Rosemary Oil	16 g
Anise Oil	3 g	Coriander Oil	13 g
Angelica Root Oil	16 g	Juniper Berry Oil	940 g
Rectified Fusel Oil	12 g		

Old Tom Gin Oil
Formula No. 3

Coriander Oil	270 g	Caraway Oil	20 g
Rectified Anise Oil	80 g	Angelica Root Oil	15 g
Rectified Juniper Berry Oil	610 g		

Port

Acetic Ether	1 oz	Tincture of Kino	4 oz
Grape Essence	4 oz	Raspberry Essence	8 oz
Vanilla Essence	4 oz		

Rose

Rose Oil (Rose Geranium Oil)	2 dr	Alcohol	1 pt

Rosoglio

Bitter Almond Oil	40 oz	Neroli Oil	½ oz
Anise Oil	25 oz	Tincture of Ambra	4 oz
Rose Oil	5 oz	Alcohol (80%)	1520 oz
Cinnamon (Ceylon) Oil	2 oz		

Rum

Butyric Ether	1 oz	Alcohol	4 oz
Nitrous Ether Spirit	1 oz		

Rum

Rum Ether	200 g	Tincture of Peruvian Bark	15 g
Ethyl Acetate	40 g	Orange Flower Water	100 g
Tincture of Cinnamon	10 g	Woodruff Essence	30 g
Tincture of Catechu	10 g	Butyric Ether	20 g
Tincture of Vanillin	10 g	Alcohol (90%)	650 g
Ethyl Formate	75 g	Rum	1000 g
Tincture of Angelica Root	2 g		

Scotch Whiskey

Creosote (Dissolved in ½ oz Alcohol)	2	drops	Or		
			Glycerin	1	oz
Acetic Acid	20	drops	Caramel	30	drops
Alcohol	4½	pt	Butyric Ether	10	drops
Water	3½	pt	Oenanthic Ether	10	drops
			Formic Ether	10	drops
			Creosote	2	drops
			Alcohol (80%)	1	gal

Scotch Whiskey Oil

Rectified Fusel Oil	510 g	Bitter Almond Oil	20 g
Cade Oil	84 g	Sweet Almond Oil	20 g
Ethyl Butyrate	445 g	Guaiacum Oil	10 g

Sherry

Nitrous Ether Spirits	15 oz	Tincture of Orange	1 oz
Oenanthic Ether	1 oz		

Slivowitz

Cognac Oil	1½ oz	Benzaldehyde	12 oz
Fusel Oil	2½ oz	Alcohol	2½ gal

Spice

Cinnamon Oil	20 oz	Lemon Oil	60 oz
Clove Oil	30 oz	Sweet Orange Oil	60 oz
Bitter Almond Oil or Benzaldehyde	60 oz	Alcohol	4000 oz

Rum Punch

Rum Essence	3½ oz	Citric Acid Solution	1½ oz
Jamaica Rum	1½ oz	Caramel Color (1:1)	8 oz
Lemon Essence	1 oz	Alcohol	160 oz
Vanilla Essence	½ oz		

To make a punch add 2 ounces of the above essence to 1 gallon cherry syrup. Dilute 1 quart of this syrup with 2 quarts of water and 1 quart of alcohol.

Rye

Caramel Color	¼ oz	Or	
Glycerin	¾ oz	Amyl Alcohol	10 oz
Wintergreen Oil	7 drops	Cognac Oil	1 oz
Butyric Ether	10 drops	Ethyl Acetate	6 oz
Formic Ether	10 drops	Anise Oil	1 oz
Vanilla Essence	20 drops	Fennel Oil	½ oz
Alcohol (60%)	1 gal	Coriander Oil	1½ oz
		Alcohol	400 oz

Tokay

Tincture of Civet	2½ dr	Tincture of Carob	16 oz
Pineapple Essence	8 oz	Grape Essence	8 lb
Raspberry Essence	8 oz		

Walnut

Coumarin	8 oz	Alcohol	5 pt 4 oz
Rose Oil	40 min	Water for	1 gal
Lovage Oil	½ oz	Caramel Color	To suit

Whiskey

Ethyl Acetate	250 oz	Anise Oil	1 oz
Ethyl Nitrate	200 oz	Juniper Oil	2 oz
Caraway Oil	1 oz	Alcohol	1000 oz

Rock and Rye Whiskey Essence

Rectified Grain Fusel Oil	340 g	Coumarin	15 g
Green Wine Lees Oil	12 g	Raisin Wine Essence	580 g
Peru Balsam	12 g	Peach Essence	8 g
Jamaica Rum Essence	12 g	Bitter Orange Extract	50 g
Vanillin	6 g	Cinnamon Oil	2½ g
Ethyl Acetate	12 g	Clove Oil	2½ g

Waldmeister

Tincture of Coumarin	150 oz	Currant Essence	20 oz
Strawberry Essence	30 oz		

Vermouth

Wormwood Oil	25 oz	Calamus Oil	2 oz
Ginger Oil	10 oz	Bitter Almond Oil	½ oz
Coriander Oil	2 oz	Alcohol (90%)	800 oz

Aromatic Waters

Cherry Laurel Water

True Fruit Extract of Wild Cherry	8 oz	Balance Water for	1 gal
Bitter Almond Oil	1 dr	Filter through Magnesia	1 oz

Orange Flower Water

Orange Flower Essence	8 oz	Filter through Magnesia	1 oz
Balance Water for	1 gal		

Rose Water

Rose Essence	4 oz	Balance Lukewarm Water for	1 gal
		Filter through Magnesia	1 oz

Peppermint Water

Peppermint Oil	2 dr	Filter through Magnesia	½ cz
Balance Water for	1 gal		

Essences for the Above Aromatic Waters

Orange Flower Essence

Neroli Oil	2 dr	Alcohol	1 pt

Rose Essence

Rose Oil *	2 dr	Alcohol	1 pt

Tinctures

Tincture of Benzoin

Powdered Gum Benzoin	6 oz	Alcohol	2 pt

Macerate for 7 days in a closed vessel. Filter and add enough alcohol to make the tincture measure 2 pints.

Use: Add a few drops of this tincture to an orange extract to make it cloudy.

Tincture of Capsicum

Oleoresin Capsicum	1 oz	Alcohol	1 pt

Let stand and filter.

Tincture of Celery

Macerate 1 pound of the sliced and cut plant, or of the bruised seeds, with 5 pints of diluted alcohol. The seeds must be bruised in a mortar. Percolate, if desired, otherwise filter.

Tincture of Coffee

Powdered Coffee	1 lb	Diluted Alcohol	3–5 pt

Macerate for several days: filter.

Tincture of Curacao

Powdered or Sliced Peel of Curacao	4 oz	Diluted Alcohol	20 oz

Macerate 24 hours in a stoppered bottle. Percolate until 1 pint is obtained.

* Rose geranium oil, being cheaper than the natural rose oil, may be used.

Tincture of Musk

Artificial Musk Powder	2 oz	Alcohol	1 gal

Tincture of Kino

Take 100 grams kino to make 1000 cubic centimeters. Place the kino in a capacious flask and pour on it 500 cubic centimeters of boiling water. Agitate the mixture thoroughly and heat it for 1 hour on a water bath, shaking it frequently. Allow the liquid to cool. Add enough recently boiled water to make the product measure 500 cubic centimeters and add 500 cubic centimeters of alcohol. Stopper the flask, set it aside in a cool place for 24 hours and then decant the mixture through cheesecloth. Preserve in a cool, dark place in tightly stoppered bottles.

Tincture of Kino

Kino in Coarse Powder	1 oz	Alcohol (60%)	5 oz

Macerate and filter.

Tincture of Orris Root

Powdered Orris Root	15 lb	Water	1 gal 7 pt
Alcohol	3 gal 1 pt		

Macerate and filter.

Tincture of Peru Balsam

Macerate 1 ounce of the balsam with 10 ounces of diluted alcohol. The latter will only dissolve a small proportion, however, the undissolved portion is reserved for the next tincture. Filter.

Tincture of Rhatany

Krameria Root in No. 40 Powder	1 oz	Alcohol (60%)	5 oz

Macerate and filter or percolate.

Tincture of Tolu

Tolu	1¾ oz	Alcohol (60%)	16 oz

Agitate occasionally until dissolved. Filter.

Tincture of Styrax

Liquid Styrax	1 oz	Alcohol	½ pt

Digest by agitation for a week and then decant the clearer portion.

True Fruit and Flavor Syrups and Fountain Supplies

In each gallon of fruit juice, dissolve 15 pounds of sugar and clarify with isinglass solution or by settling.

To each gallon add 1 ounce of powdered citric acid and sufficient color to produce the desired shade. One-tenth of 1 per cent of sodium benzoate may also be added.

Orange Syrup

Cut the oranges in half, express and strain the juice. Mix with an equal bulk of water to each gallon. Add 8 ounces of citric acid and 15 pounds of sugar. Do not heat.

If the syrup is intended for immediate consumption it may be flavored with 1 ounce of soluble orange extract, or by paring off the outside yellow rind and rubbing it in a mortar with a portion of the sugar.

Color slightly yellow with certified food orange color.

Raspberry, Strawberry and Similar Fruit Syrups

Crush the raspberries and set in a warm place until fermentation commences. Then express juice and make in the regular way, as orange syrup.

Lemon Syrup

Make in same way as orange syrup.

Vanilla Syrup

Make simple syrup by using 15 pounds of sugar to 1 gallon of water. Heat and clarify. When cold, add 2 ounces of pure vanilla extract to each gallon and tint slightly with caramel.

Simple Syrup

Take 8½ gallons of warm water, put in a stone jar, add 100 pounds of sugar, stir up and let stand overnight.

Orangeade Syrup

Syrup	1 gal	Fruit Acid Solution	2 fl oz
Orange Emulsion	1 oz	Color	1 fl oz
Benzoate of Soda Solution	1 fl oz		

Orange Syrup

Syrup	1 gal	Fruit Acid Solution	6 fl oz
Orange Extract	4 fl oz	Color	½ fl oz
Benzoate of Soda Solution	1 fl oz		

Lemon Syrup

Syrup	1 gal	Fruit Acid	6 oz
Lemon Extract	4 fl oz	Color	1 oz
Benzoate of Soda Solution	1 oz		

Lime Syrup

Syrup	1 gal	Fruit Acid	6 fl oz
Lime Extract	4 fl oz	Green Color	1 oz
Benzoate of Soda Solution	1 fl oz		

Strawberry Syrup

Syrup	1 gal	Strawberry Extract (True Fruit)	4 fl oz
Benzoate of Soda		Fruit Acid	2 fl oz
Solution	1 fl oz	Strawberry Color	3 fl oz

Anise Syrup

Syrup	1 gal	Color	Delicate red
Oil of Anise	8 drops		

Banana Syrup

Syrup	1 gal	Citric Acid	½ oz
Banana Extract	2 oz	Color	Light yellow

Blackberry Syrup

Syrup	3 qt	Citric Acid	1 oz
Blackberry Juice	1 qt		

Coffee Syrup

Syrup	1 gal	Vanilla Extract	1 oz
Coffee Fine Powder	2 oz	Caramel Color	Enough

Bring the syrup to a boil. Add the coffee and boil for 5 minutes; then add vanilla and color.

Raspberry Syrup

Make in same way as strawberry syrup.

Cherry Syrup

Syrup	1 gal	Benzoate of Soda Solution	1 fl oz
True Fruit Cherry Extract	4 fl oz	Fruit Acid Solution	3 fl oz
Synthetic Cherry F. Oils	1½ fl oz	Cherry Color Solution	2 fl oz

Ginger Ale Syrup

Syrup	1 gal	Capsicum Tincture	10 min
Ginger Extract	4 fl oz	Caramel Color	2 fl dr
Saturated Water Solution of Tartaric Acid *	1 fl oz		

* See under Sundries.

Pineapple Syrup

Syrup	1 gal	Fruit Acid Solution	3 fl oz
True Fruit Extract	4 fl oz	Color Solution	2 fl oz
Benzoate of Soda Solution	1 fl oz		

Caramella Syrup

Syrup	1 gal	Benzoate of Soda Solution	1 fl oz
Caramel Flavor	4 fl oz	Caramel Color	1 pt

Imitation Maple Syrup (Breakfast Syrup)

Syrup	1 gal	Mapleine	2 fl oz

Vanilla Syrup

Syrup	1 gal	Fruit Acid Solution	1 fl oz
Pure Vanilla Extract	2 fl oz	Caramel Color	1 fl oz
Benzoate of Soda Solution	1 fl oz		

Imitation Vanilla Syrup

Syrup	1 gal	Fruit Acid Solution	1 fl oz
Vanilla Flavor Compound No. 1	10 fl oz	Caramel Color	1 fl oz

Root Beer Syrup

Syrup	1 gal	Benzoate of Soda Solution	1 fl oz
Root Beer Extract	4 fl oz	Caramel Sugar	12 fl oz

Chocolate Syrup
Formula No. 1

Cocoa Powder	2½ lb	Water	2½ gal
Granulated Sugar	35 lb	Condensed Milk	5 lb

Bring to a boil. When cool add:

Extract Vanilla	5 fl oz

Formula No. 2

Bitter Chocolate	2 lb	Sugar Syrup (30° Bé)	6 lb

Melt the chocolate, then add the sugar syrup and boil down to desired thickness. Add 1/10 of 1 per cent sodium benzoate, if desired.

Formula No. 3

Syrup	4 qt	Vanilla Extract	½ oz
Cocoa Powder	8 oz		

Bring 1 quart of syrup to a boil and then add 8 ounces of cocoa. Stir for 10 mnutes over a slow fire. Remove from fire and add 3 quarts of syrup and vanilla extract.

Cocoa Paste

Cocoa paste is a very thick product and is a base for ice cream topping, fudge or cocoa syrup.

Sugar Syrup (34° Bé)	1 gal	Sodium Benzoate	4 g
Cocoa Powder	3 lb		

Warm the sugar and cocoa powder up together, rubbing the powder into a smooth paste. Heat as high as 190° F and add the sodium benzoate, dissolved in a small amount of water. Be sure there are no lumps.

Chocolate syrup is made by adding 5 gallons of 34° Bé sugar syrup to 1 gallon of the above formula.

Compound Sugar—Maple Syrup

Sugar Syrup	85 pt	Vermont Maple Syrup	15 pt

Corn Syrup—Cane Sugar Blend

Corn Syrup (39° Bé)	50 pt	Caramel Color	To suit
Sugar Syrup	50 pt		

Cane Sugar—Invert Syrup Blend

Invert Syrup (37° Bé)	50 pt	Caramel Color	To suit
Sugar Syrup	50 pt		

Cane Sugar and Maple Syrup Blends

15% Maple Blend

Maple Syrup	15 gal	Water	38 gal
Sugar	617 lb		

17½% Maple Blend

Maple Syrup	17½ gal	Water	37 gal
Sugar	600 lb		

20% Maple Blend

Maple Syrup	20 gal	Water	36 gal
Sugar	580 lb		

25% Maple Blend

Maple Syrup	25 gal	Water	34 gal
Sugar	545 lb		

Mix sugar with water and heat until dissolved. Add maple syrup and bring the batch to the boiling point (219° F), agitating it thoroughly. Pack containers at 170 to 180° F.

Pineapple Fountain Syrup

Pineapple Juice	⅓ gal	Sodium Benzoate (Dissolve in Juice)	3 g
Simple Sugar Syrup	⅔ gal		

Crushed Orange Fruit Syrup

Orange Pulp	50 lb	No. 100 Powdered Pectin	2 oz
Cane Sugar	50 lb	Citric Acid	4 oz
Sodium Benzoate	1½ oz	Water	1 gal

Prepare the above recipe in the same manner as for crushed pineapple fruit.

Orange or Lemon Fruit Syrup

		Water	5 pt
Lemon, Orange or Other Fruit Pulp	1 qt	Citric Acid	½–1 oz
Sugar	6½ lb	Pectin	1 oz

Thoroughly mix 1 pound of sugar with 1 ounce of pectin. Bring the 5 pints of water to a boil and add slowly, while stirring, the mixture of pectin and sugar. Then boil vigorously for 1 minute. If artificial color is desired, it may be added at this point. Now add the balance (5½ pounds) of sugar and cook until completely dissolved. Allow to cool to 180° F and add the citric acid, previously dissolved in a little water. Add the fruit pulp and stir slowly until cool. If some additional flavor is desired, it is added at this point. If a preservative is required, 3.6 grams of benzoate of soda is stirred in. Stir the finished syrup slowly while bottling. It is advisable to shake each bottle the next day before packing for shipment. The pulp will now remain in suspension for long periods.

Cherry Fountain Syrup

Sugar	7	lb	Citric Acid	¾ oz
Water	5	pt	Amaranth Food Color	Enough to color
True Fruit Cherry and Wild Cherry Extract	4	oz		

This formula will make 1 gallon of syrup.

Invert Sugar Syrup

Sugar	100	lb	Tartaric Acid	50 g
Water	44	lb		

Mix these ingredients together and boil for 30 to 35 minutes. Stop boiling when temperature reaches 230° F. This syrup has a sweetening value at least 20% greater than a similar cane sugar syrup.

Crushed Strawberry Fruit Made with Cold Pack Fruit

Fruit and Juice	45	lb	Sodium Benzoate	1⅕ oz
Sugar	55	lb	Agar-Agar	9 oz
Citric Acid	2	oz		

(1) Soak agar-agar overnight in 1 gallon of water. Then cook up with 5 pounds of sugar to 220° F. Add to (2) on cooling table and mix in with paddle.

(2) Cook juice and 50 pounds of sugar to 220° F. Then add the fruit and cook to 223° F. Dissolve sodium benzoate in 2 ounces of warm water and add to kettle, with stirring. Then transfer to the cooling table.

Preparation of Red Strawberry Color

Amaranth—Certified Food Color	2 oz	Sodium Benzoate	3 g
Ponceau 3R—Certified Food Color	1 oz	Citric Acid	½ oz
		Enough Water to Make	1 gal

Dissolve the acid and sodium benzoate separately in small amounts of water and add separately to the gallon.

Strawberry Fountain Syrup

Strawberry Juice (Drained from Cold Pack)	⅓ gal	Sodium Benzoate (Dissolve in Juice)	5 g
Simple Sugar Syrup	⅔ gal	Citric Acid (Dissolve in Juice)	¾ oz
Strawberry Color	½ oz		

Add the sodium benzoate and citric acid solutions separately to the gallon.

Walnuts in Syrup for Use at Fountain

Granulated Sugar	100 lb
Water	7 gal
Walnuts (Usually ¼ in.)	Pack in jars

Boil sugar and water to 221° F. Then cover walnuts with this hot syrup. Place cover on jar and pasteurize for 30 minutes.

Avoid iron contamination.

Coffee Syrup

Drip Coffee	10 pt	Sodium Benzoate	1 oz
Granulated Sugar	25 lb		

Dissolve the sodium benzoate in the hot drip coffee. Now add the sugar. Stir until solution is complete.

Chocolate Syrup
(Double Strength)

Cocoa Powder Natural	14	lb
Cocoa Powder Dutched	1	lb
Granulated Sugar	65	lb
Water	35	pt
Vanillin	2½	g
Oil Bitter Almonds	2	drops
Sodium Benzoate	4	oz
Water	To make 10	gal

Mix the cocoa powder with part of the sugar. Bring the water to a boil. Now slowly add the cocoa powder, mix and agitate until smooth. Then add balance of sugar. Heat around 200° F for 30 minutes. Add flavor and benzoate. When the mixture is cold, add enough water to make 10 gallons. One gallon of this added to 1 gallon of simple syrup will make 2 gallons of regular chocolate syrup.

Punch Syrup

Strawberry Juice	7½	gal	Sodium Benzoate	6	oz
Grape Juice	3⅗	gal	Citric Acid	1	lb
Loganberry Juice	4	gal	Water	5	gal
Pineapple Juice	6	gal	Grape Color	Enough	
Cherry Juice	2	gal	Sugar	30	lb

Mix all ingredients together in the cold.

Crushed Pineapple Fruit

Crushed Pineapple (No. 10 Tins)	30 lb		Sodium Benzoate	18 g
Sugar	15 lb		No. 100 Powdered Pectin	3 oz
			Water	1 gal

Mix the powdered pectin with 5 pounds of sugar. Add this to the boiling water, while stirring. Then add balance of the sugar and cook to 220° F. Now add crushed pineapple and cook to 225° F. Shut off steam and add the sodium benzoate previously dissolved in 5 ounces of water. Stir around and then transfer immediately to the cooling table.

Crushed Fruits for the Fountain

Peel, pit, and prepare the fruit.

Pass through a meat chopper having a perforated plate with medium-sized holes.

To each gallon of this add:

Sugar	8 lb	Tartaric Acid	1 oz

Place in kettle, heat to 190° F, and stir until sugar is dissolved.

To each 3 gallons of this add:

Japanese Gelatin (Agar-Agar)	2 oz

which has been soaked in cold water for 6 hours and dissolved by heat in a little water.

Stir well together. Add color to produce desired shade.

To each 100 pounds of this product add:

Preservative (Benzoate of Soda Solution)	2 oz

Nut Sundae

Chopped Walnuts, Pecans, or Almonds	Sugar Syrup (34° Bé)
	Maple or Walnut Flavor

Cover the nuts with a hot 34° Bé sugar syrup and let soak overnight. In the morning fill ½-gallon jars with the nuts, cover with a 34° Bé syrup. Add maple or walnut flavor and then sterilize 30 minutes at 180° F. Sodium benzoate, ⅒ of 1%, and caramel color may be added, if desired.

Lemonade Powder for Soft Drinks

Sugar	86 oz	Citric Acid	⅒ oz
Dry Borden's Lemon Powder	14 oz		

Color with a yellow certified food color.

The above powders are mixed and colored. Four ounces of

above powder, when mixed with a pint of cold water, will make delicious lemonade.

Lemon Beverage Powder

Sugar	64	lb	Citric Acid	2½	lb
Salt	2	oz	California Lemon Oil	2	oz
Powdered Lemon Juice	2½	lb	Lemon Coloring	As desired	

Place 48 pounds of sugar, the salt, powdered lemon juice and citric acid in a mixing bowl and blend thoroughly. Blend the lemon oil and coloring with 16 pounds of sugar and add this to the first mixture. Additional sugar may be used, if desired. Put the entire mixture through a fine sieve.

To serve, make up in the proportion of 1 pound of mixture to 1 gallon or more of water.

Orange Powder Beverage

Sugar	64	lb	Citric Acid	2	lb
Salt	2	oz	California Orange Oil	2	oz
Powdered Orange Juice	2½	lb	Orange Shade	3	oz

Place 48 pounds of sugar, salt, powdered orange juice and citric acid in machine bowl, and blend thoroughly. Blend 16 pounds of sugar, orange oil and shade carefully, and add to first mixture. Use additional sugar as required.

Mix well and sift through fine sieve. One pound produces 1 gallon or more of beverage.

Flavoring Emulsions

Lemon Emulsion

Gum Tragacanth	18 oz	Water	4	gal

Place the gum tragacanth in a kettle similar to a baker's mixing machine. Pour the water over it. Stir well with an egg whip until all the lumps have disappeared. Let stand overnight.

Next morning, squeeze the gum solution through a cheesecloth bag. Sugar bag may also be used.

Add to the homogeneous gum solution:

Glycerin	1 qt

Place the kettle under the machine and allow to mix for 5 minutes at low speed.

Now put the machine in second speed and gradually add:

Lemon Oil	4 qt

After 1 quart of the oil has been added, it is advisable to put the machine in third speed, otherwise the oil will not be taken up by the gum solution. When all the oil has been added, stop the machine, scrape the walls of the kettle with a long-bladed spatula and add 5 ounces of benzoate of soda in solution to prevent fermentation. Start the machine again in third speed for about 5 to 10 minutes.

The emulsion is then ready to be placed in bottles.

Yield is about forty-three one-pound bottles.

Lemon Oil Emulsion

Gum Arabic	13 oz
Terpeneless Oil of Lemon	20 oz
Lemon Oil	20 oz
Glycerin	40 oz
Water	To make 10 gal

Mix the gum arabic and the glycerin. Then mix in the terpeneless oil and the lemon oil. Add the water slowly with good stirring. Beat intermittently until homogeneous. Then pass through an homogenizer.

Orange Oil Emulsion

Gelatin	4 oz	Terpeneless Oil of Orange	20 oz
Water	16 lb	Orange Oil	20 oz
Cane Sugar	24 lb		
Invert Sugar	60 lb		

Dissolve the gelatin in the water. Add the cane sugar and heat until dissolved. Then add the invert sugar and mix well; homogenize.

Orange Emulsion

Make in same way as lemon emulsion.

The other flavored emulsions are generally prepared in smaller quantities. This prevents too-long storing, and a fresh product is always on hand.

Formula for Smaller Batches
(About Sixteen One-pound Bottles)

Gum Tragacanth	9 oz	Glycerin	3 pt
Water	2 gal	Oil	1 pt

Proceed according to directions for lemon emulsion.

Rye Bread Emulsion

Caraway Oil

Proceed according to directions for lemon emulsion.

Cinnamon Emulsion

Cassia Oil

Proceed according to directions for lemon emulsion.

Strawberry Emulsion

Synthetic Strawberry Oil

Proceed according to directions for lemon emulsion.

Formula for 1-Pound Bottles
Almond Emulsion

Gum Tragacanth	5 oz 5 dr	Glycerin	10 oz
Water	5 qt	Bitter Almond Oil	10 oz

Proceed according to the directions for lemon emulsion.

Vanilla Emulsion

Vanillin	3 oz	Coumarin	½ oz

Dissolve these in 3 pints of glycerin. Then emulsify with 1 pint of sesame oil.

Delacator Emulsion

Vanillin	1/4 oz	Almond Oil	1 oz
Coumarin	1/16 oz	Rose Oil	1/4 oz
Lemon Oil	1 oz	Strawberry Oil	1/4 oz
Orange Oil	2 oz	Anise Oil	1/16 oz

Dissolve these in 3 pints of glycerin. Then emulsify with 1 pint of sesame oil.

Root Beer Emulsion

Gum Arabic	17 lb	Water	6½ gal

Heat and stir until dissolved. Filter through cheese-cloth.

Formula No. 1
(To Make 10 Gallons of Concentrate)

Synthetic Wintergreen Oil	1 pt	Vanillin	4 oz
Sassafras Oil	1 pt	Coumarin	1 oz

Formula No. 2
(To Make 10 Gallons of Concentrate)

Synthetic Wintergreen Oil	2 qt	Clove Oil	1 pt
Sassafras Oil	1 qt	Cassia Oil	1 pt

Emulsification, or absorption of the oils in the gum solution, can be accomplished simply by gradually pouring the oils into the gum solution, while it is being vigorously agitated. It should be agitated for at least 10 minutes, and, as this process is going on, you will note the tendency of the gum solution to thicken. If you have an emulsifier, so much the better. An emulsifier, however, is not essential, as a good stirring with a wooden paddle or a large size cream whipper will do the work. While you are mixing this solution, about 1 gallon of sugar coloring, caramel coloring, should be added to give the concentrate a dark color. The resulting product is then made up with water to make exactly 10 gallons and then given another thorough stirring before being placed in a container. One gallon of this product will flavor 20 barrels of root beer.

Transparent Lemon Oil Emulsion

Invert Sugar Syrup (80%)	60 lb	Water	16 lb
Medium Fine Sugar	24 lb	Gelatin	4 oz

The gelatin is dissolved in water. Corn syrup is added, and the mixture heated until solution is complete. The invert syrup is then added, and the whole product is thoroughly mixed. The refractive index of this solution should be approximately 1.4720. In the preparation of the emulsion, 20 ounces of terpeneless and 20 ounces of regular lemon oil are mixed with enough of the above solution to produce 10 gallons. This mixture is then homogenized and left standing for 18 hours to allow entrapped air bubbles to rise to the surface. Glass containers are filled from the bottom outlet of the emulsion container and tightly stoppered. The resultant emulsion will be practically transparent.

Almond Emulsion

Gum Tragacanth	1/3 lb	Oil of Bitter Almond	1¾ lb
Propylene Glycol	2 lb	Water	12 pt

Add the oil to the propylene glycol and mix. Then add the gum and the water, and mix to a uniform emulsion.

Lemon or Orange Oil Emulsion

Gum Tragacanth	½ lb	Lemon or Orange Oil	2½ lb
Propylene Glycol	3 lb	Water	17 pt

Add the oil to the propylene glycol and mix. Then add the gum and the water, and mix to a uniform emulsion.

Mace or Cinnamon Emulsion

Gum Tragacanth	¾ lb	Oil of Mace or Cinnamon	2¼ lb
Propylene Glycol	3 lb	Water	17 pt

Add the oil to the propylene glycol. Then add the gum and the water and mix to a uniform emulsion.

Oil of Bitter Almond Emulsion

Gum Arabic	10 oz	Cane Sugar	8 oz
Bitter Almond Oil	2 oz	Water	120 oz

Mix the sugar and gum together. Then add it to the water, which is agitated by means of a rapid mixer. Allow to mix until a uniform emulsion results. Then add the oil slowly until completely emulsified.

True Fruit Extracts

True fruit extracts made by pressing the fruit, peelings and kernels, and then distilling with diluted alcohol, will result in an extract of richer aroma and flavor than if made from the fresh fruit alone. Add to the pressings 2 to 3 times their weight of 60 to 70% alcohol. Allow to stand overnight and distill over the following morning. Discontinue the distillation when the distillate shows about 40% alcohol. Do not distill below this percentage.

Apple

Squeeze the apples in a stone mill to pulp and press them afterwards in a fruit press.

Yield is 60 to 75% juice.

To 84 litres of strained apple juice add 16 litres of alcohol or place 10 kilograms of fresh apple peelings in 16 litres of alcohol. Allow to stand for 3 days, then press.

To the press residue add 20 litres of apple juice and allow to stand for 2 to 3 days to recover the alcohol remaining in the residue. The whole makes about 100 litres.

Apricot

Press apricots which have been previously freed from stones.

Yield is 70% juice.

To 84 litres of strained juice add 16 litres of alcohol.

Blackberry

Press only full ripe berries.
Yield is 75 to 90% juice.
To improve aroma, add 5% sugar and let ferment. Strain.

Bilberry

Press the berries between layers of willows.
Yield is 70 to 90% juice.
Add 4 to 5% sugar and let ferment a short time. Add a little alcohol.

Cherry

Use either sour cherries or a mixture of sweet and sour, but the sour should predominate.

Grind the cherries with the kernels and allow the juice to stand for 12 hours. Push the layer forming on top back into the liquid. Strain and add alcohol.

Currant

Press the fruit with the stems.
Yield is:

Red berries	75 to 80% juice
White berries	65 to 75% juice
Black berries	65 to 75% juice

The black juice must be fermented.

To the red berry juice add 10% of the black juice to obtain a more powerful and wine-like aroma.

Lemon

Press the lemons and clarify the juice at first by filtering through felt. Further clarify by adding 10 grams of dry albumen dissolved in 250 grams for each 10 litres of juice.

Heat juice to 65° C.

Filter through asbestos (Seitz Filter style).
Remove bitter taste by charcoal treatment.

Peach

Press the fruit.
Yield is 65% juice.
There is little aroma.
Aroma can be increased by adding other fruit juices.

Orange

Use only hard and sour oranges.
Press and strain.
Add a little alcohol.

Raspberry

Press the fruit.
Yield is 65 to 90% juice.
To the juice add 3% of sugar and let ferment at 18° C.
Duration of fermentation is 3 to 6 days.
Push the layer forming on top back into liquid.

To test for completion of fermentation, filter a small quantity of the juice and add the same volume of alcohol. After ½ hour standing, no cloudiness should appear, otherwise ferment longer.

Add a little alcohol to the strained juice.

Strawberry

Select the fruit carefully, using fresh fruit only. Press between willows.

Yield is 70 to 85% juice.

Add to the residue 18 litres of alcohol. Let stand for 1 day. Then decant and add to the juice.

Red Whortleberry

Press between willows.

Yield is 80% juice.

Let ferment.

The juice is used to increase the aroma of weak fruit juices.

True Fruit Aromas (Concentrates)

Extract the fruit with alcohol several times.

Shake out with pure petroleum ether, then separate and distill off the ether. The residue is the natural fruit aroma together with wax. Chill, and the wax will separate.

Press the fruit. Place in a percolator cask and cover with alcohol.

After 24 hours, allow percolation to proceed, adding alcohol until 25 gallons are obtained from 100 pounds of fruit. This is percolate No. 1.

Continue percolation until a second 25 gallons are obtained. This is percolate No. 2.

Distill the residue with percolate No. 2 and use this for treating another 100 pounds of fruit.

Percolate No. 1 is shaken out with one-tenth its volume of petrol ether. Separate. Recover alcohol. Distill off ether, using column. Dissolve the remaining oil in 5 times its quantity of alcohol. Freeze out the wax. Filter.

Strength: 100 times stronger than ordinary essence.

Use: 1 ounce to 50 gallons of syrup.

Note: Fruit juice may also be treated this way. If fruit residue and juice are worked up, 3 tons of berries yield 2 pounds of oil and 10 pounds alcoholic concentrated fruit aroma. The solvents, alcohol and ether, can be used over again.

Manufacture of Dry Shredded Coconut

The preparation of dry or desiccated coconut for domestic confectionery and bakers' consumption has developed into a very large industry. Shiploads of coconuts arrive daily from the West Indies to be manufactured into a highly commercialized product. The coconuts come packed in burlap bags, containing anywhere from 80 to 120 fruits depending on their size. When the coconuts arrive at the factory, the nuts are placed in steam cabinets and subjected to a temperature of 212° F for ½ hour. The object in placing these nuts in the steam cabinet is to loosen up the hard shell from the meat. The nuts are opened by means of a small tool similar to an oyster knife. The nut meat itself is covered with a tough brown skin which is removed by skilled workers. The nut meats are then thoroughly washed by passing them under water sprays and are then conveyed to the shredding machines. A belt conveyer serves the battery of shredders, carrying the product to a galvanized iron tank truck. In this way the product is relayed to the sweetening operation nearby.

Rotating copper drums are used for incorporating the sugar. The shredded coconut is tumbled around for 20 to 30 minutes. Refined cane sugar is used for sweetening. A small quantity of salt is added. Glycerin is incorporated to render the shreds less brittle, and, when the tumbling is completed, the shredded coconut is transferred over the driers. The product enters the driers with about 45% moisture. It is warmed and dried by addition of conditioned air and finally cooled. The maintenance of a proper moisture content is very important to assure no infestation with insects. After the drying operation, the product is screened and separated into four grades, differentiated by the length of the shreds. The product is then piled up in bins for blending and equalizing the moisture content. Desiccated coconut is marketed in barrels, tins and cardboard boxes, and, because of its high oil content, glassine paper liners should be used.

Beverages

The essential principles are extracted by:

1 Infusion with hot spirit, if the aromatic principle is not too volatile, the ingredients being first moistened with boiling water

2 Maceration with cold spirits, which may require hours, days, or even months, according to the solubility of the essential principles

3 Digestion, or prolonged infusion, or maceration, in hermetically closed vessels, when the aromatic principles are extracted only with difficulty by cold alcohol

4 Distillation, when only the natural essential oils or other volatile flavoring constituents are required

The following plants should always be distilled by steam: wormwood, anise, balms, caraway, citronella, fennel, juniper, hyssop, lavender, melilot, mint, orange, rose, sage and thyme.

Aromatic tinctures are prepared by filtration or percolation.

Percolation

Moisten the powdered drug or mixed drugs with a sufficient quantity of the menstruum to render it evenly and distinctly damp. Transfer it to a percolator and, without pressing the powder, allow it to stand well covered for 6 hours. Then, unless otherwise stated, pack it firmly and pour on enough of the menstruum to saturate the powder and leave a menstruum above it. Close when dripping begins. Cover the percolator. Macerate for 24 hours. Let drip slowly.

Clarification or Collage

It is best to avoid this with fining materials but often it is necessary. The principal materials and the proportions used are as follows:

Albumen (White of Egg)
  1 egg for every 10 litres
Gelatin
  30 grams per 100 litres
Isinglass
  10 grams per 100 litres
Milk
  1 litre boiled in 100 litres with 15 grams Alum
Animal or Vegetable Carbon

Tannin and Gelatin Combinations
Bentonite
Alum and various filtering powders are also employed.
For colored liquids, filtration is preferable to collage.

Bitters

They are prepared by maceration without distillation and are considered to be tonic and stomachic and to improve the appetite when taken in moderation.

A bitter made according to the following formula, which represents the proportions necessary to make 20 gallons, may be considered typical:

Gentian	6	lb	Juniper Berries	1	lb
Cinnamon	5	lb	Cloves	½	lb
Caraway	5	lb			

Macerate in 7 gallons of 60% alcohol for 10 to 12 days. Strain and filter. Add syrup consisting of 10 pounds of sugar to 13 gallons of water and color with cochineal.

English Bitters

Outer Rind of Dried Orange Peel	5 oz	Carduus Benedictus	8 oz
Cinchona Bark	6 oz	Centaury	8 oz
Gentian	6 oz	Wormwood	8 oz
		Orris Root	4 oz

Digest with 12 gallons of proof spirits and 12 pounds of sugar.

Orange Bitters

Ground Orange Peel	5	lb	Saccharin (1:300)	¼	oz
Ground Coriander Seed	2	lb	Alcohol	8	gal
Ground Lemon Peel	1	lb	Distilled Water	12	gal

Macerate 14 days with a portion of the alcohol and water (enough to cover the drugs). Percolate with the remainder. Filter, if necessary.

Angostura Type Bitters

Gentian Root	4	oz	Orange Peel	1	lb
Calisaya Bark	10	oz	Alkanet Root	1	lb
Canada Snake Root	10	oz	Caraway Seed	1½	oz
Virginia Snake Root	10	oz	Cinnamon	1½	oz
Licorice Root	10	oz	Cloves	½	oz
Yellow Bark	10	oz	Catechu	2	oz
Allspice	10	oz	Coriander Seed	2	oz
Dandelion Root	10	oz	Nutmeg	2	oz
Angostura Bark	10	oz	Wormwood	2	oz
Cardamom Seeds	6	oz	Mace	1	oz
Tolu Balsam	4	oz	Red Sanders Wood	1¼	lb
Galanga	4	oz	Turmeric	8	oz
Rhubarb	4	oz			

Macerate for 15 days in 50 gallons of proof spirits. Add 30 pounds of honey and filter.

Cordials

	Ordinaire	Demifine	Fine	Superfine
Alcohol	6½ gal	7¼ gal	8¼ gal	9¼ gal
Sugar	27½ lb	55 lb	82½ lb	110 lb
Water	17½ gal	15 gal	12 gal	9 gal

Benedictine

Angelica Root	2	lb
Angelica Seed	1¼	lb
Cloves	6	oz
Balm Herbs	2	lb
Cinnamon Bark	5	oz
Mace	5	oz
Arnica Flowers	5	oz

10 lb of Herbs need 4½ gal of Menstruum
6½ lb of Herbs need 3 gal of Menstruum

Menstruum

Water	3	gal
Alcohol	1½	gal
	4½	gal

Macerate for 2 weeks. Decant and press.

Peppermint Oil	1 oz	Lemon Oil	2 oz

Add to the above.

Use 1 ounce of the infusion for the gallon of Cordial.

Example

To make 5 gallons of Superfine Benedictine according to the scale just given we use:

Alcohol	2¼ gal	Infusion	5 oz
Sugar	27½ lb	Tartaric Acid Solution	Trace
Water	2¼ gal	Liquid Egg Color	To suit

Benedictine Type

Bitter Almonds	40 g	True Saffron	6 g
Powdered Nutmeg	4500 g	Sugar	2000 g
Vanilla Extract	120 g	Boiling Milk	1000 cc
Powdered Cloves	2 g	Alcohol (95%)	2000 cc
Sliced Lemon	2 g	Distilled Water	2500 cc

Mix. Let stand for 9 days, agitating occasionally. Filter.

Benedictine Type

Cloves	2 g	Angelica	25 g
Nutmeg	2 g	Genepi of the Alps	25 g
Cinnamon	3 g	Calamus	15 g
Balm	25 g	Cardamom	50 g
Peppermint	25 g	Arnica Flowers	8 g

Macerate for 2 days in 4 litres of alcohol with 3 litres of water added. Distill and take 4 litres. Make syrup using 4 kilograms of sugar and 2 litres of water, and bring the whole up to 10 litres with water. Color and filter.

Chartreuse

Bitter Orange Peel	1¼ lb	Calamus Root	10 oz
Ginger Root	5 oz	Iva Herb	10 oz
Cinnamon Bark	3 oz	Menstruum (See under Benedictine)	2 gal
Zedoary Root	15 oz		

Macerate 2 weeks. Decant, press and filter. The infusion is improved by the addition of:

Coriander Oil	1 oz	Thyme Oil	1 oz
Angelica Oil	1 oz		

Make the finished product as directed under Benedictine. Color green or yellow.

Chartreuse (Trappistine)

Large Absinthe	20	g
Angelica	40	g
Mint	80	g
Cardamom	40	g
Balm	30	g
Myrrh	20	g
Calamus	20	g
Cinnamon	4	g
Cloves	4	g
Mace	2	g
Alcohol	4.5	litres and 3 litres of water added
Sugar	3.75	kg and 2 litres of water added

Macerate for 2 days, distill and rectify.

Add syrup and color green or yellow. Bring up to 10 litres with water.

This was originally prepared at one of the Trappist monasteries in Italy.

Anise Cordial

Anethole	7 fl dr	Alcohol	8 pt
Fennel Seed Oil	80 min	Syrup	5 pt
Bitter Almond Oil	16 drops	Water	16 pt

Mix the oils and anethol with the alcohol. Then mix the syrup with the water. Mix the two and filter.

Anisette

Anise Seeds	4 oz	Alcohol	½ gal
Coriander Seeds	1 oz	Water	3 qt
Fennel Seeds	1 oz		

Macerate for 5 to 6 days. Distill over 7 pints. Sweeten with 2½ pounds of sugar. Strength is 30% by volume.

Brandy
(Apricot, Cherry, Ginger, Peach)

These brandies are made by macerating the fresh fruits for prolonged periods or mixing the fresh fruit juice with brandy and sweetening with sugar. Essential oils are also used.

General Formula

Extract	¾–1 oz	Alcohol	1 gal 1 pt
Tartaric Acid	½ oz	Water	1 gal 1 pt
Syrup	¾ gal		

Strength is about 38% by volume.

Cocktails

The cocktail is a compounded drink much used in America. The spirit basis is brandy, gin, whiskey, or other spirits and their different varieties. Depending upon the kind of flavoring used, the various kinds of cocktail are numerous. A bottle of brandy cocktail is made as follows: ½ brandy, ¼ water, a wineglass of bitters, curacao, and gum syrup. Whiskey and gin cocktails may be made by the same recipe, substituting these spirits for brandy.

Manhattan Cocktail

Vermouth	½	wineglass	Bitters	10	drops
Whiskey	½	wineglass	Curacao	6	drops
Syrup	30	drops	Lemon Peel	1	small strip
Angostura			Little Ice		

Martini Cocktail

Gin	½	wineglass	Syrup	6	drops
Vermouth	½	wineglass	Orange Bitters	12	drops
Rock Candy			Lemon Peel	1	small strip

Gin Rickey

Gin	1 wineglass
Lemon or Lime Juice	1 dessertspoon
Carbonated Water	Enough to fill glass
Ice	

Liqueurs

Creme de Menthe

This is one of the most popular liqueurs and is typical of the class of cremes, which are highly sweetened and usually flavored with only one characteristic ingredient, e.g., creme d'anis, cocoa, cafe, menthe, noyan, orange, vanilla. The inferior grades are made by flavoring plain spirit with essential oil without distillation.

Peppermint	600 g	Sage	10 g
Balm	40 g	Orris Root	10 g
Cinnamon	20 g	Ginger	15 g

These are macerated with 5 litres of alcohol and 3 litres of water distilled and sweetened with 3.75 kilograms of sugar and made up with water to 10 litres. Color green.

General Formula

Extract 1 ounce per gallon. Start with less and taste, then add. Strength of extract varies.

Syrup	1 gal 1 pt	Alcohol	1 gal 7 pt
Water	1 gal 7 pt		

Color green.

Curacao

Outside Yellow Rind of Orange	1 lb	Ground Saigon Cinnamon	½ oz
Ground Bitter Orange Peel	½ lb	Ground Nutmeg	¼ oz
		Alcohol	2 gal

Soak for 2 weeks. Filter. Add 7½ pounds of sugar dissolved in ½ gallon of water. Add a trifle of caramel color to tint.

Curacao Liqueur

Tincture of Fresh Orange Peel	1	oz	Orange Oil	2	dr
Tincture of Tangerine Orange	1	oz	Alcohol	12	oz
			Water	10	oz
			Syrup	8	oz

Mix, macerate for a few days, and filter.

Kümmel

General Formula

Caraway Oil	½	oz	Water	4	gal
Capsicum Tincture	¾	oz	Tartaric Acid	½	oz
Syrup	1	gal	Alcohol	5	gal

Cheap Quality

Caraway Oil	4	oz
Fennel Oil	7½	dr
Bitter Almond Oil	15	drops
Syrup	40	lb of Sugar in 12 gal of Water
Alcohol	7	gal
Fine with Alum		

Superior Quality

Caraway Seeds	4	lb	Florentine Orris Root	2	oz
Fennel Seeds	¼	lb			

Macerate with 10 gallons of alcohol and 8 gallons of water. Distill. The first part of the distillate is set aside because of its rough character, and about 8 gallons are then distilled over for making the fine kümmel. The last 3 or 4 gallons are used for inferior grades. To the middle fraction a syrup made with 60 pounds of sugar in 10 gallons of water is added, and the whole well mixed in an open heated vessel. The liquor is then cooled and made up to 20 gallons with water. It may be fined with isinglass.

Kümmel

Caraway Oil	30 drops	Spirit of Nitrous Ether	30 drops
Peppermint Oil	3 drops	Sugar	72 oz
Lemon Oil	3 drops	Alcohol	96 oz
Acetic Ether	30 drops	Water	96 oz

Dissolve the oils and ethers in the alcohol and the sugar in the water. Mix and filter.

Maraschino

Alcohol	2 gal	Orange Flower Water	½ pt
Bitter Almond Extract	2 dr	Sugar	7½ lb
Raspberry Syrup	½ pt	Water	½ gal

Mix alcohol and almond extract. Dissolve sugar in water. Add orange flower water and raspberry syrup. Then mix and filter.

Peppermint
(Eau des Chasseurs) (Sportsman's Cordial)

Peppermint Water	1 pt	Sugar	¾ lb
Gin	1 pt		

Ratafia

Cherry Juice	20 litres	Sugar	13.5 kg
Alcohol	20 litres	Bitter Almond Water	4–8 litres

Vermouth

The principal varieties are the French and the Italian. The basis in each case is white wine flavored with various herbs. Inferior kinds are made from brandy or other spirits sweetened and flavored with essential oil.

The French variety is less sweet and stronger in bitter principles, particularly wormwood, than the Italian.

Italian Vermouth

White Wine	100 litres	Holy Thistle	125 g
Coriander	500 g	Elecampane Root	125 g
Rinds of Bitter Oranges	250 g	Little Centaury	125 g
Powdered Orris Root	250 g	Germander	125 g
Elder Flowers	200 g	Chinese Cinnamon	100 g
Red Cinchona	150 g	Angelica Root	65 g
Calamus	150 g	Cloves	50 g
Large Absinthe	125 g	Galanga	50 g
		Nutmeg	50 g
		Cassia	30 g

Digest for 5 or 6 days. Draw off the liquor. Add gelatin and allow to stand for 15 days. Strength is 27% by volume.

French Vermouth

Wormwood	4 oz	Centaury Leaves	4 oz
Gentian	2 oz	Germander Leaves	4 oz
Angelica Root	2 oz	Nutmegs	15 oz
Blessed Thistle	4 oz	Sliced Oranges	6 oz
Calamus	4 oz	Alcohol	9 pt
Elecampane Root	4 oz	Sweet White Wine	20 gal

Macerate for 15 days and filter.

French Vermouth

Cardamom	½ oz	Peppermint	4 oz
Cinnamon	½ oz	Centaury Herb	4 oz
Cloves	½ oz	Balm Leaves	8 oz
Coriander	2 oz	Angelica Root	8 oz
Fennel	2 oz	Gentian Root	1 lb
Star Anise	2½ oz	Alpine Wormwood	7 lb

Percolate with:

Wine Brandy	6 lb	Sweet Wine	16 lb
Alcohol	7 lb		

After 3 days collect, without expressing, 10 pounds of extract.

Vermouth

General Formula

Cardamom	½ oz	Peppermint	4 oz
Cinnamon	½ oz	Centaury Herb	4 oz
Cloves	½ oz	Balm Leaves	8 oz
Coriander	2 oz	Angelica Root	8 oz
Fennel	2 oz	Gentian	16 oz
Star Anise	2½ oz	Alpine Wormwood	7 lb

Macerate with 1½ gallons of alcohol and 3 gallons of water for 15 days. Filter off. This is the extract.

(A) French Vermouth

Syrup	2½ gal
Extract	½ gal
Tartaric Acid	6 oz
Alcohol	5 gal
Water	for 20 gal
Caramel Color	Small quantity

Strength is 25% by volume.

(B) Italian Vermouth

Syrup	5 gal	Alcohol	5 gal
Extract	1 qt	Water	for 20 gal
Tartaric Acid	5 oz	Caramel Color	To suit

Strength is 25% by volume.

Whiskey

Rye

Proof Spirit	50 gal
Pelargonic Ether *	2 oz
Pear Oil	1 oz
Wintergreen Oil (Dissolved in Alcohol)	10 drops
Acetic Ether	4 oz
Clove Oil (Dissolved in Acetic Ether)	4 drops

* Pelargonic ether appears to be identical with oenanthic ether.

Scotch

Alcohol (95%)	46 gal	Creosote	5 drops
Scotch Whiskey	8 gal	Acetic Acid	2 oz
Water	18 gal	Pelargonic Ether	1 oz
Honey	3 lb	Ale	1 gal

Irish

Spirits	30 gal	Creosote (Dissolved in Acetic Acid)	4 drops
Irish Whiskey	5 gal	Pelargonic Ether	1 oz
Old Ale	½ gal		

Brandy

Cologne Spirits Proof	40 gal	Tannin	¼ oz
Cognac Oil **	⅙ oz	Brandy Essence *	1 oz
Burnt Sugar Coloring	1½ pt	Alcohol	1000 oz
		Water	600 oz

Rum

Rectified Spirits	6 qt	Water	2 qt
Jamaica Rum	22 qt	St. John's Bread	1½ oz
Rum Essence ***	1½ oz	Raisins	1½ oz
Vanilla Essence	⅒ oz		

Or

Proof Spirits	40 gal	Sugar Coloring	½ pt
Rum Essence	½ pt	Sugar Syrup	1 qt

* Brandy Essence	Grape Oil	5 oz
	Acetic Ether	4 oz
	Allspice Tincture	1 oz
	Gall Tincture	3 oz
	Alcohol	100 oz
** Cognac Oil	Amyl Alcohol	5 oz
	Oenanthic Ether	5 oz
*** Rum Essence	Butyric Ether	15 oz
	Acetic Ether	2 oz
	Vanilla Tincture	2 oz
	Violet Essence	2 oz
	Alcohol	90 oz

Gin

Corn Spirits	80	gal	Water	35	gal
Turpentine Oil	1	pt	Caraway Oil	½	oz
Juniper Oil	8	oz	Sweet Fennel Oil	¼	oz
Salt	21	lb	Cardamom	8	oz

Distill over 100 gallons.

SYRUP TABLE

Per Cent by Weight (Brix)	*Degrees Baumé*	*Pounds Sugar in 1 Gallon at 60° F*	*Per Cent by Weight (Brix)*	*Degrees Baumé*	*Pounds Sugar in 1 Gallon at 60° F*
27	15	2½	51.5	28	5¼
29	16	2⅝	53.5	29	5½
31	17	2⅞	55	30	5⅝
33	18	3	57	31	6
35	19	3¼	59	32	6¼
36	20	3½	61	33	6½
38	21	3⅝	63	34	6¾
40	22	3⅞	65	35	7
42	23	4	67	36	7¼
44	24	4¼	69	37	7½
46	25	4½	71	38	7⅞
48	26	4¾	73	39	8⅛
50	27	5	75	40	8½

This table is based on using 100 pounds of sugar and sufficient water to make any of the below desired strengths.

PER CENT SUGAR IN SOLUTION AT A GIVEN TEMPERATURE

Per Cent Sugar	*Per Cent Water*	*Temperature in ° F*	*Per Cent Sugar*	*Per Cent Water*	*Temperature in ° F*
67.25	32.75	220	85.20	14.80	239
75.20	24.80	225.5	85.80	14.20	240.8
80.00	20.00	230.8	86.50	13.50	242.6
82.20	17.80	233.6	87.20	12.80	244.4

SIMPLIFIED TABLE FOR MANUFACTURING SYRUP

Bé Wanted	*Pounds of Sugar*	*Added Water*		*Syrup Obtained*		*Bé Wanted*	*Pounds of Sugar*	*Added Water*		*Syrup Obtained*	
		Gallons	Ounces	Gallons	Ounces			Gallons	Ounces	Gallons	Ounces
40	100	3	119	11	67	29	100	10	46	17	107
39	100	4	56	11	123	28	100	11	20	18	87
38	100	4	113	12	52	27	100	12	7	19	74
37	100	5	41	12	114	26	100	13	3	20	70
36	100	5	106	13	36	25	100	14	8	21	69
35	100	6	49	13	118	24	100	15	20	22	83
34	100	6	119	14	61	23	100	16	44	23	102
33	100	7	68	15	8	22	100	17	82	25	13
32	100	8	23	15	84	21	100	19	8	26	72
31	100	8	111	16	47	20	100	20	79	28	13
30	100	9	74	17	10						

The following table will yield approximately 10 gallons of syrup corresponding to the desired Bé.

Bé Wanted	*Weight of Sugar in Pounds*	*Amount of Water*		*Yield, Gallons*	*Bé Wanted*	*Weight of Sugar in Pounds*	*Amount of Water*		*Yield, Gallons*
		Gallons	Ounces				Gallons	Ounces	
20	35	7	48	10	31	60.5	5	48	10
21	37	7	16	10	32	63.5	5	32	10
22	40	7	0	10	33	66	5	0	10
23	42	6	112	10	34	69	4	96	10
24	44	6	80	10	35	72	4	80	10
25	46	6	64	10	36	75	4	48	10
26	49	6	48	10	37	77	4	16	10
27	51	6	16	10	38	82	4	0	10
28	53	6	0	10	39	85	3	96	10
29	56	5	96	10	40	86	3	48	10
30	58	5	80	10					

The following table uses a 33° Bé syrup, from which a syrup of lower concentration can be made.

Bé Syrup Wanted	*Use 33° Bé Syrup*	*Water to Add in Ounces*	*Amount of Syrup Made*
20	1 gallon	110	1 gallon 110 ounces
21	"	97	" 97 "
22	"	85	" 85 "
23	"	74	" 74 "
24	"	64	" 64 "
25	"	55	" 55 "
26	"	46	" 46 "
27	"	38	" 38 "
28	"	30	" 30 "
29	"	23	" 23 "
30	"	17	" 17 "
31	"	11	" 11 "
32	"	6	" 6 "

The following table uses a 36° Bé syrup, from which a syrup of lower sugar concentration can be made.

Bé Syrup Wanted	*Use 36° Bé Syrup*	*Water to Add in Ounces*	*Amount of Syrup Made*
20	1 gallon	140	2 gallons 12 ounces
21	"	126	1 gallon 126 "
22	"	112	" 112 "
23	"	100	" 100 "
24	"	88	" 88 "
25	"	78	" 78 "
26	"	68	" 68 "
27	"	59	" 59 "
28	"	51	" 51 "
29	"	43	" 43 "
30	"	36	" 36 "
31	"	29	" 29 "
32	"	22	" 22 "
33	"	16	" 16 "
34	"	11	" 11 "
35	"	5	" 5 "

Caramel Sugar Color

Simple Sugar Syrup	10 lb	Ammonia Water (28° Bé)	15 g
Ammonium Carbonate	25 g		

While bringing the sugar syrup to a boil, add the ammonia water with stirring. Now add the ammonium carbonate, previously dissolved in a little warm water. Continue heating syrup until the temperature reaches 350 to 450° F. Stir the contents continuously throughout the heating period.

THE PRINCIPAL STAGES OF SUGAR BOILING

Syrup	at 220° F
Thread	" 225
Pearl	" 230
Blow	" 235
Soft ball	" 245
Hard ball	" 255
Soft crack	" 280
Hard crack	" 312
Caramel	" 350

CHAPTER X

FRUITS AND VEGETABLES

Fruits

Frequently, fruit may be of a very attractive appearance and yet may be of poor quality and have defects and blemishes hidden beneath the surface. To the inexperienced user of fruits and vegetables, the text which follows should be invaluable, from the standpoint of judging quality, and should serve as an aid to economy.

Apples

The market supply of apples comes from many distant states as well as from nearby districts, the largest shipments coming from Washington, New York, Virginia, Idaho, California, and Oregon. The Gravenstein, a late summer and early fall apple, is the only California apple generally found in eastern markets.

Firm apples of good color and flavor are desirable. Immature apples lack color and are poor in flavor. Over-ripe apples yield to slight pressure, and the flesh is often soft, mealy and lacking in flavor. The condition in which the flesh of an apple is mealy or brown is known as internal break-down. Such apples are usually wasteful and should be avoided. This condition is not always apparent on the surface but becomes noticeable when the soft mealy flesh yields to pressure. Internal break-down may be due to various factors such as: over-ripeness, delay in placing in storage, holding too long in storage, unfavorable storage conditions, etc. Summer and fall varieties are more or less susceptible to this defect, particularly at the end of the seasons.

Apples, which have been slightly frozen but have received

no other injury, usually recover their healthy condition. Their qualities as food are not affected, but their keeping qualities are somewhat impaired. When apples have been severely injured by freezing or bruising, the skin appears to be brown and water-soaked, and is often tough and leathery because of the thick layer of flesh adhering to it. The flesh is brown at point of injury and may be either dry and mealy, or watery and mushy, depending on the severity of the injury. Severely frozen apples do not recover their earlier healthy conditions; they continue to deteriorate and, in addition, are very susceptible to decay organisms.

Apricots

Apricots are usually picked when slightly immature in order that they may reach the market in good condition. The finest quality and best flavor are found only in fruits that have ripened on the tree. Because of their extreme perishability, such fruits are difficult to ship and are usually found only in markets adjacent to the district in which they are grown.

Well-matured apricots are plump, fairly firm, uniformly golden-yellow in color, with a juicy flesh, and of good flavor. Immature apricots are usually greenish-yellow in color and the flesh is too firm. By the time they get to market, more or less shriveling is evident and the typical apricot flavor is lacking.

Avocados or Alligator Pears

The avocado, known as the salad fruit of the tropics, is marketed to a considerable extent under the trade names of Flavocado and Calavo, the former being a contraction of Florida avocado and the latter a contraction of California avocado. Avocados vary greatly in shape, size, color and appearance. The shape may vary from spherical to pear- or bottle-shape. The size may range from 5 or 6 ounces to 2 or 3 pounds. The skin may be almost paper-thin and relatively smooth; or it may be thick and smooth; or it may be leathery, rough and shell-like. In color, the fruits may vary from green to a dark mahogany or almost black. No constant relation has been found between color and flavor or between color and texture of flesh.

The softening of firm avocados may be hastened by keeping the fruit in a warm humid place for 2 or 3 days. Softening may be retarded by keeping the fruit in a cool dry place at a temperature not below 42° F. Exposed to the average summer temperature, firm avocados will soften in 4 to 5 days.

Heavy medium-sized avocados, which have a bright fresh appearance and are fairly firm or are just beginning to soften, usually are the most desirable. After a little practice, the feel of the fruit will indicate this condition. The texture of the flesh should be soft, buttery, or marrowlike; such fruit usually has a high oil content and a rich, delicate, nutlike flavor. Avocados whose flesh does not have a buttery smoothness may have a flat watery taste.

Bananas

Bananas are one of the few fruits that reach their best quality and flavor after being harvested in the green state. If allowed to ripen on the plant, the fruit is insipid and of poor flavor. A banana is said to be full when the individual fruit is plump or well-developed; it is said to be thin when it is poorly developed.

With bananas, ripeness is indicated by the color of the skin. The fruit is of a green color when shipped to market. Then it is ripened in storage rooms and takes on a yellow or red color depending upon variety. Good eating-quality is indicated when the solid red or yellow color of the banana is tinted with brown and flecked with small brown specks. At this stage, the flesh is mellow, and the flavor is fully developed. Bananas with tips still green, or those which are of a solid yellow or red color from stem to tip, have not developed their full quality for eating, but are good for cooking.

Poor condition is indicated by a badly discolored skin and a soft, mushy, sometimes discolored flesh. In some instances, the skin may be entirely brown or almost blackened, and yet the flesh may still be in prime eating condition if it is fairly firm and not discolored.

Bananas that have become too cold will not ripen properly, will not develop the bright yellow color of prime fruit, and will usually be of poor flavor.

Blackberries, Dewberries, Loganberries and Raspberries

Quality in this group of fruits is indicated by a bright, clean, fresh appearance combined with a solid full color and a plumpness of the individual berry. Good berries should be free from dirt, trash, moisture, and adhering caps.

Over-ripe berries usually are very dull in color, soft and sometimes leaky. Such fruit is wasteful and is generally not desirable. Natural break-down, decay, bruising, and crushing, may cause berries to be soft and leaky.

Wet or leaky berries should be avoided. A leaky condition is indicated by stained containers and by the general appearance of the fruit. Sometimes the stains on the containers are plainly evident, but at other times they are not seen until the container is tilted so that the side is exposed. Leaky soft or damaged berries are not always seen in a casual examination for they may be at the bottom or in the center of the container.

Decay can be easily detected by the presence of molds on the surface of the berries. Berries with the caps attached may be immature since the caps usually adhere firmly to immature fruits but not to mature fruit.

A berry that has a number of cells or drupelets that are green or off-color when the rest are of the normal ripe color will not have as good flavor.

Blueberries and Huckleberries

Blueberries and huckleberries that are plump, of fresh appearance, clean, dry, free from leaves and trash, fairly uniform in size, and of a deep, full color throughout the lot, are usually of good quality. Ripeness is indicated by the color, which may be blue, black, bluish-black or purplish. The berries may be covered with more or less bloom depending on the variety. Generally, the distinction between blueberries and huckleberries is on the basis of the seed, blueberries having small inconspicuous seeds and huckleberries having very prominent seeds which make them less desirable for consumption.

Decay is usually indicated by the presence of molds. Freedom from moisture is essential to good quality berries. Moisture may be caused by natural break-down, decay or some form of mechanical injury and should be looked for carefully. Over-ripe fruit has a dull, lifeless appearance and is often soft and watery. Berries held long after picking have a similar appearance and may be more or less shriveled.

Cherries

Sweet cherries are grown mostly in the western states but are also grown to some extent in the eastern states. The important varieties are: (light fleshed) Napoleon, Royal Ann; (dark fleshed) Black Tartarian, Bing, Republican, Lambert, Windsor and Schmidt.

Sour cherries are produced mostly in the eastern and Great Lake states. The important varieties are: Montmorency, English Morello and Early Richmond.

Sweet cherries are used primarily for eating fresh, whereas sour cherries are used primarily for culinary purposes, including sauce and pie making. The season extends from May through August.

Good quality in cherries is indicated by bright fresh appearance, plumpness and good color. Well-matured cherries are plump, fairly firm, well colored for the variety, juicy, and usually have a well-developed flavor. Immature cherries are usually smaller than the average ripe cherries found in the container. They are frequently hard and of poor color, lack juice and are likely to be very acid. Over-mature or stale fruits are generally soft and dull in appearance. They may be more or less shriveled and may be leaky. Such fruit is wasteful and may prove to be costly. A close examination should be made for worm injury.

Decay in the form of small brown circular spots is sometimes found. Cherries containing such spots should be avoided. Decay and other forms of damage are often indicated by damp, stained and leaky boxes. Cherries that have been bruised or otherwise mechanically injured are not desirable. The quality may be affected through molds which develop readily at the point of injury.

Cranberries

Cranberries vary in size and color according to the variety. They are rarely sold to the consumer under variety name. The most common types on the market are the rather large bright red fruit and the small darker kinds, which are somewhat sweeter than the lighter colored berries. Cranberries are generally sold by weight and are on the market from fall until the end of the winter.

A fresh plump appearance, combined with high luster and firmness, indicates good quality in cranberries. Poor quality and condition are indicated by shriveling, dull appearance and softness. Cranberries that show moisture are usually leathery, tough and sticky, and the flesh may be discolored.

Figs (Fresh)

Figs are produced mainly in California, the southeastern states and Texas. Because of their highly perishable nature, only a small percentage of those produced are shipped fresh.

The important varieties grown in the southeastern states are: Brown Turkey, Brunswick, Celeste, and Magnolia. The Lob, Inger, Kadota, Mission, and Adriatic are the important varieties grown in California.

Fresh figs must be fully ripe to be of good quality. A ripe fig is fairly soft to the touch and will vary in color from a greenish-yellow to purplish or almost black, depending on variety. Varieties differ in size; therefore, size cannot be used as a method of judging maturity. Ripe figs sour and begin to ferment quickly, and fermentation proceeds rapidly. A characteristic odor is noticeable when figs begin to sour. Bruised or mechanically injured fruit should be avoided, for such fruit breaks down very quickly.

Grapefruit

Grapefruit is received on the markets from Florida, Texas, California, Arizona, Porto Rico, and Cuba, including the Isle of Pines. One of the most important varieties is Marsh (Marsh Seedless), which is practically a seedless fruit. Grapefruit often has a discolored appearance caused by the normal yellow color being overcast by a reddish-brown or reddish-yellow color. Such discoloration is known as russeting. Grapefruit is called bright when the surface of the fruit shows very little russeting, and russet when most of the surface of the fruit shows considerable russeting.

Grapefruits of good quality are firm, but springy, to the touch, not soft, wilted or flabby. They are well-shaped and heavy for their size. Fruits heavy for their size are usually thin-skinned and contain more juice than those that have a coarse skin or are puffy or spongy.

Generally speaking, most of the defects found on the grapefruits in the market, such as scale, scars, thorn scratches, and discoloration, are minor in nature; they affect appearance but not eating-quality.

Grapes

Grapes usually found in the market are of two very distinct types—the American, grown mainly in the eastern and central states, and the European or vinifera, grown chiefly in California. The muscadine type is grown in the South Atlantic and Gulf states but is not shipped to any extent.

Western grapes, shipped largely from California, usually have a higher sugar and solids content, and in many cases are larger than the eastern type. The skin and pulp of the western type do not separate, but the seeds separate readily from the pulp. Some varieties of western grapes were planted for table use and others for juice. The more popular varieties of western table grapes are: Flame Tokay, Malaga, Emperor, Cornichon, and Sultanina (Thompson Seedless).

Eastern grapes are sold indiscriminately for both table and juice purposes. The eastern type is shipped from many sources, but the largest supplies are received from New York, Michigan, and Pennsylvania. The most common varieties are Concord, Catawba, Moore Early, Worden, Niagara, and Delaware. The Concord is used most extensively for juice purposes.

To be of good quality, table grapes should have a general appearance of freshness. They should be mature, and the individual berries should be firmly attached to the stems. Mature grapes are usually plump.

For juice purposes, maturity is the most important factor of quality. No consideration need be given to compactness of the bunches, or to shattering of the berries from the stems, provided the berries are not shriveled or dried.

Lemons

California produces most of the lemons grown in this country. In addition to the California supply, many have until now been received from Italy. The California lemon is usually larger and smoother than the Italian.

Deep yellow-colored lemons are usually relatively mature and are not as acid as those of the lighter or greenish-yellow color. They are also generally thinner-skinned and may have a relatively larger proportion of juice, but they are not as desirable since lemons are used for their acid flavor.

Limes

Limes that are green in color and heavy for their size are the most desirable. Deep yellow-colored fruits do not have the desired acidity.

Limes often become mottled with irregular-shaped spots of a

purple to brown color. Sometimes the whole fruit turns brown. This is the result of a defect known as scald. Such fruit has a poor appearance and brings a lower price than sound fruit, but in many cases the flesh is unaffected, although, occasionally, a moldy taste can be detected immediately below the spots.

Oranges

The oranges from California and Florida constitute by far the greatest part of the market supply. Some importations are received, as well as shipments from Texas, Arizona, Alabama, Mississippi, and Louisiana. The leading varieties shipped from California are the Washington Navel and the Valencia.

The Washington Navel season begins in November and continues to about May, when the Valencia season starts and continues throughout the summer and fall.

The greatest supply of Valencias begins in June and continues to about July, when it slackens; the supply grows heavier again in September, continuing until about November. Tangerines, Satsumas, King, and Mandarin types and varieties come chiefly from the Gulf states. The California orange is usually of a deeper color than the Florida fruit. The Valencia has from one to five seeds, whereas the Washington Navel is seedless. Generally, the skin of the Washington Navel is thicker and may be rougher than that of the Valencia, and the segments are more easily separated.

Florida and Texas oranges are often designated as bright or russet, depending upon the extent of the russeting or discoloration of the surface. The russeting does not affect the flavor. California oranges are not subject to russeting.

Oranges of the best quality are firm, heavy, have a fine-textured skin for the variety and are well-colored. Such fruits, even with a few surface blemishes, such as scars, scratches and slight discolorations, are preferable to oranges that have a badly creased skin, or are puffy or spongy, and light in weight.

Puffy oranges are likely to be light in weight, lacking in juice and of generally poor quality. Exceptions occur in the tangerines, of which Dancy is the principal variety, Satsumas,

King and Mandarin types and varieties. These oranges are usually thin-skinned, and are usually oblate or decidedly flattened at the ends. The skin is easily removed. There is little coarse fibrous substance between the skin and the flesh, and the segments of the fruit separate readily. The flavor is distinctive; the aroma is pungent and pleasant. Because of the looseness of the skin these oranges are likely to feel puffy. Consequently, judgment as to quality should be based mainly on weight and the deep yellow or orange color of the skin.

The fruit is sound when shipped, but sometimes decay develops before the fruit reaches the consumer. When present, decay usually exists in the form of soft areas on the surface of the fruit, which appear to be water-soaked; these areas may be covered by a mold. In the early stages of development, the skin in the affected area may be so soft and tender that it breaks easily under pressure. Oranges that have been mechanically injured should be carefully examined. Decay may be present at the point of injury, and decay organisms may easily find entrance to the flesh of the fruit. Wilted, shriveled, or flabby fruit is sometimes found, and such oranges are not desirable. Age or injury often causes these conditions.

Peaches

There are two classes of peaches—white-fleshed and yellow-fleshed; both classes have varieties that are clingstones and freestones. The freestone varieties of both classes are most popular for dessert, home canning and general culinary use. The clings are not so popular for dessert but are used to some extent for slicing and home pickling, and chiefly for commercial canning.

Peaches for home canning should be firm and ripe. Generally, the yellow-fleshed midseason and late varieties, such as Elberta, Salwey, J. H. Hale, Late Crawford, and Smock are preferred, although sometimes personal preference calls for a white-fleshed peach, such as Belle, Hiley, and Champion.

Quality in peaches is indicated by the general appearance and firmness of the flesh. A peach of fine quality should be free from blemishes, have a fresh appearance and a ground color that is either whitish or yellowish, sometimes combined with a red color, or blush, depending an the variety. The red color or blush alone is not a true sign of maturity; the flesh should be firm.

Pears

The Bartlett, a midseason pear, is one of the leading dessert varieties and is used extensively for domestic and commercial canning. Clapp Favorite (an early-season pear), Bosc, Comice, and Seckel (midseason pears), Anjou and Winter Nelis (late-season pears) are some of the varieties used for dessert purposes. In the greater part of the central and southern states, none of these can be grown successfully because of pear blight. The coarser varieties—Keiffer, Garber, and Le Conte—are somewhat resistant to blight and furnish large supplies of cheap fruit of good size, extensively used for canning, preserving and fresh stewed fruit.

Pears are normally picked when slightly immature and are ripened in a cool dark place. The flesh of storage-ripened pears is usually of fine texture, whereas that of tree-ripened fruit is often coarse and, in some varieties, gritty.

Pears that are firm, or fairly firm, but not hard, free from blemish, clean, not misshapen, wilted or shriveled, are generally of good quality. Some varieties are in prime condition while still a green or greenish-yellow color; others may be yellow and yet be too immature for eating.

Pears that are soft or that yield readily to pressure at the base of the stem are usually mature and ready for immediate consumption, but are too ripe to keep for future use.

Pineapples

The only production of pineapples in continental United States is in Florida. The chief part of the crop is imported from Porto Rico, Cuba, Bermuda, Costa Rica and the Bahamas.

They are picked in a slightly immature state so that they will reach the markets in good condition.

Color and odor are the factors indicating quality in pineapples. A ripe pineapple in good condition has a fresh clean appearance and has a distinctive dark orange-yellow color combined with a decidedly fragrant odor. The eyes are flat and almost hollow. Usually the heavier the fruit is in proportion to its size, the better the quality, provided the fruit is mature. Pineapples may lose moisture, particularly if held in a relatively dry atmosphere. The fruit remains firm as it dries out, if not affected by decay, but it shrinks in size, and the outer surface develops a poor and darkened color.

Plums and Prunes

The leading dessert varieties of prunes are: Italian Prune and Agen, or French. The leading dessert plums are Burbank, Bradshaw, De Soto, Golden Drop, Jefferson, Reine Claude (Green Gage), Diamond, Wickson, Tragedy, Santa Rosa, and Kelsey.

For culinary purposes the leading varieties are: Lombard, Golden Drop, Grand Duke, Italian Prune, and Reine Claude (Green Gage). Varieties of the Damson type, because of their tart flavor, are especially desirable for jam and jelly making.

A prune is a variety of plum which is suitable for drying purposes. Such plums can be dried without removal of the pit and without fermentation of the flesh. A ripe fresh prune can be separated from the pit like a freestone peach, but a plum cannot be opened in this way. Ripe fresh prunes are superior for canning.

Plums and prunes of good quality are plump, clean, of fresh appearance, full-colored for the variety and soft enough to yield to slight pressure. Fruit that is shriveled usually has flesh of a leathery texture and is generally poor in flavor. Plums and prunes that have growth cracks should be examined carefully since these cracks, unless well-healed, provide means by which decay organisms may enter the fruit.

Quinces

Most of the quinces found on our markets are produced and sold locally, but some shipments are made from California and New York. Most varieties have a heavy fuzz over the outer surface of the fruit. Usually this is rubbed off when the fruit is packed, and the general appearance is improved. Sometimes this is not done, and then in the later handling some of the fuzz is removed, leaving the fruit, even though it is of good quality, with a generally poor and often dirty appearance.

Quinces are in season in the fall and are usually sold by measure, for preserving, flavoring and jelly making. Since the entire fruit including the core is used for jelly, wormy fruit causes great waste in trimmings.

Quinces of good quality are firm to hard, free from blemish and of a greenish-yellow or golden-yellow color. Immature quinces are hard and green, very astringent and, when cooked, lack the peculiar quince flavor. Usually such fruit wilts and shrivels if kept for any length of time.

Strawberries

Large strawberries, because of their size and generally attractive appearance, are preferred for dessert purposes, but size is no indication of flavor, which varies with the variety. There are many varieties of strawberries, those grown locally or in nearby districts being preferable, especially for canning, preserving, etc.

The varieties commonly found in markets are: Klondike, Missionary, Howard 17 (Premier), Dunlop, Aroma, Chesapeake, Marshall, Clarke, and Blakemore, a new variety which is gaining rapidly in popularity.

Quality in strawberries is indicated by the general appearance. Strawberries should be of a fresh, clean, bright appearance, have a full solid red color, be free from moisture, dirt and trash, and the cap should be attached. Small misshapen berries or nubbins usually are of poor quality and flavor, and often have a small, hard, green area.

Over-ripe strawberries, or those that are not fresh, have a dull lusterless appearance, are sometimes shrunken and are likely to be wet or leaky; such berries are wasteful. The presence of leaky or damaged strawberries is often indicated by a stained container. Decay is easily detected through the presence of mold on the surface of the berries. It may be found anywhere in the container and is not always evident in the top layers.

Strawberries without the caps should be carefully examined, since they may have been roughly handled, or may be over-mature. Such berries are likely to break down rapidly and be wasteful.

Vegetables

Artichokes

Two very different vegetables are called artichokes. One is known as the French, or globe; the other is known as the Jerusalem artichoke. The French, or globe, artichoke is a plant allied to the thistle. The edible part is the large unopened bud which, if not cut from the plant, opens into a flower. The globe artichoke is very perishable and becomes tough with age.

The Jerusalem artichoke is a type of plant related to the sunflower and is grown for the thick, potato-like underground tubers. These tubers, which constitute the edible part, are now little used except occasionally for pickling and as a substitute for potatoes.

A compact, heavy, globular, plump globe, or French, artichoke, which yields slightly to pressure and which has large, tightly clinging, fleshy leaf scales of a good green color, is the most desirable. Freshness is indicated by the green color which, with age or injury, becomes brownish.

Over-maturity is indicated when the artichoke is open or spreading. The center may be fuzzy and of a dark pink or purple color, and the tips of the scales are hard. When cooked, the flavor is strong and the scales are tough and woody.

Beets

The early, or new-crop, beets are usually marketed in bunches, three to five to the bunch, with the tops attached, either full size or cut back to 6 or 8 inches in length. Occasionally, small quantities of the new crop are shipped loose to local markets; in such cases, the tops are cut back to within an inch or two of the beet. The fresh green tops of bunched beets make excellent greens; leaves that are discolored and ragged, with tough stems, are undesirable for this use.

The late-crop beets are usually marketed with the tops removed. These beets are suitable for storage. Those that have not been washed are likely to keep well if storage conditions are good. Medium-sized beets of the late crop are less likely to be tough or woody than are the large or very small ones. Beets that are marketed with the tops removed are usually sold by weight and in some instances by measure.

Good beets should be smooth and free from blemish. Beets that are rough or ridged or that have deep growth cracks are wasteful and may be tough and woody. It is expected that beets will have some soil on them, but those that are caked with it are difficult to clean.

Soft flabby or shriveled beets are wasteful and usually poor in flavor. Examination may reveal decay, which generally appears as rot. Bunched beets that have remained too long in the field and have become tough or woody can often be recognized by a short neck covered with deep scars, or by several circles of leaf scars around the top of the beet. Only the early or smaller leaves will have been shed by a tender beet. Fresh condition of the leaves alone should not be considered as an indication of the quality of the beet. The roots or beets may be of excellent quality, but the tops may be badly wilted or otherwise damaged.

Beet Tops

Beet tops are usually supplied to markets from local sources. They consist chiefly of the plants that have been pulled out to thin the rows and, generally, the whole plant is used. Beet tops are usually sold by weight.

Beet tops of good quality are young, fresh, tender and clean. Old, coarse, heavy-veined, heavy-midribbed leaves are usually tough and generally undesirable. Flabbiness and wilting are the factors of condition that most frequently affect beet tops, but freshness can be restored if the tops are not old and the condition is not too far advanced. Occasionally beet tops may be found to be in a slimy condition; such stock should be avoided.

Cabbage

There are several market types of cabbage: the pointed, Danish, domestic, savoy and red. The pointed type is the familiar early or new cabbage and includes the varieties that normally develop conical or pointed heads. The Danish type includes varieties maturing late which normally develop hard tight-leaved compact heads. A head of Danish-type cabbage, even after being trimmed, appears tight and smooth-leaved around the basal portion and, when viewed from the stem end, is circular and regular in outline. The domestic type includes varieties of cabbage that normally develop heads, either round or flat in shape, and less compactly formed than those of the Danish type. This type includes early, midseason and medium-late varieties. The savoy types are finely crumpled-leafed varieties which are roundish or drumhead in shape and are usually late in maturing. The red type includes all red varieties.

Early or new cabbage shipped from the southern sections during the winter months is frequently harvested before the heads are firm. It is not trimmed so closely as the late or main crop and can be readily identified by the conical, or pointed, head and comparatively smooth deep green leaves. New cabbage that has not been properly handled soon loses its freshness. Freshness of appearance of the late or main-crop cabbage is not such an important factor of quality as in the early crop. Late cabbage out of storage may be trimmed down to perfectly white heads by the retailer and still be good.

Prime heads of cabbage should be reasonably solid, hard, or firm and heavy or fairly heavy for their size. They should be closely trimmed, that is, the stems should be cut close to the head, and all except three or four of the outer or wrapper leaves should be removed. Early or new cabbage usually is not so solid or firm as cabbage of the late or winter crop. Soft or puffy heads, although edible, are usually of poorer quality than those that are hard or firm.

Carrots

There are two crops of carrots, early and late. The early or new carrots are usually marketed in bunches, three to five to the bunch, with the tops attached. The late crop is marketed with the tops removed. Carrots of the early crop, because they are usually harvested before reaching full maturity, are generally smaller, and of a brighter color and milder flavor than those of the late crop.

Late-crop carrots are generally grown to full maturity and, for this reason, are generally of deeper color and of more pronounced flavor, but are sometimes coarser in texture than those of the early crop. In some varieties, the heart may be tough and woody. The late crop is usually stored and marketed during the winter. Carrots are generally washed before being offered for sale but will often be found with considerable soil adhering to them. This does no harm, but it mars the appearance and may possibly hide some defect.

Under present conditions of production and marketing, bunched carrots are on the market practically the whole year and offer strong competition to the late or storage-crop topped carrots.

Good quality carrots are firm, clean, fresh in appearance, smooth, well-shaped and of good color. Usually, but not always, poor color of carrots is associated with poor quality. The tops of bunched carrots should be fresh and green. The condition of the tops is an indication of quality, but it is not always a trustworthy one, since the tops may be damaged, and the roots may still be in prime condition.

Carrots that are wilted, flabby, soft, or shriveled are undesirable, since they usually lack flavor. Those that are excessively forked or pronged, rough, or have deep growth cracks, are wasteful. Carrots showing excessively thick masses of leaf stems at the point of attachment, or neck, usually have undesirably large cores or hearts. Decay is easily noticed, since it usually appears as soft or watersoaked areas which may be more or less covered with mold.

Corn

A slender ear of corn may have just as much kernel as a large coarse ear, owing to the variation in the size of the cob. Corn found in the markets is of two classes, sweet and field. The latter is usually an early-maturing variety of field corn and is known, in some sections as roasting ears. It may be as tender as sweet corn, but it never has a notably sweet taste.

The ears of sweet corn are usually smaller than those of field corn, and the husks are generally darker green in color with ribbonlike ends which hang free and give the ear a somewhat ragged appearance. The husks of field corn do not have ribbonlike ends; they taper and cling tightly at the top of the ear. The color of sweet corn may be either white or yellow, depending on the variety; the field corn sold for cooking is usually white, but may be yellow.

A good ear of corn is one that has a fresh green husk and a cob that is well-filled with bright, plump, milky kernels that are just firm enough to offer slight resistance to pressure. Dry, yellowed or straw-colored husks are an indication of age or damage. Corn heats rapidly when packed for shipment. Heating causes the yellowing and drying-out of the husk, as well as the toughening, discoloration, loss of flavor and shriveling of the kernels. Corn that is too immature is unsatisfactory, since the kernels of immature corn are very small and soft and, when cooked, lack flavor. Worm injury is not serious when confined to the tips, since the injured portion usually can be removed with little waste, but it is more objectionable if it occurs along the side of the ear.

Corn should be cooked as soon as possible after being picked as it loses flavor very rapidly.

Cucumbers

Cucumbers for slicing purposes should be firm, fresh, bright, well-shaped and of good color. The flesh should be firm, and the seeds immature. Withered or shriveled cucumbers should be avoided, since their flesh is generally tough or rubbery and somewhat bitter. Over-maturity is indicated by a generally overgrown puffy appearance. The color of over-mature cucumbers is generally dull and not infrequently yellowed, the flesh is rubbery and tough, the seeds are hard, and the flesh in the seed cavity is almost jelly-like. Cucumbers in this condition are not suited for slicing but are excellent for certain kinds of pickles. Some varieties are of solid green color when mature enough for slicing, but usually there is a little whitish color at the tip, with a tendency to extend in lines along the seams. These lines advance from pale green to white and, finally, to yellow, with age.

Decay, when present, usually appears anywhere on the surface as a dark sunken irregular area.

Onions (Dry)

Two general classes of onions grown in the United States are found on the markets, the large, mild-flavored and the medium-sized, strong-flavored varieties. The former are the Bermuda and the Spanish, or Valencia, types and the latter the domestic, or American type.

The Bermuda, a flat early type, is found on the market from March to June. Usually two varieties are available, the white and the yellow, although, occasionally a red type may be seen. The most popular sizes are those about 2½ to 2¾ inches in diameter.

The Spanish or Valencia type is a large mild sweet onion which may be either light yellowish-brown or white in color. It is usually globular or somewhat oval in shape.

The domestic or late-crop onions are practically all globular and have yellow, red, or white skins; occasionally brown-skin varieties are seen. They usually keep well and are found on the market the entire year. The white varieties are usually milder in flavor than the others.

Imported onions are available during most of the year. The Spanish variety is very mild, large in size, oval in shape, yellow-skinned and has sweet flesh. The Egyptian onion is usually darker in color and stronger in flavor than the Spanish.

Boilers is the term used for domestic and Bermuda-type onions that range from 1 to 1⅝ inches in diameter. Pickling onions are those that do not exceed 1 inch in diameter. White varieties of this size are frequently in demand, particularly in the fall. Large onions of the Spanish type are sold by the unit; those of the Bermuda and domestic types are usually sold by weight. Sometimes the pickling sizes are sold by measure.

Bright, clean, hard, well-shaped, mature onions with dry skins are usually of good quality. Onions in which the seed stem has developed are undesirable. Usually the neck of such stock is thick and a tough woody condition of the stem or neck is noticeable; the tough stem extends from the very base of the bulb, causing much waste.

Decay generally appears as a rot attacking either the outer scales or the scales in the center of the bulb and may penetrate the bulb from the neck to the base. Moisture at the neck is an indication of decay.

Misshapen onions are sometimes found on the markets. The most common of these forms are known as splits, or doubles, and bottle necks. These terms are self-explanatory and refer to shape only. Misshapen onions are objectionable mainly because of the possible waste in their preparation for table use.

Onions (Green), Leeks, and Shallots

The green bunched onion may be any early onion, usually white, that was pulled when it had reached the required size but before maturity. The leek, a plant similar to the onion, has broad dark green leaves and a straight, thick, white neck about 1 inch or more in diameter. It has an agreeable, but not strong, flavor and is used cooked or raw, usually for flavoring. The shallot is a species related to the onion that grows in oblong clusters which are pulled before maturity. The mature shallot is used chiefly for flavoring.

The green onions, leeks and shallots, as they appear on the market, are merely blanched leaf bases or incompletely developed bulbs, together with the green portions of the leaves. They are usually sold in bunches. The green onion differs from the shallot in that the onion is somewhat bulbous at the base, whereas the shallot is practically a straight stem. The leek, much larger than the shallot, has a rounded but not a bulbous base.

Green onions, leeks and shallots of good quality have green fresh tops, and medium-sized necks which are well-blanched for at least 2 or 3 inches from the root and which are young, crisp and tender. Bruised, yellowed, wilted, or otherwise damaged tops are not attractive and may indicate poor quality or damaged necks. The wilting and yellowing of the top may indicate age and flabby, tough, fibrous necks. This condition can be ascertained by puncturing with the thumb nail and twisting. Bruised tops are unimportant if they can be trimmed without waste or without spoiling the appearance for table use.

Peppers

Two types of peppers are found on our markets, the sweet type and the pungent or hot type, to which the chili or cayenne peppers belong.

Sweet peppers are usually of the bell or bull-nose type. They are shipped usually when mature but still green in color. They change from a deep green to a bronze-red or bright red color as they continue to mature. Sweet peppers are offered for sale in both the green and the red stages of maturity. The varieties vary in shape. Some are long, somewhat slender and pointed, others are short, chunky and wide. The latter type are more desirable for stuffing.

The pungent or hot varieties are found on the markets in either the green or the red stage. They vary from the small chili to the large sizes approaching the bell type. As a rule, they are not so thick-fleshed as the sweet pepper and are generally used in the red stage. The chili and cayenne peppers are sometimes threaded on strings and dried before being offered for sale. Sometimes the whole plant is pulled up with peppers attached. It is dried and then sold.

Peppers, to be of good quality, must be mature, firm, well-shaped, thick-fleshed, and of good color and fresh appearance. Immature peppers are usually soft, pliable, thin-fleshed, pale in color and have undeveloped seeds. In mature peppers, the seeds are hard and the pepper may yield to slight pressure, but it should not be shriveled, limp or pliable. Shriveling or softness may indicate age or immaturity. Peppers that have been kept too long are usually in this condition and are often of poor color.

Constricted, crooked, deformed peppers, provided they are of good quality otherwise, are objectionable only because of their limited uses and the possibility of waste. Peppers that have surface blemishes should be examined carefully. Such stock may prove to be wasteful, particularly if the blemish passes through the fleshy wall. Sometimes peppers are found on which there is a bleached discolored area that may be sunken or that may resemble a water-soaked blister. Such peppers should be avoided because they decay rapidly.

Potatoes

Potatoes of the same variety may be had as either new or old stock during the late spring or early summer.

New or early potatoes are dug before they reach full maturity and are marketed as soon as dug. Because of the slight immaturity and the warm weather at the time they are dug and shipped, they will not stand rough handling and are more subject to injury than are those of the late crop. The skins are easily broken and often present a feathery appearance.

Potatoes of the late crop are usually more mature when dug.

Most of this crop is stored and shipped during the winter and spring. Immature late potatoes are not often on the market; they may be detected by their feathery skins and frequently by a discolored appearance.

Until a comparatively short time ago washing or brushing dirt from potatoes was not considered profitable. Such cleaning is now becoming a commercial practice in some sections, but it adds to the cost of preparation for market.

Potatoes that are sound, smooth, shallow-eyed and reasonably clean are usually of good quality. Varieties vary considerably as to shape, size, color of the skin and cooking qualities. It is difficult for even the most experienced buyer to identify them. A mixture of varieties, which are not similar as to cooking quality, is undesirable. It is often economical to buy and cook a small sample before buying in quantity. Medium-sized potatoes are usually the most desirable for general use, but selection on the basis of size should be governed by the use for which they are intended. Dirty potatoes are unattractive, but the presence of dirt does not detract from the eating-quality. However, wilted, leathery, discolored potatoes should be avoided, since they may have been dug up too early or injured by some other means.

Occasionally, both new and old potatoes show a green color on some part of the surface. This condition is known as sunburn. It is usually caused by long exposure to light which may have occurred in the field or in storage. Sunburned potatoes should be avoided as they usually have a bitter taste that makes them quite inedible.

Potatoes may sometimes have a hollow center known as hollow heart. The size of the cavity may be very small and cause no appreciable waste; or it may be very large, causing considerable waste. Another serious defect, known as blackheart, is sometimes found. This is a black, often slimy, center in the potato. It is usually the result of poor storage or shipping conditions. Hollow heart and blackheart are defects that can be detected only in cutting. Hollow heart is more likely to be found in large potatoes, but it may be present in potatoes of smaller sizes. Blackheart is very objectionable, particularly in potatoes used for baking.

Potatoes injured by freezing are sometimes found on the market during the winter. Bad cases are indicated when the potato is wet and leaky, or, when cut across, it may show a black ring just within the outer surface. In such potatoes, the flavor is usually affected, and the flesh turns dark in cooking.

Decay is one of the most serious defects. It may appear as either a wet or a dry rot, which may affect both the surface and the interior flesh. Sometimes the decay is so slight that it can be cut away with little waste. Another defect is caused by wireworms. Affected potatoes show numerous small perforations which may be so deep as to cause appreciable waste.

In the late spring or early summer, old potatoes may have a shriveled appearance, may be soft and spongy or may even be sprouted. They are very wasteful and may not cook satisfactorily.

Badly formed or misshapen potatoes are sometimes offered for sale. They should be considered from the standpoint of the possible quantity of waste in preparing them for use.

Sweet Potatoes

There are two general types of sweet potatoes—those having dry, somewhat mealy, flesh when cooked, and those having a moist flesh and high sugar content. The latter are commonly, but incorrectly, called yams on the markets.

The most common varieties of the dry type are Big Stem Jersey, Little Stem Jersey and Yellow Jersey. These usually have a yellowish fawn-colored skin and a very light yellow flesh. The moist-fleshed varieties have skins that vary from whitish to a reddish color and flesh that varies from light greenish-yellow to a reddish or orange tint. The varieties vary in shape, and there is considerable variation of shape within the varieties. Nancy Hall and Porto Rico are the most important varieties of this type.

Sweet potatoes are either shipped as an early crop or are

stored for use during the winter and, therefore, are found on the markets practically throughout the year. Maturity is of little importance, except that the early crop may be more readily bruised and skinned than those shipped later in the season. Sweet potatoes are very perishable when abused and are easily injured by careless handling. A mixture of varieties is not desirable because of the difference in flavor and color of the flesh and because of the lack of uniformity in cooking qualities.

Good sweet potatoes are smooth, well-shaped, firm and of bright appearance. The most common defects are decay, misshapenness and growth cracks. Badly misshapen potatoes and those with growth cracks are undesirable only from the viewpoint of waste in preparation.

Sweet potatoes affected by decay are objectionable, because the decay usually spreads rapidly and imparts a disagreeable flavor to the potato, even if the decayed portion is removed before cooking. Decay may appear either as a soft wet rot or as a dry, shriveled, discolored, sunken area, usually at the ends of the potato, but it may appear anywhere on the surface. Another form of decay may appear as greenish, almost black, circular spots, varying from small to large. At times, the spots are irregular, occurring in bruised and injured places.

Sweet potatoes sometimes are marked with small, dark, clay-colored spots which may unite and form large dark blotches. The spots are only skin deep and affect the appearance but not the flesh. Sweet potatoes that appear damp should be carefully examined. They may have been badly handled or frozen, or decayed specimens may be present.

Tomatoes

Tomatoes that are to be shipped long distances are picked when still green, but mature enough so that they will ripen properly after arriving at their destination. Such tomatoes have a dark, clear, shiny, green, body color, and the blossom end usually has a white or yellowish cast. They are generally firm and have a polished surface. Immature fruit usually appears shriveled or is angular in form and has a whitish-green color.

Tomatoes can now be had throughout the year. In addition to the supplies received from distant sources, many markets are supplied with locally grown stock, either greenhouse or field-grown, according to the season. Tomatoes that have been picked green and ripened in the markets are less juicy than those ripened on the vine; therefore, the very best for fresh table use are to be had from the home-grown crop.

Good-quality tomatoes are mature, firm but not over-ripe, fairly well formed, plump, smooth, of good color and free from blemish.

There are many defects in tomatoes which are serious only from the standpoint of waste in preparing them for the table. Catfaces or scars around the blossom end are typical of the defects within this class. Tomatoes that are rough or irregular in shape may not be attractive in appearance, but, for certain purposes, they can be used with little waste.

Tomatoes that have been attacked by worms are very objectionable, especially if the worm has bored deeply. Those having growth cracks will seldom keep long but are fit for immediate consumption. Puffy or watery fruit is usually of poor flavor and is wasteful. Puffy tomatoes are usually angular in shape.

Turnips

Early turnips are usually sold in bunches with the tops attached, although sometimes the tops are cut back to about 6 or 8 inches. The early turnips are usually much smaller than those of the late or main crop, which is marketed with the tops removed. The early crop is marketed as soon as harvested, but much of the late crop is stored for shipment during the winter. Early or new turnips are shipped during the winter months, while the late or winter crop is still on the markets.

Turnips that are smooth, firm, with few leaf scars around the crown, and with very few fibrous roots at the base are usually of good quality.

The condition of the tops of bunched turnips is an indication of quality. The tops should be fresh, green, young and turgid. Yellowed or wilted tops of bunched turnips may indicate damage of some kind, possibly caused by long keeping.

Fruit Juices

Many fruits, when fully ripened, are rich in juice, which, when pressed out, makes a delicious drink.

These fruit juices can be bottled for fountain or household purposes. Grape juice makes a good unfermented beverage, and so does a blend of sweetened blackberry and raspberry juices. In the last few years prune juice and pineapple juice have become popular in the home.

Fruit juices are rich in vitamins and mineral salts, and are considered very nourishing. At the same time they make a good beverage of the non-intoxicating type.

The Manufacture of Fruit Juice

Cloudy juice frequently presents a serious problem to the manufacturers of fruit juices. Such juices may be sparklingly clear when bottles are filled but, upon standing a short while, will again become cloudy. A fruit juice tastes best when it is first pressed from the fresh fruit. Treatment that is added sometimes injures its delicate taste and aroma. The important thing to retain in a bottled juice is its pleasing taste. The methods of clarifying fruit juice, so that no further cloudiness takes place in the bottled juice, should not remove too much of the natural taste and body. Making the juice clearer improves the physical appearance, but it does not improve the flavor.

Fresh fruit juices will spoil quickly, either through fermentation or molding, if nothing is done to prevent the growth of yeasts and molds. Yeasts and molds are usually killed by temperatures between 160 to 170° F. If the juice is immediately transferred to hermetically sealed containers and then sterilized, it will not spoil, as long as the container remains closed. The containers should be pasteurized for 35 minutes at a temperature around 175° F. Too high a temperature of pasteurizing, or too long a period of heating, injures the fresh fruit flavor and imparts a cooked taste. Juices, held in cold storage at 32° F, may begin to spoil after 6 months but, if maintained in a frozen condition below 32° F, will keep indefinitely. The flavor, as well as the color, however, may be slightly impaired.

Chemical preservatives are also commonly used to preserve fruit juices. Sodium benzoate is allowed to the extent of 1/10 of 1% and must be declared on the label. Sulfurous acid is also used but is efficient for only short time preservation. The sulfurous acid oxidizes to sulfuric acid. After several weeks practically all of the sulfurous acid is gone, especially if used with juice in wood-filled containers. Bottles and corks can be sterilized or treated with antiseptics, so that the perfect filtered juice will not become infected with yeasts or molds after the bottles are filled.

Bottles can be rinsed either in a solution of 25 parts per million of chlorine water and then allowed to drain, or in 0.025% solution of sulfurous acid (250 parts per million). Corks also may be immersed in the latter chemical solution and then used after a short while.

Filtering the Juice

Cloudy juice containing suspended particles and other impurities can be clarified by the following methods:

1 Filtration. This is most commonly used and most satisfactory.
2 Centrifuging. This is clearing by centrifugal force.
3 Clarifying materials, such as gelatin, tannic acid, isinglass, egg albumen, casein, talc, bentonite, calcium sulfate are used.

In the third process, the clarifying material is added to the juice, coagulates the suspended cloudy substances, and then finally carries them down to the bottom of the container. These clarifiers have both advantages and disadvantages. Gelatin, when used by itself, is partly soluble in juice. If used in conjunction with tannic acid, some of it may remain in solution if there is insufficient tannic acid to combine with it.

When used alone, it has the other disadvantage of removing tannins, in this way removing the natural color and flavor of the juice. When the proportion of tannin and gelatin is correct, a flaky precipitate immediately forms. This precipitate, as it settles, carries down with it a pectinaceous slime, bacteria and gummy substances. The quantity of tannin naturally present in fruit juices varies, and in order to determine the correct amount of gelatin and tannin to add, it is necessary to make trial precipitations with small quantities of juice. To each of three jars add 1 quart of juice and 10 cubic centimeters of a 1% solution of tannic acid, mix thoroughly, and then add to each jar, respectively, 5 cubic centimeters, 10 cubic centimeters and 15 cubic centimeters of a 2% gelatin solution. Mix the juice again and then allow to stand. After 2 hours, the jar in which the best clarification has taken place is selected.

The fruit juice in the tank is now clarified by the addition of tannin and gelatin in the same proportion.

Clearing the Juice with Chemical Combinations

In using a combination of chemicals for juice clarification, the reaction yields a heavy rapidly settling precipitate, which carries down with it the pectinaceous and gummy colloids. The chemical precipitate formed must be insoluble in the fruit juice and should also be harmless. Such combinations are:

(1) Potassium aluminum sulfate and sodium carbonate
(2) Potassium aluminum sulfate and dibasic sodium phosphate

The combination of (1) yields a dense white precipitate of aluminum hydroxide. Dissolve 4½ ounces of potassium aluminum sulfate in 1 pint of water and then stir it into 50 gallons of juice. Now dissolve 3 ounces of sodium carbonate in 1 pint of water and add it with stirring to the same 50 gallons of juice. Allow the precipitate to settle. Siphon out the clear juice.

The combination of (2) yields a heavy flocculent precipitate of aluminum phosphate and aluminum hydroxide. Use by dissolving 3 ounces of potassium aluminum sulfate in 1 pint of water and stir it into 50 gallons of juice. Now dissolve 3$\frac{3}{10}$ ounces of dibasic sodium phosphate in 1 pint of water and add it with stirring to the same 50 gallons of juice. Allow the precipitate to settle. Siphon out the clear juice.

Flow Sheet in Manufacture of Juice

A The fruit is sorted or examined on broad cotton or rubber belts, which slowly move past workers who remove bruised or blemished fruits.

B Washing equipment is then used, similar to the rotary tomato washer, to improve the quality of the fruit before crushing. Berries, which cannot stand rough treatment, must either be rinsed by hand or washed under a gentle spray of water.

C Crushing the fruit is the next step. The most satisfactory type of crusher for general use is that commonly used for apples under the name of apple grater. The grater usually consists of a heavy steel cylinder, geared to revolve about 2400 times a minute, and grooved to receive knives, which are six or more pieces of tempered steel about $\frac{3}{8}$ inch thick and having one edge cut to form teeth $\frac{3}{8}$ inch square. The knives, adjustable by set screws, are usually set so that they project about $\frac{3}{16}$ of an inch. One side of the cylinder housing is made up of the concave, consisting of three or four heavy curved iron plates, each attached by its top to the housing, and adjusted at the bottom by a heavy spring. When properly adjusted, such a grater is much more efficient than any roller crushing device.

For grapes, the best crusher consists of two corrugated or fluted metal rollers which revolve close together and towards each other, carrying downward between them and crushing the grapes that are fed into a hopper above. Connected with the crusher is a stemmer consisting of a horizontal metal cylinder with a perforated bottom, through which the grapes are forced by revolving paddles. The stems cannot pass through these openings, and are thrown out at the end of the crush.

D Pressing the juice from the pulp is the next operation. The press, known as the rack and cloth press or hydraulic press, will give a higher yield and a clearer juice than with the basket press. In the rack and cloth press, the crushed fruit is built up in layers in heavy burlap cloths between racks made of wooden slats. Pressure is applied by means of a ram operated by a hydraulic pump.

In basket presses, the cloths and racks are not used. The crushed fruit is held in a strongly reinforced wooden basket of cylindrical shape, which rests on the press floor. Pressure is applied by lever and screw in small presses, and by hydraulic pressure in the larger presses.

The crushed fruit may be heated up first in a steam-jacketed kettle, before pressing, if so desired.

E Pasteurizing fruit juice, used for bulk, is the next operation. Proteins and gums are coagulated, which otherwise would cloud out in the bottle. Grape juice is usually sterilized and stored several months to permit separation of cream of tartar before bottling. A very common and effective pasteurizer consists of a block tin or Monel metal coil surrounded by hot water. The juice flows through this pipe and can be heated to any desired temperature, depending upon the surrounding water. A thermometer at the juice outlet is necessary for temperature control. The temperature of the juice should be around 175° F when it comes out of the pipe. The hot juice is delivered into containers for settling.

F Filtering the juice improves the appearance of the product. The simplest filter is the bag filter, which consists of a conical heavy duck or felt bag. The rate and effectiveness of filtration may be increased by the addition of infusorial earth, or asbestos pulp to the juice before filtration. Bag filters commonly hold 10 gallons of juice at each filling.

For large scale production the juice is filtered through an ordinary plate and frame filter press, using infusorial earth as a filter aid. The filter aid and water are mixed together in the kettle and then pumped through the filter. In this way the filter bed is built up. A new type of filter equipment has been introduced. This machine contains a series of round filter plates which are chromium or silver plated. The plates are arranged so that a round sheet of asbestos fits into a pair, which is separated by a special grid, and which permits the filtered juice to flow to the outlet. The juice is pumped through by means of a very small motor. The juice which then goes through this layer of asbestos is crystal clear but not sterile. When the layer of asbestos becomes clogged, the filtration slows down. The plates must then be removed, and the used asbestos replaced with fresh asbestos. Absolutely sterile juice can be had by filtering the product through a bacteria-removing filter.

Concord Grape Juice and Its Manufacture

The Concord grape is well recommended as the standard for juice manufacture. The Clinton as a rival lacks the foxy flavor of the Concord. The Concord grape is very juicy and the juice has a rich red color. Practically 95% of commercial grape juice is prepared from the Concord grape.

Preservation Methods

There are a number of methods which may be used to preserve the juice. The following are recommended:

A Pasteurization of the unfermented juice.

B Holding juice at a temperature of 28 to 30° F. The argols will precipitate out at this low temperature. The juice must be pasteurized, however, when ready to be sold in bottles.

C Freezing storage is used but slightly. Too much of the tartrates will precipitate out in this method, removing some of the desirable ingredients of the juice.

D Filtering the juice through germ filter presses. A sparklingly clear juice is obtained and there is no question that the germs are filtered out.

E Storage of the juices under a high pressure of carbon dioxide.
F Concentrating the total solids by evaporation or freezing. The high total solids will prevent further spoilage.
G Adding sulfur dioxide or sodium benzoate to the juice.

The process of juice manufacture may be divided into six distinct manipulations:

A Crushing and stemming the fruit
B Heating the crushed fruit
C Pressing the heated fruit
D Sterilizing and bottling juice for storage
E Siphoning the juice
F Bottling and pasteurizing the trade juice

Crushing and Stemming the Fruit

The crusher consists of two rollers of metal, usually bronze, revolving at high speed. The rolls are set so as not to crush the stem or seed. The stemmer is a cylindrical revolving drum inside of which are arranged revolving fingers that separate the berries from the stems. The crushed berries and liberated juice pass through the openings in the side of the drum and fall through to the cooker while the stems are picked out of the drum by the fingers.

Heating the Crushed Fruit

The liberated juice is run into heating kettles which may be steam-jacketed or double boilers or equipment commonly called cookers. The heating temperature should be between 140 and 145° F. Heat until the bright purple color is diffused throughout the mass. Heating too long or too high a temperature will result in a harsh bitter taste.

Pressing the Heated Fruit

The hot grapes are placed on heavy burlap cloths which are on racks, and which can be folded over and built up on top of one another. Any number of presses of various types is available. The hydraulic press is the one generally used in the factory. The grape solids on the cloth are usually referred to as cheeses. As the number of cloths or cheeses increase there is a flow of juice called the free run juice, since it is the juice which is recovered by the weight of the cheese itself, without additional pressure. The free run juice may amount to 50 to 55% of the entire quantity of available juice in the fruit. Pressure is gradually applied until all of the juice has drained out of the cheeses.

Cold pressing of the juice is utilized only where light colored juices are wanted and for juice intended for fermentation, making light colored wines.

Sterilizing and Bottling the Juice for Storage

The juice should be strained again to remove coarse particles, and then conducted into the sterilizers. These may be aluminum or steam-jacketed kettles. The juice is heated to destroy all living organisms. The sterilizing temperature ranges from 175 to 190° F. The lowest possible temperature without risk of not destroying all living organisms is the best. It is inadvisable, however, to go much lower than 180° F.

The precipitating vessels in which the argols or cream of tartar separates out may be glass carboys or jugs of five-gallon capacity. They are filled to a few inches from the top with the hot sterile juice. The bottles should be sterilized and warmed up in advance. The necks of the receptacles are sealed with corks of good quality and prepared by dipping into heated paraffin, which closes the pores and canals. The sealed vessels containing the juice are placed in vaults. In about 4 months or more, the tartrates, tannins and solid particles form a heavy sludge on the bottom. To obtain the best precipitation the temperature should be kept as low as possible, as close to 32° F as can be maintained without danger of freezing.

Siphoning the Juice

The juice should be siphoned over only during the winter months. The remaining sediment is poured on several thicknesses of burlap stretched on a frame, and the juice is allowed to drain. This juice should be sterilized and be allowed to precipitate out again. The hot juice when placed in the carboy has a nice red color. As it cools, it becomes turbid, and when it clears becomes reddish purple. The hot juice has a harsh taste, whereas the clear juice is pleasant in taste and aroma. As the juice in the carboy cools, the cream of tartar separates out, until the solution contains the normal amount of cream of tartar.

Bottling and Pasteurizing the Trade Juice

The pasteurizing temperature of juice in the bottle ranges from 160 to 175° F. If the bottles are filled cold it may take about 10 minutes for the juice to come up to the temperature. It should then be pasteurized for about 35 minutes and cooled off in 10 minutes. If filled hot from 25 to 30 minutes of pasteurizing temperature is sufficient. The heating time will vary, depending upon the size of bottle. Pints and quarts should take around 55 minutes. Large containers should be filled hot before pasteurizing. Bottled juice should be stored in a cool dark place. It should not be stored at a lower temperature than that which was employed to precipitate out the cream of tartar.

Removal of Argols or Cream of Tartar from Grape Juice

The removal of crude cream of tartar takes place at low temperatures and over an extended period of time. If too much is removed from the juice, it lacks snap, and tartaric acid should be added to make up for it. This method is a long-drawn-out process, but there are a number of other methods for hastening this crystallization, some of which are as follows:

A Freezing the juice and then removing the cold juice from the sludge formed during the freezing

B Storing the juice in suitable containers, just above the freezing point of the juice, which is 28 to 30° F

C A method patented in 1931 calls for the addition of calcium acid malate, lactate, or acid phosphate to the juice.

Apple Juice Cider

Apples used in the manufacture of cider must be carefully washed and sorted before grinding and pressing. It is very important that all dust or other dirt particles be thoroughly removed in order to prevent fermentation. The apples should be cleaned in a large container of clean water, stirred around for a while, and then taken out by hand for further inspection. Many modifications may be used, but it is necessary to secure sound clean fruit for pressing.

When the fruit has been sorted and washed, it can be crushed between two rollers or by grating the apples with a cylindrical grater. The second method is the most efficient.

The pressing out of the juice may be accomplished in several ways. The barrel press is considered the poorest method because the resulting juice is poor, and, when the necessary force is used, the juice spurts out of the filtering cloth. The power presses of the hydraulic type are the most efficient. They have a great capacity, are inexpensive to operate, and a greater yield is obtained from the same amount of fruit as compared with the barrel press. A ton of apples should yield about 175 gallons of juice.

The residue remaining on the filter cloth after juice has been extracted is called pomace. In most cider mills this represents waste. A ton of apples will yield about 700 pounds of wet pomace. This can be pressed again at the end of several hours, and the juice added to the first pressing. Another good idea is to moisten the pomace with the juice of the first pressing and then repress. The residual pomace may then be dried and sold for its pectin value for making imitation jellies, or used as cattle feed.

Clarification of Cider

Apple juice is cloudy after extraction, and a common method for cleaning and clarifying it is heating the juice in order to coagulate the proteins and gums, cooling and then filtering until clear. Another method consists of mixing thoroughly ½ to

1% by weight of refined infusorial earth with the juice and pumping through the filter press. Two or more filtrations may be necessary. Generally, freshly pressed fruit filters with difficulty because the filter becomes clogged with a gummy deposit. By removing a large portion of the colloidal suspension of starches and gums, the filtration becomes easier and more rapid. The following are some brief methods for this removal:

Sedimentation

The pressed cider should be stored in clean barrels or tanks for about 10 to 15 hours, and the clear juice siphoned off without disturbing the sediment.

Centrifuging

A high-speed centrifuging machine can be used for separating the particles from the cider. Sometimes this machine clogs quickly, and a cream separator may be used instead.

Gelatin, Tannin, and Enzymic Treatment

Flash Heating

In this process the juice goes through a series of coils which are steam-jacketed. During its passage through, the temperature may go as high as 185° F. The time required for this operation may be about 1 to 3 minutes. A considerable amount of gummy substance is coagulated in this way, and this makes the juice much easier to filter.

Heating

Fresh cider, when heated to 175° F, forms considerable coagulum. The juice should be kept at 175° F for at least 10 minutes and then cooled very quickly to room temperature. After settling, the clear liquid is siphoned over, without disturbing the sediment.

Pasteurization

Apple juice may be pasteurized in a number of ways. Flash pasteurization consists in heating the juice to 185° F for a few seconds and then cooling quickly, or the juice is kept in the pasteurizer for 20 minutes at about 175° F. Pasteurization may also be done in bottles or other suitable containers. Bottle pasteurizers may be home-made wooden troughs fitted with steam coils, or a live steam pipe; or the bottles may be placed in basket conveyors, which carry them through water baths increasing in temperature until the bottles are pasteurized and finally through cold water baths until they are properly cooled. The temperature recommended is 175° F for 20 minutes. Yeast cells are destroyed at 140 to 150° F for a few minutes. Yeast spores are more resistant and require a higher temperature of 175° F for 20 minutes. Juices having higher acidity need be pasteurized only to 160 to 165° F. Regardless of what method is used to pasteurize the product, it is necessary to cool promptly in order to avoid a cooked taste. The sterilization of bottles and caps before filling will greatly reduce loss from fermentation and mold growth.

Chemical Methods of Preserving

Sodium benzoate may be used to preserve cider. The amount necessary is 1¼ ounces to 10 gallons of cider. This chemical is more toxic to yeast than to molds. Maximum toxicity is obtained with higher percentage of acid. The amount of this preservative must be declared on the label, e.g., "Contains 1/10 of 1% of Sodium Benzoate."

Sulfurous acid is more toxic to molds than to yeasts. By the addition of 0.10% of sulfurous acid (20 grams of 5% solution of sulfurous acid per liter of cider) to cider, it may keep for a year, if stored at a temperature of about 60° F. The federal government will tolerate 0.0350% of sulfur dioxide in the product offered for sale. This must absolutely be declared on the label, as some states do not allow its use at all.

Carbon dioxide has also been used to preserve fruit juices. In one process, the cider is placed in a specially constructed iron keg, and the air is exhausted, making a vacuum. The cider is then carbonated at a pressure above 60 pounds per square inch. The cider in the keg is exhausted of its carbon dioxide, a vacuum is once more created, and then carbon dioxide pumped in a second time. It is claimed that this method destroys the microorganisms which cause spoilage.

Cranberry Juice

The two methods of expressing the juice are heat extraction and cold pressing. In the heat method, use 6½ gallons of water to 100 pounds of berries and heat for 10 minutes or longer, to soften. Express the juice which should be bright red. The skin and seeds can be either strained or filtered out. The yield is about 8 gallons per 100 pounds of fruit. To make the juice palatable, dilute with 1½ to 2 parts of water and sugar enough to test 15° Brix. Juice in the bottles should be pasteurized at 185° F for at least 30 minutes.

A cold extract of cranberry juice is made by grinding in a food chopper, and allowing to set overnight before pressing through a hydraulic press. Approximately seven gallons of juice are derived from 100 pounds of berries. Clarification of this juice can be accomplished with the use of pectinol and filtering with earth. The juice blends well with pineapple, grapefruit and apple juice. Of course, only the pure juice can be correctly termed cranberry juice.

Red Raspberry Juice

This juice possesses a very pleasing flavor and is made in almost the same manner as grape juice. Cuthbert, Miller and Brandywine varieties are used.

The juice yields, obtained by crushing and pressing the berries, are high, averaging from 70 to 85% of the fruit. It is necessary to press slowly and to use double cloths. The juice should settle clear, preferably without filter aids. Pasteurization at 175° F in bottles is a satisfactory method of preservation.

Black Raspberry Juice

Berries upon being pressed without previous heating, yield 60 to 75% juice. Pressure must be applied very gradually for good yield. The juice is not noticeably injured in flavor or color by pasteurization. The berries may also be pressed after heating. The sterilized juice is rather acid and requires the addition of sugar to make it palatable.

Strawberry Juice

Best results are secured by grinding the fruit first, and then applying pressure very gradually. The yields range from 60 to 85%. The juice can be sterilized without injuring the color, but there is generally marked injury to the flavor. Color and flavor change rapidly during storage, the bright red color changing to a dull brownish-red tone. Disagreeable flavor also develops during storage.

Red Currant Juice

Without previous heating of the berries, juice yields vary from 65 to 70%. When the berries are pressed with heat, the yield is generally about 73 to 80%. There is a slight loss in flavor when heated, but not in color. The sterilized juice will keep well in cold storage at from 32 to 35° F. The juice is best prepared by cooking until soft and then pressing. The juice is then freed from sediment and sterilized in glass.

Black Currant Juice

It is necessary to heat the fruit before pressing to secure a satisfactory yield and quality of juice. The yield of heat-pressed juice ranges from 68 to 78%. The flavor and color keep well for a long period of time after sterilization.

Blackberry Juice

Cooking before pressing increases the yield, aroma and flavor of the fruit. It is necessary to apply the pressure gradually in order to prevent the pulp from passing through the cloths. Yields, cold pressed, range from 67 to 70%; hot pressed, from 75 to 80%. There is very little loss of flavor or color in sterilization.

Loganberry Juice

This juice has become very popular. It has a deep red color, a very rich flavor and high acidity. Thoroughly ripe fruit gives the best juice, while under-ripe berries are light in color, excessively tart and astringent in taste. The juice is best extracted by heating the berries in the kettle to about 150° F and then pressing on the hydraulic press. The juice should then be cooled and filtered through a felt bag, subsequently through cotton pulp filter. This juice is rich in pectins and gums, and difficulties may be encountered in filtering it. In bottles, the juice is pasteurized at 175° F for 40 minutes.

The filtered loganberry juice may be bottled undiluted and unsweetened, diluted and sweetened, or sweetened but undiluted. Sugar helps to retain the fresh berry flavor and to prevent the development of a bitter astringent taste.

Cherry Juice

This juice is of recent introduction. The English Morello, and Montmorency cherries from New York, Michigan and other states are the varieties used.

Without previous heating, the yield, after pressing, ranged from 70 to 80%. The fruit is washed and ground to a coarse pulp and then pressed. The juice is heated rapidly to 190 to 200° F and cooled. The juice will be cloudy and must be clarified before bottling. The distinctive color and flavor are well retained after sterilization. Juice pressed containing the kernels is slightly better than juice without the crushed kernels, because it possesses flavor derived from the kernels.

Pineapple Juice

The pineapple used for this purpose should be field-ripened. The fruit is first peeled and then cored. It is then shredded and pressed, the resulting juice is strained and then heated to 170° F in stainless steel-jacketed kettles, filled at once into plain cans, and held for 20 minutes before cooling. In some factories, the warm juice is poured into cans which are then closed under vacuum. The canned juice is then sterilized in tanks of hot water for 30 minutes at 180° F. In other cases, the juice is flash pasteurized, and poured into cans which are closed under vacuum. Heating the juice causes a slight change in flavor. Keeping the juice in cold storage at 32 to 35° F after sterilization will help to retain a distinctive, satisfactory, pineapple flavor.

Orange Juice

In spite of extensive study of orange juice, a successful method for retaining the fresh orange flavor has not yet been found. Sterilizing the juice injures the flavor, which gradually deteriorates when the juice is kept at ordinary temperatures. In cold storage, the flavor is well retained.

The two methods for making the juice involve the pressing of peeled oranges in screw presses or pressing the orange halves down a rapidly revolving burr. Removing the peels before pressing results in a juice deficient in orange flavor. With the burr method, more of the oil goes into the juice and it therefore has a better flavor.

The Valencia oranges yield a juice which is far better than that from Navel oranges. The juice from both undergoes a distinct but slight change in flavor when sterilized at 80° C. If kept at 32 to 35° F, no flavor change occurs for a few months. At ordinary temperature, the change in flavor is rapid, accompanied by darkening of color. It has been found that by keeping the air from contact with the surface of the orange juice the color and flavor changes are not so pronounced. Orange juice must be frozen quickly and deaerated, or the dissolved air must be removed from the juice previous to freezing. This may be done by treating with a high vacuum in a closed tank or by allowing it to flow over a surface while exposed to a high vacuum.

In canning the juice, it is first screened, deaerated, and then flash pasteurized quickly for 5 seconds at 205° F. It is then quickly cooled to 180 to 190° F and immediately poured into cans which are closed under 12 to 14 inches of vacuum and cooled. Carbonating and sterilizing the juice in carbon dioxide permits satisfactory flavor and color retention.

Lemon Juice

This juice is prepared by cutting each lemon into two or more pieces, placing the fruit in cloths between racks and pressing it. The yield ranges from 35 to 40%.

A large proportion of oil is included, but is removed by passing the juice through a separator, which at the same time removes some of the suspended material.

The juice can be sterilized by heating at 70° C for 30 minutes. There is very little loss of flavor. When kept at a low temperature, it retains its rich lemon flavor for many weeks, and, if frozen quickly, it will keep for a long time. Eventually, however, a peculiar flavor similar to a lime flavor seems to develop. The exclusion of all air from this juice is important; but even this exclusion is not successful in controlling the change of flavor.

Prune Juice

Prune juice is made by extracting dried prunes with water until practically all of the soluble substances have been removed. The prunes are washed clean and then soaked in hot water for 1 to 4 hours at 200° F. The soaking time depends upon the body and toughness of the prune. The liquid from the first soaking is used to extract the soluble material from a second, and finally from a third lot of an equal amount of prunes. In the meantime, the first lot of prunes is extracted a second, third, or even a fourth time, for an hour each time at 200° F. All of the extractions are then combined. Each lot of prunes may be extracted 4 to 5 times. Fifty pounds of prunes should yield approximately 15 gallons of juice, with about 18% solids. The juice may be filtered through several layers of cheese-cloth or through a pulp filter. The concentration of the juice may be controlled by boiling or dilution with water.

Prune juice in bottles, quart size, should be pasteurized at 190° F for 35 minutes.

Grapefruit Juice

The fruit is washed and cut into halves or quarters. The fruit should not be shredded, for the more the juice comes in contact with the peel, the more bitter will the flavor of the finished product be. The fruit is pressed gradually, and the yield generally ranges from 70 to 100 gallons of juice from a ton of grapefruit. The juice is strained through copper or bronze gauze, and then given a preliminary pasteurization at 170 to 185° F. Best results are obtained by flash pasteurization, in this way avoiding any chances of overheating. When ready for bottling or canning, the juice can be clarified by the addition of kieselguhr and then by filtering. If the juice exceeds 1% in acidity, a slight addition of sugar will improve the taste. For final bottling or canning, the juice is heated once more, poured into the containers and sterilized at 170 to 185° F for 30 minutes.

Concentrated Fruit Juices

Fruit juices can be concentrated in a number of ways, but concentrating in vacuo is the most desirable. Concentrating in open kettles results in loss of flavor and color. Concentrating by the spray process has been used successfully for making powdered lemon or orange juice. Spray-drying of grape juice has been accomplished with some degree of success. Concentrating by freezing preserves the aroma and color of the fruit. The fruit juice is then separated from the ice crystals.

Sweetened juice applies to fruit juice to which has been added 50% or less cane sugar.

Fruit syrup is the name applied to fruit juices to which has been added 50% or more of cane sugar, and must contain at least 33⅓% of pure fruit juice.

Fruit juice concentrate is the name applied to fruit juices which have been concentrated by evaporation of the excess water in a vacuum or under atmospheric pressure or by freezing and the subsequent separation of the ice and concentrated juice. Most apple concentrates made by evaporation in the open kettle possess a cooked taste.

Orange Fruit Concentrate

Orange Juice Concentrate, 72° Brix	1	gal
Sugar Syrup, 60° Brix	5	gal
Orange Oil, California Cold Pressed	¼	oz

One and one-half ounces of this syrup is added to a 6½-ounce bottle and carbonated, or added to an equivalent amount of cold water.

Lemon Fruit Concentrate

Lemon Juice Concentrate, 72° Brix	1	gal
Sugar Syrup, 60° Brix	20	gal
Lemon Oil, California Cold Pressed	¾	oz

Use this syrup in the same manner as above.

The Clarification of Fruit Juices by Enzymes

Fruit juice, when first pressed, contains finely suspended matter. Some of it may settle out on standing; much of it can be removed by filtration or by centrifuging. It sometimes requires several such treatments to completely remove the colloidal material, which is the real cause of cloudiness. These treatments are expensive and time consuming. By destroying the colloids chemically, fruit juices can be quickly clarified.

The abundant colloidal substances in fruit juice are pectin, small amounts of protein, tannin and starch. By adding the enzyme pectinase to fruit juices containing pectin, hydrolysis of the pectin takes place, and by adding the enzyme diastase, the starches are converted into soluble sugars.

The enzyme pectinase is most efficient at a pH of 3. The rate of enzyme action or breaking down of the pectin into soluble forms depends upon time of reaction, temperature of juice, pH, and the quantity of enzyme used. In the case of cloudy cider, the enzymes may be added to the juice as soon as pressed. If the temperature of the tank is maintained around 70° F enzymatic action and clarification can be accomplished overnight. At lower temperatures, it may take as long as 48 hours for complete hydrolysis of the pectins. At higher temperatures, around 110 to 135° F, enzyme action is complete in several hours. If there is no appreciable loss of color or flavor, this method of clarification is satisfactory.

Some juices naturally clarify quicker than others with the aid of enzymes. Different varieties of grape and apple juice require separate methods of enzymatic clarification. Strawberry juice clarifies easily, cherry juice with difficulty. Currant juice contains considerable pectin and more precipitate is formed than in other juices, but still it filters easily after clarification. It is difficult to set up a fixed formula for the enzymatic clarification of juices. The following methods, however, give some idea of how to treat pectin syrups and fruit juices.

*Using Clarase * for Clarifying Pectin Juice and Fruit Juices*

1 Preparatory to enzyme treatment for clarification purposes, thorough initial cooking of the pomace or the fruit is a necessary prerequisite, as the starch must be thoroughly gelatinized before enzyming in order to produce a subsequently starch-free product. After the pressing operation, the pectin liquor or fruit juice should be cooled to between 77 and 86° F.

2 Settling the juice before enzyme treatment would be a decided advantage, as this permits much of the suspended solids to settle out. The use of a preservative, such as sulfur dioxide, during settling has no harmful effect on Clarase during the subsequent digestion.

3 For treating average pectin juice, 1 pound of Clarase will be sufficient to clear the starches from between 10,000 and

* Clarase is supplied by Takamine Laboratory, Inc., Clifton, N. J.

15,000 pounds of thoroughly cooked juice, at a temperature of from 77 to 86° F in an hour or less.

For the treatment of apple juice, quince juice, or crab apple juice, 1 pound of Clarase will be sufficient for between 6000 and 8000 pounds, under the same conditions.

4 For the amount of juice to be treated, calculate the quantity of Clarase required. Weigh out the Clarase in dry form and proceed to dissolve it in water of about 80° F. Pour just a little water at first on top of the dry Clarase and stir into a paste. Add a little more water until the Clarase is thoroughly wet but still in the form of a thick paste, after which more water can be added and within a short time the material will all pass into solution.

5 Pour this Clarase solution into the pectin or fruit juice to be treated, which at this point must have been cooled to between 77 and 86° F, and stir thoroughly throughout the pectin or fruit juice. Digest until the iodine test shows a red or reddish-brown color. Traces of blue indicate undigested starch and digestion should be continued until complete conversion is accomplished; otherwise more Clarase may be required for the particular fruit juice being treated. It is customary to make preliminary laboratory tests on a small scale for each batch of juice before digesting on a large scale. Such preliminary trials will indicate the exact amount of enzyme required in each instance and the time required for the digestion.

6 After a red or reddish-brown color is obtained, the pectin or fruit juice is ready for final filtration and should be heated to about 125° F or over to stop the enzyme action, and higher still if it is to be filtered at once. Filtration must be conducted by a filter capable of removing fine colloidal particles. The most thorough digestion possible to obtain will be nullified completely by the use of inadequate filter-equipment; whereas with a proper filter, the juice is brilliant and sparklingly clear.

7 Pectin liquor so filtered is then ready for concentration, pasteurizing or bottling. Fruit juices so treated are ready for making up into jelly. They are free from starch and other suspended matter and will produce brilliant jellies which will maintain their clarity indefinitely.

*Using Protozyme PX * for Clarifying Pectin Juice and Fruit Juices*

The following method is recommended for the removal of starch and protein from pectic solutions:

Quantity used: 0.1% of Protozyme on a basis of weight, or approximately 4½ to 5 pounds of Protozyme to 500 gallons of juice. This quantity may be increased or decreased to suit individual requirements, condition of the juice and time allowable for the reactions. About an hour to an hour and a half before the batch is ready for treatment, weigh the required amount of Protozyme into a suitable container. Add water (½ gallon to every pound of enzyme used). Allow the mixture to stand at least an hour, stirring occasionally. Squeeze through cheese-cloth, using only the liquid so obtained.

Note: In treating pectic solutions it will not be necessary to strain off the solid matter if you conclude your operation with a final filtration of the pectic liquor prior to concentrating. However, it is advisable in any case to make a solution of the Protozyme with water to facilitate the enzyme going into solution more rapidly.

When all the juice from each batch has been accumulated, adjust the temperature to 120° F. Then add the enzyme liquid to the juice; if convenient, agitate for 2 or 3 minutes. Stop the agitation and do not disturb the liquor in any way until the desired reactions are completed. It will not be necessary to apply any heat to the liquor during the treatment as the reactions will be completed before the temperature takes a material drop.

Time of Action: The reactions should be complete in 40 to 60 minutes. The end point of the starch reaction is indicated by the appearance of a reddish-brown color when a small amount of dilute iodine solution is added to a sample of the juice. Previous to the removal of the last trace of starch, samples will show the characterictic blue color when treated with the iodine solution.

* Jacques Wolf and Co., Manufacturers of Protozyme PX, Passaic, N. J.

After the desired reactions have been obtained, raise the temperature of the batch to 170° F rapidly, maintaining this temperature for 10 minutes. The acting powers of the enzymes are then destroyed, after which the liquid can be allowed to settle in the tanks, or, if the operations are concluded with a final filtration, the batch can be filtered at once.

Note: If acid is used in the extracting water, the temperature of the liquor must be adjusted to 90° F before adding the enzyme. The destructive temperature can also be lowered to 125° F when acid is used. Optimum pH is 3.4 to 3.8 at 120° F. However, if the pH is adjusted to 3.2 to 2.9, the temperature for the reactions must be lowered to 90° F.

*Pectinol A * (Soluble) for Clarifying Apple Cider*

Enzyme clarification of cider is covered by U. S. Patent 1,932,833.

Directions for Use: Add Pectinol A (Soluble) to the cider to be clarified as quickly after pressing as possible. Use the proportion of 1 pound of Pectinol A (Soluble) per 100 gallons of cider (maximum). Allow to stand 5 to 6 hours at room temperature or, better, overnight if possible. Any convenient receptacle such as a barrel, vat or tank is suitable, depending on the amount of juice to be clarified.

Pectinol A (Soluble) goes readily into solution when sifted directly into the juice with gentle stirring.

To produce a polished sparkling juice, filter in the usual manner after clarification is complete.

Pectinol A (Soluble) is not a preservative. If the juice is to be bottled for later sale, it should be preserved by pasteurization. Heating to 170° F in the bottle and holding for 20 minutes is recommended.

Fruit Juice and Approximate Solids

Fruit juice used in jelly making should contain not less than the approximate percentage of total solids shown below.

* Röhm & Haas Co., Philadelphia, Pa. (Manufacturers).

The total solids is determined by immersing a Brix hydrometer into a cylinder filled with the juice at room temperature.

Juice	*° Brix or Per Cent Solids*
Apple	12 –13
Blackberry	8 – 9
Blueberry	14
Cherries (sweet)	17 –20
Cherries (sour)	15 –17
Crab apple	13 –15
Currants (red)	10 –11
Currants (black)	13 –15
Grapefruit	10 –12
Grape	14 –23
Lemon	11 –12
Lime	9 –11
Loganberry	11 –12
Orange	12 –15
Peach	13 –14
Pineapple	13 –14
Plum	10 –12
Quince	12 –13
Raspberry (black)	11 –13
Raspberry (red)	9 –11
Strawberry	5½– 7

CHAPTER XI

CANNED GOODS

Specifications for Canned Goods *

Canned goods are commonly packed in 5 sizes of cans, the commercial designation and average net contents of which are:

No. 1	12 oz.
No. 2	1 lb. 4 oz.
No. 2½	1 lb. 13 oz.
No. 3	2 lb. 3 oz.
No. 10	6 lb. 9 oz.

California Canned Fruit Grades

Adopted by Canners League of California

GENERAL DESCRIPTION OF GRADES

Superlative Quality—The top grade to be designated *"Fancy"* and packed in extra heavy syrup.

Fine Quality—The grade between the top and the Standard Grade to be designated *"Choice"* and packed in heavy syrup.

Good Quality—The medium grade to be designated *"Standard"* and packed in medium syrup.

Second Quality—Fruit packed: In 10% syrup to be designated *"Second."* In Water to be designated *"Water."* For Bakery Use to be designated *"Pie"* and *"Solid Pack Pie."*

Heavy Syrups to be used on Apricots, Peaches and Plums, and, beginning with 10% sugar by weight on Seconds, to increase in steps of 15% to each grade.

Light Syrups to be used on Pears, Cherries and Grapes, and, beginning with 10% sugar by weight on Seconds, to increase in steps of 10% to each grade.

* Reprinted by permission, from *Canning Trade Almanac*, Baltimore, Md., (1938).

Either the Brix or Balling Scale is used on hydrometers or saccharometers to indicate the percentage by weight of sugar going into the solution, the terms "percentage" and "degree" being synonymous when using these instruments.

It is to be understood in connection with the systems of nomenclature herein adopted that the grades of the different varieties of fruits are to follow, as closely as is *practicable,* these general specifications together with the detailed specifications for each grade and variety; and it is to be understood that the words or expressions descriptive of grade are used in the following definitions with the significance which they generally carry among the American canned foods trade. Also, while goods for any grade must not go below the qualifications fixed for such grade, nevertheless, in comparing a delivery of the pack of one season with a delivery of the pack of another season, it must be borne in mind that a great variation in the quality of the raw material one season as compared with another is possible and such variation will necessarily be reflected in the quality of the pack of each grade of the varieties affected, one season as compared with another.

The same specifications as to quality of grade apply to all sizes of cans.

Unless otherwise specified, fruit listed below are understood to mean:

Apricots—Approximate halves, unpeeled fruit
Pears and Peaches—Approximate halves, peeled fruit
Cherries, Plums, Grapes—Whole unpeeled fruit

MAXIMUM NUMBER OF PIECES PER CAN

APRICOTS	3" 8 oz.	3¼" 8 oz.	Picnic	No. 1 Tall	No. 2	No. 2½	No. 10
Fancy	9	10	12	16	19	24	86
Choice	11	12	15	20	24	30	108
Standard ..	13	15	18	26	31	42	151
* PEARS AND PEACHES							
Fancy	4	5	6	8	9	12	43
Choice	5	6	7	10	12	15	54
Standard ..	7	7	9	13	15	21	76

* Pears and Peaches in the 8-oz. Cans (3 inch and 3¼ inch) not otherwise specified may be either halves or quarters.

ROYAL ANNE CHERRIES							
Fancy	31	33	41	57	67	85	306
Choice	38	41	51	71	83	105	378
Standard ..	46	50	62	89	106	145	522
FIGS							
Fancy	9	10	11	17	21	30	110
Choice	11	12	14	21	26	38	140

MINIMUM NUMBER OF PIECES PER CAN

All Grades

	3" 8 oz.	3¼" 8 oz.	Picnic	No. 1 Tall	No. 2	No. 2½	No. 10
Apricots			(No minimum)				
Pears and Peaches	3	3	3	4	4	6	18
Royal Anne Cherries			(No minimum)				
Figs	3	3	4	7	8	12	40

MAXIMUM VARIATION IN THE NUMBER OF PIECES PER CAN

No lot or parcel shall vary more than the following:

	3" 8 oz.	3¼" 8 oz.	Picnic	No. 1 Tall	No. 2	No. 2½	No. 10
APRICOTS							
Fancy	3	3	3	4	5	6	21
Choice	3	3	3	5	6	7	25
Standard ..	4	4	4	6	7	8	29
PEARS AND PEACHES							
Fancy	2	2	2	3	3	4	14
Choice	2	2	2	3	4	5	18
Standard ..	3	3	3	4	5	6	22
ROYAL ANNE CHERRIES							
Fancy	6	6	8	12	14	16	58
Choice	8	8	10	15	17	20	72
Standard ..	16	16	20	30	35	40	144
FIGS							
Fancy	3	3	3	5	6	7	25
Choice	4	4	4	6	7	8	29

Unless otherwise specified, the maximum variation in the number of pieces per can, with the exception of cherries and grapes, shall in no case exceed 25% of the high count in any can in any lot or parcel; that is, the low count shall not be more than 25% below the high count in a particular parcel. In the case of cherries and grapes, the maximum percentage, unless otherwise specified, shall in no case exceed 20%. The number of units specified as the minimum of a range shall be the integer nearest to the exact quantity. For example, if the maximum count be 10 halves, the minimum will be 7; if the maximum count be 13 halves, the minimum will be 10. The above counts may be shown on the label as: Approx. 7 to 10 Halves and Approx. 10 to 13 Halves.

Yellow Cling Peaches

Fancy—In Syrup, 55% Sugar When Packed.—Fruit to be of very high color, ripe yet not mushy, and free from blemishes serious for the grade, halves uniform in size and very symmetrical.

Choice—In Syrup, 40% Sugar When Packed.—Fruit to be of high color, ripe yet not mushy, and free from blemishes serious for the grade, halves uniform in size and symmetrical.

Standard— In Syrup, 25% Sugar When Packed.—Fruit to be of reasonably good color and reasonably free from blemishes serious for the grade, halves reasonably uniform in size, color, and degree of ripeness, and reasonably symmetrical.

Seconds—In Syrup, 10% Sugar When Packed.—Trade designation "Seconds." Fruit to be tolerably free from blemishes serious for the grade, halves tolerably uniform in size, color, and degree of ripeness.

Packed in Water.—Trade designations "Water." Fruit to be tolerably free from blemishes serious for the grade, halves tolerably uniform in size, color, and degree of ripeness.

Packed for Bakery Use.—Wholesome fruit unsuited for above grades. Trade designation "Pie": Minimum drained weight 76 ounces. Trade designation "Solid Pack Pie": Minimum drained weight 92 ounces.

Figs

Fancy—In Syrup, 55% Sugar When Packed, to Cut Out a Minimum of 32%. Fruit to be of very fine color, ripe yet not mushy, free from blemishes serious for the grade, and uniform in size; woody stems removed.

Choice—In Syrup, 45% Sugar When Packed, to Cut Out a Minimum of 27%.—Fruit to be of fine color, ripe yet not mushy, free from blemishes serious for the grade, and uniform in size; woody stems removed.

Split Figs—Quoted as Fancy or Choice, to conform to the Fancy or Choice specifications except for the fact that the figs are split.

Seconds—Figs that do not conform to the foregoing specifications.

Prepared Prunes in Syrup

Prepared Prunes in syrup to be properly matured, clean, sound fruit, dried in accordance with the usual commercial practice; to be whole, separate, and of good color, odor and flavor. The count, after packing and after draining in the standard manner, not to vary in any one parcel in excess of 10 to the pound; thus, for example, the count in any one parcel to be within a range of 20 to 30, 30 to 40, 40 to 50, etc. Cans to be well filled with fruit and packed with sugar syrup testing not less than 20% sugar when packed. The minimum drained weights thirty days after packing to be as follows:

3″ "Eight Ounce"	6¼ Ounces
3¼″ "Eight Ounce"	6¾ Ounces
Picnic	8 Ounces
No. 1 Tall	12 Ounces
No. 2	15 Ounces
No. 2½	22 Ounces
No. 10	79 Ounces

At time of packing, the weight usually runs 4 to 5% less than the above and increases gradually for thirty days or longer.

Canned Fruits

Cherry

About 250 varieties of cherries are grown.

Oxhearts (black and white) *Tartarian* *Bigeran*	Are best for eating
Morello *Montmorency* *Early Richmond*	Are cooking cherries (for pies, puddings)

Royal Anne cherries are considered the best canned cherries.

Apricots

Closely related to the peach, and originally came from Armenia. They are grown extensively in Japan and China, and also in California. The kernel is used in making *"Kernel Paste,"* a substitute for almond paste.

Grades	*Park*	A fleshy fruit
	Blenheim	High in Color

Apricots are sold in canned form and as dried fruit. The so-called Spanish apricot has a higher acidity than the domestic and the latter is therefore preferred in preparations where an excess of acid would be harmful (See Apritop).

Pears

Although a great many different pears are grown in America and Europe, the *Bartlett* and *Kieffer* varieties are commonly found in the canning industry.

Peaches

Originally grown in China, are now cultivated in Europe, Africa, Asia and also in America. In general there are two kinds of peaches

Free, in which the pulp separates readily from the stone, and

Clings, in which the pulp adheres strongly to the stone.

Peaches are dried and canned. The canned peaches are packed under the following trade designations:

Yellow Cling Peaches	Fine appearance and color
Sliced Yellow Cling Peaches	
Yellow Free Peaches	Inferior in appearance but of peculiar, much preferred flavor

The dried peaches are of the yellow free variety.

Nectarine

A variety of the peach. Smaller in size and with a smooth skin.

Muscat Grapes

A very sweet, spicy-tasting grape.

Plums

The plum is a native of Asia, brought to Europe 100 years ago, and shortly after to America.

Varieties	The Cherry plum in Europe
	The Japanese plum
	The Chinese plum
	The Wild Goose plum in America
	The Sand plum in America
	The Damsons in England
	The Green Gage in England
	The Mirabelles in England

Fruits for Canned Fruit Salads

Apricots	Halves
Bartlett Pears	Halves or quarters
Yellow Cling Peaches	Halves, quarters, sixths, or eighths
Pineapple	Sectors
Maraschino Type Cherries	White, pitted, colored red with U. S. food colors

Apples

The tins shall be well filled with ripe, firm winter apples in water.

Apple Butter

Apple butter shall be made from whole, sound, fresh, ripe fruit, with no cores, peelings, or stems, free from preservatives and glucose. The finished product shall contain not less than 55% of solids.

A. Canned Soups

1. Consomme
   A clear soup, made from beef broth and a combination of the juices of several vegetables, usually those of carrots, turnips, onions, leeks, and parsley.
2. Chicken Gumbo
   A soup made from chicken and tomato stock, to which is added chicken meat, rice, and okra (gumbo). The latter gives the soup its peculiar flavor.
3. Julienne
   A soup made from beef broth, small peas, shredded carrots, turnips, celery, leeks, and cabbages. A medium thick soup.
4. Mock Turtle
   A dark colored, rich, thick soup. Made from tomato and meat juice, thickened with flour, and flavored with spices and herbs.
5. Tomato Soups

Tomato Bouillon	Made from tomato juice and beef broth.
Tomato Soup	Cooked down tomato juice, flavored with salt, sugar and spices. Thickened with flour.
Tomato Cream Soup	Made from tomato juice and added cream.

6. Mulligatawny
   Meaning *"pepper water."* A hot and spicy soup, made from chicken stock and ground tomatoes, flavored with curry powder.

B. Canned Vegetables

Beans

State Beans	Pea bean.
California Navy Beans	In *"small"* and *"large"* size.
Michigan Navy Beans	An easy and quick-cooking bean.

Kintoki Beans	From China and Japan; used principally in chili con carne.
Pinto Beans	A freely speckled bean. Very tender.

Beans are put up in four different ways:

Beans with Plain Sauce	Molasses, sugar, salt, water.
Beans with Tomato Sauce	Molasses, sugar, salt, tomato pulp, spices.
Beans with Pork and Plain Sauce	Slices of pork and plain sauce.
Beans with Pork and Tomato Sauce	Slices of pork and tomato sauce.

String Beans

There are two different kinds of string beans:

The flat pod bean and

The round or refugee bean.

Both can be wax or green beans.

No. 2 can shall contain not less than 13 ounces after draining.

No. 10 can shall contain not less than 4 pounds after draining.

Bean Sprouts

These are sprouted beans of the *Mung* variety, a bean indigenous to China, but now also grown in the United States. It is a small, round bean.

"*Bean sprouts*" are used in chop suey and other Chinese food preparations.

Lima Beans

Lima beans, if canned, shall be medium sized, fresh green or white limas, free from undeveloped, overripe, or hard beans.

No. 2 tins shall contain not less than 13 ounces after draining.

No. 10 tins shall contain not less than 4 pounds after draining.

Peas

There are two kinds of peas:

1. The early June peas, which mature in June. They are round, of smooth skin, and are the more expensive kind.
2. The sweet peas, which mature in July. They are irregular in shape, mostly oblong, and have a wrinkled skin.

The peas themselves are distinguished by their size. According to the mesh of the screen through which they pass, they are classified as:

No. 1; No. 2; No. 3; No. 4; No. 5; No. 6

Peas are put up in a brine consisting of water, sugar, and salt.

A No. 2 can shall contain not less than 12 ounces of peas after draining.

A No. 10 can shall contain not less than 4 pounds 8 ounces of peas after draining.

"Alaska" and "Winner" peas are trade names for early June peas, not indicating their origin.

Asparagus

There are three kinds of asparagus:

The white asparagus—the spear is kept under the ground.

The green asparagus—the spear grows above the ground.

The peeled asparagus—the spear is freed from its skin.

There shall be about 40 stalks per No. 2½ can.

Carrots

Carrots are put up in cubes with a brine like peas, and together with peas under the name of Carrots and Peas.

Okra

Okra is the young pod of the okra or gumbo plant.

The pods have from five to twelve sides and are from two to ten inches long. They contain numerous small seeds.

Okra is used principally in gumbo soup.

Dried Beans

Black Lima	Ceci
Black Eye	Fava (a large bean)
Black Turtle	Lima

Grades: Extra Choice, Recleaned, Choice Recleaned, Sample.

Marrow	Roman
Pea	Soya
Red Kidney	Yellow Eye

Tomatoes

Canned tomatoes must be clean, sound, ripe, fairly evenly colored, well-peeled, and cored. The can, upon being opened and drained over a sieve of ¼-inch mesh for 2 minutes, shall show not less than 50% of the contents above the screen.

Tomato Puree

May be made from either strained tomatoes * or trimming stock,** but when made from the latter the label must carry an explanatory legend equivalent to "Made from trimmings."

The customary requirements for tomato puree are:

Specific Gravity for Fancy Goods 1.045
Specific Gravity for Heavy Goods 1.050

Succotash

Succotash is a mixture of 75% corn and 25% lima beans.

Corn

Canned corn must be well cleaned and must be free from silk, bits of cob, and other foreign material. It should be of New York, Maryland or Indiana grading.

A brine consisting of salt, sugar, and water is added.

* Strained tomatoes. Obtained by straining sound, ripe tomatoes, either raw or cooked, through a screen that removes skins and seeds.

** Obtained by straining sound peeling, pieces and trimmings from ripe tomatoes through a screen that removes skins and seeds.

"Shoe-Peg" Corn, is corn on the cob minus the cob.

No. 2 tins shall contain not less than 13.5 ounces of corn after draining.

No. 10 tins shall contain not less than 4 pounds 8 ounces after draining.

Spinach

Canned spinach must be solid-packed, from fresh green stock, and must be free from grit. No. 3 shall contain not less than 1 pound, 9.5 ounces after draining the tin for 2 minutes over a ⅛-inch mesh screen.

Pimentos

Pimentos are large, sweet, red peppers that have been peeled and put up in their own juice.

Dried Fruits

Apricots

See under Canned Fruits.

Citron

The citron grows on a tree, similar to the orange or lemon tree, which is cultivated in China, Persia, Greece, Palestine, Italy and California. The fruit is an oblong, 5–9 inches long, resembling a lemon, but with a very thick peel and a smaller amount of acid pulp. The peel, thick and spongy, is used in the preparation of candied or preserved citron, and is also used in the dried state as citron peel. The drained citron peel is sugared, but is not coated with sugar, as the candied kind is.

Dates

Mesopotamia produces about 80% of the world supply of dates, the fruit of the date palm, which grows in bunches of 20 pounds and more. The palm is cultivated in western Asia, northern Africa, southern California and Arizona. Dates contain about 55% of sugar and 7% of protein; they are of high nutritive value.

Varieties

Hallawi Date	From Persia. Firm in texture, bright golden in color.
Khadrawi Date	Also from Persia. Less firm in texture, dark in color.

Both varieties are alike in flavor.

Fard Date	A dark colored, firm Arabian date.

Figs

The fig is largely cultivated in the Mediterranean countries. It looks like a pear when fresh and is harvested in June and again later in fall. Only the fruit of the second harvest can be dried and shipped.

Varieties

Smyrna Figs	Large, sweet, white fruits. Shipped in small cases, boxes, or baskets, pressed flat. They are considered the best variety.
Adriatic Figs	Shipped from Greece, squeezed together and placed on strings.
Calimyrna Figs	Grown in California from stock brought from Smyrna, Turkey.
Istrian Figs	Small, very sweet. Do not keep well. Shipped mostly in barrels or cases.
Mission Black	A black fig, cultivated in the Southern States.

Prunes

Prunes are dried plums. (See under Canned Fruit.)

California, Oregon and Italy take the foremost place in producing prunes.

There are three ways of curing prunes:

I. Sun-dried.
II. Curing in evaporators.
III. Partially cooking and then drying either by method I or II.

Varieties

California or French Prunes	Very sweet
Oregon or Italian Prunes	Tart
Oregon Petites	Sweeter than Oregon

Prunes are graded into ten sizes:

20– 30s
30– 40s
40– 50s
50– 60s
60– 70s
70– 80s
80– 90s
90–100s
100–110s
110–120s

Raisins

Raisins are dried grapes.

There are several special grapes, which are best adapted for making raisins:

I. The muscatels or sun-dried raisins. From Malaga and Valencia.
II. The Valencias or dipped raisins. Dipped in potash lye, containing lavender and rosemary, and in olive oil.
From Valencia, called Denis.
From Turkey, called Smyrna.
From California, called Valencias.
III. The Sultana, the small, tart, seedless raisins from Turkey, Greece, and Persia.

IV. The Thompson raisin, a true seedless, very sweet raisin from California.

V. The Puffed Raisin, seedless raisins, not sticky.

Size Grades

1. Crown — Raisins that at time of boxing will not pass through a screen with opening smaller than $^{12}/_{32}$ inch.
2. Crown — Raisins that at time of boxing will not pass through a screen with opening smaller than $^{17}/_{32}$ inch.
3. Crown — Raisins that at time of boxing will not pass through a screen with opening smaller than $^{21}/_{32}$ inch.
4. Crown — Raisins that at time of boxing will not pass through a screen with opening smaller than $^{21}/_{32}$ inch.
5. Crown }
6. Crown } Hand sorted jumbo, generally sold in clusters.

Currants

Currants are the dried fruit of the small black Corinth grape grown in Greece. The currant growing in the United States is the fruit of a bush like the gooseberry and is a different fruit altogether.

Grades

Vostizza
Patras
Amalias
Pyrgos

Patras and Amalias are most commonly imported in the U. S. A.

Tamarinds

Tamarinds are the fruits of a tree cultivated in the East and West Indies and in Africa. The pulp found in commerce is mostly used in medicine as a purgative. But the raw fruits are also used in the manufacturing of chewing tobacco and the extract prepared from the fruits is often added to beverages, where a tart taste is preferred. The East Indian tamarind is considered the best.

Canned Dessert Apples

Northern Spy, Northwest Greening, Arkansas, Stark Apples, Winesap and Stayman, produce a satisfactory dessert product. The following process is recommended:

1 Apples are washed, peeled, cored, trimmed and cut into eighths or sixteenths or may be diced or sliced into rings.
2 They are then placed in a 10% sugar syrup and deaerated under a vacuum of 24 inches of mercury or more for 10 to 15 minutes.
3 Place pieces in cans and cover with a 35% sugar syrup, a small part of which may be replaced with dextrose or honey.
4 The cans are now vacuumed, sealed and pasteurized at 212° F for 10 minutes in a No. 2 can for diced apples. Rings and slices are pasteurized at 15 minutes at this temperature.

Canning of Fruits and Vegetables

The aim of any method of food preservation is to keep food in good condition for those seasons when fresh fruit cannot be obtained.

Methods of canning have progressed so successfully during the past few years that, almost without exception, all foods fit for consumption can now be preserved. The particular method of preservation for long time storage to be used for a particular food should be carefully considered. Food should be stored under refrigeration, dried, cured or pickled. Under proper and satisfactory conditions, beets, cabbage, carrots, celery, onions, parsnips, pumpkin, potatoes, sweet potatoes, turnips and apples may be satisfactorily used at later periods.

Cold storaging of fruits and vegetables is cheap and easily

accomplished at the present time. Small cold storaging plants have sprung up throughout the country and lockers for depositing foods for cold keeping may be rented. It should be borne in mind, however, that if the temperature in the cold locker is too low, fruit or vegetable products are apt to become spoiled. The average freezing points at which the following fruits and vegetables will remain palatable are listed below.

Fruits and Vegetables	*Average Freezing Temperature in ° F*
Apples	28.50
Cherries	27.81
Grapes	28.16
Oranges	28.03
Peaches	29.41
Plums	28.53
Strawberries	29.93
Blackberries and raspberries	29.40
Potatoes	28.92
Tomatoes	30.40
Sweet corn	28.95
Onions	30.00
Lettuce	31.20
Carrots	29.57
Peas	30.00
Cauliflower	30.00
Celery	29.73
Cabbage	31.18

Freezing injury does not always occur when fruit or vegetable products are exposed to temperatures at, or below, their actual freezing points. Under certain conditions, many of these products can be undercooled below the true freezing temperature and, when warmed up again, no apparent injury is noted. Certain products may actually be frozen below their freezing temperatures without injury, whereas some products are injured by chilling at temperatures above their actual freezing point. In view of these facts, the possible danger of freezing injury must always be considered even though the temperature is somewhat above the freezing point, especially, if the food is exposed for a long period of time to cold temperatures.

Spoilage Prevention

Successful canning prevents spoilage by: (1) establishing conditions whereby it is impossible for the food to come in contact with anything which might cause spoilage; (2) destroying everything already present in the food which might cause it to spoil. This is accomplished by packing the food in a suitable container, applying sufficient heat to destroy any and all spoilage agents present, making the container air-tight, and storing under proper conditions. If canned food has been properly packed, heated, sealed and stored, it will keep indefinitely without spoilage. Canning powders and preservatives should never be used, as they are not necessary and may be harmful. The use of vinegar or lemon juice, formerly recommended in the canning of vegetables by the hot water-bath method, is not advisable because the method is not dependable and the flavor of the product is frequently unsatisfactory.

Spoilage Agents

The spoilage of canned food is caused either by the growth of micro-organisms (molds, yeasts and bacteria) or by chemical agents. Some micro-organisms are found almost everywhere, in the air, in the soil and on the surface of food. It is the failure to kill these organisms or to prevent their contact with food which causes all microbial spoilage. In view of this fact, it is important to recognize the kinds, causes and dangers of spoilage, and to understand how spoilage can be prevented.

Molds

The familiar fuzzy growth produced by molds is easily recognizable. Molds usually begin their growth upon the surface of the canned food and penetrate slowly to the center. If the mold is present in small amounts, it may be removed from the surface, and the remainder of the food may be eaten with safety, as there is little evidence that moldy food causes illness. If a heavy growth of mold is present, however, the food should be discarded, as the flavor and the texture will be unfavorably affected.

Most molds require air. They grow rapidly in warm, dark places and in the presence of sugar, acid and moisture. For this reason, jellies, preserves, fruit juices and tomatoes are likely to be affected. Acid foods, such as fruits, when canned with a large amount of sugar, are particularly susceptible. Although molds are easily destroyed by heating for a short time at the temperature of boiling water, and sometimes even at a lower temperature if the heat is continued for a sufficiently long time, spoilage may take place if the food is heated in an open kettle and then transferred to hot jars, sealed and stored. The molds from the air may contaminate the food as it is being transferred, or the molds may be in the jars if they have not been properly sterilized. Molds will not spoil preserved food if the jars containing the food are heated in a boiling water-bath long enough to kill the molds and are then adequately sealed.

Yeasts

Yeasts are micro-organisms which often cause canned fruits and fruit products to ferment. Fermented products usually have a peculiar appearance, a sour taste and an alcoholic odor. One of the products of fermentation is carbon dioxide. The gas accumulates in the container and, even though there may be no outward sign of spoilage, there will be an outburst of gas or a spurt of liquid when the container is opened. If sufficient gas accumulates, it may break the seal of a glass jar so that the contents leak out, and occasionally a jar may blow up. Spoilage of this kind in foods, canned in tin, causes the ends of the cans to round outward. The trade term for such cans is swells. Any gas would produce the same result.

Tomato products and fruit products may be spoiled by yeasts which grow in the presence of acid and sugar. Yeasts are destroyed, even more easily than are molds, by heating for a short time at the temperature of boiling water. They are more easily killed in unsweetened fruit juice than in juice containing a large amount of sugar. Like molds, yeasts may cause the spoilage of a food if it is heated in an open kettle and then transferred to hot jars, sealed and stored. There is no danger of fermentation, however, if the jars containing the food are heated in a boiling water-bath long enough to kill the yeasts and are then adequately sealed and stored.

Bacteria

Bacteria are the chief sources of trouble in canning because they may exist in resistant forms, called endospores. These are not always destroyed at the temperature of boiling water and may develop later, when the canned food is stored, and cause spoilage.

Bacteria are easily killed by heat in the presence of acid; therefore, such acid foods as fruit products and tomatoes, when canned by the boiling water-bath method, are not subject to bacterial spoilage. However, non-acid foods, e.g., meat and vegetables other than tomatoes, may be heated in containers in a boiling water-bath for as long as 6 hours without destroying the spores of certain bacteria. The reason for spoilage can usually be traced either to inadequate heating or to the absence of an air-tight seal. Inadequate heating usually occurs in the following ways: heating in a boiling water-bath, i.e., a steamer (212° F is about the highest temperature which the food reaches by this kind of heating), instead of in a pressure cooker; heating at too low a pressure when a pressure cooker is used; or heating for too short a time. To prevent the spoilage of meats and non-acid vegetables, a pressure cooker must be used, since only in this way is a temperature (from 240 to 250° F) obtained which is sufficiently high to kill the spore forms of the bacteria.

Bacterial spoilage causes a number of conditions which may result in cloudiness, softness, putrid and disagreeable odors, gases, acid or other off flavors and toxins.

Cloudiness and Softness

Cloudiness and softness alone do not necessarily indicate spoilage, but food with such an appearance should not be tasted until it has been boiled actively for 10 minutes. If an odor develops during boiling, the food should be discarded without tasting.

Putrefaction

Food which has begun to putrefy gives off a disagreeable odor and should be discarded without tasting.

Gases and Swells

Certain bacteria produce gas which, like the gas from yeast fermentation, may cause: the ends of tin cans to bulge; containers to set free an explosive puff of gas or spurt of liquid when they are opened; jars to leak; or jars to blow up. Food from such containers should be regarded with suspicion. If bacterial spoilage has taken place, the product usually has a sour or putrid odor and, sometimes, a foamy appearance. Any food showing such a spoilage should be discarded without tasting. If the food does not show spoilage, it should be heated. If the food neither looks nor smells unusual after it has been heated, it should be boiled actively in an open vessel for 10 minutes, and then tasted. If the food is wholesome, the swelling probably was due to overpacking and not to the formation of gases during spoilage. If fruit is canned in tins, the acid present may have acted on the tin to produce a harmless gas with no harmful effect on the product.

Acids and Flat Sour

Other bacteria produce acids which give the canned food a sour taste. No gas is formed, however, and the ends of the spoiled cans remain flat. There may be no evidence of spoilage until the container is opened; then the food usually has a sloppy appearance, a cloudy liquid, an unpleasant odor and a sour taste. Such food should be discarded. Peas, asparagus, corn, lima beans and string beans are particularly susceptible to this type of spoilage. The damage is usually caused by thermophilic, or heat-loving, bacteria, which require heat for their development. Such spoilage can be reduced to a minimum by packing the food into the jars quickly, heating the jars promptly after they are packed, cooling them immediately and thoroughly after they have been heated and storing them in a cool place.

Toxins and Botulism

Some bacteria during their growth produce extremely poisonous substances called toxins. One of the most dangerous of the toxin-producing organisms is Clostridium botulinum, an organism found in the soil, which causes botulism, usually a fatal disease. Unfortunately, signs of spoilage, which can be detected by casual observation are not always evident when the fatal toxin has been produced. Sometimes a gas is formed together with the toxin, so that swells are produced. The food may have a dark or mushy appearance and a cheesy or putrid odor. Food showing evidence of such spoilage should be discarded without tasting, for one taste may prove fatal.

The toxin is rare but may be present in other forms of bacterial spoilage. It has been found in home-canned meats, in soup mixtures and in such non-acid vegetables as asparagus, corn, greens, peas and string beans which have been canned at the temperature of boiling water. Acid foods, such as fruit products and tomatoes, are not subject to this type of spoilage because the organism does not develop and produce its toxin in an acid medium.

The bacteria produce heat-resistant endospores which are harmless themselves, but which, if not destroyed during the heating period, develop in the cans during storage, and produce the poison. To insure the destruction of the spores, meat and non-acid vegetables must be canned in a pressure cooker, for this is the only method by which a temperature high enough to kill the spores can be obtained. The canning of meats and non-acid vegetables by methods other than the pressure-cooker method is unsafe and cannot be recommended. These foods should be preserved by methods other than canning, if pressure cooking is not possible. Because of the danger of poisoning from toxins, the Bureau of Home Economics, at Washington, D. C., recommends that all home-canned meats and vegetables be removed from the can, brought to a boil, and boiled actively in an uncovered vessel for 10 minutes before they are tasted, to destroy the toxin if it is present. The treatment is, of course, applied only to food that appears wholesome; other food should be immediately discarded.

Chemical Agents

Chemical changes in a container may be of two kinds. They may result from a reaction between the food and the material of which the container is made, or from enzyme action. When fruit is canned in tin, the acid may attack the tin and produce hydrogen gas, with the production of hydrogen swells. The food, although it may be bleached, will otherwise have a normal appearance and will be wholesome. When corn is canned in tin, discoloration may take place because of chemical action. In order to avoid this action as much as possible, the product should be canned in glass or in enamel-lined containers recommended by the manufacturer for that particular product.

The most noticeable effect of enzyme action is discoloration. If the food is in a glass jar, the top surface may first turn brown, then black. Much discoloration produces a food of very poor quality and, at best, the taste is somewhat unfavorably affected. The discolored portions should be discarded, although the remainder of the food may be used if no objectionable odor develops after it has been boiled for 10 minutes. The development of any foreign odor indicates that other spoilage has taken place, in which case the food should be discarded without being tasted.

To prevent this type of damage, the canned products should be stored in a cool place, since most chemical reactions are retarded at lower temperatures.

Canning Methods

In general, commercial canning methods for all food products are similar, but minor variations are required by different foods. Prompt delivery of the raw products is of the greatest importance, since most vegetables and fruits change or deteriorate rapidly after gathering. So important is the matter of freshness that canneries are usually located near farms or orchards. Peas, corn, asparagus and green beans are all seriously injured by standing. Berries must be collected in shallow lug boxes to prevent bruising. The milk collected for the evaporated product must be even fresher and cleaner than that generally delivered to city markets, and fish must be brought to the canneries by special boats employed solely for the purpose.

On arrival at the cannery, the loads of vegetables, fruit or fish are first given a general examination by the superintendent. The next step is to sort for condition and size. This is usually done by machinery. After the product is shelled, husked, peeled or trimmed, as its nature may require, it passes to machines where it is thoroughly washed. Some products are then placed in the cans after preliminary heating or cooking; others are packed cold and then passed through a steam chamber known as an "exhaust box" in order to heat the contents before the can is closed. Mechanical fillers are usually employed.

The filled cans are passed under a machine called a capping machine, or double seamer, which places a cover on the can and clamps it on by a rolled seam so that it is air-tight. The cans are then carried mechanically by a conveyor to the process kettles, where the product is cooked, either in boiling water or under pressure. The time and temperature of the cook, or process, depends on the nature of the food product. Because many products are extremely perishable and must be expeditiously handled if they are to be of the best quality, the aim of the modern canning plant is to get the product into the can and cooked just as quickly as possible. In modern plants the product is usually in the cans within a very short time after the raw material has been received. The delicious flavor of many canned foods is due to this quick handling, and to the fact that the raw products are gathered at the best stage of ripeness. In most canneries, after the preliminary handling, the product is not touched by hand.

Sorting, Cleaning and Trimming

In an endeavor to prepare a properly finished product, various methods for sorting, cleaning and trimming have been devised. Practically every product now canned goes through these three processes.

Where the nature of the product permits, the sorting is done by machinery, but in many cases it must still be done by hand. The sorting consists of a series of processes; general sorting before the product is washed and cleaned, and a final sorting before the product goes into the cans. The purpose is twofold. It removes the damaged or inferior material in the raw product as it arrives at the cannery, and it separates the material into various sizes. The latter might more properly be called grading for size.

The most ingenious machines have been perfected to wash and clean fruits, vegetables, fish, etc., before they are finally put into cans. Many of these machines are so arranged that a continuous stream of water is sprayed on the product, which, in many cases, is mechanically scrubbed by brushes. Such products as peas, beans, fruits, etc., after a thorough washing are passed over long tables or traveling belts, where imperfect vegetables or fruits and any foreign materials are removed by hand. This insures a clean sound product for canning.

Blanching

Certain vegetables and fruits are given a preliminary cooking in hot water before they are packed in the cans. This treatment is known as blanching and is varied in degree, according to the size and maturity of the fruits or vegetables so that even though the final sterilizing process is the same for a given product and size of container, the contents will all be cooked to the proper degree of tenderness. The blanching treatment not only softens the raw product so that the cans may be properly filled, but also serves as a preliminary cooking treatment and removes gummy substances from the outside of certain vegetables, e.g., peas and beans, which would make the liquor in the can more or less cloudy.

Filling

While fruits and fish, except salmon in tall cans, are placed in cans by hand, vegetable and milk cans are generally filled by mechanical fillers. Hand-filling is necessary with the former to secure the proper fill without crushing or breaking up the food. Hand-filling is checked by weight, and mechanical fillers are generally designed to deliver a definite volume into each can.

When syrup or brine is used, or oil for sardines, it is added by mechanical fillers, adjusted to deliver a certain volume to each can.

Preheating or Exhaust

The next step in the canning process, employed chiefly when products are placed in the can cold, is to give the cans a preliminary heating, or exhaust. This is done by passing the filled cans on a conveyor through a steam chamber or hot water tank for 5 to 10 minutes. Products such as corn, which is placed in the can while hot, and peas, which are covered with a hot brine, are ordinarily not exhausted. Tomatoes, which are packed cold, are generally exhausted. The purpose of the exhaust treatment is to produce a partial vacuum in the can after it is closed and finally cooled, so that the ends of the can will be flat or concave and will not bulge.

In recent years, machines have been devised which produce a partial vacuum in the cans by sealing them in a chamber from which the air has been partially exhausted.

Closing

The open-top can is closed by an automatic closing-machine known as a double-seamer. The edge of the cover and the flange of the can are first clamped together and then rolled under. To make the seam air-tight, a gasket is used, composed of paper or of a rubber compound spread on the edge of the cover.

Processing

Success in canning depends largely upon the *process,* that is, proper sterilization by heat. A knowledge of the principles of sterilization and sanitation is something that every successful canner must have and must put into practice. The increase of such knowledge is one reason why canned foods, as a class, are second to none in cleanliness and wholesomeness. The commercial canner has the advantage of producing his own raw materials or securing them directly from the farmer or grower. He is also equipped to process or sterilize canned foods by methods best adapted to each product.

Acid products such as tomatoes and fruits can be readily sterilized at temperatures not over 212° F, but most vegetables, meat, fish and milk require a considerably higher temperature to sterilize them within a reasonable time.

Four types of equipment are used for processing or sterilizing canned foods. They are: the open-bath cooker; retort or pressure cooker; continuous cooker; and the agitating cooker.

The open-bath cooker is simply a large tank or trough containing water, heated by open steam coils, in which the cans are immersed, usually in metal crates or by means of a mechanical conveyor. In such cookers, the temperature cannot exceed 212° F, the boiling point of the water.

The retorts, or pressure cookers, are made of steel boiler plate and have a lid which can be fastened down air-tight. After the cans are placed in the retorts, the retorts are closed and steam under pressure is admitted. For some products, the temperature is carried up to 250° F, and the period of heating, to complete sterilization, varies from 20 minutes to 3 hours, depending on the product and the size of the containers. Either vertical or horizontal retorts, which have steam and cold water connections and are equipped with thermometers and pressure gages, may be used.

A continuous cooker is one provided with a conveyor system which carries the cans through the cooker, making it continuous in operation. Either hot water or steam may be used for heating, and continuous cookers of special construction are made so that heating can be done under pressure if desired. By regulating the speed of the conveyor or changing the point of exit for the cans, the cooking time can be closely regulated.

Agitating cookers are similar to continuous cookers, but are provided with an arrangement for turning the cans in the cooker, which makes cans of semi-solid products heat through more rapidly.

Cooling

As soon as the sterilizing process is complete, the cans are cooled. Cooling is essential for many products, to prevent over-cooking, and is done either by water, or by spreading the cans out in the warehouse so that the air circulates freely among them. In water cooling three methods are commonly used: the retorts may be filled with cold water before the cans are removed; the cans may be conveyed from the cooker through a tank of cold water; or they may be placed under sprinklers.

Syrup and Brine

In canning fruits and vegetables, as previously mentioned, brine or syrup is usually added, as a certain amount of liquid is usually necessary in the canning process to secure proper sterilization and to retain the desired flavor and consistency. The syrup or brine is, therefore, as essential a part of the product as the fruit or vegetable itself.

With the exception of the water or pie grade of fruit, which is packed in water and sold largely to be manufactured into other products, such as pies, fruit syrups and preserves, all grades of canned fruit are packed with syrup of varying degrees of strength. The strength of the syrup in the finished product is not the same as that of the added syrup, because of the gradual diffusion of the juice of the fruit into the syrup. As the amount of juice present and of syrup added vary somewhat with the condition of the fruit and fill of the can, it is impossible to tell exactly the strength of the original syrup, but it is possible to make an approximate estimate.

In commercial practice, syrups for canning are made up to a definite sugar strength, as measured by degrees upon the Brix scale, and tested by a hydrometer adjusted to give direct readings on that scale. For example, a 20° syrup, Brix scale, is prepared by dissolving 20 pounds of sugar in 80 pounds of water to make 100 pounds of syrup.

Syrup in the canned fruit may be tested by the hydrometer, just as plain sugar syrup may be. The hydrometer reading will then depend not only on the amount of sugar added in the form of syrup, but also on the amount of sugar and other soluble substances naturally present in the fruit. It will indicate approximately the total percentage of sugar in the liquid contents of the can.

Most vegetables are canned with the addition of a certain amount of brine, which serves the same purpose as syrup in canned fruits. The use of brine in canned foods is subject to the same restrictions that govern the use of syrup. The brine is a weak solution of common salt (1½ to 2%) or of salt and sugar, the choice being determined by the character of the product and the flavor desired. Some vegetables, e.g., sweet potatoes, tomatoes and pumpkins, do not require the addition of any brine.

Grades and Qualities

Raw products for canning, particularly fruits and vegetables, are usually graded for size or quality, or for both. This is one of the first steps in the canning operation. Quality grading is often closely associated with the size of fruits and certain vegetables, and it is an interesting fact that in the case of fruits, the larger sizes go into the higher quality grades, while the reverse is true of vegetables such as peas, beans and beets. Grading for size is largely done by machinery, and grading for color, defects, etc., is done by hand and eye.

Fruits

The designation of the three grades of quality used by fruit canners in this country for table fruit are, in order of quality: fancy, choice and standard. In addition, there are two lower grades packed in very light syrup or in water and variously designated by such terms as *seconds, natural, water,* or *pie* grade. The lower grades usually consist of undersized or irregular fruit, or of fruit which is either too ripe or not ripe enough for table fruit but is, nevertheless, sound and wholesome.

A good idea of the quality of canned fruit may be obtained from the general appearance of the product. Generally speaking, the quality of the fruit governs the style of its preparation. A heavier syrup is used on the higher grades, and in the case of the large fruits, like peaches, pears and apricots, the larger and more uniform fruit goes into the higher grades. With certain of the smaller fruits, the quality of the fruit in the higher grades does not differ materially from that in the lower grades, the difference being largely in the amount of sugar used. The lower grades are usually canned with very little sugar, or in water. A cheaper, but wholesome, article is furnished in this way to meet the great demand for the manufacture of pies and various fruit products. The quality of fruit used in the lower grades is subject to much greater variation than that in the higher grades.

The grading of peaches illustrates very well the principle governing the grading of the large fruits. For fancy grade, perfect stock in the best state of ripeness is used and handled with particular care. The fruit must be large enough to take not more than 12 pieces (halves) to fill a No. 2½ can; a 55° Brix syrup is added. For choice grade, the fruit must be nearly perfect but need not be so large, the limit being 15 halves per can; a 40° Brix syrup is used. The limit for standard grade is 21 halves of good but less uniform fruit and the strength of the syrup is 25°. The so-called *second* grade calls for sound fruit of fairly uniform size and ripeness, without any definite size limits, packed in 10° Brix syrup. The *pie* grade simply requires sound, wholesome, fruit packed in water in large cans, chiefly for the manufacture of other food products. In the lower grades, the fruit is of smaller size, may be a little underripe or over-ripe, may lack uniformity and, though sound, may contain some blemishes.

The smaller fruits are usually graded only for quality, which is based largely on uniformity of color, size, flavor and freedom from damaged fruit.

Vegetables

There are three grades of quality for canned vegetables in commercial use, the grades generally being fancy, extra standard or choice, and standard. Other designations are also used, based upon the size of the material, such as the number of stalks in a can of asparagus or the size of peas.

The grading of vegetables varies for different products. Grading of peas is carried to an extreme, so that a canner may put up fifteen or more grades of the same variety. This is due to the usual grading of peas into five or six sizes and three or four quality grades. On the other hand, such products as pumpkins and sweet potatoes are packed in only two grades, fancy and standard.

The condition of the food product is the chief factor in determining vegetable quality grades. The composition of the brine or liquor is independent of the grade, although in a fancy grade of corn, for example, more sugar is usually added in the brine than in the standard grade.

Fancy grade calls for materials of the highest quality and flavor, uniform in color and size, handled with special care. Standard grade is prepared from sound material, usually about average run of the crop, so that it may lack uniformity in size, color and ripeness. The extra standard or choice grade is intermediate between fancy and standard, and applies to a product that falls slightly below the rigid requirements of the fancy grade, but is still better than standard.

Other Products

Evaporated and condensed milk are not graded but must comply with the government standards in fat and total solids content.

Canned meats and meat products are not graded, but are subject to the rigid inspection and regulations of the Federal Meat Inspection Service. This includes also canned stews and soups with a meat base.

Canned salmon and tuna are graded for species and color of flesh in the can, while sardines are graded for size according to the number of fish per can. Oysters and shrimp are graded only for size.

Methods of Canning Typical Products

Vegetables

Asparagus

The canning of asparagus differs considerably from the canning of most vegetables in that a large amount of hand labor is involved. Promptness at every step of the harvesting and canning process is especially necessary to prevent deterioration and loss of flavor. The plants are set in rows and covered with a deep layer of soil so that long stalks are sent out by the plant in order to reach the surface. The stalks or portions of stalks that are not exposed to light are white, and these are generally preferred to the green stalks that grow above the ground. The stalks are harvested by hand, washed and cut to uniform length in the field and then hauled to the cannery.

Before being canned the asparagus is sorted into grades, both as to color and as to size of stalk. It is then blanched and packed into cans by hand, both round and rectangular cans being used, brine is added, and the cans are sealed and processed.

Beans

The canning of beans has grown to very large proportions in recent years. They are packed with plain sauce, with tomato sauce, with pork, or with both pork and tomato sauce. Dried white pea beans or navy beans of the grade known as choice hand-picked are generally used; these are grown principally in Michigan, New York and Wisconsin.

The beans are first soaked in tanks of cold water from 10 to 15 hours, with frequent changes of water. Sometimes a short blanching of the beans helps to secure prompt and uniform swelling. The beans are then blanched in hot water for several minutes, a blancher of the same type as the pea blancher being commonly used. Some beans are also cooked in jacketed kettles or ovens. Such a dry heating process is necessary to entitle the product to the term *baked beans,* under the ruling of the United States Bureau of Chemistry.

Plain sauce for canned beans is prepared from molasses,

sugar, salt and water. Tomato sauce is made from the same ingredients, with the addition of tomato pulp and spices. In canning beans with pork, a piece of pork is placed in each can before it is filled with beans and the sauce is then added. Finally the cans are sealed and processed under pressure.

Red kidney beans are canned to a considerable extent by similar methods.

Green and Wax Beans

Great improvement has been made in the varieties of green and wax beans grown for canning with regard both to color and to elimination of strings.

The beans are hand-picked while young and tender and are brought in boxes to the cannery, where they are graded for size and snipped mechanically. The larger sized beans are also cut into even lengths by machine. No. 1 grade are beans which pass through a $\frac{3}{16}$-inch opening, while the largest, or No. 5, grade are too large to pass through a $\frac{5}{16}$-inch opening.

Hand sorting is often necessary with wax beans to remove defective beans.

Lima Beans

Lima beans have long been cultivated in this country for canning and drying as well as for direct table use. The lima bean is believed to have originated in South America. Two types are cultivated in this country, the bush or dwarf varieties and the long-vine or pole varieties. The young green beans are the highest grade of product, as the beans become more starchy as they mature, lose their green color and are less attractive in appearance.

Canned lima beans prepared from dried California beans are also packed to a considerable extent. They are required to be labeled to distinguish them from green canned beans.

Lima beans are harvested on the vines and the beans are separated by means of vining machines like those used for shelling peas. The beans are mechanically graded for size and are picked over by hand to remove foreign matter and separate green from white beans. The beans are then washed, blanched and packed with hot brine in cans, which are sealed and processed under pressure.

Succotash

Both fresh and dried lima beans are largely used with sweet corn in preparing canned succotash. Where fresh beans are used, varieties of corn and beans which mature together are selected and planted for the purpose. The preliminary preparation of the beans is the same as already described. After the corn and beans are mixed, the product is treated as in the case of corn. The usual proportion of lima beans in succotash is 20 to 30%.

Corn

Sweet corn for canning is grown in the northern part of the United States, from the Atlantic Coast to the Missouri River valley. In the eastern states the crops are generally grown by farmers on contract, while many of the western canners raise their own corn. The effort of canners is to develop a tender fine-flavored sweet corn. The seed is especially selected and grown in regions which have been found to be best suited for producing the finest quality. The principal varieties of sweet corn canned are Evergreen, Country Gentlemen, Crosby and Golden Bantam.

Canned corn is put up principally in two forms, whole-grain style and cream style. In the former, the kernels are simply cut from the ears after husking, whereas in the latter, the upper portion of the kernel is cut off and the creamy contents of the lower portion are forced out by means of scrapers, which leave a large portion of the fibrous part of the kernels on the cob. Both types of product are freed mechanically from silk and pieces of cob and are packed in cans with the necessary amount of hot brine, to which sugar is usually added. The contents of the cans are heated nearly to the boiling point and, after the cans are sealed air-tight, they are cooked and sterilized under pressure in such a way as to affect the appearance of the product as little as possible.

Hominy

Canned hominy is a convenient form of vegetable and breakfast food and is a staple article of diet in camps. Lye hominy was a staple food of the early settlers, who used lye prepared from wood ashes to soften and remove the corn hulls. For canning, lye hominy is prepared from white corn, which is screened, washed and treated with hot lye solution for a sufficient time to loosen the tough hulls, which are removed in a mechanical hulling machine. The hulled corn is then thoroughly washed to remove lye, cooked till tender and packed in cans. Weak brine is added and the cans are sealed and processed under pressure.

Beets

Only red varieties of garden beets are used for canning, and they are planted so as to reach the desired small size in the fall. This ensures a maximum of tenderness and prevents the development of large, coarse-fibered beets.

In the cannery, the beets are graded for size and thoroughly scoured mechanically. They are then steamed in retorts to facilitate peeling, and are peeled and filled into cans by hand. Hot water or weak brine is added and the cans are sealed and processed under pressure. Enamel-lined cans are used to prevent bleaching of the red color. Large beets, i.e., beets over 2 inches in diameter, are cut into pieces before being canned, and are sold as cut beets. Small and medium size beets are canned whole or sliced.

Peas

Peas are one of the staple canned vegetables. The quality of the product has been continually improved by the selection of types of peas grown especially for canning, and by the development of better machinery and equipment for handling them promptly in very large quantities. In modern canneries peas are not touched by the human hand from the vine to the can, except defective peas which are removed at the picking tables as described later.

Peas are successfully grown for canning in nearly all the northern states and are of two types, the smooth, early pea or Alaska type and the sweet, wrinkled type. After they are canned, the Alaska peas may generally be distinguished from the other types by their more uniformly round appearance.

Peas in the same pod vary considerably in size, from the big ones in the middle down to the small ones at each end. Therefore, the canner must judge by experience the best time to gather the peas to secure not only the largest yield, but also the most desirable size and degree of tenderness. To spread the canning season out and prevent the peas from maturing all at once, the planting is extended over several weeks so as to give a fairly uniform supply that can be conveniently handled.

When peas in the field have reached the proper stage, they are harvested, vines and all, and hauled to the cannery or vining station, where the peas are separated by a special type of threshing machine known as a viner. The shelled peas are next freed from leaves, pieces of pod and loose dirt by passing them in front of a blower or fanning mill and over vibrating slotted screens. The peas are also washed under sprays of cold water, this washing process taking place either before or after they pass through the graders, which are a series of large horizontal cylinders perforated with round holes, graduated in size, and which separate the peas into five or six different sizes. Although some very tender peas are canned without being graded, it is generally more satisfactory to grade them to achieve uniformity. By separating the different sizes, the canner is able to vary the method of preparation for each size to yield the most satisfactory product.

After being washed, the peas are carried on moving belts past pickers, who sort out defective peas or foreign materials that may have escaped in the previous treatments. The peas are next blanched or given a preliminary cook in a large cylinder revolving in a tank of hot water. The peas are fed continuously through this blancher and the length of cook is adjusted by varying the rate of revolution of the cylinder. After another washing, the peas are conveyed to fillers, machines which automatically measure and fill the peas into the cans and add hot brine. The cans are promptly sealed and processed in retorts under pressure.

Canned peas are graded for quality, as well as for size, so that it is not unusual for a canner to produce as many as twelve or fifteen grades of peas. The terms fancy, extra standard and standard are used to describe quality grades, while the expressions petit pois, extra sifted, sifted, Marrowfat and telephone apply to sizes. The term *early June* was originally used to designate a certain sized pea, but it is now generally applied to any peas of the early, smooth type.

Pimientos or Sweet Peppers

Until a few years ago, canned pimientos were imported chiefly from Spain, but pimientos of equally good quality and of fine red color are now produced in the south and in California. After being delivered at the cannery, the peppers are usually graded for size. Two methods of coring and peeling are used. By one method, the whole peppers are immersed in hot oil to loosen the skin and, after being cooled, they are peeled and cored by hand. In the other method, the peppers are cored by cutting off the stem end and removing the core and seeds, either by machine or by hand. The peppers are then passed on a conveyor through a roasting furnace, which parches the skin and makes it easy to remove. The skins are removed by passing the roasted peppers through cylindrical washers, and the peeled peppers are then sorted and packed in cans, exhausted, sealed and processed.

Pumpkin and Squash

Pumpkin and squash are canned to a considerable extent and are handled in the same way. In recent years decided improvement has been made in the methods of canning. The pumpkins on arrival at the cannery are washed, stemmed and cut into pieces with cleavers or by slicing machines. The seeds are then removed by washing under strong sprays or by a mechanical shaker of perforated metal, through which the seeds, but not the pieces of pumpkin, can pass. The cut pumpkin is then either placed in crates and cooked in closed retorts or is steamed in large covered tanks until thoroughly softened, after which it is conveyed to a cyclone, or horizontal cylindrical sieve with inner paddles revolving at a high rate of speed, which removes pieces of skin and coarse fiber. If the resulting product is too thin and watery, it is thickened by pressing or draining to the desired consistency and may then be passed through another cyclone with finer mesh to give the product a smooth body. It is then placed in the cans while hot, sealed and processed under pressure. Pumpkins and squash are now generally packed in enamel-lined cans.

Sauerkraut

Sauerkraut is cabbage which is subjected to a process of fermentation that converts the small amount of natural sugar in the cabbage to lactic acid, and gives the product the acid flavor characteristic of sauerkraut. The cabbage is trimmed mechanically, shredded, and packed tightly in large vats or tanks and sprinkled with dry salt. If it is not too cold (70 to 85° F), the fermentation soon starts and proceeds for 2 to 4 weeks, depending on the temperature, after which the product is ready for packing. The sauerkraut is placed in cans by hand, covered with hot water or brine, and then exhausted, sealed and processed in boiling water.

Spinach

Spinach for canning is grown in the spring and the fall, as cool weather is necessary for the production of a tender and succulent product. On arrival at the cannery, the spinach is picked over by hand to remove coarse stems and leaves that are discolored or too mature. It is then washed in a cylindrical washer under strong sprays; revolving paddles are often used to stir up the leaves and insure removal of all dirt and grit. The washed spinach is blanched in hot water for a few minutes and placed in cans while hot. Brine is next added and the cans are exhausted, sealed and processed under pressure.

Sweet Potatoes

Sweet potatoes are a valuable food, rich in starch and sugar, but, unlike white potatoes, spoil easily in ordinary storage. Consequently, canned sweet potatoes, although a comparatively new product, have met with popular favor by making this vegetable available throughout the year.

When received at the cannery, sweet potatoes are graded for size, as subsequent treatment depends to some extent on the size of the potatoes. They are next placed in crates and steamed in retorts for sufficient time to loosen the skins and cook the potatoes. As soon as the potatoes are cool enough to handle, the skins are stripped off by hand and the potatoes are packed tight into cans while still hot. The filled cans are given a long exhaust, and are then sealed and processed at 212° F.

Tomatoes

Tomatoes are canned in nearly all sections of the United States, and the average annual output for the past 10 years has been about 11,800,000 cases (equivalent to about 590,000,000 pounds). In addition to being one of the most important canned vegetables from the standpoint of annual output, the tomato is foremost of the vegetables in the variety of products which it yields; these include canned tomatoes, tomato pulp, tomato puree and tomato paste. Other canned products of which tomatoes form an essential ingredient are catsup, chili sauce, tomato soup, tomato sauce, okra and tomatoes, baked beans, spaghetti with tomato sauce and sardines in tomato sauce.

Canned tomatoes are customarily packed and designated under three different grades: fancy, choice or extra standard, and standard. The difference in these grades depends largely upon the brightness and uniformity of the color and the proportion of whole tomatoes or large pieces contained in the can. Special varieties of tomatoes have been developed by careful selection and breeding, which are distinguished for their bright color, size and comparatively smooth exterior.

Tomato canneries are located in the center of tomato-growing districts and the canner usually buys his tomatoes from the grower by contract. In this way, it is possible to secure tomatoes of the best quality for canning purposes and to assure their prompt delivery to the cannery in the best possible condition.

Tomatoes are delivered in baskets or boxes to the cannery and are washed in a mechanical washer; spray washers supplying jets of water under pressure are now largely used for this purpose. From the washer, the tomatoes are passed through a scalder, where they are subjected to the action of live steam for a few seconds, to loosen the skin and make the subsequent peeling easy. They are then generally given a short spray of cold water and are carried to the peelers' tables, either in pails or on conveyor belts. Tomatoes are usually sorted, peeled and trimmed by hand, but mechanical tomato peelers have also been used to some extent in recent years.

After the tomatoes are peeled they are delivered in buckets to the fillers. The sorting and filling of fancy tomatoes is usually done by hand, while tomatoes of the standard grade are usually placed in the cans mechanically. The filled cans are then passed through the exhaust box, after which they are sealed and processed.

Tomato Paste and Puree

These are more or less concentrated tomato products, especially convenient for use in the preparation of various dishes. The tomatoes for making tomato paste or puree are washed, scalded, sorted and trimmed. They are then run into a machine called a cyclone, which consists essentially of a cylinder made of fine screening with a revolving inside paddle, which separates the skins, cores and seeds.

The liquid juice and pulp is then concentrated by boiling in large kettles, which may be done under vacuum for the more highly concentrated paste. It is finally run through a finisher, a machine similar to the cyclone but with a finer mesh screen, and then packed in cans. Cans holding from 5 to 6 ounces are used for tomato paste and those holding 10 ounces to 6½ pounds, for tomato puree.

Fruits

Apples

In canning apples, several methods are used, the nature of the canned product varying accordingly. The varieties usually canned are those best adapted for cooking purposes, fairly uniform in shape, mildly acid and of fine texture. The canning method largely in use up to a few years ago consisted of peeling, trimming and coring the apples, packing them in cans, exhausting the cans for a few minutes, adding hot water, and sealing and processing at about 212° F.

A number of changes have been effected by the introduction of efficient peeling, coring and slicing machines, and by the practice of immersing the peeled fruit in warm water, or in a weak salt solution, to prevent discoloration and to use up the oxygen naturally present in the cells of the fruit, which may cause trouble later. Air is sometimes removed from the fruit by immersing it in water in a retort and then producing a high vacuum in the retort.

For solid-pack canned apples, which have gained considerable popularity in the past few years, the apples, after the preliminary preparation described above, are placed in a warm, weak salt solution for 30 to 40 minutes. This treatment is for the purpose of getting rid of the oxygen contained in the cells of the fruit, which would otherwise cause darkening of the fruit when heated. The apples are then blanched for a few minutes with steam, which makes them soft and flexible so they can readily be packed solid in the cans, which is the next step. The cans are then exhausted, sealed and processed.

Apple sauce is prepared from peeled, cored and trimmed apples which are thoroughly cooked with sugar and then run through a mechanical sieve or cyclone.

To a limited extent, apples are canned whole, after being peeled and cored.

Apricots

Canned apricots are almost entirely a Pacific coast product, as in no other part of the country has the cultivation of this fine table fruit reached such a large scale development. The raw fruit is gathered and handled in much the same manner as peaches.

The apricots are first mechanically graded for size, then washed, trimmed if necessary, split and pitted. Few canned apricots are peeled, for the skin is so tender that peeling is not necessary. The pitted fruit is then packed into cans by hand, the packers separating the pieces according to quality. The large, uniform, bright-colored pieces go into the fancy grade, the choice grade is a little smaller, while the standard grade is still smaller and may show slight blemishes and lack some uniformity in color. The small, irregular, but sound fruit goes into the two still lower grades, viz., second and pie grades. The syrups added in packing these grades are 55°, 40°, 25°, 10° Brix and water, respectively. The cans are then thoroughly exhausted, sealed and processed at 212° F.

Apricots that are over-ripe and too soft for canning may be cooked down in kettles, with or without sugar, to form apricot pulp, which is packed in large cans for manufacturing purposes.

Berries

Blackberries, raspberries, loganberries and strawberries are canned by very similar methods. For transportation to the cannery, berries are picked either in the shallow quart boxes used in the retail trade, or in shallow trays that are, in turn, packed in crates with space between each tray to prevent the crushing or bruising of the fruit.

The berries are picked over, sorted and washed, either by being conveyed through a series of sprays or by hand dipping. Red raspberries are usually not washed, as they are very tender and break apart easily.

The cleaned fruit is then packed in cans by hand, hot syrup is added, and the filled cans are exhausted, sealed and processed at about 212° F. Syrup containing 10 to 70% sugar is used, depending on variety and grade. Enamel-lined cans are generally used in canning red berries, to preserve the color and flavor.

As blueberries are gathered by the use of special hand rakes, they are mixed with leaves and stems, which are separated out by passing the fruit through a fanning mill. The berries are also picked over by hand to remove green fruit, leaves and stems. They are then cooked with live steam and the hot product is packed into cans and sealed up.

Many blueberries are packed in No. 10 cans in water for bakers' use, as well as in No. 2 cans in sugar for the table. They are usually not processed after sealing, the acid character of the fruit preventing spoilage when the cans are filled hot.

Cranberries

A number of firms in New England and New Jersey can cranberries in the form of either cranberry jelly or cranberry sauce. The berries are gathered in the cranberry bogs by the use of special rakes, and upon arrival at the cannery they are cleaned, sorted, washed and cooked in steam-jacketed kettles with the desired amount of sugar or syrup. The cooked product then passes through a mechanical pulper or cyclone to remove the skins, and the strained sauce or jelly is placed in cans and the cans are sealed.

Cherries

Cherries used in canning are of two general classes, the sweet cherries packed on the Pacific coast and in New York state, and the red sour cherries of the northeastern and central states. The Royal Anne, a white cherry, is by far the most important sweet cherry canned, though black cherries, chiefly Bings and Lamberts, are packed to some extent. The principal varieties of red sour cherries canned are the Montmorency and the Early Richmond. For table use, sweet cherries are packed in syrup and they are nearly always canned with the pits, whereas sour cherries, used largely for pies and preserves, are generally pitted. The syrup added runs from 10 to 70% sugar, depending upon the variety and grade.

Sweet cherries are stemmed and graded for size before being canned. The quality grades, fancy, choice, standard, etc., depend on the size, freedom from blemishes and the strength of syrup used. Sour cherries are graded only according to uniformity of color and strength of syrup.

The stemmed fruit is hand-picked to remove defective fruit and then washed. Sour cherries are pitted by means of a special machine, consisting of a horizontal revolving cylinder on the circumference of which are rows of small pockets with a hole in the bottom of each. As the cylinder revolves, the cherries drop into these pockets from a hopper, and the pits are punched out through the holes by a set of plungers. The fruit is then packed into cans with syrup or, in the case of sour cherries in No. 10 cans, with water, and the cans are exhausted and processed in boiling water.

Figs

Figs are canned to a considerable extent in some of the southern states and in California. The varieties used for canning differ from those prepared as dried figs. The canning procedure is simple, the fruit being generally left unpeeled and packed in a medium syrup. Put up in this way, the natural appearance and flavor of the fruit are preserved to a remarkable degree.

Canned figs for table use are graded in three sizes, blanched in hot water and packed in cans by hand. Hot syrup is added, and the filled cans are then exhausted, sealed and processed in boiling water. Fruit that is sound but over-ripe or mechanically damaged is cooked in kettles without syrup and packed solid in cans for the manufacture of preserves and pies.

Fruit Salad

This canned product is filling a need for a ready-prepared salad and fruit dessert. As packed on the Pacific coast, five fruits are used, viz., peaches, pears, pineapples, apricots, and cherries prepared from white cherries and colored in Maraschino style. The fruit used is generally of choice grade, though slight blemishes in the larger fruits are of no importance, as they are removed in trimming and slicing.

The fruit used for this product is usually packed in season in No. 10 cans and mixed as required. The cans are packed by hand and a syrup of about 30° Brix is added. The filled cans are exhausted, sealed and processed at 212° F.

Grapefruit

Canned grapefruit, a new product, has assumed considerable production proportions in Florida and Porto Rico. It supplies this popular fruit in a convenient form for breakfast service or for salads. The grapefruit is peeled by hand, and some canners are using a lye process, similar to that used for peaches, to remove the white inner layer that covers the fruit. The fruit is then thoroughly washed and the hearts or segments of fruit are separated from the protecting membrane and packed in cans by hand. Hot syrup is added, and the cans are exhausted, sealed and processed in hot-water tanks.

Olives, Ripe

Olives of the mission and Manzanillo varieties are used largely in canning ripe olives in California and Arizona. In order to be sufficiently firm to withstand the treatment necessary before canning, the olives must be picked as soon as ripe, but before they become soft and tender. They are first graded for size, eight sizes being commonly used. The olive naturally contains a high oil content, which distinguishes it from other fruits both in character and use as a food. It also contains a bitter principle, which must be removed to make it palatable, and for that purpose it is treated with a weak lye solution. The olives are then held in several changes of water until all lye is washed out. Then the olives are cured in brine of gradually increasing strength until 3 to 4% brine has been reached. Gradual treatment is necessary to preserve the original form of the olives and to avoid shriveling. The olives are then packed in glass or tin containers of various special sizes and processed under pressure.

The ripe olive industry of California is now operating under special regulation by the State Board of Health. All canners of olives are obliged to take out licenses, and a system of thorough inspection and control is provided from the raw materials to the finished product.

Peaches

The output of canned peaches in this country exceeds that of any other fruit. Canning is done chiefly in California, Michigan and New York. The varieties of peaches used for canning differ markedly; in the east, the Freestone varieties are the most common, while Clings are largely used in the west.

The method of grading peaches has already been described in connection with the subject of quality and grades. Great care is taken to prevent bruising the fruit in picking it and hauling it to the cannery, and to get it into the cans as promptly as possible.

The first step in the cannery is that of halving, pitting and trimming, which is done by hand by the use of special knives for each operation. Machines for splitting both the fruit and pit have been tried out experimentally. Peaches are now usually peeled by the lye process, which is entirely a mechanical one. The pitted fruit is either sprayed or immersed for ½ to 1 minute in hot lye, which completely removes the skin without injuring the fruit. All the lye is removed from the fruit immediately afterward by thorough washing under sprays or in running water. Should any trace of lye be left after this washing, it would be neutralized by the acid of the fruit. Complete removal of the lye is necessary to prevent darkening of the fruit.

After passing the fruit on to sorting belts, where defective or improperly trimmed or peeled fruit is removed by hand-sorting, the peeled fruit is sorted for size by means of a grading machine, which consists simply of a series of sieves with openings of standardized sizes. The smallest fruit is used in the lower grades, which have been described elsewhere, and the larger sizes for choice and fancy grades. Some of the largest fruit is also used for preparing canned sliced peaches. Peaches are also canned whole in syrup.

The graded fruit is carried on conveyor belts, one or more lines for each grade, to the filling tables, where the fruit is again washed and then packed into cans by hand. A certain amount of sorting is also done at the filling tables to remove defective or improperly graded fruit.

The filled cans are next conveyed to the syruping machines, where syrup of various strengths or water is added, according to the grade of fruit desired. The cans are then exhausted, sealed and sterilized at a temperature of 212° F.

Pears

As a table fruit the Bartlett pear is preferred above all other varieties of pears, as its fine flavor is well preserved in canning. Pears of the Kieffer variety are also extensively canned.

On account of the irregular shape of pears and the ease with which they may be bruised and discolored, the canning operations require an unusual amount of hand labor. The fruit is hand picked and the peeling and coring are done by hand with special knives. Expert labor is necessary in order that the fruit may be well peeled, uniformly split and carefully cored. The fruit is graded at the time of picking and coring, and is kept submerged in water or weak brine after it is peeled, to keep discoloration at the minimum.

The pears must be packed into the cans by hand, as it is necessary to layer the larger sizes to obtain the proper fill. The fancy and extra-fancy grades call for six to twelve perfect pieces for a No. 2½ can. The lower grades, choice, standard, second and pie, call for a smaller less perfect fruit in lighter syrup or water. The syrups used are 40°, 30°, 20° and 10° Brix respectively.

Pineapple

Pineapple canning is now largely restricted to the Hawaiian Islands, as canneries located at the source of fruit production are assured of prime material. On the Hawaiian Islands, canners aim to cut the fruit in as ripe a condition as possible and to get it into the cans within 36 hours.

The fruit is first trimmed, top and butt, and then run through a sizing and slicing machine, which cuts the fruit to slightly less than the diameter of the can and in cross-sections so that eight pieces will make the proper fill. The core is removed during the same operation. The slices are inspected as they pass along a belt to the filling tables. Those which are perfect, with the eyes fully developed, of fine texture, and good color are classed as extra grade; those having irregularities in color or slight defects in cutting become standard; and torn or broken pieces become *broken slices*. The rest of the peeled and cored fruit, too irregular for slices, is packed as crushed and grated pineapple. The pineapple, though naturally high in sugar, requires a heavy syrup for table use.

Plums and Prunes

Both the yellow and purple varieties of plums are canned fresh. In the first type are the green gage and yellow egg, and in the second are the Lombard, Damson and Italian prune. The fruit is stemmed, washed and, in some cases, graded for size. It is then packed in cans by hand, hot syrup or water is added, and the cans are sealed and processed at 212° F.

Dried prunes are also canned to a considerable extent, either with or without syrup. As the cooking is done after they are sealed in the can, the flavor of the fruit is preserved to a remarkable extent.

Sea Foods

Clams

The little neck or hard clam is canned to some extent on the New England coast, the long neck or soft clam on the South Atlantic coast, while the razor clam is the basis of the minced clam industry that has assumed considerable proportions on the North Pacific coast. Clam bouillon or nectar and clam chowder are also well-known canned products.

Clams for canning are washed and the shells are then opened by steaming or immersing them in hot water. The meats are separated from the shells and trimmed by hand by clipping off the necks or siphons. Razor clams, which are larger, are also split, trimmed and passed through a meat chopper. A measured quantity of clams is placed in each can, hot brine or juice from the steamed clams is added and the filled cans are exhausted, sealed and processed under pressure.

Crabs

The canning of the Pacific crab in Alaska and Washington has developed to a considerable industry within the last few years. The blue crab is also canned in a small way on the Virginia coast.

After being washed, the crabs are boiled or steamed. The edible portions are then separated from the shell and viscera and are packed with a little salt in parchment-lined cans, which are exhausted, sealed and processed under pressure.

Fish Flakes

This product has been canned on the New England coast for many years, with fresh cod and haddock as the basis. The fish are dressed and cleaned and are then held in brine for a time, depending on the size of the fish. After this they are cooked in retorts with live steam till tender, when the bones are separated from the meat, which is packed by hand into cans. The cans are exhausted, sealed and processed under pressure.

Lobsters

On account of the comparative scarcity of lobsters and the great demand and high prices paid for them in the large city markets, they are no longer canned in the United States, the industry being carried on only in Canada and Newfoundland. The so-called *spiny lobster* or *rock lobster,* a species of crawfish, is also canned in the south.

Oysters

Oysters were one of the first food products to be canned in this country, the industry having been started in Baltimore to develop a larger market for the famous Chesapeake Bay oyster. On account of great improvements in the handling and shipment of raw oysters in recent years, the demand for raw oysters largely absorbs the supply of northern oysters and canning has chiefly been confined to the South Atlantic and Gulf coasts.

Oyster canneries are located along the coast and receive their supplies by boat. The oysters are unloaded into cars of slatted ironwork construction, which are run into a steam chamber for a few minutes to kill the oysters and open the shells. The cars are then run out into the packing room, and, when sufficiently cool, the oysters are shelled by hand by use of special oyster knives. The oysters are washed thoroughly and packed into cans, brine is added, and the cans are sealed, processed under pressure and cooled promptly under water.

Salmon

The Pacific salmon is one of the most important marine foods produced on this continent, and the canning of salmon is carried on from the coast of Northern California to Bering Sea, Alaska.

Five varieties of salmon are canned, grading being done almost entirely by variety and color of the flesh. In the following list the different varieties are given in the commonly accepted order of quality. It will be noted that several names are given to each variety, the name depending on the district in which the fish are caught.

Red, Sockeye or Blueback
Chinook, King or Spring
Medium Red, Cohoe or Silverside
Humpback or Pink
Chum or Dog

The production of canned salmon depends on the run of fish and varies greatly from year to year. The fish are caught at the time they are returning to fresh water to spawn, and several ingenious devices are used to catch them commercially, including traps, purse seines, gill nets, fish wheels and trolling. The fish are unloaded from the boats at the cannery by means of conveyors, from which they are fed to the iron chink, a machine that removes the heads, tails, fins and entrails.

They are then thoroughly cleaned and washed by hand and fed to the cutting machine, which cuts them crosswise into pieces of the right size to fit the cans. Nothing is packed in the cans with the salmon except salt, which is placed in mechanically. The fish is packed in cans either by hand or by machinery. Hand labor is generally employed for filling flat and oval cans, while the tall can is filled mechanically. After inspection of the cans for proper fill, the covers are clamped on loosely, and the cans are exhausted, sealed and sterilized under pressure. Because of the oily nature of the fish, the cans are finally cleaned in a lye bath and rinsed, and when cold they are tested for leaks.

Sardines

Under the definition of the Bureau of Fisheries, any small fish of the herring family, properly prepared, are entitled to be called sardines. In Europe, the pilchard and sprat are canned as sardines, in Maine, the young herring, and in California, the Pacific sardine. Either a frying or smoking process is used for cooking and drying European sardines, while in this country only frying and steaming processes are employed for that purpose.

In Maine, the sardines are caught principally in traps or weirs, heart-shaped enclosures erected in bays or coves. The fish are conveyed to the canneries in special sardine boats, from which they are carried by running water in sluices to the pickling tanks, where they are held for 1 or 2 hours in strong brine. They are next spread by machine on wire trays, or flakes, and are then either cooked in steam chambers and dried in drying rooms, or are first dried and then fried in oil. After cooling somewhat, the fish are packed by hand in cans, oil is mechanically added and the cans are sealed and processed.

In California the packing of sardines is closely connected with the canning of tuna, and only when the latter are scarce do the canners devote themselves to sardine canning. Although small California sardines of fancy quality are put up to some extent by frying and packing in oil, a majority of the fish are of larger size and are packed in oval cans in tomato sauce. The larger fish are first sealed in a revolving perforated drum provided with water sprays, then beheaded and eviscerated. After being held in brine till they are firm and of proper flavor, the fish are dried, fried and packed in cans by hand. Sauce is added, and the cans are exhausted, sealed and sterilized under pressure.

Shrimp

The canning of shrimp is carried on in the South Atlantic and Gulf states, where two styles of pack are put up, the wet pack and the dry pack. Larger shrimp are preferred for canning purposes. They are brought to the cannery in ice or are iced immediately on arrival. When canning begins, the shrimp are peeled, that is, the head and shell are removed by hand. The peeled shrimp are then thoroughly washed in fresh water and cooked or blanched by immersion in boiling brine. After the shrimp have cooled and drained, they are inspected to insure thorough peeling and are packed into parchment-lined cans to prevent discoloration. For wet pack shrimp, some brine is added. The cans are sealed and processed under pressure, and promptly cooled under water.

Tuna

Tuna is canned in southern California, four varieties being used. The season usually extends from June to September, and various methods are used for catching the different varieties. Hook and line are used for albacore and striped tuna, and purse seines for the blue-fin and yellow-fin tuna.

The fish are cleaned and washed before being delivered to the canneries, into which they are carried on conveyors. Along these conveyors are sprays of water which wash the fish a second time. Then they are placed on racks and run into steam chambers, where they are cooked for 2 to 4 hours. When thoroughly cooled, the fish are skinned and the flesh is easily separated by hand into four longitudinal sections, two of dark and two of white meat. Both dark and white meat are canned, but the white meat is in much greater demand. After the meat is cut into proper lengths, it is packed in cans, and salt and vegetable oil, usually cottonseed, are added. The cans are then exhausted, sealed and processed under pressure.

California tuna is also put up Italian style, that is, more heavily salted and packed in olive oil.

Meats

Canners of meats and canned meat products, being engaged in interstate business, operate under the Federal Meat Inspection Service. This inspection covers examination not only of the animals before and after slaughter, but also of the portions used in canning, other ingredients and the method of labeling. Rigid sanitary requirements are enforced by the inspectors.

Such a large variety of meat products is canned that it is impracticable to describe in detail the methods used. In general, all meat for canning is carefully trimmed and freed from bone and other inedible portions. It is then given a preliminary cooking, and is either cut up in pieces to fit the cans, or ground, or chopped up and mixed with other ingredients to produce such products as potted meats, sausage and hash.

After the cans are filled, stock produced in cooking the meat is added to certain products and the cans are exhausted, sealed and sterilized in retorts under pressure.

In the case of some products packed in glass, the jars are sealed in a vacuum chamber and the exhaust treatment is omitted.

Milk

In the early days of the industry, the canned milk output was confined to the sweetened product, which depends for its preservation on the addition of a large amount of cane sugar in the condensing process. Since that time, the canning of unsweetened evaporated milk has grown to a very large industry, and mechanical equipment has been developed for every stage of the process.

Sweetened condensed milk is milk from which a large part of the water has been removed by heating in vacuum pans, and to which a considerable proportion of cane sugar has been added. Cane sugar, in the proportion present in the finished product, acts as an efficient preservative.

The government standard for sweetened condensed milk requires a minimum of 28% milk solids and 8% milk fat, and for the unsweetened product not less than 25.5% milk solids and 7.8% milk fat.

Milk for canning is subjected to constant supervision when delivered at the cannery, to ensure cleanliness, freshness and proper fat content. Before being tested, it is strained, weighed and mixed in large glass-lined tanks, and if found acceptable, it is drawn off into heating tanks for pre-heating. It is then pumped into vacuum pans or evaporators and heated by closed steam coils. Here the milk is concentrated to the desired degree, which is checked by test samples. From the vacuum pans the milk passes to the coolers, a set of horizontal pipes through which cold water circulates and over which the milk flows in a thin layer. The cold milk then passes through a homogenizer, to reduce the size of the milk fat globules and to prevent separation of the fat in the cans. It is held in large glass-lined storage tanks from which it is delivered to mechanical can fillers. After being filled, the cans are sealed and, in the case of unsweetened milk, are processed under pressure.

Soups

Because of their quality and convenience, the production and consumption of canned soups has increased very rapidly. Two classes of soup are made, light and condensed. Light soups can be served without the addition of water or milk and include chiefly beef, mutton, veal, chicken and clam broth, and bouillon. They are usually packed in 8, 13, 32 ounce and No. 10 cans.

Condensed soups require the addition of an equal volume of water or milk. Among these are: asparagus, beef, bouillon, julienne, mock turtle, mulligatawny, mutton, ox tail, celery, consomme, chicken, pea, pepper pot, printanier, tomato, tomato-okra, chicken gumbo, clam bouillon, clam chowder and vegetable. Condensed soups are usually packed in No. 1 cans, although beef, clam chowder, mutton, ox tail, tomato and vermicelli soups are also packed in No. 10 cans.

Vegetable soups, such as asparagus, celery, pea and tomato soups, are packed during the season when fresh vegetables are available. The packing of others that require meats and materials available throughout the year is not restricted to any one period. Federal laws require the inspection of meat and meat products, and as many soups contain considerable meat stock, the production of these is under government inspection.

Specialties

Chili Con Carne

This Mexican dish has attained considerable popularity in the southern states, and quite recently the canning of the product has been taken up to some extent. Being a meat product, it is subject to government supervision.

The red beans and chopped beef are cooked and mechanically placed in cans and a special sauce containing Mexican chilis and other seasoning is added. The cans are sealed and processed under pressure.

Tamales

This is another popular Mexican dish which is now being canned in the southwest. It consists of cooked meat, usually chicken or beef, enclosed in corn meal, which has been cooked up with the meat stock.

Other Specialties

A number of other specialties, such as plum pudding, brown bread, whole wheat bread, chicken a la king and lobster a la Newburg, are put up to a limited extent.

Storage and Care of Canned Foods

Many kinds of canned foods will keep for years without appreciable deterioration, if stored in a cool dry place. Some foods, particularly acid products, slowly attack the tin container and gradually suffer a loss of color and flavor. The higher the temperature of storage, the more rapidly such action takes place. In such acid foods, the action of the acid on the tin plate may cause a leak in the can or gradual development of sufficient gas to cause bulging of the ends of the cans, even though the contents remain sound. The presence of air or other oxidizing substances in the can hastens such action, and recent improvements in the methods of packing these products, such as a more thorough exhausting or driving out of the air and better sealing, are greatly reducing troubles of this kind. However, it is well to recognize that such products are somewhat perishable and should be treated accordingly.

Rusting can be prevented by storing canned foods in a place that is not damp and where the temperature is not subject to frequent and wide fluctuations. If the cans are colder than the surrounding atmosphere and the latter is heavily charged with moisture, water will condense on the surface of the can and rust spots may start that will gradually corrode through the tin plate until the can is perforated. If cans must be stored in a cellar that is not dry, it is best not to place them directly on the floor but a few inches above it.

While freezing does not affect the wholesomeness of canned foods, it injures the appearance and flavor of some foods. Frequent freezing and thawing are much more injurious to the appearance of the foods than allowing them to remain frozen continuously. Overheating in storage is also bad, and is particularly injurious to foods of an acid character, like fruits, for reasons above stated.

In short, stocks of canned foods should be kept only in dry storerooms or cupboards, where the temperature is fairly uniform and, if possible, not over 70° F, preferably lower. Exposure of canned foods to freezing temperatures should also be avoided.

CHAPTER XII

JELLIES AND PRESERVES

Jellies, Preserves, Jams and Marmalades

Jelly

Fruit jelly is a transparent or translucent substance, which while warm is fluid or semifluid, but becomes stiff on cooling. The jelly owes its flavor, color and taste to the juice of the fruit and cane sugar. The property of forming the gel is due to the pectin derived from the fruit. The percentage of pectin in fruit is highest just before the fruit becomes fully ripe. When equal quantities of fruit juice and sugar are heated together, the pectin gelatinizes the mass on cooling.

Pure fruit jellies usually contain 65% or more of soluble solids. The United States Department of Agriculture, Food and Drug Administration, defines fruit jelly as follows: The semisolid, gelatinous product made by concentrating to a suitable consistency the strained juice or the strained water extract from fresh fruit, from cold pack fruit, from canned fruit, or from a mixture of two or of all of these, with sugar or with sugar and dextrose.

It is impossible to give the exact amount of pectin and fruit acid needed in making jellies, inasmuch as the quality of the natural pectin and natural acid of the fruit varies so widely. Use more or less pectin, depending upon the natural gel strength of the fruit juice. The same holds true for the acid.

Directions

Use the juice made from fresh fruits and, if it is not perfectly clear, filter through felt filters, a wine press, or through a hydraulic press.

Cane Sugar	50 lb	
Fruit Juice	50 lb	
No. 100 Powdered Pectin	2 oz to 8	oz (or about 50 oz Liquid Pectin*)
Powdered Citric Acid	1 oz to 2⅘ oz	

Add fruit juice to kettle and, while bringing it to a boil, add 45 pounds of sugar with stirring. Mix the powdered pectin with the other 5 pounds of sugar in a dry pan, then slowly add the pectin-sugar mixture to the kettle. Stir the boiling contents until the pectin mixture becomes uniformly distributed. Now boil continuously and rapidly until a sample gels, normally at around 217 to 222° F. Judge the finish by the drop from the paddle. First it will run, then string, then break from the paddle in sheets.

Dissolve the acid in about 4 ounces of warm water and add it to the batch just before pouring the jelly into glasses or containers. Do not make too large batches of pure jelly at one time for pouring into glasses, because some pure fruit jellies, like currant, start to set in the kettle, if not removed quickly.

For barrel and pail jellies, more pectin and acid have to be used, according to the consistency wanted. Cook fruit juice, sugar and pectin up to around 220 to 223° F, depending upon the heaviness of product desired. Now let the batch in the kettle cool down to about 150° F. Some copper kettles are constructed with cold water inlets and outlets, especially for the purpose of rapidly cooling off the product. The barrel is now filled with 50 gallons of jelly. The calculated amount of liquid acid is added quickly to the barrel, stirred in vigorously with a garden-type hoe, the blade of which has two large holes in it. Just as soon as the jelly grabs or sets, quickly remove the hoe. Jelly made this way should age for 2 or 3 days before shipment. For a 30-pound pail of jelly the required amount of liquid acid is added to the pail. The cooled-off jelly in the kettle is then allowed to run quickly into the pails. The jelly should grab or set within several minutes.

* A method for making this concentrated liquid pectin is described on page 730.

Manufacture of Liquid Pectin

Dried Apple Skins or Apple Pomace	200 lb	Water	250 gal

Add water to a 350-gallon wood tank with cover. Bring water to a boil. Add the dried skins. Now boil vigorously, having live steam come through several perforated pipes extending up and down the tank. Allow to boil for about 40 minutes. Now run the cooked batch onto the hydraulic press and collect the liquid. Approximately 180 to 190 gallons of liquid pectin juice result from the above formula. It may read from 3 to 4° Brix. Concentrate this juice in the kettle to 9 to 10° Brix. This is now the liquid pectin product to be used in all jelly, jam and preserve formulas outlined in this book.

Low Cost Sugar-Fruit-Pectin Jellies

Acid Coagulant for all Jellies Except Imitation Jellies

Tartaric Acid	48 oz
Citric Acid	16 oz
Water Enough to make a total of	1 gal acid solution

Use 2 to 4 ounces of this prepared liquid acid for each 30-pound pail of jelly. If jelly is too firm or leathery use less acid. If jelly is not firm enough use more acid, provided you do not make the jelly too tart.

Apple-Raspberry Jelly
Formula No. 1

Liquid Pectin	50 gal	Red Raspberry Juice	4 gal
Cane Sugar	600 lb	Black Raspberry Juice	5 gal

Add the 50 gallons of pectin juice and the 600 pounds of sugar to the kettle. Cook with stirring to 219° F, or, if a heavier product is desired, cook to a higher temperature. Close the steam valve and now add the 9 gallons of fruit juices. Mix the contents. Allow the jelly to cool down to around 150° F. Pour the required amount of liquid acid into 30-pound pails, and then let the jelly run in quickly.

Formula No. 2

Liquid Pectin	45 gal	Red Raspberry Juice	4 gal
Cane Sugar	375 lb	Black Raspberry Juice	4 gal
True Fruit Raspberry Flavor	6 oz		

This jelly is to be made in the same way as described for Formula No. 1.

Raspberry-Pectin Jelly

Water	30 gal	Amaranth Color (4 oz Color Dissolved in 1 Gal Water)	12 oz
No. 100 Powdered Pectin (Mix with Sugar)	5 lb	True Fruit Raspberry Flavor	20 oz
Cane Sugar	260 lb		
Corn Syrup	140 lb		

Place water in kettle and bring to a boil. Now add gradually the pectin-sugar mixture with stirring until the solution is complete. Add the balance of the sugar and cook to 214° F. Now add the corn syrup and the color, and cook to 220° F. Close the steam valve and then add the flavor. Cool the jelly as recommended and then run into 30-pound pails, with the required amount of liquid acid already added.

Orange-Pectin Jelly

Water	15 gal	Egg Color Solution (3¾ oz Tartrazine and ¼ oz Orange I in 1 gal Water)	4 oz
No. 100 Powdered Pectin (Mix with Sugar)	2 lb	Orange Oil (Dissolved in 8 oz Alcohol)	5 oz
Cane Sugar	130 lb		
Corn Syrup	70 lb		

Make this orange-pectin jelly in the same manner as described for apple-raspberry.

Pectin-Sugar Jelly

Water or Fruit Juice or Combination of Both	60 lb	Flavor	As desired
Cane Sugar	100 lb	Color	As desired
No. 100 Powdered Pectin (Mix with 5 lb Sugar)	1 lb	Prepared Acid Solution	¾ oz

Bring water to boiling in kettle. Mix 1 pound of powdered pectin with 5 pounds of sugar and add slowly to boiling liquids (water or juice or both), stirring the added pectin until a uniform solution is obtained. Add the balance of 95 pounds of sugar to the kettle, stir, and cook up to not less than 217° F and not higher than 221° F, depending upon how heavy the jelly should be. The steam is now turned off and the desired color and flavor are added. Cool the jelly down to around 150° F, and then run into 30-pound pails with the acid previously added to it.

The above formula represents a product known as pectin-fruit jellies. Flavor combinations like apple-currant, apple-strawberry, apple-raspberry and apple-grape can be made as above described. Liquid pectin can also be used instead of the powdered pectin. Approximately 3 gallons of liquid pectin are equal to 10 to 16 ounces of No. 100 powdered pectin.

Apple-Mint Jelly

Apple Juice	12 gal	Mint Flavor	1 oz
Cane Sugar	120 lb	Green Color	Enough to suit
Liquid Pectin	4 gal		

Cook apple juice, sugar and liquid pectin to 217 to 219° F. Turn steam off. Allow to cool off somewhat and then add color and flavor. Add enough acid solution to glasses or other receptacles to make jelly set.

Mint Flavor

Oil of Spearmint	1 oz	Alcohol	Enough to make 1 qt
Oil of Peppermint	1 oz		

Mint Color

Tartrazine Food Color	¾ oz
Guinea Green B Food Color	¼ oz
Water	Enough to Make 1 gal color solution

Directions for Extracting Fruit Juices for Jelly Making

Heat, extraction and filtration are essential steps in making fruit juices. The object of extracting the juice from the fruit is to obtain the pectin, natural coloring, acids, salts and flavoring needed for jelly making. Firm fruits require longer cooking to soften the cellulose and to obtain a maximum yield of fruit juice, while soft fruits require short cooking.

Examine the fruits carefully, wash them, and remove decayed parts. Firm fruits may be cut into small pieces to allow for quicker softening while cooking and to obtain a higher yield of juice. For firm fruits use around 1 pint of water for each pound of fruit. For grapes and berry fruits, the water at first may be left out entirely and the fruit allowed to cook up in its own juice. Too much water will give a thin juice which will require later concentration. Too little water will give a cloudy juice, which may also have the tendency to form a jelly. Cooking of the fruit longer than is absolutely necessary results in loss of fruit flavor, darkening of color, extracting bitter substances and destroying some of the pectin. The boiling time for apples is about 15 to 20 minutes, for citrus fruits about 1 hour, and for berries from 2 to 4 minutes.

The juice is separated from the pulp by straining through jelly bags, by pressing through a wine or hydraulic press, or by allowing the cooked batch to settle and pouring off the juice. A second and third extraction may be made on some fruits such as apples, quinces, currants, grapes and raspberries, by cooking up the extracted pulp with more water.

In extracting the juice from grapes and berries, crush the fruit and then add 1 pound of water for about every 5 pounds of fruit. Heat fruit and water until tender, and then strain or press out the juice.

Quince Jelly

Quinces require long cooking to become tender. The quince flavor is exceptionally strong. Combining the quince with tart apples not only improves its flavor but also results in a jelly of better texture.

The quince is cut up into small pieces, sufficient water added, and it is cooked until tender. When the fruit is cooked tender drain off the juice and use it alone or combined with other fruit juices. Use about two-thirds as much sugar as fruit juice and make jelly in the usual manner.

Loganberry or Gooseberry Jelly

Use sound, slightly underripe fruits. Examine fruit carefully and remove decayed parts. Wash the fruit. Add ½ pint of water for every 5 pounds of fruit. Heat the water and fruit slowly to the boiling point and boil for about 3 minutes. Drain the juice the best way and make a second extraction by covering the pulp with water and boiling for another 3 minutes. Combine the two extractions. Usually two-thirds as much sugar is required as fruit juice in making this jelly.

Blueberry Jelly

Add 1 pint of water for each quart of berries, bring to a boil and heat for 5 minutes. Drain the juice, then make a second extraction by covering the pulp with sufficient water, and cook slowly for 10 minutes. Drain the juice and then combine the two extractions. Concentrate the combined juice by rapidly boiling to 1 pint for each original quart used.

Currant Jelly

Add 1 pint of water for each quart of fruit used. Boil slowly in a covered kettle for 6 to 8 minutes. Strain and then return pulp to kettle. Add same amount of water, cover kettle, and cook slowly for 10 minutes. Let stand 10 minutes and strain. Combine the two extractions and concentrate by rapid boiling to receive 1 pint of juice for each quart of fruit originally used. Usually 1 pound of sugar to each pint of juice is used for making jelly.

Orange Jelly

Water	85 lb
Citrus Pectin (100-Grade)	18 oz
Concentrated Orange Juice (72% Soluble Solids)	15 lb
Sugar	100 lb
Citric Acid Solution (50%)	20 fluid oz

Put the water into the kettle and turn on the steam. When it has reached a temperature of 160 to 180° F, add the pectin thoroughly mixed with 8 times its weight of granulated sugar. Allow the mixture to come to a brisk boil, stirring occasionally. Boil vigorously for about ½ minute.

Add the fruit juice and the balance of the sugar. Boil as vigorously as possible to a temperature equivalent to 219 to 220° F. Turn off the steam and immediately add the acid solution. Mix thoroughly.

Draw off the hot jelly and fill containers immediately. Finish the filling and capping while the jelly is still above 190° F, or sterilize the capped jars.

The yield should be approximately 170 pounds of finished jelly at 65% soluble solids.

Apple Jelly

Wash and clean the fruit. Slice or cut into small pieces. Add 1 pint of water for each pound of fruit used. Boil in covered kettle for 15 minutes. Strain the juice. Return pulp to kettle. Add same amount of water and cook slowly for 15 minutes. Strain. Combine the two extractions and concentrate by boiling so that 1 pint equals 1 pound of original fruit. Usually ¾ pound of sugar to 1 pint of apple juice will be found satisfactory for jelly making.

Raspberry Jelly

Wash and clean the fruit. Add 1 pint of water for each quart of fruit. Cover the kettle and boil for 5 minutes. Strain the juice. Return pulp to kettle and add same amount of water. Cook slowly for 10 minutes. Strain the juice. Combine the two extractions and then concentrate quickly by boiling to 1 pint for each quart of original fruit. Add ¾ pound of sugar for each pint of juice for jelly making.

Grape Jelly

Use firm ripe grapes. Add 1 pint of water for every 2 pounds of fruit. Remove stems and green and decayed fruits. Boil slowly for 10 minutes. Strain the juice. Return pulp to kettle, cover with water and cook slowly for 10 more minutes. Strain juice. Combine the two extractions and concentrate the juice so that 1 quart is the equivalent of 2 pounds of fruit. Use 1½ pounds of sugar for each quart of juice. Over-ripe grapes may be used if apple jelly stock is used measure for measure. In this case, 2 pounds of sugar for each quart of mixed fruit juice will make a satisfactory jelly.

Strawberry Jelly

A satisfactory jelly can be made by taking a mixture of apple and strawberry juice. The strawberry is low in pectin and will not form a jelly. Apple jelly stock is made by taking equal parts of apples and water and cooking together until tender, and then straining the juice. The strawberries should be hulled, and the decayed fruit removed. The berries should be slowly cooked for about 5 minutes, using ½ pint of water for each quart of berries. All possible juice should be squeezed out. To each pound of this juice add 1 pound of apple jelly stock, and then concentrate this combination to about 1½ pounds. Add ¾ pound of sugar per pint of concentrated juice for jelly making.

Cherry Jelly

Cherry juice makes an excellent jelly when combined with an equal measure of apple jelly stock. Sweet as well as sour cherries can be used. The cherries are covered with water in a kettle and heated to boiling and cooked until tender. The juice is then squeezed out. Concentrate the juice so that each pint is equal to 1 pound of original cherries.

Cranberry Jelly

Put 100 pounds of cranberries in a kettle with 6 gallons of apple juice. Bring to a boil and cook 10 minutes. Express the juice and place in kettle with 50 pounds of sugar. Cook to 217° F until jelly forms.

Combination Jellies

As seen from the above jelly formulas, some fruits will not make jellies unless mixed with apple jelly stock. These fruits contain insufficient pectin or acid to produce a good product.

The raspberry, blueberry, blackberry and grape will make jelly by themselves, but with the slight addition of apple juice they make a better product. Half-grown, green apples always give best results, especially when mixed with the berry or grape juices.

Bakers' Quick Fruit Jelly
(Unsweetened)

Water	8	oz
Agar-Agar	1/4	oz
Fruit Juice	16	oz
Cane Sugar	1/4	oz or more

Mix the sugar and the agar well. Bring the water to a boil. Add the sugar-agar mixture and, when dissolved, remove from the fire. Add the fruit juice. When cooled to room temperature, it will gel.

Any fruit juice can be used in the above formula. Stewed fruits or berries can be used to make a jam, or milk to make a custard.

Pectin in Fruit

Most fruits are lacking in natural fruit pectin and fruit acid. Apples, grapes, lemons and currants are rich in pectin and acid, whereas strawberries, plums, raspberries and apricots are average or poor. The following fruits are considered as having a high percentage of pectin:

Apples	Grapes	Oranges
Cranberries	Guavas	Pomegranates
Currants	Lemons	Quinces
Gooseberries	Limes	Tomatoes

Fruits having an average pectin content are:

Apricots	Blueberries
Plums	Raspberries
Loganberries	Strawberries
Blackberries	

Fruits having a low pectin content are:

Bananas	Pears
Cherries	Rhubarb
Pineapples	Figs
Peaches	

To be able to make commercial jellies, preserves or jams, pectin and acid must be added. Pectin is available either in powdered or liquid preparations. Powdered pectin is made from apples or citrus peels.

Preparation of Liquid Pectin from the Powdered Pectin

A stock solution of liquid pectin can easily be made as follows:

No. 100 Powdered Pectin	1 lb	Cane Sugar	5 lb
		Water	2 gal

Mix the powdered pectin and the sugar in a dry pan. Add this mixture slowly and with stirring to the 2 gallons of boiling water. When solution is uniform shut off steam. Make final volume 2 gallons. One pint of liquid pectin is now equal to 1 ounce of powdered pectin.

A perfect jelly should be clear, transparent, should quiver, hold its form when cut and should be sufficiently firm and not mushy. It should not be syrupy, gummy or sticky. To obtain satisfactory jelly, the pectin, acid and sugar must be adjusted in proper proportions. Varying proportions will naturally result in different jellies.

Helpful Facts on Jelly and Preserve Manufacture

1 Jelly forms best at a pH of 3.1 to 3.4. Above 3.4 the jelly has a tendency to form more slowly, sometimes not at all. Below 3.1 the jelly may be too acid and *weep* or *water-out.* This process is called syneresis. Watering-out may happen because of lack of solids, poor pectin or too much sugar.

2 Before use, powdered pectin should always be mixed with cane sugar, and then added gradually to hot water. The less solids there are in this hot water, the easier it will be for the

pectin to dissolve. Too much solid substance, like sugar syrup, may form lumps with the pectin.

3 Jellies and preserves should be cooked quickly and cooled quickly. This precaution will result in better taste, flavor, color and texture.

4 Liquid pectin made from dried apple skins, pomace, cores, pieces, chops or wet apple skins is not intended to replace fruit juice in making pure jellies. It can be used for making imitation jellies, jams or preserves. It can be used in pure jellies and preserves only where there is a natural deficiency of pectin in the fruit.

5 Juice extracted from berries, currants and grapes averages 80 to 85% of the weight of the fruit taken. From solid fruits, the average is 70%.

6 Elderberry, currant, loganberry, Columbian raspberry and black raspberry juices are used to intensify color of jellies and preserves.

7 Too much pectin makes jelly leathery. Too much acid toughens the jelly and will carry more sugar as a result. If excess acid is used, watering may take place due to hydrolysis of pectin.

8 Pectin testing: To find out whether fruit juice contains enough pectin, put equal parts of juice and alcohol in an evaporating dish. If the content of pectin is sufficient for jelly making, the whole will settle to a gelatinous mass. If a loose mushy mass forms, use less sugar than fruit by weight.

9 Jelly for filling glass goods is automatically poured from filling machines. If the jelly is poured while cold, it should be pasteurized at 180° F for 30 minutes. If the jelly is poured while still hot, pasteurizing for 20 minutes is satisfactory.

*Method for Using Clarase * in the Treatment of Fruit Juices to Produce Sparkling and Permanently Clear Jellies*

There is a growing demand for sparkling clear jellies, especially those produced from apples, quinces and crab apples. The appeal to the purchaser is greatly enhanced and the man-

*** Takamine Laboratory, Inc., Clifton, N. J.**

ufacturer who produces these finer jellies will have a better opportunity to obtain a larger share of the available business.

Clarase is a dry powder which maintains its effectiveness for an indefinite period when kept properly closed against air at a temperature not above 100° F and not near steam pipes, heating units, etc. It is completely soluble in water, and should be dissolved and then added to the liquid to be treated.

The steps in the Clarase process of clarifying apple juice, quince juice and crab apple juice, as well as similar juices high in starches, are as follows:

1 Maceration and cooking of the fruit
2 Pressing the juice
3 Cooling the juice to 85 to 104° F
4 Adding Clarase and digesting for one hour, more or less, as occasion may require
5 Thoroughly filtering the juice through suitable filter
6 Using, canning or concentrating the juice as occasion may require

The amount of Clarase to be used is proportionate to the amount of starches in the fruit juice. These starches vary considerably in different localities and depend upon the variety and ripeness of the fruit from which the juice is pressed. It may be said, however, that on an average, from 6000 to 8000 parts (by weight) of apple juice, quince or crab apple juice can be treated with one part of Clarase (by weight).

The temperature is very important as the juice is normally acid, and, in an acid medium, the temperature must be kept low to effect the most efficient results.

After the juice has been pressed from the cooked fruit, it should be run through three or four sugar bags to remove as much of the suspended solid substance as possible. Then it should be cooled to the proper temperature of between 85 and 104° F either by natural cooling or by running cold water or brine through coils built into the tanks. After cooling, the juice is ready for the next step, the treatment with Clarase. From the amount of juice to be treated, calculate the quantity of Clarase required.

Weigh out the Clarase in dry form and proceed to dissolve it in water at about 80° F. Pour just a little water at first on top of the dry Clarase and stir into paste. Add a little more water until the Clarase is thoroughly wet but still like a thick paste with no surplus water in the container. The very fine Clarase powder would float on top of any surplus water and would dissolve slowly, so the water must be added slowly in small amounts and thoroughly mixed all the while to get all particles thoroughly wet. After that more water can be added, and within a short time the material will all pass into solution.

Pour the water solution of Clarase made in this way into the fruit juice to be treated, which at this point must already have been cooled to between 85 to 104° F, and stir throughout the juice so that the enzyme is thoroughly dispersed. Note the time and digest for one hour. Take a sample after one hour and test with iodine to note the starch color, which at this stage should begin to be quite brown or reddish-brown, indicating the conversion of the starches into dextrins and maltose. If the iodine reaction still indicates the presence of undigested starch by a blue color, the reaction should be continued for a sufficient number of minutes to accomplish complete conversion, as indicated by a reddish-brown color with iodine. If complete conversion is not indicated, more Clarase should be used for the particular juice being treated. It is not unusual to make preliminary laboratory tests on a small scale for each batch of juice before digesting on a large scale. Such preliminary trials will indicate the exact amount of enzyme required in each instance.

The solution is now ready for final filtration and should be heated up to 160° F or a little higher, as this stops the enzyme action and makes the filtering operation much easier. A successful type of filter frequently used is the plate filter using Filter-cel as an aid to filtration. Care must be taken to avoid any of the aid passing through the filter with the juice. The fruit juice so filtered is then ready to be made up into jelly right away, or it may be placed in 5-gallon cans and pasteurized for cold storage until needed, or it may be concentrated by evaporation for future use.

No change is required in the method of using fruit juice that has been treated with Clarase. The Clarase treatment converts all colloidal starches into soluble sugars so that the undesirable effect of starch, in producing cloudiness in the finished jelly, is entirely eliminated. The conversion of the starch also frees any other colloids which may have been held together by the starches, so that all of these particles will be removed by the filtering operation, which is therefore much more effective. The resulting juice is brilliant in appearance and makes jelly which is correspondingly sparkling and maintains this property instead of becoming cloudy and dull as would otherwise be the case.

Imitation Jelly

Imitation jelly can be made with corn syrup, cane sugar, artificial or genuine fruit flavors, coal tar certified food colors, liquid or powdered pectin, and with acids such as lactic, citric, tartaric or phosphoric.

Approximate Basic Formula

Concentrated Liquid Pectin (As Described under Manufacturing of Liquid Pectin)	55 gal
Corn Syrup (42° Bé)	85 gal
Red Color	3 pt
Sodium Benzoate (Dissolved in 1 qt of Warm Water)	2 lb
Acid	Sufficient to make jelly set
Flavor	As desired

To the warm solution of pectin in the kettle, add 85 gallons of corn syrup. Mix with an agitator. Now add color, flavor and sodium benzoate, and, when the solution is thoroughly mixed, run into pails or barrels. The temperature of the batch should be around 150° F. If pail goods are desired (30 pounds of jelly to a pail) add 4 ounces of phosphoric acid solution to the pail and then fill with the jelly. In about 1 minute the jelly will set. If barrels are desired, fill the barrel with the jelly first, and then add the required amount of phosphoric acid, with vigorous up-and-down motion with a paddle. In about 1 to 2 minutes the jelly should be firm. If a barrel holds 51 gallons add 68 ounces of the prepared acid.

The jelly pails should be closed the day following the filling. If jelly is low in solids it may ferment or develop surface molds.

Prepared Phosphoric Acid Solution

Phosphoric Acid (50%)	3 pt	Water	2 pt

Use 4 ounces of this solution to each 30 pounds of jelly. More or less may have to be used, depending upon the quality of pectin and sugar ratio.

Prepared Red Color Solution

Amaranth Certified Food Color	5 oz	Ponceau 3R Certified Food Color	15 oz
		Water	5 gal

Dissolve the color in the water.

Glucose Jelly

Water	15 gal	Red Color (As Prepared)	¾ pt
Corn Syrup (42° Bé)	300 lb	Prepared Phosphoric Acid Solution (to Each 30-lb Pail)	3–4 oz
No. 100 Powdered Pectin	2¾ lb	Sodium Benzoate (Dissolve in 1 pt Water)	5 oz
True Fruit Raspberry Flavor	10 oz		

Mix the powdered pectin with about 10 pounds of sugar, and then add slowly to the boiling water with stirring. Now add the corn syrup and heat up to 221° F. Shut off steam and then add color, flavor and sodium benzoate. Let the batch cool down to around 150 to 175° F, and then fill containers.

Preserves, Jams and Marmalades

The manufacture of preserves, jams and marmalades is conducted on a large scale. The equipment consists of large, open, steam-jacketed copper kettles, lined with tin, or vacuum kettles. Sound fruit and sugar should be used, and foreign substances like corn syrup and food colors must not be used in making pure products.

Preserves and jams are made by cooking to a suitable consistency properly fresh fruit, cold pack fruit, canned fruit, or a mixture of two or all of these with sugar, or with sugar and dextrose, with or without water. In its preparation not less than 45 pounds of fruit are used to each 55 pounds of sugar or of sugar and dextrose. A product in which the fruit is whole or in relatively large pieces is customarily designated a preserve, rather than a jam, according to the definition given by the Department of Agriculture.

The name jam is generally applied to preserved fruits which have lost their shape in cooking with sugar.

A good jam or preserve is bright in color, may or may not have jelly-like consistency, and the flavor is characteristic of the fruit from which it is made. It should not be oversweet. Fruit and sugar are the essential ingredients in making a jam or preserve but, if a jelly-like consistency and tartness are desired, both pectin and acid are used.

Directions

Fruit and sugar should be weighed carefully and both heated in the kettle. Cooking should be rapid in order to retain maximum natural color, natural flavor and taste. When the batch begins to get heavy in the kettle while cooking, stir the contents to avoid side-burning. When the proper temperature has been reached, 220 to 224° F, transfer contents immediately to cooling table, or pour into jars, seal them, turn them upside down to sterilize covers. To thicken up the preserves or jams, either powdered or liquid pectin can be used.

High Quality Pure Preserves

Fruit	50 lb
Cane Sugar	50 lb
No. 100 Powdered Pectin	3–4 oz or 1 gal Liquid Pectin
Fruit Acid Solution	2–4 fluid oz
Water	3 gal

Add to kettle 3 gallons of water and bring to boil. Add 45 pounds of sugar and cook to 220° F. In a separate kettle, cook up with ½ gallon of water a mixture of the powdered pectin and 5 pounds of sugar. Add this liquid pectin to the kettle containing the sugar. Also add the juice of the fruit and cook again to 220° F. Now add the fruit and acid, and cook quickly to 221 to 223° F, or until the product gives a good sheet test on the paddle or wide blade. Now transfer quickly to cooling table.

If the product does not sheet off the paddle at 223° F, it usually is an indication of insufficient pectin, and more should be added. Cooking above 223° F in order to thicken the product may sometimes result in the product crystallizing out and having an oversweet and caramelized taste.

When frozen or cold pack fruit, packed with sugar, is used, the amount of sugar and fruit in the above formula will vary as follows:

	2 Plus 1 Fruit	3 Plus 1 Fruit	4 Plus 1 Fruit
Use this amount sugar	25 lb	33 lb	37½ lb
Use this amount cold pack fruit	75 lb	67 lb	62½ lb

Allow fruit to thaw out completely before making preserves.

Average Quality Pure Preserves

Fruit	45 lb
Cane Sugar	55 lb
No. 100 Powdered Pectin	3–4 oz or 1 gal Liquid Pectin
Fruit Acid Solution	2–4 fluid oz
Water	3 gal

Make these preserves in the same way as above.

When frozen or cold pack fruit, packed with sugar, is used, the amount of sugar and fruit in the above formula will vary as follows:

	2 Plus 1 Fruit and Sugar	3 Plus 1 Fruit and Sugar	4 Plus 1 Fruit and Sugar
Use this amount sugar	32½ lb	40 lb	44 lb
Use this amount cold pack fruit	67½ lb	60 lb	56 lb

Ginger Preserve

Drain the ginger well and then cut it up. Place in cold water in a steam pan, gently bringing to a boil and simmering for 20 minutes. Place in sieves in drain. Transfer to a cold syrup of 4 pounds of sugar to each gallon of water, and allow to stand until the next day. Transfer all to steam pan, gently bring to boil, and simmer for 15 minutes. Then place in a clean dry tub and allow to stand until the next day. Run off the syrup into the steam pan and add 3 pounds of sugar to each gallon of syrup. Stir well and bring to boil. Return this syrup to the ginger in the tub and allow to stand until the following day, then place in sieves to dry. Roll in sugar and shake out the loose sugar through a coarse sieve. Then spread out to dry.

Imitation Jams or Preserves

These products are deficient in fruit and therefore must be labeled as imitation. The percentage of ingredients should be declared on label.

Formula No. 1

Fruit	35 lb	Fruit Acid	
Sugar	55 lb	Solution	4–6 fluid oz
Liquid Pectin	3 gal		

Formula No. 2

Fruit	25 lb	Fruit Acid	
Sugar	55 lb	Solution	4–6 fluid oz
Liquid Pectin	4 gal		

Formula No. 3

Fruit	15 lb	Fruit Acid	
Sugar	55 lb	Solution	6–8 fluid oz
Liquid Pectin	5 gal		

With cold pack fruit, the above should be prepared as follows:

Formula No. 1

52½ lb 2 plus 1 Fruit	and	37½ lb Sugar
47 lb 3 plus 1 Fruit	and	43 lb Sugar
44 lb 4 plus 1 Fruit	and	46 lb Sugar

Formula No. 2

37½ lb 2 plus 1 Fruit	and	42½ lb Sugar
33 lb 3 plus 1 Fruit	and	47 lb Sugar
31 lb 4 plus 1 Fruit	and	49 lb Sugar

Formula No. 3

22½ lb 2 plus 1 Fruit	and	47½ lb Sugar
20 lb 3 plus 1 Fruit	and	50 lb Sugar
19 lb 4 plus 1 Fruit	and	51 lb Sugar

Preparation of Frozen Fruit

The preserve industry uses large quantities of cold pack or frozen fruits. This makes all-year manufacturing of preserves a reality. Fruit may be frozen with or without sugar. Berries used for frozen pack are completely ripened. They are washed and graded for size. Foreign matter and imperfect fruit are removed. Before freezing, berries and sugar should be mixed, and the sugar should be in complete solution before freezing. The fruit in barrels is frozen quickly in order to retain the maximum color, flavor and texture. It takes 2 to 3 days at 0° F to complete the freezing operation. The barrels are then stored at 12 to 15° F until ready for shipment. Fruits are packed in the ratios of 2 parts fruit and 1 part sugar, 3 parts fruit and 1 part sugar, and 4 parts fruit and 1 part sugar, or all fruit without sugar.

The varieties of strawberry most used are the Marshall, Missionary, Blakemore, and Klondike. The bulk of raspberries used are Red Cuthberts and Black Columbians.

Blueberry Jam

Blueberries	45 lb	Cane Sugar	55 lb

Cook the blueberries until soft, then put through strainer or pulper. Combine the pulp and sugar and boil the mixture until it is thick and clear.

Apple-Blueberry Jam

Blueberry Pulp	20 lb	Cane Sugar	55 lb
Tart Apples (Peeled and Cut)	25 lb	Grated Lemon Rind	4 oz
		Citric Acid	2 oz

Combine the above mixture and cook up until thick and clear.

Cherry Preserves

The cherries should be washed, stemmed and pitted. Combine 45 pounds of the cherries with 55 pounds of sugar and cook together in kettle as directed under making of jam.

Currant Jam

Wash the currants and remove the stems. Combine 45 pounds of fruit with 55 pounds of cane sugar and cook together in kettle as directed under making of jam.

Damson Plum Jam

Wash the plums, then cover with water in the kettle and cook the fruit until the skins are soft. Cool and then remove the pits in a pulper. The jam should be cooked up in the ratio of 45 pounds of fruit to 55 pounds of cane sugar, as directed under making of jam.

Grape Jam

The grapes should be washed and stemmed. Then press out the pulp from the skins. Now cook the pulp with the addition of about 10% water, and then press through sieves, or put through a pulper to remove the grape seeds. Add the skins to the pulp if desired and then take 45 pounds of pulp and 55 pounds of cane sugar and cook in kettle to proper consistency as described under making of jam.

Apple-Peach Jam

Peaches (Cut into Small Pieces)	20 lb	Tart Apples (Cut into Small Pieces)	25 lb
		Cane Sugar	55 lb

Boil the above mixture until it is thick and clear. Follow directions as given under making of jam.

Peach-Orange-Pineapple Jam

Peaches (Cut into Small Pieces)	30 lb	Cane Sugar	55 lb
Crushed Pineapple	10 lb	Grated Orange Rind	8 oz
Lemon Juice	2½ lb	No. 100 Powdered Pectin	1–4 oz
Orange Juice	2½ lb		

Combine the above mixture and cook together as directed under making of jam.

Peach-Orange Jam

Crushed Peaches	40 lb	Cane Sugar	60 lb
Orange Juice	5 pt	No. 100 Powdered Pectin	1–4 oz
Lemon Juice	5 pt		
Grated Orange Rind	12 oz		

Combine all of the above ingredients and cook together in the kettle as directed under making of jam.

Plum Jam

Plum Pulp and Juice	300 lb
Pitted Prunes	100 lb
Cane Sugar	275 lb
Concentrated Liquid Pectin	20 gal (Or 2¾ lb No. 100 Powdered Pectin)
Citric Acid	1 lb
Caramel or Sugar Coloring	3 lb
Water	5 gal

Bring the plum juice and water to a boil, and then add the cane sugar. While it is still boiling stir the sugar until dissolved, and then add the plum pulp and the prunes, which have been previously cooked until very soft and then put through the meat grinder, making them into a paste. Then add the concentrated liquid pectin. If powder is used, mix it at first with about 5 times its own weight in sugar. Now add the citric acid dissolved in water. Cook the contents in the kettle until the temperature reaches about 223° F. If the sheet test is satisfactory, add the caramel color, mix in thoroughly, and quickly transfer to the cooling table. If sheet test is poor, continue cooking to a higher temperature.

Imitation Raspberry Jam
Formula No. 1

Raspberry Fruit and Juice	200 lb
Cane Sugar	400 lb
Concentrated Liquid Pectin	32 gal (Or 4 lb No. 100 Powdered Pectin)
Water	20 gal
Citric Acid	2 lb
Red Color	If desired

Bring the water to a boil and then add the sugar. Continue the boiling until the temperature registers 220° F. Now add the pectin, continue the boiling, and then add the raspberry fruit and juice. Then add the citric acid dissolved in about 1 pint of warm water. Continue the boiling until the temperature reaches about 223° F, or until the proper sheet test is had. Transfer quickly to cooling table.

Formula No. 2

Raspberry Fruit and Juice	110 lb
Cane Sugar	200 lb
Corn Syrup	75 lb
Concentrated Liquid Pectin	15 gal (Or 3 lb No. 100 Powdered Pectin)
Citric Acid	¾ lb
Red Color	If desired

The above imitation jam is made the same way as No. 1. The corn syrup is to be added after the cane sugar.

Imitation strawberry jam is made in the same way as described for imitation raspberry jam, except that strawberry fruit is used.

Imitation Apricot Jam
Formula No. 1

Apricot Pulp	120	lb
Cane Sugar	250	lb
Concentrated Liquid Pectin	7	gal (Or 2½ lb No. 100 Powdered Pectin)
Citric Acid	½	lb
Water	15	gal

The above apricot jam is made in the same way as described for imitation raspberry jam.

Formula No. 2

Evaporated Apricots	50 lb
Cane Sugar	85 lb
Concentrated Liquid Pectin	5 gal (Or 12 oz No. 100 Powdered Pectin)

Soak the fruit overnight in sufficient water and, in the morning, put through the grinder. Add fruit and sugar to the kettle, and while it is boiling add the pectin. Cook up to about 224° F or until the batch shows a good sheet test. Transfer immediately to the cooling table.

Imitation Prune Paste or Lekvar

Prunes *	250 lb
Cane Sugar	100 lb
Corn Syrup	200 lb
Concentrated Liquid Pectin	10 gal (Or 2 lb No. 100 Powdered Pectin)
Sugar Coloring	4 pt
Water	10 gal
Citric Acid	8 oz

* The prunes have pits in them and are rather tough. The pits must be removed and the flesh cooked until softened. The procedure is to allow the prunes just covered with water to soak overnight. In the morning transfer to kettle, add more water, bring to a boil and cook until the prunes are soft. Then put the prunes through the pulper, which removes the pits and pulps at the same time. Use this prune paste for the above.

Bring the water to a boil in the kettle and then add the sugar Stir until dissolved. Then add the corn syrup, liquid pectin, prunes, acid and color. Cook to a pretty high temperature, around 224 to 228° F, or to the consistency desired. Transfer to cooling table.

Orange-Pineapple Jam

Crushed Pineapple	400 lb	
Orange Peels *	90 lb	
Cane Sugar	500 lb	
Liquid Concentrated Pectin	15 gal	(Or 5 lb No. 100 Pectin)
Citric Acid	1 lb	
Water	10 gal	

Bring water to boil in kettle and then add the crushed pineapple, sugar, orange peel, liquid pectin and acid. Cook quickly to 223° F, or until proper sheet test. Then transfer to cooling table.

Orange Flavored Pineapple Jam

Crushed Pineapple	450	lb
Cane Sugar	300	lb
Orange Oil (Dissolved in 16 oz Alcohol)	12	oz
Sodium Benzoate	10	oz
Tartaric Acid	6	oz
Orange Color	Enough to suit	
Agar-Agar **	1¼	lb
Water	20	gal

* The orange peel for this formula must be prepared as follows:
Soak the fresh orange peel overnight. In the morning cook the peel in the kettle covered with sufficient water, and change the hot water at least 3 times to remove the bitterness, and to render the peel soft and tender. Then put the peel through the grinder.

** Let the agar-agar soak overnight in 5 gallons of water. In the morning add it to the kettle, bring to a boil and then add the 25 pounds of cane sugar. Cook this mixture until the temperature registers 222° F. Use this mixture of agar and sugar and work into the pineapple batch on the cooling table.

Bring the water to a boil. Add 275 pounds of sugar and cook to 220° F. Then add the crushed pineapple and, while the batch is cooking rapidly add the sodium benzoate dissolved in a little water, the color and dissolved acid. Cook to 222 to 224° F. Transfer to the cooling table, and here work in the solution of agar-agar first into the batch until uniformly mixed. Then work the orange flavor into the batch on the cooling table.

Orange Marmalade

Cane Sugar	100 lb	
Oranges (Use ¾ Oranges and ¼ Lemons)	50 lb	
No. 100 Pectin	6–8 oz	(Or 2 gal Concentrated Liquid Pectin)
Water	5 gal	

Wash the fruit. Cut in half, remove juice with a reamer, and filter or strain the juice. Slice orange and lemon peel very thin. Cook up with water in kettle until soft and tender. This may take as long as 1 hour. Mix the cooked peel with the strained juice, boil slowly, then sprinkle in slowly the pectin-sugar mixture, or the liquid pectin. Then add the balance of the sugar, and cook rapidly to 224° F or until the proper sheet test.

Orange Marmalade

Cane Sugar	67 lb	
Orange Pulp and Skins	33 lb	
No. 10 Powdered Pectin	4 oz	(Or 100 oz Concentrated Liquid Pectin)
Citric Acid	2 oz	

Mix together the 4 ounces of pectin and 1 pound of sugar. Cook together balance of sugar and pulp. Add pectin mixture or pectin solution. Stir, and cook quickly until batch sheets off paddle. Shut off steam. Add solution of citric acid. Transfer to cooling table.

Citrus Pulp Marmalade

Considerable quantities of citrus pulp preserved with sulfur dioxide are used for the manufacture of marmalade which may be a promising product for domestic consumption. The barrels of grapefruit and orange pulp are treated with a sufficient amount of sulfur dioxide to prevent the product from becoming spoiled or fermented. The following recipe is suggested for making the jam,

Sulfured Grapefruit Pulp	18 7/10 lb	Sulfured Orange Pulp	9 3/10 lb
		Sugar	65 lb

A simple method of manufacturing is to add the pulp to plenty of water in the steam-jacketed kettle and boil the mixture vigorously to remove practically all of the sulfur dioxide. It may be necessary to add additional pectin in order to bring about a good set of the marmalade. The faster the jam can be cooked, the better and more desirable the product that will result. The remaining sulfur dioxide in the product should not exceed 0.005% otherwise it may impart a brassy taste.

Marmalade Flavoring From Orange Peels

Orange Peel and Pulp	6 lb	Standardized Invert Sugar	10 lb
Water	15 lb		
Corn Syrup	10 lb		

This formula is designed to use the orange peel and pulp remaining after the juice has been extracted at soda fountains or restaurants. The product can be used for flavoring rolled and cast cream centers, nougat fudge bar candy centers, jellies and other confectionery products.

Dice or grind the peel. Add the water and allow to boil until soft. This requires about ½ hour. Add the corn syrup and standardized invert sugar. Cook the batch slowly to 222° F. Transfer at once to a container, place the cover in position and allow to set for 24 hours.

Apple-Raspberry Juice

Apple Juice, Natural, Baldwin, Northern Spy, Cortland Blend	76 lb	Black Bristol Raspberry Juice	10⅕ lb
		Purple Raspberry Juice	10⅕ lb
		Sugar	3⅘ lb

Freeze the whole raspberries with 10% added sugar. Press in fall in an ordinary apple juice press immediately before pressing apples. Blend freshly pressed apple juice with raspberry juice, strain, deaerate, flash pasteurize, pour hot into No. 2 tall (30 oz) cans lined with "berry" enamel or into hot bottles, close and lay on side for 3 minutes. Cool in cold water.

Keeping-Qualities of Jams and Jellies

All jams in open packages, unless artificially preserved, should contain from 65 to 70% sugar in the finished product. In air-tight packages, which are given a special process, the product may be kept successfully with less sugar. The percentage of sugar in a jam or jelly may be readily calculated from the amount of sugar used and the amount of jam produced, allowing about 3% more for the sugar in the fruit and inversion of the cane sugar.

The most satisfactory package for jams and jellies is the air-tight closure. Jams are best poured hot, sealed at a temperature of 170° F or more, and inverted or rolled in such a way as to force the small amount of air in the jar through the hot mass. Another method is to pour hot, allow to cool, vacuumize the jar, and pasteurize at a low temperature.

Jellies are best poured hot, and sealed at a temperature not lower than 170° F, or, they may be poured at lower temperatures, allowed to set, vacuumized, and then pasteurized.

Units of 60 to 80 pounds of jellies and 100 to 150 pounds of jams will give better products than larger amounts. The quicker the operation is finished and the preserve cooled off, the better are the flavor and color. The simplest cooler is a copper or plain wooden table about 8 feet long by 3 feet wide and 6 inches deep. This is placed on a slight incline and the jam being spread in a thin layer cools very quickly. The size and slope can be varied to deliver the jam into the filling receptacle at the required temperature.

Crystallization in jams can be prevented by regulating the amount of invert sugar present.

After boiling, the jam should be analyzed for invert sugar, and if the amount does not fall between the limits of 25 to 30%, either the acidity or the time of boiling should be adjusted so that more or less inversion takes place.

For soft fruits the finishing temperature is usually about 223 to 224° F.

For hard fruits the temperature is a little lower, about 220 to 221° F.

Chemical Composition of Fruit, Juice, Cold Pack Fruit, Fruit Jellies and Preserves

In manufacture of jellies and preserves, either the fresh or frozen fruit may be used. The cold pack or frozen fruit may also be packed with added sugar. When 3 parts of strawberries are mixed with 1 part of sugar, and then frozen, full reliance is placed on the packer, as to the correct amounts used, etc. The chemical composition of the berry or the pack cannot be guaranteed and is generally not discussed by the merchant. The food chemist is able to determine only by chemical analyses the amount of fruit or fruit juices used in the manufacture of preserves or jellies.

The discussion of the composition of fruit juices, fruit jellies or preserves in the following chapters may be accepted as authentic because a great deal of this work has been done in the author's laboratories. The analyses of these jellies and preserves represent pure products. However, it must be recognized that the chemical composition of fruit varies, depending upon the crop, soil, drought, fertilizer and especially upon the location from which it is derived. Therefore the composition of the finished product cannot always be the same.

COMPOSITION OF FRUITS *

Fruit	*Per Cent Total Solids*	*Per Cent Water-Insoluble Solids*	*Per Cent Total Invert Sugar*	*Per Cent Ash*	*Per Cent Alcohol Precipitate*	*Per Cent Pectic Acid*	*Per Cent Acid as Citric*	*Per Cent Malic Acid*	*Per Cent Tartaric Acid*
Apple	17.25	2.50	11.50	0.33	0.76	0.32	—	0.57	—
Apricots	16.50	1.79	8.50	0.75	0.80	0.46	0.50	0.80	—
Blackberries	15.57	7.23	7.00	0.44	0.62	0.32	0.96	—	—
Black raspberries	20.00	9.00	6.50	0.65	0.65	0.35	0.95	0.04	—
Cherries	16.25	1.38	8.22	0.49	0.20	0.10	—	0.92	—
Crab-apples	18.22	3.22	10.90	0.41	1.25	0.72	—	1.25	—
Currants	16.00	5.85	6.20	0.57	0.68	0.41	1.90	0.08	—
Cranberries	15.00	3.22	4.80	0.28	0.82	0.52	1.25	0.90	—
Figs	21.00	3.22	15.25	0.56	1.00	0.68	0.08	0.15	—
Gooseberries	12.00	3.00	4.50	0.44	0.72	0.42	0.50	1.50	—
Grapes (Concord with much skin)	16.00	1.25	12.00	0.48	0.58	0.32	—	0.35	0.87
Loganberries	15.54	6.07	3.68	0.50	0.72	0.36	1.65	0.15	—
Peaches	13.50	1.65	9.00	0.50	0.98	0.45	0.20	0.50	—
Pineapples	15.60	1.10	12.50	0.43	0.10	0.05	0.73	0.20	—
Plums (Damson with practically all skin)	14.50	1.48	8.00	0.43	1.22	0.78	—	2.14	—
Quinces	17.00	3.22	7.25	0.46	1.00	0.62	—	1.10	—
Raspberries	14.00	6.00	4.32	0.46	0.78	0.32	1.30	0.05	—
Strawberries	10.00	2.70	5.36	0.47	0.52	0.33	0.87	0.20	—

* Examined in Shirley Laboratories. New York.

COMPOSITION OF FRUIT JUICES*

Fruit	*Per Cent Total Solids*	*Per Cent Total Invert Sugar*	*Per Cent Ash*	*Per Cent Citric Acid*	*Per Cent Malic Acid*	*Per Cent Tartaric Acid*	*Per Cent Protein*
Apple	12.98	11.32	0.29	—	1.17	—	0.10
Blackberries	15.60	8.40	0.40	0.92†	—	—	0.38
Apricot	11.54	7.40	0.51	—	0.84	—	0.51
Cherries	17.56	11.36	0.48	—	0.64	—	0.48
Currant	11.55	6.90	0.48	2.11	—	—	0.34
Grape	16.80	14.50	0.38	—	—	0.98	0.28
Grapefruit	12.17	6.40	0.41	1.28	—	—	0.32
Loganberry	11.00	6.50	0.48	1.89	—	—	0.60
Lemon	12.00	4.89	0.28	6.50	—	—	0.32
Quince	12.50	8.98	0.49	—	1.20	—	0.30
Orange	14.00	10.00	0.46	1.35	—	—	0.50
Peach	10.65	6.88	0.46	—	0.98	—	0.41
Pineapple	17.00	11.45	0.50	0.70	—	—	0.20
Raspberries	10.39	6.02	0.45	1.67	—	—	0.34
Black raspberries	12.17	7.51	0.55	1.38	—	—	0.36
Strawberries	7.71	4.70	0.43	0.71	—	—	0.26

* A. L. Winton and Kate Barber Winton, *Structure and Composition of Foods*, New York, John Wiley & Sons, 1935.
† Isocitric.

COMPOSITION OF COLD PACK FRUIT*

Fruit	*Per Cent Total Solids*	*Per Cent Total Invert Sugar*	*Per Cent Water-Insoluble Solids*	*Per Cent Ash*	*Per Cent K_2O in Ash*
100% Strawberry	13.41	7.36	1.93	0.33	38
100% Strawberry	15.67	9.82	3.35	0.48	42
3 + 1 Strawberry	38.20	30.20	2.00	0.33	32
2 + 1 Strawberry	40.52	37.10	2.70	0.28	33
100% Raspberry	13.20	5.15	5.20	0.38	42
100% Raspberry	14.57	5.36	8.82	0.46	34
12% Added Sugar To Raspberry	27.82	17.92	6.16	0.43	38
3 + 1 Raspberry	35.20	27.32	4.80	0.31	30
100% Loganberry	12.54	3.68	6.07	0.50	42
100% Blackberry	15.24	7.09	5.87	0.42	38

* Examined in Shirley Laboratories, New York.

PURE FRUIT JUICES—ALKALINITY * OF ASH AND PROTEIN CONTENT

Juices	Degrees Balling or Brix	Per Cent Ash in 100 Grams: Total Ash	Soluble Ash	Insoluble Ash	Alkalinity in Ash †: Total Alkalinity	Soluble Alkalinity	Insoluble Alkalinity	Acid Content Grams per 100 Cubic Centimeters	Per Cent Protein
Apple juice	15.95	0.2999	0.2356	0.0639	37.70	27.30	3.40	0.83	0.15
Grapefruit juice	12.85	0.3204	0.2404	0.0800	42.58	23.82	18.76	1.98	0.32
Lemon juice	12.10	0.2860	0.2231	0.0629	40.96	26.32	14.64	6.12	0.32
Navel orange juice (late)	17.10	0.4986	0.3994	0.0992	87.10	60.90	26.20	0.83	0.65
Valencia orange juice	14.75	0.5149	0.4341	0.0808	78.50	55.40	22.10	1.40	0.58
Pineapple juice (canned)	17.70	0.5080	0.3593	0.1487	40.90	34.20	6.70	1.07	0.15
Grape juice (bottled Concord)	17.85	0.2188	0.1499	0.0769	27.70	23.20	4.50	1.40	0.25
Black cherry (California)	19.50	0.5554	0.3060	0.2494	50.90	44.70	6.20	0.96	1.10
Loganberry juice (early)	14.20	0.3984	0.2747	0.1237	41.40	35.20	5.90	2.49	0.85
Strawberry juice	14.15	0.4874	0.3421	0.1453	26.40	23.10	3.30	1.12	0.42
Raspberry juice	12.45	0.5894	0.4336	0.1558	40.30	35.20	5.10	1.04	0.46
Royal Anne cherry juice	20.10	0.5101	0.4094	0.1007	50.20	44.50	6.20	0.69	1.20
Navel orange juice	11.50	0.328	0.273	0.055	83.70	56.80	26.90	1.19	—

* Cubic centimeters N/10 HCl per 100 gram sample
† H. Aref and W. V. Cruess, *Fruit Products J.* Vol. 3, (1937).

COMPOSITION OF JELLIES *

Fruit	Per Cent Total Solids	Per Cent Total Invert Sugar	Per Cent Ash	Per Cent P_2O_5 in Ash	Per Cent K_2O in Ash	Per Cent† Alkalinity of Ash	Per Cent Acid	Per Cent Protein (N × 6.25)
Apple	66.28	63.05	0.18	9	35	10	0.87‡	0.06
	63.92	60.00	0.15	7	32	8	0.48‡	0.02
Apple-currant	70.32	66.92	0.20	8	38	11	0.70‡	0.18
	65.47	62.70	0.17	10	36	12	0.41‡	0.12
Apple-raspberry	67.25	64.10	0.20	8	30	10	0.76‡	0.18
	67.85	66.40	0.16	9	34	9	0.58‡	0.10
Blackberry	72.07	67.72	0.26	12	35	13	0.62**	0.22
	64.15	61.82	0.18	15	28	9	0.50**	0.15
Apple-strawberry	67.85	64.22	0.17	12	33	8	0.68‡	0.18
Apple-blackberry	69.20	66.82	0.18	16	38	12	0.72‡	0.16
Apple-mint	65.40	64.10	0.14	6	26	8	0.52‡	0.05
Currant	69.40	68.00	0.24	16	42	16	0.92**	0.20
	65.10	63.00	0.19	12	36	10	0.72**	0.16
Crab-apple	66.00	61.80	0.20	9	38	14	0.45‡	0.12
Grape	67.60	65.60	0.18	7	29	8	0.50§	0.22
	66.60	63.30	0.15	11	32	7	0.60§	0.15
Plum	69.20	66.01	0.24	9	42	11	0.62‡	0.20
Raspberry	66.10	62.32	0.22	14	38	12	0.86**	0.20
	67.22	64.00	0.16	16	26	14	0.42**	0.18
Strawberry	69.28	65.18	0.23	12	38	16	0.50**	0.12
	63.76	60.25	0.18	18	33	12	0.42**	0.15

* Examined in Shirley Laboratories, New York
† Cubic centimeters normal acid required to neutralize the alkali in 1 gram of ash
‡ As malic acid
** As citric acid
§ As tartaric acid

COMPOSITION OF PRESERVES AND JAMS *

Fruit	Per Cent Total Solids	Per Cent Water-Insoluble Solids	Per Cent Ash	Per Cent P_2O_5 in Ash	Per Cent K_2O in Ash	Per Cent† Alkalinity of Ash	Per Cent Acid	Per Cent Total Invert Sugar
Apricot	70.15	1.48	0.36	10	33	14	0.82**	64.96
Blackberry	72.83	3.42	0.23	12	42	10	0.71‡	65.89
(seedless)	70.95	0.15	0.16	9	28	10	0.53‡	67.72
Cherry	76.22	0.91	0.20	11	40	14	0.73**	71.20
Grape	72.56	0.42	0.17	10	28	10	0.42§	67.00
Grape	71.60	1.20	0.25	9	38	7	0.59§	65.60
Grape	70.40	0.50	0.23	8	33	6	0.53§	67.70
Loganberry	73.83	2.58	0.23	13	42	13	0.85‡	67.36
	72.26	2.12	0.17	8	33	8	0.97‡	65.64

Fruit	Per Cent Total Solids	Per Cent Water-Insoluble Solids	Per Cent Ash	Per Cent P_2O_5 in Ash	Per Cent K_2O in Ash	Per Cent† Alkalinity of Ash	Per Cent Acid	Per Cent Total Invert Sugar
Peach	70.72	0.85	0.23	8	38	8	0.52**	67.70
Pineapple	73.92	0.86	0.22	12	33	10	0.58‡	67.45
Plum (Damson) ..	67.06	0.82	0.29	9	28	14	0.85**	62.40
	73.29	0.61	0.20	10	38	10	0.74**	68.56
	63.80	0.59	0.18	10	40	12	0.48**	59.20
Raspberry	69.15	2.74	0.24	14	35	15	0.93‡	64.15
	73.20	2.18	0.19	12	38	12	0.72‡	68.10
	71.60	2.44	0.26	10	33	14	1.00‡	65.56
	70.59	3.58	0.17	10	44	10	1.04‡	62.20
Strawberry	73.90	0.65	0.21	15	43	16	0.89‡	67.00
	68.05	1.10	0.18	13	38	12	0.76‡	64.48
	70.32	1.72	0.23	12	52	14	0.75‡	64.88
Marmalade .... (70% orange) (30% lemon)	66.40	2.40	0.25	18	40	16	0.65‡	56.80

* Examined in Shirley Laboratories, New York
† Cubic centimeters normal acid required to neutralize the alkali in 1 gram of ash
** As malic acid
‡ As citric acid
§ As tartaric acid

Sulfuring of Fruits

The sulfuring of fruits has been practiced since ancient times. The active compound employed is sulfur dioxide generated by burning sulfur in the presence of water. Fruits are treated with sulfur dioxide for the purpose of preserving color, flavor, nutritive values, to prevent enzymatic oxidation, and to repel insect infestation during drying and storage.

When sulfur burns in the air, it combines with oxygen to form sulfur dioxide, which in a pure state is a colorless gas but reacts with water to form sulfurous acid.

$$\underset{\text{Sulfur}}{S} + \underset{\text{Oxygen}}{O_2} \longrightarrow \underset{\text{Sulfur dioxide}}{SO_2}$$

The sulfur dioxide, being soluble in water or fruit juice, forms sulfurous acid.

$$\underset{\text{Sulfur dioxide}}{SO_2} + \underset{\text{Water}}{H_2O} = \underset{\text{Sulfurous acid}}{H_2SO_3}$$

Sulfurous acid, often referred to merely as sulfur dioxide, is used as a preservative in a variety of foods. In this country the food manufacturer is allowed 350 parts per million (0.035%) of sulfur dioxide in the finished food product. This is the amount tolerated by the federal government in interstate commerce. Although practically all states allow this percentage, there are notably a few exceptions where no sulfur dioxide is permissible in most foodstuffs. The purpose of using sulfur dioxide is to prevent foods, both in liquid and dried form, from fermenting or spoiling. For example, if dried fruit were stored without sulfur dioxide in it, it would lose color, flavor and become infested with insects faster than if it contained the preservative. The amount of sulfur dioxide which should be incorporated in the fruit, so that it will be preserved, varies with the particular fruit; for example, apricots, peaches and nectarines should have at least 2000 parts per million, pears 1000, bleached raisins 800, apples 1500, cherries in sulfur brine 3000, grapefruit peel, either diced pieces or halves, 1500, figs 1500, orange and lemon halves 1500, Cuban, Porto Rican and Singapore pineapple 2000 parts per million.

The method of sulfuring fruits consists simply of exposing them, in trays in a closed chamber, to an atmosphere of burning sulfur. For fruits packed in a brine, e.g., cherries, peels, pineapples, etc., the sulfurous acid is added directly in a calculated amount to the barrel of fruit. There is any number of methods and procedures employed in sulfuring fruits, depending upon the operating conditions and the part of the country in which the preservative is employed.

When fruits are exposed to the fumes of burning sulfur the absorption of sulfur dioxide is great. However, a considerable proportion of the gas escapes during the first 4 hours in drying of the fruit. It is important to dry sulfured fruit rapidly, so as to retain the maximum amount of sulfur dioxide; one way to insure this is to use dry heat in the process of driving off the moisture.

The sulfuring of fruits has played an important part in balancing demand and supply, and in stabilizing prices during those months when the fresh fruit is unobtainable. The law, regarding the permissible amount of sulfur dioxide in the finished fruit preparations, is constantly being enforced. This places a responsibility upon the food manufacturer for removing the greater part of sulfur dioxide present in the fruit when it arrives at the factory. It is not an uncommon procedure for a food manufacturer, in order to remove the sulfur dioxide, to allow sulfured fruits to remain submerged in water overnight, and then let the wash water run off. When this procedure is followed, almost 10% of the fruit itself is dissolved out and lost.

Preservation of Fruits with Sulfur Dioxide Brining

Surplus fruit supplies and certain varieties unsuitable for sale as fresh fruit are finding an important and growing outlet through the use of a solution of sulfur dioxide as a temporary preservative, in conjunction with calcium compounds in some instances.

All products stored in the sulfur dioxide brine are used for further manufacture and the sulfur dioxide is largely removed in each case.

Advantages of Sulfur Dioxide Brining

Some of the advantages of the process are: it is inexpensive; it is a quick method of handling the fruit in bulk; little equipment is required; the barrel in which the brining is done is suitable for bulk shipment; the preservative can be removed; the preservative is effective for a period of 1 to 2 years which is ample time to allow for storage, shipment and manufacture.

In addition to preventing spoilage of the product, the brining treatment may be used to bleach and/or harden certain fruit, depending on the type and use for which it is intended. In practically all cases, calcium compounds are used as the hardening or firming agents to prevent the collapse of the cells. Finely divided calcium hydroxide, such as ordinary hydrated spray lime, is advantageously used in the preferred method described below although calcium carbonate (whiting, chalk) can be used. In the alternate methods described, either material may be used.

Although formulas often specify lime and calcium carbonate interchangeably as to amount, 1 pound of lime is actually equivalent in hardening effect to about 1⅓ pounds of the carbonate (whiting or chalk).

With raspberries and such fruits as apples, apricots, blackberries, gooseberries, loganberries, peaches, plums, prunes, rhubarb, etc., which are badly mashed or almost unrecognizable as individual fruits when found in jam, no attempt is made to harden the fruit, and calcium compounds are omitted from the sulfur dioxide brine. In the case of these fruits, the general object of the process is to temporarily preserve the fruit pulp in suitable form, in this way overcoming the peak load of jam plants at harvest time, and facilitating the transportation of the pulp to export markets. Sulfur dioxide is used as the basic preservative, as this chemical is volatilized by heat and so passes off during the jam making process.

On the other hand, in brining strawberries, calcium is supplied in various compounds to act as a hardening agent to prevent mushing. Properly processed strawberries should remain whole and should be sufficiently firm to resist gentle pressure by the fingers. Strawberries that are not firm will break down in transportation.

Black currants are typical of fruits which are inclined to harden if treated with sulfur dioxide without a preliminary boil. Therefore, this product is heated in a kettle with water until most of the fruits are broken before it is placed in the barrel.

The Brining Operation

In the preferred method of carrying out the brining operation, the barrel containing some water or a portion of the brine is placed on the scale and the proper amount of fruit added through the open end. The barrel head is inserted and the barrel moved to the brining station. The appropriate amount of calcium compound, preferably hydrated lime, is poured through the bung hole by means of a funnel, and then the sulfur dioxide solution is added to fill the barrel. Most conveniently, the solution is prepared by adding the sulfur dioxide, from a measuring chamber equipped with a gauge glass, directly into a stream of water at a mixing valve. The solution then flows through a rubber hose, the end of which may be inserted in the bung hole. The filled barrel can immediately be closed by driving in the bung. If calcium carbonate is used instead of lime, carbon dioxide will be released and care must be taken to prevent loss of sulfur dioxide and brine due to bubbling out of the carbon dioxide. Obviously, the bung cannot be immediately driven in tight under these conditions.

In the alternate method of brining, the sulfur dioxide is added to a vat or other large vessel containing the calcium compound in a suspension. The sulfur dioxide reacts with the calcium compound to form insoluble calcium sulfite. Since the lime or calcium carbonate is largely insoluble and the product first formed is insoluble, the sulfur dioxide must be added at a comparatively slow rate if excessive loss from the open container is to be avoided. Later in charging the barrels with the brine, some additional loss is encountered. When the sulfur dioxide solution is charged directly into the barrel containing the lime, the bung can be immediately inserted and no loss of sulfur dioxide need occur.

Preservation of Strawberries

Firm strawberries to be packed in sulfur dioxide brine should be picked when they have become red over three quarters of their surface and one quarter is still whitish green. Soft strawberries should be picked when they are three quarters ripe with the shoulders and bodies a red color and the tips, or noses, green. If the fruit is not to be shipped in brine but is to be preserved for later use at the same plant, it can be harvested in a riper condition.

Berries must be clean. Sand, adhering straw, decayed fruit and foreign material should be removed by washing.

Calcium compounds are used in the brine solution for strawberries, to harden the tissue so that there will be whole berries in the finished product. Either calcium carbonate or hydrated lime may be used. The following tabulation shows the proper quantities of these ingredients as well as the amount of sulfur dioxide required to properly brine a barrel of 350 pounds of strawberries. Different quantities of hardening agents are shown for various stages of maturity of the fruit.

Stage of Maturity	*Whiting*	*Hydrated Lime*	*Sulfur Dioxide*
All green	2½ oz	2 oz	13 oz
or shoulders	3¾ oz	2¾ oz	13 oz
Red with green tips			
Ripe-red but not soft	4½ oz	3¼ oz	13 oz
Fully ripe	5 oz	3¾ oz	13 oz
Ripe fruit in bruised condition	6 oz	4½ oz	13 oz

It is important that the proper quantity of water is added. The barrel should not be filled completely with fruit and solution. In any of the formulas used, only about 3 to 4 gallons of water or brine should be added. This variation depends on the season and quality of the berries. Soft wet berries do not require any extra water, whereas berries that are dry and firm may receive the full amount. When barrels of processed berries, which have received the proper amount of water, are examined, the berries and liquid are at approximately the same level.

The most satisfactory method of brining the fruit is to add the 350 pounds of drained, washed berries to the barrel, head the barrel and add the chemicals through the side bung as explained above. This method provides less chance of loss of sulfur dioxide.

Preservation of Raspberries and Other Mashed Fruits

The brining of raspberries is described as an example of brining in which no effort is made to maintain the original shape of the fruit. In this classification are all fruits that are badly mashed when found in jam. These have been listed earlier. No calcium compounds are required since no hardening of the fruit is necessary.

In barreling fruit, a total of 13 ounces of sulfur dioxide and 3¼ gallons of water per barrel has been recommended. About one-quarter of the brine is placed in the open-ended barrel and berries which been picked over to remove leaves, etc., are added until the barrel is about half full, using constant agitation to insure a uniform distribution of the solution. Another quarter of the sulfur dioxide brine is added and then the remainder of the 400 pounds of fruit. The third-quarter portion of the brine is added and the head inserted, after which the final quarter of the brine is added through the bung.

Although use of a mixing valve and hose for introducing the sulfur dioxide brine directly into the barrel is possible, it may prove easier in some cases to use the mixing valve and sulfur dioxide measuring chamber to prepare a brining solution in an auxiliary tank and measure it into the barrel as required in spite of the probable greater loss of sulfur dioxide. With fruits that are easily mashed, more trouble results from lack of distribution of the sulfur dioxide brine than from any other cause.

Preservation of Black Currants

This fruit is an example of a product which hardens if preserved in sulfur dioxide brine without the pre-treatment to soften the fruit. The suggested procedure is as follows:

The currants are picked without the stems and are passed through a fanning mill to remove leaves and other foreign material. The currants are placed in large kettles with water, using about 50 to 75 pounds of water to 350 pounds of currants. The fruit is heated with constant stirring to temperatures below the boiling point (190 to 200° F) until all of it is broken, and the pulp is smooth. It is then cooled to 160° F and is ready for barreling. The fruit pulp and the brine are added through the bung openings on the sides of the barrels. The brine solution is made up, using 10 ounces of sulfur dioxide for 2½ gallons of water. One-third of the acid is added to the barrel before any pulp is introduced, another third of the brine is added when the barrel is half full and the final third is added at the top. Four hundred pounds of fruit pulp are used for each barrel. The barrel is bunged immediately.

Preservation of Apple Flesh for Candying

Apples for this product must be used while quite firm. The following varieties have yielded satisfactory fruit for candying for the bakery trade: Grimes Golden, McIntosh, Jonathan, Delicious, Winesap, Newtown, Winter Banana, Wagener, Rome Beauty, Northern Spy, Stayman Winesap, Greening and Spitzenburg.

The apples are peeled, cored and sliced in sixteen radial sectors. All pieces of core should be removed. The sectors are kept in 2% salt brine until a sufficient quantity has been obtained to fill a barrel. The prepared fruit is then placed in the barrel and lime and sulfur dioxide solution added as in the preferred cherry brining method. The amount of each of these materials must be determined depending on the volume of brine which can be added to each barrel. The brine should be made up to contain 5 pounds of lime and about 8⅓ pounds of sulfur dioxide per 100 gallons. These quantities are slightly smaller in each case than those required for brining of sweet cherries.

Handling and Storage of Fruits Preserved in Sulfur Dioxide

Newly packed barrels of fruit should be inspected daily for leaks and any deficiency of brine should be made up for with fresh solution. A quarter turn of the barrels every day during the first 5 days of this inspection will aid the curing process by preventing pocketing of brine solution. It is a good practice to store the brined barrels in the coolest common storage available. Storage in direct sunlight is very undesirable. It is suggested that a complete record is kept on each barrel of brined fruit. The barrel number, quantity of sulfur dioxide, quantity of water, quantity of lime, maturity of fruit, the number of hours from picking to brining, etc., are all facts that will probably prove useful at some later date.

Sulfur Dioxide Limits and Methods of Analysis

The strength of the sulfur dioxide solutions used for brine should be accurately known. This is important in order to know that the fruit will be properly preserved and is necessary when preparing fruit pulp for export, so that the tolerance named by the importing country is not exceeded.

The British tolerances for sulfur dioxide in fruit and fruit pulp are:

Cherries	3000 parts per million
Strawberries and raspberries	2000 " " "
Other fruit	1500 " " "

In the case of preserved fruit that is to be stored, it is well to watch the sulfur dioxide content and add more sulfurous acid if the percentage of SO_2 falls below 2000 parts per million.

BRINING OF CITRON, WATERMELON, LEMON, ORANGE AND GRAPEFRUIT PEEL AND CHERRIES

Brining simply means placing the raw product in a solution of salt. Certain chemical changes take place, which preserve the fruits so that they can be used months or even years later. It also makes it possible to permit the shipping of raw materials from the source of supply to any destination or food factory. In some instances sulfurous salts are also used with the salt, as has previously been discussed.

Preparation of Stock Sulfur Dioxide Solution

The sulfur dioxide solution is prepared as follows:

Place the cylinder on a scale and weigh it. Then subtract from the gross weight 20 pounds. This 20 pounds of gas should be allowed to run into a barrel filled with 45 gallons of water. When the entire amount is in, the scale will be in true balance. The cylinder of gas is to be connected properly with rubber tubing conducting the gas into glass tubing, which goes through a hole on top of the barrel, and reaches down to the bottom. The time required for the gas to be dissolved in the water is about 10 hours. The strength of the solution in the barrel will now average between 4 and 5% of sulfur dioxide. Mark this solution *stock solution.* By diluting with water, any desired strength may be attained.

Citron

The citron may be brined with sea water, as it is done in foreign countries. The addition of the salt water at first sets up a fermentation in the citron. This has its advantages, for it removes bitterness through the opening of the pores and also develops flavor. After several weeks of equalization between the citron and liquid, the salt content will be about 8%. This concentration of salt is sufficient to preserve the citron for months and sometimes years until made into the candied product. Instead of sea salt, ordinary salt can be used for brining. Add the salt at the rate of 1% a day, giving the citron a chance to undergo a slight fermentation. When the brine as well as the citron shows about 8% salt, the process is complete.

Citron Melon

Citron melon is also brined. While it is called a substitute for citron, it is really not. Citron melon develops molds quickly in a weak salt solution. Before adding salt, add about 3 pints of 4 to 5% stock solution of sulfur dioxide to each 100 pounds of citron melon. The salt is added in the same way as suggested for regular citron.

Watermelon

The skin is removed from the rind which is then cured in brine. To each 100 pounds of rind add about 14 pounds of salt and 2 pints of sulfur dioxide solution. The addition of sulfur dioxide prevents softening of the rind and also the growth of molds. The brine in which the rind is finally stored should contain at least 8% salt. The rind is used for candying purposes.

Lemon and Orange Peel

Lemon and orange peel are preserved in a salt brine. To each 100 pounds of lemon or orange peel add about 15 pounds of salt. Dissolve the salt in enough water to cover 100 pounds of peel. If the barrel is closed, it should be rolled several times in order to mix the salt solution. A very small amount of sulfur dioxide stock solution may be added. This addition will soften up the white or rag on the peel, which is an advantage later on in candying the peel. For long storing, the peel should contain at least 8% salt by weight.

Grapefruit Peel

After the juice and pulp are removed from the fruit, the peel is collected by companies who are in the brining business.

The white or rag of the grapefruit peel is heavier, thicker and tougher than either the lemon or orange rag. The peel is ripened in a salt brine solution by slow additions of salt. During the first few days, it is advantageous if a slight fermentation sets in.

This will tend to soften the tough white rag on the peel. The final salt concentration in the peel should test about 8%. A quart of sulfur dioxide stock solution is added to each 100 pounds of peel to prevent softening action of wild bacteria, and also growth of molds.

Other fruits packed in a weak solution of salt and varying percentages of sulfur dioxide are sliced pineapples, pears, figs, kumquats, apricots, peaches and plums.

Cherries

The cherry most commonly used is the sweet cherry such as the light-colored Royal Anne, or Waterhouse, and also the black cherries, such as the Bing, Lambert, Black Republican, Tartarian, Chapman, and others. The best cherries for making a high quality Maraschino style cherry are the Royal Anne, Bing, and Lambert. Small to medium-sized cherries are usually preferred, while cherries with stems attached, used for cocktails, are generally made from larger cherries.

Other cherries which are sulfured or bleached for brining purposes are the sweet Napoleon, Windsor and Smith varieties. These cherries grow mainly in the States of Michigan, New York and Ohio.

Sour cherries are now also sulfured in large quantities for making into a Maraschino style cherry. These cherries grow abundantly in Michigan, New York, California, Oregon, and a few other states. As a rule, cherries which are to be bleached are picked just before they become fully ripe. If cherries are picked when too mature, the pits will stick to the fruit and make difficult pitting. Also, if pickled when over-ripe, the flesh is tender and will not absorb the color properly. When the cherries have remained in the sulfur solution for several weeks, they are sorted or graded according to quality and size.

Cherries are bought either by size or count. The size specifications are designated by millimeter measurements. There are at present four standard sizes:

Dipping cherries	10–14 mm
Small cherries	14–18 mm
Medium cherries	18–22 mm
Large cherries	22–26 mm

By count is meant the number of cherries required to fill a one-gallon container. The cherries are graded in a motor-driven, perforated, rotary cylinder screen, and the various sizes are caught in barrels under the machine.

Bleaching Cherries

Cherries that have been bleached to a yellowish-white shade in a sulfurous acid (sulfur dioxide) solution are known as *cherries in brine.* The word brine means water impregnated with salt, or a combination of salt with other chemical ingredients. Brining methods are used to preserve foodstuffs, which themselves may be a complete food, like fish products, or foods to be processed into a finished product.

Brined cherries are prepared in California, Washington, Oregon, Utah, Michigan and New York, also in Italy and France. The brine used for cherries consists of water impregnated with sulfurous acid, known as the preserving solution. The fresh cherries are placed in barrels and then covered with a 0.60% solution of sulfur dioxide.

The object in sulfuring the cherries is to arrest decay of the cherries temporarily, to bleach them to a uniform pale amber color, and also to give the fruit firmness. The bleaching chemical is sulfurous acid (sulfur dioxide) or sulfite salts.

Calcium carbonate or calcium oxide, which has the tendency to harden the tissue of the fruit, may also be added.

The sulfurous acid is prepared by conducting sulfur dioxide gas from the cylinder into a barrel filled with water. About 20 pounds of the gas, slowly dissolving in 45 gallons of water, will give a solution testing about a 4 to 5% solution of sulfurous acid. This solution is the starting point for making sulfur solutions. By adding 5 gallons of the above 4 to 5% solution to 45 gallons of water, and then adding to this 1½ to 2 pounds of calcium carbonate or calcium oxide, there is obtained a solution to be used for bleaching the cherries. Usually 220 to 250 pounds of cherries are placed in the barrel and covered with this diluted sulfur water solution. Another method of bleaching cherries is to add to each barrel about 2 pounds of sodium metabisulfite, first dissolved in water. Also add to each barrel 6 ounces of milk of calcium carbonate or calcium oxide. Barrels best suited for brining are paraffin-lined, made of the best quality spruce or oak. The cherries should be stored in a cool warehouse for at least 60 days before use. For the first few days, it is advisable to roll the barrels around 3 times a day, especially if unslaked lime is used. The brine should also be examined for its sulfur dioxide content. At no time should it be allowed to run under 2500 parts per million or 0.25%. The barrel should always be full of brine, and leaky barrels should be coopered immediately. The cherries are considered bleached when they have attained a slight yellowish tint, and the fruit becomes toughened. Normally this takes 4 to 6 weeks. After this process, the cherries are stemmed and graded into sizes in a mechanical grader built of stainless steel or Monel metal.

Pitting the Cherries

The next operation is the removal of the cherry pits. Practically all cherries are pitted by hand; with a small pitting spoon similar in design to the well-known peach-pitting spoon. Pitting by machine is also practiced, but it is necessary for the cherries to be of uniform size. These cherries are called *special machine-pitted cherries.*

Cherries pitted by one machine regardless of size are called common machine-pitted cherries. In this type of machine, many cherries are crushed, resulting in an inferior product.

There can be no definite statement as to what the pitting loss may be. It depends upon many factors, such as: whether the cherries are of the free-stone or cling-stone type, whether they are firm or soft, and whether the pits are large or small.

With high-quality fruit, the average pit and stem loss should range about as follows:

Large fruit	15%	Small fruit	20%
Medium fruit	18%	Dipping fruit	24%

Shrinkage of original fruit, after brining, is about 10%.

Maraschino Cherries in Syrup and Alcoholic Beverages

The name Maraschino cherries is applied to all red cherries flavored with wild cherry or maraschino cherry flavor. In a United States decision, the term *Maraschino cherries* is defined as cherries preserved in true Maraschino cordial, prepared by fermentation and distillation from Marasca cherries. Cherries, preserved in syrup and cherry flavor, must therefore be labeled Maraschino style cherries.

Maraschino cherries are used in cherry-flavored fountain syrups, ice cream, garnishing for fruit cups, cocktails, bakery products, canned fruit salads and confectionery.

For Maraschino cherries to be used in canned fruit, all of the sulfur dioxide must be removed. Any percentage left in the fruit may attack the tin lining of the can. In removing sulfur dioxide from the cherries it is customary to place the pitted cherries in large wooden troughs. The sulfur dioxide is leached out by a stream of cold running water. The Food and Drug Administration allows 350 parts per million of sulfur dioxide in the finished Maraschino cherry. The state of Pennsylvania does not allow any at all. After the sulfur dioxide has been leached out the cherries are then transferred to steam-jacketed kettles and cooked in several changes of water until tender. They are then dyed.

Coloring Cherries

Cherries are dyed with pure food certified colors. Cherries to be used in canned fruit salads are dyed with an acid fast dye that will not bleed and discolor the other fruit. Erythrosine is used for the purpose. The dye can be used in a water solution and then heated with the cherries. A 0.025% solution of erythrosine in water is used or the cherries can be dyed in a weak solution of sodium bicarbonate and then the color is set fast with a citric acid solution. In this method, therefore, a 0.025% solution of erythrosine is prepared and made alkaline with 0.05% of sodium bicarbonate. The fruit and dye are set aside for several days. The dyeing can be hastened if fruit and dye solution are kept at about 140° F. When the dye has penetrated through the cherries, pour off the dye solution and then cover the cherries with a ½ of 1% citric acid solution. After several days the dye is set. The acid solution is then drained off, and the cherries are now sweetened with a 20° Bé sugar syrup.

The other three permissible food colors used for dyeing cherries for cocktails, Maraschino type cherries, or garnishing purposes, are Ponceau 3R, Ponceau SX, and Amaranth. Ponceau 3R, however, is the preferred color.

Cherries sold in glass bottles may contain the usually tolerated amount of sulfur dioxide, which is 350 parts per million, or 1/10 of 1% sodium benzoate, artificial color, acid and bitter almond flavor. The usual final syrup in the bottles may be about 12 to 20° Bé.

Good results are obtained by making Maraschino cherries for bottles in the following way. Leach the cherries with cold water, and for 200 pounds of fruit use about 1½ to 2 ounces of Ponceau 3R certified food color. After the cherries have been dyed, they are covered with a 22° Bé sugar syrup. Also, add separately, first, 7 ounces of sodium benzoate and then 4 ounces of citric acid. At the end of 10 days add another 50 to 60 pounds of cane sugar and allow to soak in the fruit.

The cherries are then flavored with 3 ounces of oil of bitter almonds dissolved in 3 ounces of alcohol.

Bulk cherries are sold in different sizes:

600 count	Number of cherries in 1 gallon
700 count	
800 count	
900 count	
1000 count	
1200 count	Dipping cherries used primarily for cordial chocolate-coated cherries.
1300 count	
1400 count	

A 50-gallon barrel of cherries contains 300 pounds of cherries and approximately 20 gallons of syrup.

Cherries are packed in 3-, 5-, 8-, 16-, 28-ounce glass jars, and also in ½- and 1-gallon jars. The 3-, 5-, 8-, 16-, and 28-ounce jars are pasteurized at 170° F for 40 minutes. The ½- and 1-gallon jars are pasteurized at 180° F for 1 hour.

With some exceptions, the state laws governing the sale of Maraschino style cherries conform with the Pure Food and Drug Act of 1906. The exceptions are: Pennsylvania which allows no sulfur dioxide; Wisconsin which allows no sodium benzoate; and New York which requires a statement of amount of sulfur dioxide to be placed on label, 8-pt. brevier type.

TYPES OF CHERRIES (STANDARD TRADE TERMS)

Whole cherries	Cocktail cherries
Whole and broken cherries	Orange-flavored cherries
Orchard run cherries	Lemon-flavored cherries
Minced cherries	Peppermint-flavored cherries
Crushed cherries	Spiced-flavored cherries
Sliced cherries	Rum-flavored cherries

Maraschino Cherries in Brandies and Wine Syrups

Fruits preserved in alcoholic beverages are very popular. The methods of packing cherries described in this chapter may also be used for other fruits, such as apricots, peaches, Damson plums and pears. The percentage of sugar present in the finished product may vary. The amount of spirituous ingredients, such as brandy, wine, or rum, depends wholly on what the consumer likes. The use of alcohol alone, artificially flavored, is a poor substitute for brandies or wine, since straight alcohol has the tendency to toughen the fruit and impart a harsh disagreeable taste. For processing Maraschino style cherries in wine, fortified wines like sherry or angelica are recommended. The wine should be aged for at least several years.

The Maraschino cherries used in spirituous liquors should average about 18 to 20 millimeters in size. They should not contain any sulfur dioxide, since the taste is objectionable. The cherries are packed in pint bottles, and 3 to 5 ounces of brandy or rum are added. The balance of the bottle is then filled with a 30° Bé cane sugar syrup. It is advisable to pasteurize the product for 45 minutes at 175° F. Cherries packed in wines are prepared by adding 4 or 5 ounces of sherry or angelica to a pint of fruit, and to the balance of the bottle a 30° Bé cane sugar syrup is added. These bottles should also be pasteurized. There is no loss of alcohol content during the pasteurization period. Sufficient head space should be allowed in the bottles for expansion during pasteurization.

Candied and Glazed Fruits

Candied fruits are prepared by replacing the water in the fruit with sugar until impregnated with a high sugar content. In manufacturing this class of merchandise, the fruit must be handled very carefully and large amounts must not be cooked at one time. The use of certified food colors is permitted, if declared on the label. It is very seldom that a preservative is used, unless the manufactured product is purposely treated with less sugar.

Candied fruits are used extensively in the confectionery and baking industries. In fruit cakes, they add flavor as well as eye appeal. In the candy industry, they are coated with chocolate and make desirable confections.

Cherries

Glazed or candied cherries are manufactured from the same raw material that is used for making the Maraschino style cherries. The fruit when picked should be hard, ripe, and of uniform color. The cherries should be washed and the blemished ones removed. The cherries are then put in barrels and bleached with sulfurous acid solution or chemical salts, as previously described.

Leaching Out the Sulfur Dioxide

Bleaching of the fruit to a light yellowish tint makes a good base for the red color, which is used later on in the process. After the cherries have been bleached the sulfur dioxide content may run as high as 0.40%. The sulfur dioxide is removed by washing the cherries with cold water, in the original barrel, or by transferring several barrels of cherries to a large wooden trough, and allowing a continuous slow stream of water to come in from the bottom of the trough and to run out over the top. When the sulfur dioxide has been washed out so that only about 0.05% remains in the fruit, the fruit is ready for the cooking process. The cherries are transferred to steam-jacketed kettles, covered with water, and cooked for about 20 minutes until they become tender. They are then cooled rapidly with cold running water. The fruit is then transferred to three- and five-gallon crocks or enamel pails, or to twenty-gallon wooden tubs, and is ready to receive the sugar syrup and color.

Heating Fruit with Syrup

The sugar syrup is prepared as follows:

Granulated Cane Sugar	100 lb
Water	18 gal
Citric Acid	2 oz
Certified Red Color	½–¾ oz

The sugar, water, acid and color are brought to a boil. Cover the cherries with this hot syrup. The syrup should read around 20° Bé. The red color may be Ponceau 3R, Ponceau SX, or a combination of both. Use ½ to ¾ ounce of color to every 200 pounds of fruit. The next day, transfer fruit and syrup to the steam-jacketed kettle, and add enough sugar syrup to bring the Bé up to 22°. Boil fruit and syrup for 1 minute and replace in containers. The following day, pour fruit and syrup into the kettle, add sufficient sugar syrup to bring Bé up to 24°. Again boil fruit and syrup for 1 minute and transfer hot fruit and syrup back into the containers. The next day, transfer fruit and syrup back into the kettle, add enough sugar syrup to raise the Bé up to 30°. Boil contents of kettle for 2 minutes. Now return the cherries and syrup to their containers and let stand for 48 hours. The cherries and syrup are now returned again to the kettle, and sugar syrup is added to raise the Bé to 33°. Bring the contents to a boil, and continue for 2 minutes, then return to their containers. Allow fruit and syrup to stand for 48 hours. Then transfer fruit and syrup to kettle and reinforce the sugar syrup to 36° Bé with corn syrup. Bring contents to a boil and continue the boiling for 2 minutes. Replace contents of kettle into their containers, and let stand for 48 hours. Now pour fruit and syrup once more into kettle and reinforce the sugar syrup to 39° Bé with corn syrup. Bring contents to a boil and continue for 2 minutes. Transfer the contents into their containers, and let the candied cherries remain there for at least 4 days, or even longer, before they are used. When the cherries are to be shipped, transfer fruit and syrup into the kettle and bring contents to a rapid boil. Then immediately scoop up the cherries on a perforated ladle and let excess syrup fall back into the kettle. Now place the cherries on draining screens and allow to remain overnight, during which time the excess syrup will drain off. The fruit will then be dry and ready for packing into containers.

Pineapple

Glazed pineapple slices or fingers are made from fresh pineapples or from fruit packed in a weak solution of sulfur dioxide and salt. Fresh pineapples arrive in May, June and July, whereas pineapples in brine are available the year round. Fruit in brine is packed in Singapore and Cuba.

The fresh pineapples are difficult to handle. They are peeled, sliced and then stamped out into proper sizes. The fruit is high in acid, is tough and fibrous, and should be processed quickly, since fermentation sets in very rapidly.

Transfer the sliced pineapple fingers, both fresh and canned, to barrels. Cover the fruit with cold water and add ½ pint of sulfurous acid stock solution. Let stand for 24 or 48 hours. A very slight fermentation may set in. This is beneficial, since it softens the tough fibers of the fruit and aids uniform absorption of sugar later on in the process. The fruit is then transferred to the steam-jacketed kettle, covered with cold water, parboiled until tender, about 20 minutes. The hot water is then allowed to run off, and the fruit is covered with cold water until chilled. This restores a certain amount of crispness to the fruit.

Heating Fruit with Syrup

Transfer the pineapple slices or fingers to shallow three- or five-gallon crocks, and cover with a hot 20° Bé cane sugar syrup, acidulated with 1/10 of 1% of citric acid. The next morning, transfer the fruit and syrup to a ten-gallon tilting kettle, then add enough cane sugar, and boil the batch, until the syrup registers 20° Bé hot. Return fruit and syrup to crocks. The following day repeat the procedure, boil until the syrup registers 22° Bé hot. The next day repeat same procedure, boil until the syrup registers 25° Bé hot. Repeat same procedure on the following day, this time boiling until the syrup registers 28° Bé hot. Allow the fruit to remain at rest for 48 hours.

Transfer again to the tilting kettle and, for every 15 pounds of fruit, add 4 pounds of corn syrup, 42° Bé. Boil for about 1 to 1½ minutes. Return the fruit to the crock and allow to stand for 48 hours. Now transfer to the kettle, add another 4 pounds of corn syrup, and boil as outlined above. Again return the fruit to the crock. At this point, the syrup should register around 41° Bé hot. If it does not, continue the process of concentrating the syrup, after 48-hour rest periods, as suggested above. The pineapple should remain in the final syrup at least 4 days, before being drained for packing. The longer the fruit remains in the final syrup, the plumper it becomes and the more additional weight it takes on. The ratio of cane sugar to corn syrup used above is 3 to 1. The minimum amount of total sugar solids necessary to prevent spoilage or fermentation of candied pineapples is 75%. A thoroughly cured product will contain as high as 85% total solids. Pineapples should always be thoroughly cured. If they are not, they may sweat, become tacky, shrink, and lose their glaze coating. If sugar crystals form on top of the crocks, after complete processing of the pineapples, this is an indication that too much cane sugar has been used. If this occurs, heat the contents of the crock and add a small amount of corn syrup, or acidulate with $1/10$ of 1% of citric acid during the cooking in the kettle.

Draining Fruit from Excess Syrup

The pineapple slices or fingers are drained from their final heavy syrup by transferring the contents of the crocks into the steam-jacketed kettles. The steam is turned on, and contents are brought to a quick boil. As soon as the fruit is rendered loose from the syrup, remove the fruit with perforated ladles and spread over draining screens. Allow to drain and dry overnight in this manner.

Glazing Pineapples

To glaze pineapple slices or fingers, the following procedure is recommended. Place cane sugar and water in the ratio of 2 to 1 into the kettle and cook the mixture up to 235° F. Now shut off the steam. Add small amounts of the drained pineapple to the kettle. Rub the syrup up and down on one side of the kettle, and, when the syrup starts to crystallize, bring the fruit under this area, and then, with a perforated ladle, transfer to draining screens. The fruit is straightened out by working quickly with forks. The glazing should take place in a cool room, free from steam vapors. Allow to dry out on the screens for at least 24 hours before packing.

Citron Halves (Long Process)

Citron halves come packed in a brine solution, containing about 8% salt. The salt is used to preserve the rind for a long period until the manufacturer makes it into a candied product. The salt must be washed out completely. Transfer the fruit to large open wooden troughs or barrels and allow a steady stream of cold water to run just over the sides of the receptacles. Continue this operation until practically all of the salt is removed. Then cook the fruit in the kettle, using several changes of water, to remove bitterness and make it tender. The practical test for tenderness is to take a piece of stiff copper wire and press it into the fruit. If it penetrates with very little resistance, the fruit is sufficiently tender.

After the fruit has been cooked with hot water, chill it thoroughly with cold water to restore crispness.

Heating Fruit with Syrup

Now prepare a 22° Bé hot cane sugar syrup, and add the citron to it. Boil the contents for 1½ minutes, and then transfer the citron to trough or to 25-gallon wooden tubs. The citron should be just covered with the hot syrup and pressed down. Remove the excess syrup, if any. The next day transfer citron and syrup to the kettle, and for every 100 pounds of citron add 20 pounds of cane sugar. Bring to a boil and continue boiling for 1½ minutes. Replace in receptacle. For the next 2 days repeat the same operations. Now allow the citron and syrup to rest for 48 hours. Then transfer the batch to the kettle and, for each 100 pounds of citron, add 15 pounds of corn syrup 42° Bé. Bring to a boil and continue for 2 minutes. Replace in the receptacles. At the end of 48 hours transfer citron and syrup to the kettle, add another 15 pounds of corn syrup, bring to a boil, and continue boiling for 2 minutes. Replace in receptacles. Allow to rest for 48 hours. Now transfer again to the kettle, and, for each 100 pounds of citron, add another 10 pounds of corn syrup. Bring the contents to a boil and continue boiling for 2 minutes. Replace in the receptacles and let rest for at least 6 days before draining. The final syrup should register around 41° Bé hot. If it does not, then at the end of 48 hours, transfer syrup to the kettle and boil until it does. Cover the citron with this hot syrup. The longer the citron remains in the last syrup, the better the product. It becomes more translucent and takes on additional weight. When the rind is ready to be drained, transfer the syrup to the kettle. When it is boiling add small amounts of the citron, stirring them around in the hot syrup, and then placing them on the draining screens. If the citron is to be glazed, cook up cane sugar and water to around 240° F. Shut off steam and place the citron halves in this concentrated syrup. Leave in for about 30 seconds and then carefully transfer to draining screens.

Citron Cubes ¼ Inch (Short Process)

The citron used in this process may be either the Italian or Puerto Rican variety. The Italian citron has a superior flavor. Without preliminary washing of the citron with cold water, the rind is put right through the dicing machine. In this process, the dicing machine slices and cubes the peel in one operation. If the diced peel is ¼-inch size, it can be very satisfactorily processed in the quick way. If the cubes are larger, the longer process is recommended.

When dicing citron, be sure to remove stems, stones, pieces of wood, etc., as these impurities will break the knives of the machine. The diced citron should now be washed with cold running water. Remove practically all of the salt in this fashion, and be guided by the taste of the cubes. Then transfer to the kettle, cover with cold water, bring to a boil, and continue boiling for about 2 minutes. Let this hot water drain off, and repeat the operations of cooking the rind until tender and neutral-tasting. From one to three hot washes are usually necessary. Drain the wash water each time. When the citron is ready, cover with a 32° Bé syrup. The contents of the kettle are now brought to a boil. Continue cooking, but not too vigorously, from 1¼ to 1½ hours, to finish the batch, from the time of adding the cold syrup to its completion. The rind is taken out of the kettle and spread thinly over draining screens. If a dry product is desired, place the screens in the drying room, about 120° F, for about 2 hours; otherwise overnight draining is sufficient. The syrup used in this process is made with 70 parts of cane sugar and 30 parts of corn syrup dissolved in the correct amount of water, until it shows 32° Bé.

Orange, Lemon, and Grapefruit Peel, Halves Long Process

These citrus peels come packed in a salt brine. The salt must be washed out with cold running water. The method used here can be the same as that recommended for citron halves. The peel is then cooked in the kettle, with several changes of water. During these operations, the salt is further reduced, the bitterness is removed, and the peel is made tender enough to be processed successfully with sugar syrups. Two methods are suggested for syruping or sugar-curing the peels. One is to add increasing percentages of hot syrup over the peel, and the other is to cook peel and syrup together in the kettle. The first method takes longer, but the product looks better.

The following procedure may be used with the orange, lemon or grapefruit halves. Boil the peel with a 20° Bé cane sugar syrup for 1 minute, then transfer to crocks or wooden half tubs. The next day place the syrup in the kettle and, for every 15 pounds of peel that the syrup covers, add 3 pounds of cane sugar. Bring this syrup to boiling and continue for ½ minute. Cover the peel gain with this concentrated syrup. Repeat this operation twice, 24 hours apart. Again transfer syrup to the kettle and, for each 15 pounds of peel that the syrup covers, add 4 pounds of corn syrup, 42° Bé. Bring to the boiling point and continue boiling for 1 minute. Then cover the peel with this syrup. Allow a 48-hour rest period. Then again reinforce the syrup with another 4 pounds of corn syrup and proceed as above. The peel should be allowed to rest for at least 4 days before packing, and the syrup should register not less than 40° Bé cold. If it does not, boil up the syrup for 1½ minutes every 48 hours. Cover the peel again, and proceed as before. The longer the peel remains submerged in the syrup, the better and plumper the product will be. When the time comes for packing, the peel is drained from the syrup by cooking both peel and syrup in the kettle, and then spreading the peels out on draining screens.

If a glazed peel is desired, it must be submerged in a special sugar syrup, prepared as follows. Take 2 parts of cane sugar and 1 part of water, and boil to 238° F. Now close the steam valve. Add the peels to be glazed to the kettle, paddle them around for about ½ minute, and then place carefully on draining screens for drying.

The second method recommended for these peels is practically the same as for the citron. Follow the instructions for the citron. The only difference is that the final syrup in the citrus peels should be 40° Bé cold.

Orange, Lemon, and Grapefruit Peel
Short Process

In the following method, the peel is first put through the dicing machine, then transferred to the kettle in which the excess salt and bitterness are washed out with hot water. The cubes can be ¼, ⅜ or ½ inch in diameter. These methods can also be used for sliced orange, lemon and grapefruit.

Pieces or halves of orange, lemon or grapefruit peel can be used for dicing. No preliminary washing with cold water is necessary.

Orange and Lemon Peel

Take 200 pounds of the diced orange or lemon peel and transfer to a 75 to 80 gallon steam-jacketed kettle. Add cold water until peel swims around freely. Bring the contents of the kettle to a boil. Boil for about 5 minutes and then let the hot water run off. Repeat this operation as many times as may be necessary to wash out the salt and bitterness. Usually three to five cooks are sufficient. The peel should also be tender. When the peel has been sufficiently washed, chill it with cold water to restore natural crispness. Now let this cold water run off, and fill the kettle with a 34° Bé syrup, prepared by mixing 8 parts of 34° Bé cane sugar syrup with 15 parts of 34° Bé corn syrup. To every 10 gallons of this syrup add 1 ounce of dissolved citric acid. The contents of the kettle are now brought to a boil, and boiled for about 10 minutes, paddling occasionally with a wooden oar. The hot syrup should now register between 20 and 22° Bé hot. If it does not, then add water to bring it down or more syrup to bring it up to around 20 to 22° Bé. With this adjustment made, continue boiling until the syrup registers 20° Bé hot. The cooking should be so controlled that it will take about 35 to 45 minutes from start to finish. The peel is then transferred to draining screens and allowed to drain for at least 1 day. The peel can be dried if desired, in the drying room. To insure against fermentation and mold, the total solids in the product should not be less than 72%. Sodium benzoate may be added, the proportion being 1½ ounces to each 100 pounds of finished peel. Its presence must be declared on the label.

Grapefruit Peel

Grapefruit peel may be had either in pieces, halves, or strips. It may even be diced. All types come packed in a brine solution. Proper selection of peel is important, otherwise a mushy product will result. Without preliminary washing with cold water, the peel is put through the dicing machine. Transfer about 200 pounds to the kettle, and cover with cold water. Bring the contents of the kettle to a boil and continue boiling for 45 minutes. Then let this water drain out. Repeat this operation several times, until the salt and bitterness are completely removed. This is accomplished after three to five washings. The peel should also be tender. If a colored (red or green) peel is desired, it can be colored in the last wash water. One ounce of red color (¾ part Ponceau 3R and ¼ part Amaranth) is sufficient to color 200 pounds of peel and ¼ ounce of green color (¾ part Tartrazine and ¼ part Guinea Green B) is sufficient to color 200 pounds. The color can be dyed in by maintaining the contents of the last hot rinse at a temperature around 180 to 200° F. The color is usually taken up by the peel in about 30 minutes. The coloring operation is complete as soon as the white part of the peel has been penetrated. The peel can also be colored overnight in wooden tubs, if desired. The peel is now covered with a sugar syrup solution as outlined under orange. Add sufficient syrup so that the product swims about freely. Now bring the contents of the kettle to a boil and continue boiling for 10 minutes. The Baumé of the hot syrup should now read around 20 to 22°. If not, make the adjustment as outlined under orange. Also, if more color is desired on the peel, this is the time to add it. The cooking is then continued until the syrup reads 28° Bé hot. Try to regulate the cooking so that the syruping process is completed within 35 to 45 minutes. The peel is then drained while hot, on draining screens.

Semi-Finishing Process

It is possible to semi-process grapefruit peel at the source of origin. The best procedure is to select pieces of peel which are clear in color and without russet spots. The peel should be taken from the grapefruit in quarters. The next step is to place the peel in a large steam-jacketed kettle, cover with water, and bring to a slow boil, until the peel is sufficiently tender so that it may be perforated with the finger-nail.

The peel is then taken out, cooled, and cut into the desired shapes. Cubing to the size of about ¼ inch makes a very attractive piece. After this has been done, the cubes should be placed in a clean barrel, with the head off. The barrel is then filled with water and left standing for about 3 hours.

By means of a hose or pipe, which reaches to the bottom of the barrel, the water is changed, and the barrel is left standing again for 3 hours. This process is repeated until the extreme bitterness of the peel has been neutralized to a stage where a slight part of the bitterness remains.

At this point, the cubes are removed from the barrel and the water is eliminated from them, either by centrifugal force or some other desirable means. After this has been accomplished, the cubes are subjected to a high-speed glaze process, placed in barrels, covered with sufficient syrup, and shipped to their destination, where the draining and finishing process can take place.

In this shape, the peel can be readily used for fruit mixes. As a matter of fact, if treated with a little oil of orange, it will almost taste better than the orange peel itself.

Melon Rind, ¼ Inch Diced (Short Process)

The rind of the watermelon or citron melon is usually used. The rind is firm, tasteless, translucent and can be satisfactorily dyed, especially red. While being processed with sugar syrup, it becomes naturally green in color. Dice the rind without washing out the salt. Transfer 200 pounds to the kettle, cover with water and bring to a boil. After boiling about 10 minutes, drain off the hot water. Repeat this operation several times until the rind is neutral-tasting. Chill the rind with cold water after the hot washes. Now cover the product with a 34° Bé sugar syrup, made by mixing 50 parts of 34° Bé cane sugar syrup with 50 parts of 34° Bé corn syrup. Bring contents of the kettle to a boil, and continue boiling for 15 minutes. The syrup should now register about 22° Bé hot. If it does not adjust it accordingly. The cooking is continued, with occasional stirring, until the syrup registers 28° Bé hot. Regulate the cooking so that it will take 35 to 45 minutes for complete syruping. The hot rind is now drained and dried in the drying room. The rind may also be colored red or green in the same way as outlined under grapefruit peel.

Watermelon Rinds

The watermelon rind may be prepared by removing the green peel on the outside up to the white flesh. The adhering red portion of the watermelon on the inside should also be removed until a smooth white finish results. A rapid method of removing the green pigment is to heat the rind to 145° F in a 10% lye solution containing ½% wetting agent, for 5 to 7 minutes. The rind is then packed in a 10° Bé salt solution (11% brine) and the daily loss of salt renewed for 7 days. On the fifteenth day, the brine is completely replaced, and 0.25% sulfur dioxide added.

Another method of preservation is to cover the rind with a 10° Bé salt solution, and at the end of 15 days, remove all of the salt solution and replace only with 0.5% sulfur dioxide and 0.05% calcium carbonate. During the fifteen-day standing period, a fermentation sets up which is beneficial.

To remove the preservative solution, the rind is soaked in fresh water for 12 hours to remove salt, and then boiled vigorously in fresh water 1½ to 2 hours to drive off sulfur dioxide.

Watermelon Rind Preserves

Preserved Rinds	5 lb	Water	2 gal
Sugar	6 lb	Sliced Lemons	3
Dextrose	1 lb		

Bring the sugar, water and drained rind to a boil and cook until clear and translucent. This may take from 1 to 2 hours. The temperature of the syrup should register 222 to 224° F when finished. Preserved watermelon rinds can also be put up in a form of special candied and pickled products.

Salting of Vegetables

One of the oldest methods of preserving and making foods is by the use of table salt. A high percentage of salt will usually stop spoilage whereas a small amount will bring about certain favorable changes. Yellow or pitch pine wood should never be used for processing salted vegetables. This sort of lumber will impart undesirable flavors to the product. Granulated, flake or rock salt can be used. However, granulated salt is preferred.

Preparation of Vegetables

All vegetables should be washed. String beans may be salted, whole or in pieces. Green tomatoes, cucumbers, beets, turnip and beet greens, spinach, etc., should not be cut up. Peas and green lima beans should be shelled, and corn should be husked before being salted.

Dry Salt Method of Fermentation

After draining off the excess water from the washed vegetables, 2½ to 5% of salt based on the weight of the vegetables is used. This salt should be as evenly distributed between layers of vegetables as possible. When a vegetable does not have sufficient water of its own, it may be necessary to add a brine solution in addition to the added dry salt. The tops of fermentation vats or vessels require protection from possible dirt and insect contamination. Clean muslin cloth or stainless steel wire mesh may be used, and it should be weighted down. As the fermentation begins, it will be indicated by gas bubbles rising to the surface. This usually starts within 18 to 48 hours and depends a great deal on the temperature. During the warm seasons, the fermentation should be completely finished within 7 to 10 days, whereas during cool seasons, it may continue as long as 2 to 4 weeks. When fermentation is completed, try to store the finished product at a temperature around 50° F. By taking this precaution, losses due to acidic fermentation and moldy tastes will be avoided. The barrels or containers should also always be full of brine solution, otherwise oxidation causing bad flavors, or a mold growth may occur. Spinach, cabbage and greens such as beets and turnips can be fermented using the method just described.

Sauerkraut

In the manufacture of this product, each 100 pounds of shredded or sliced cabbage is treated with 2½ pounds of salt. The shreds are packed in 1-inch layers with salt sprinkled in between. Even distribution is important in order to obtain a uniform product. Under proper conditions, fermentation takes place between 65 and 70° F. The kraut should be ready in 10 to 14 days. The developed acidity may run as high as 2.4% of lactic acid.

Canning Kraut by Cold Pack Method

Heat the kraut in its own juice to 110 to 130° F. Fill clean jars, seal and then pasteurize at 175° F for 35 minutes and 45 minutes, for pints and quarts respectively. Try to cool the pasteurized jars as quickly as possible after pasteurization; otherwise, the product may tend to soften and darken. A small amount of sodium bisulfite, if added to the kraut before pasteurization, will help to retain a light color. However, the presence of this chemical should be declared on the label.

Vegetable Salting Method Without Fermentation

The use of heavy dry salt, very carefully distributed throughout the vegetables, is preferred by some manufacturers to the use of a heavy brine solution. In this method, the process is rather simple, and requires little work.

For each 100 pounds of washed vegetables, 18 to 25 pounds of salt are used. Place a layer of vegetables about 1 inch thick in the bottom of a clean container. Sprinkle over with salt. Add another layer of vegetables with more salt, and continue in this way until the receptacle becomes full. The salt should be so equally distributed that the fixed amount will be sufficient. Care must be exercised to get equal distribution between salt and vegetables. The container should be filled to about 3 inches from the top, covered with clean muslin and held down with a suitable weight. After 24 hours if the vegetables are not covered by their own brine, make a brine by dissolving 1 pound of salt in 2 quarts of water, and cover the vegetables. Air bubbles arising to the surface are not due to fermentation but are the result of pressure in the receptacle forcing the air and gases out of the vegetables. After the bubbling stops, the heavily brined vegetables are ready for storage.

Salt Removal from Vegetables

The salt can be removed from tanks by using the flowing water method or by changing the water every hour or so. The use of warm water around 160 to 180° F will reduce the salt-removal time. The addition of small amounts of sugar and lemon juice to the freshened salted vegetables is beneficial.

Vegetable Fermentation with Heavy Salt Brine

Vegetables such as corn, cucumbers and green beans, which may be kept for a long time, call for a high percentage of salt. These vegetables, like others, may be processed at first with a weak solution of brine, which permits fermentation and produces a desirable flavor. The amount of salt is then increased. This insures the preservation of the vegetables. Vegetables brined by this method are referred to as *salt stock* vegetables. They are stored in outside tanks and are able to withstand all kinds of weather without spoilage.

Brining Method

The vegetables are washed and excess water is drained off. They are then covered to within 3 inches of the top of the receptacle with clean cheese-cloth or muslin and weighted down. The brine is prepared by dissolving 1 pound of salt in 1 gallon of water. After the vegetables are covered with this brine, add 1 pound of dry salt for every 10 pounds of vegetables on the top of the receptacle. This is done in order to make up for the water present in the vegetables. Fermentation should start within 24 to 48 hours. At the end of a week, ½ pound of dry salt for each 100 pounds of vegetables is added.

This salt may be added on the surface of the container. It will dissolve and become equalized in a short time. Continue to add salt at this same rate for the next 10 weeks, at the end of which time the vegetables should be completely fermented and cured. If properly prepared, these vegetables should be sound and good to use for long periods of time.

Salted Vegetables and Storage Care

Salted vegetables are subject to attack and spoilage by yeasts, bacteria, molds and insect life. Loss of brine solution and oxidation also may occur. Yeasts consume the natural acids in the vegetables. When this takes place, the product becomes mushy, putrid and also takes on a scum growth. The method of prevention is keeping the air away from the surface of the brine. Be sure that the vats or barrels are always full of brine, and where necessary, add additional brine made with 2 pounds of salt to 1 gallon of water.

SOLUTION OF SULFUR DIOXIDE IN WATER (SCOTT) 15° C

Per Cent SO_2 By Weight	*Degrees Bé*	*Specific Gravity*
0.1	0.07	1.0004
0.2	0.14	1.0008
0.3	0.21	1.0012
0.4	0.28	1.0016
0.5	0.36	1.0020
0.6	0.43	1.0024
0.7	0.50	1.0028
0.8	0.57	1.0032
0.9	0.64	1.0036
1.0	0.72	1.0040
1.5	1.09	1.0070
2.0	1.46	1.0100
2.5	1.82	1.0125
3.0	2.19	1.0150
3.5	2.55	1.0180
4.0	3.07	1.0210
4.5	3.44	1.0235
5.0	3.80	1.0260
5.5	4.17	1.0285
6.0	4.53	1.0310
6.5	4.90	1.0335
7.0	5.27	1.0360
7.5	5.63	1.0385
8.0	5.99	1.0410
8.5	6.36	1.0435
9.0	6.73	1.0460
9.5	7.10	1.0485
10.0	7.46	1.0510

pH OF SO_2 SOLUTIONS *

Molar Strength	*Per Cent Strength*	*pH*
1.00	6.41	1.05
0.50	3.20	1.10

pH OF SO_2 SOLUTIONS * (*continued*)

Molar Strength	*Per Cent Strength*	*pH*
0.33	2.14	1.20
0.25	1.60	1.28
0.20	1.28	1.32
0.10	0.64	1.47
0.05	0.32	1.62

* Determined by La Motte Chemical Products Co., Baltimore, Md.

SODIUM BISULFITE SPECIFIC GRAVITY OF AQUEOUS SOLUTIONS AT 15° C

Specific Gravity	*Degrees Bé*	*Per Cent $NaHSO_3$*	*Per Cent SO_2*
1.008	1	1.6	0.96
1.022	3	2.1	1.3
1.038	5	3.6	2.2
1.052	7	5.1	3.1
1.068	9	6.5	3.9
1.084	11	8.0	4.8
1.100	13	9.5	5.7
1.116	15	11.2	6.8
1.134	17	12.8	7.8
1.152	19	14.6	9.0
1.171	21	16.5	10.2
1.190	23	18.5	11.5
1.210	25	20.9	12.9
1.230	27	23.5	14.5
1.252	29	25.9	15.9
1.275	31	28.9	17.8
1.298	33	31.7	19.6
1.321	35	34.7	22.5
1.345	37	38.0	23.6

CHAPTER XIII

PLANT SANITATION AND SANITARY FOOD CONTROL

DISCUSSION OF PLANT SANITATION

A very serious problem regarding rodent infestation exists in all food manufacturing plants. There are no short cuts or quick ways to bring about clean and sanitary conditions. The employment of trained cleaning crews, plenty of hot water, detergents, brushes, brooms, mops and other cleaning aids is necessary. This manner of good housekeeping will result in considerable rodent control. If there is evidence of the presence of rodents, locate the source so that their quarters may be destroyed. Poisoning is recommended as a means of rodent control outside of the plant only. Traps should be set for mice and rats inside of the plant, in runways and along the walls. They may be baited with a small dab of peanut butter, and this may be good for 2 to 4 weeks.

The presence of even a single rodent hair in a finished food product is now regarded by the Pure Food and Drug Administration as sufficient evidence of filth to cause condemnation of the product. This is really a serious matter, and expert care should be exercised in examining bags of salt or sugar which may have been attacked by rodents which leave behind pellets and hairs. When sacks of sugar are emptied into the syrup, kettles or tanks, this filth may drop in. There is also the possibility of rodents walking over equipment, pipes and beams in the plant, and as a result, the dropping of pellets or hairs and dirt into equipment and the food materials undergoing processing.

Insect infestation may be controlled by cleaning out the breeding places and using plenty of hot water and detergents. One gallon of 5% DDT can be sprayed over 1000 square feet. The killing effects of DDT may last about 3 months, providing there is not too much washing or steaming of the areas sprayed. It is practically impossible and, in most cases, unnecessary to manufacture foods to a point where they are completely free of bacteria. However, it is important that the bacterial content be kept at such low levels that the by-products do not lower the quality of the food. Bacteria control is accomplished by maintaining the raw products at low temperatures, handling foods in clean containers and equipment, using clean tables and belts, and working with clean hands and clean clothing. Water between 120 to 140° F with an added detergent provides the best cleaning action. It coagulates protein wastes and saponifies fats. Washing should be followed by a rinse of warm water to flush off water and suspended dirt. Chemical sanitation should be used in places hard to get at with hot water. The best compounds are hypochlorites and quaternary ammonium compounds both of which are non-toxic, free from residual odors, fast acting and have the additional property of spreading easily over the entire equipment surface. Some of the things that should be checked occasionally are dirt on beams over uncovered tanks, condensation on or sweating of ceilings, paint peelings on ceilings and walls, spider webs, and piles of trash, rubbish and spoiled food.

Roccal

Roccal is a new chemical sanitizing agent for the food industries based upon an entirely new principle of bactericidal action. It contains no phenol, iodine, mercury or hypochlorites. It is relatively non-poisonous when taken by mouth, practically odorless and, in proper dilutions, non-irritating to the skin. It exerts a powerful antiseptic and germicidal action. In dairy plants, 1 ounce to 5 gallons of water is recommended for sanitizing all equipment. The equipment should be thoroughly cleaned with a good detergent prior to sanitizing with Roccal solution. For inhibiting mold growth, 1 ounce of Roccal to 2 gallons of water is suggested as a sanitizing agent. All surfaces should be cleaned with a good detergent prior to sanitization. For an average solution of Roccal in any operation associated with the production of food, a concentration of 1 ounce to 4 gallons of water is recommended.

Hypochlorite Solutions

Sanitation in the bottling plant is of exceptional importance. First of all products must meet rigid public health requirements. Secondly they must be free from the molds, yeasts and other micro-organisms that cause spoilage. To solve this dual problem means real profit protection for the bottler.

Bottlers agree that ordinary cleanliness alone is not the answer to this problem. Bottling equipment must be made more than just clean, as judged by the naked eye; it must be made clean from the bacteriological point of view. The bottler's sanitation problem is solved only when ordinary cleaning is followed by sterilization through the use of an agent that kills micro-organisms and destroys the organic matter on which they grow.

Steam and scalding water sterilize effectively when used correctly. But it is hard to keep water or steam at the right temperature when it reaches the equipment. There is always the danger that inaccessible places will not be heated enough to kill the bacteria and that the moisture and warmth will actually aid bacterial growth.

Hypochlorite solutions have won wide recognition as highly effective, safe and easily applied sterilizing agents. Health authorities and bottlers everywhere find that a reliable hypochlorite product is the practical and economical way to guard against spoilage losses.

Sanitation HTH Points the Way to a Practical Sanitation Plan

A dry, free-flowing powder, Sanitation HTH provides dependable hypochlorite sterilizing solutions when simply added to water.

It comes packed in sealed cans, and there is no danger of waste or breakage. It always reaches the user full strength and, due to its exceptional stability, assures the user of effective sterilizing solutions at all times.

HTH is easy to use. First prepare a 2% sodium hypochlorite stock solution from HTH as follows:

1 Into a 20-gallon crock about ½ full of lukewarm water, empty the entire contents of a 5-pound can of Sanitation HTH and stir for 2 or 3 minutes.
2 Add slowly 4 pounds of light soda ash or 8 pounds of modified soda and continue stirring for 2 or 3 minutes.
3 Add water up to the 20-gallon level and stir until thoroughly mixed. Allow solution to stand until sediment has settled.
4 Cover the stock solution and store in a cool place away from the direct rays of the sun.

This is a sodium hypochlorite solution containing not less than 2% available chlorine.

The following table gives directions for preparing sterilizing solutions for various applications around the bottling plant: *

Solution A 100 PPM	1 pint 2% stock solution to 20 gallons water	For *dipping* and *rinsing* (after thoroughly cleaning) articles that have smooth surfaces (metal, glass, tile, etc.) not subject to heavy contamination
Solution B 200 PPM	1 pint 2% stock solution to 10 gallons water	For *rinsing* and *flushing* (after thoroughly cleaning) articles not subject to heavy contamination, but with surfaces difficult to clean
Solution C 300 PPM	¼ pint (4 fluid ounces) stock solution to 2 gallons water	For *wiping* and *spraying* (after thoroughly cleaning) metal, tile and other smooth surfaces
Solution F 600 PPM	½ pint (8 fluid ounces) stock solution to 2 gallons water	For *wiping* or *spraying* rough surfaces and general disinfection of premises
Solution K 5000 PPM	1 quart 2% stock solution to 1 gallon water	For combatting active outbreaks of mold

* Note: For most effective treatment for killing bacteria, apply chlorine solutions only to surfaces which have been thoroughly cleaned.

Syrup System

Syrup Jars

After use, wash syrup jars thoroughly and rinse with clear potable water. Shortly before using again, do one of the following:

1 Fill a jar with water and add 1 pint of the 2% stock solution of HTH for each 10 gallons of water (solution B). Stir thoroughly and allow to stand about 15 minutes, then drain and rinse with potable water. The same solution may be pumped into other jars for similar sterilizing action.

2 With an ordinary spray gun, spray the entire inner surface of each jar with solution C, then rinse with potable water.

Syrup Lines, Mixer and Filler

After using, flush the system with warm water. Before each new run make up enough solution B in the mixer to flush the entire system. Pump the solution through the lines and filter into an empty storage jar. From this jar, let the solution flow through the syrup lines to the filler heads. Collect the partially spent solution and use it to disinfect floors, tables, drains, etc.

Rinse the system thoroughly with potable water. Make sure none of the HTH solution is trapped in pumps, filters, etc.

Filter Cloths

Wash filter cloths free of syrup in warm or hot water. Just before using, soak the cloths 15 minutes in Solution B, then rinse in potable water.

Water System

Filters

1 Stone Filters: Clean filter with scraper in usual way. Close all outlet valves and open waste line. Add enough solution B to thoroughly saturate the stone and allow to stand 30 minutes. Open drain and wash filter free of HTH solutions with potable water.

2 Paper Filters: Flush the interior of filter chamber with solution B. Drain, then rinse with potable water. Sterilization may also be effected by spraying chamber interior with Solution C.

3 Sand Filters: Reverse the flow of water and clean the filter. Add enough of solution B to thoroughly saturate the sand and allow to stand 30 minutes. Drain the HTH solution, then flush with potable water. All water filters should be treated regularly once each week.

Water Storage Tanks

Water tanks should be emptied and cleaned at regular intervals. Treatment with chlorine at these cleaning periods will help prevent algae and slime growths and the conditions that encourage bacterial growth. After the tank has been cleaned, fill it with water, adding ½ pint HTH stock solution for every 1000 gallons of water. Let the solution stand in the tank over Sunday. Drain, then rinse with potable water until the chlorine odor disappears.

General

Bottle Washing

Modern bottle washing machines using a 3% alkali solution and water above 120° F are generally recommended, and are required by many health authorities. In emergencies, hand-washed bottles may be sterilized effectively by brief immersion in solution B, followed by careful rinsing with potable water.

Bottle Caps

In cases where crowns have been exposed to contamination, they may be sterilized by immersing in solution B. Before feeding to crowner, drain, then rinse with potable water.

Refrigerating System

Slime and algae often accumulate on the ammonia condenser tubes of refrigerating systems, lowering the cooling rate and the efficiency of the equipment. Use of HTH in the cooling water will remove this slime and prevent further accumulations. One quart HTH stock solution per day for each 100 gallons of water in the condenser system should be used at the start. The quantity can then be varied as required to keep the tubes free of slime.

Tables, Shelves, Floors, Walls, etc.

After washing with warm water and cleaning powder, disinfect these surfaces with HTH solutions that remain after treatment of syrup lines, etc., or by spraying with solution C.

Laboratory Apparatus

After washing, immerse flasks, ladles, graduates, pipettes, etc. briefly in solution B, then set aside to drain and dry.

Methyl Bromide Fumigation

Methyl bromide, a relatively recent fumigant, has become valuable to food manufacturers. It has been found effective against a wide variety of insects, and has the properties of penetration which make possible the destruction of certain sheltered pests, mites and other internal feeders. As with other fumigants, its deadly action is delayed for a time after fumigation. This period may be several days with some species and will vary according to the excess above the minimum lethal dosage. Methyl bromide is widely applied for the fumigation of plant material from quarantined areas. Its use has been approved by both federal and state agencies for vegetables, fruits, potatoes and sweet potatoes. It is also used for storage-infesting insects in flour, feeds, dried fruits, nutmeats, beans, peas, spices, chestnuts, macaroni products, dehydrated vegetables, chocolate and dairy products.

Properties of Methyl Bromide

Methyl bromide (CH_3Br) is a colorless, odorless, volatile liquid with a specific gravity of 1.732 at 0° C and a boiling point of 40.1° F. It is a gas at ordinary temperatures, in which state, it is approximately 3.3 times as heavy as air. It is soluble in most organic solvents, but is only slightly soluble in water.

Protection to Operator

Methyl bromide is injurious to all forms of animal life. Proper caution should be observed when handling it, and continued exposure to low concentrations is especially to be avoided. An operator should use an approved gas mask when fumigating. Information about the right mask to use, etc., will be submitted by the manufacturer of methyl bromide.

Types of Fumigation Chambers

Successful methyl bromide fumigation can be done in almost any type of enclosure, such as, fumigant vaults, drums, tents or tarpaulins, storage rooms, vans or refrigerated trucks and freight cars. A tent or tarpaulin fumigation can be performed best by using cloth that has been coated with ethyl cellulose, vinyl resin or other similar coatings. The ordinary canvas duck tarpaulins apparently do not retain methyl bromide sufficiently to insure satisfactory fumigation.

Dosage Schedules

Dosage schedules for methyl bromide fumigation are composed of three elements each equally important: amount of methyl bromide, exposure period and temperature. It has been found that generally ½ pound of methyl bromide and ½ hour of exposure have about the same value, so that any modification of a dosage schedule can be either in dosage or in exposure in those proportions. Methyl bromide is more effective at higher than at lower temperatures, and the change in effectiveness is at a fairly constant rate for any given insect or environment. Available data indicate that a 5° to 10° F change in temperature is equivalent to ½ hour of exposure or ½ pound of dosage. This ratio of course changes as the schedules approach dosages of less than 1 pound, exposures of less than 1 hour, or extreme temperatures.

This fairly constant rate of change has made it possible to establish series of dosage schedules ranging from low to high temperatures all of which have approximately the same effectiveness on a given insect and the same degree of tolerance from the plant host. In instances where the schedules are near the minimum lethal amount, the series have small temperature differences. Where the schedules of a series are considerably over the minimum, they are fewer with greater temperature differences.

Fumigation schedules developed to date are too numerous to be discussed here. At 70° F the minimum schedules for accessible insects appear to be about 2½ pounds for 1½ hours (or 2 pounds for 2 hours).

For some insects imbedded in vegetables, fruits and nuts, it is sometimes necessary to use dosages of 4 to 5 pounds for 3 to 5 hours, but many succumb to 2 pounds for 2 hours at 70° F.

For the fumigation of grain or inert materials, a schedule of 1 pound for 24 hours at prevailing temperatures is generally recommended.

Successful fumigation has been done with methyl bromide from 0 to 95° F.

Method of Applying Dosages

The dosage of methyl bromide is released within the chamber through a closed system from the outside. It can be released either by spraying through an ordinary spray nozzle into the air stream in front of the blower, by allowing it to evaporate from a shallow pan, or by volatilizing it through a heated coil.

Where the calculated dosage of methyl bromide is 1 pound or more, it can be measured by weight. To do this, the cylinder of methyl bromide is placed on platform scales and accurately weighed. The desired amount is then subtracted from the total, and the scales are set at that point. The gas is slowly released, and, when the weight bar again balances, the valve is closed. In this case, the cylinder of gas is connected directly with the fumigation vault by light copper or plastic tubing. Near the cylinder, the tubing should be arranged in a spiral coil of two or three turns to reduce to a minimum its influence on the weight of the cylinder.

Special glass measures, or dispensers, have been devised to measure small quantities of methyl bromide. One type is inserted in the line between the supply cylinder and the vault. By use of valves the desired amount is admitted to the measure, which is graduated in cubic centimeters. The methyl bromide remains as a liquid in the measure, since it is still under pressure. To release, valves on the other side are opened, and its own pressure forces the dosage into the vault.

Another type of dispenser holds the contents of a one-pound can, and the dosage is removed by subtraction. A third type delivers the entire contents of a can into the vault and is useful where the amount of fumigant desired is in multiples of 1 pound.

These dispensers can be procured from the Arrow Products Company, Carlstadt, N. J., The Dow Chemical Company, Midland, Mich., or Bobbink and Atkins, East Rutherford, N. J. No attempt has been made here to make a fully complete list, and no discrimination is intended or implied against firms whose names are not listed.

Calculation of Dosage

When the dosage of methyl bromide is calculated by weight the following formula is used

> Divide the volume of the vault in cubic feet by 1000, then multiply this quotient by the dosage rate as expressed in pounds per 1000 cubic feet. For example, for a dosage of 2.5 pounds per 1000 cubic feet in a 1400-cubic-foot vault it would take $\frac{1400}{1000} \times 2.5 = 3.5$ pounds.

When a dosage of methyl bromide is to be measured, the following formula is used

> Multiply the index of the desired rate by the capacity of the vault in cubic feet. For example, for a dosage of 2 pounds per 1000 cubic feet in a 31-cubic-foot vault, $0.52 \times 31 = 16.12$ cubic centimeters. The basic index is 0.26 *cubic centimeters per cubic foot per pound per 1000 cubic feet.*

Rate per 1000 cubic feet	Index
1 pound	0.26 cubic centimeters per cubic foot
1.5 pounds	0.39 cubic centimeters per cubic foot
2 pounds	0.52 cubic centimeters per cubic foot
2.5 pounds	0.65 cubic centimeters per cubic foot
3 pounds	0.78 cubic centimeters per cubic foot
3.5 pounds	0.91 cubic centimeters per cubic foot
4 pounds	1.04 cubic centimeters per cubic foot

Effect of Humidity

The efficiency of methyl bromide apparently is not impaired by the presence of moisture; in fact, better results are often obtained under conditions of high humidity. Enough water to form a protective film, however, will sometimes reduce the expected efficiency.

Circulation and Venting

Methyl bromide gas will stratify if no circulation is provided in the fumigation chamber, and an overdosage or lengthened exposure is necessary to offset that condition.

If fumigation is done at prevailing temperatures, a short period of circulation during and following the application of methyl bromide is sufficient to mix the gas and air thoroughly. If temperature control is necessary, circulation should continue for the duration of the exposure. Otherwise heat stratification will occur. If rapid penetration is desired in a short exposure period, continuous circulation will probably be desirable.

The fumigation chamber should not be entered, except by men wearing gas masks, until the gas in the free air spaces has been exhausted through the venting system. This can be determined with the halide test lamp. Gas will continue to be present in the containers or load for a period of 1 hour or more but it is generally safe to handle such material if the exhausting continues, and this will keep a drift of fresh air moving toward the rear of the chamber.

Caution

Since methyl bromide is known to be injurious to human beings, persons not familiar with handling fumigants are cautioned against using it until they have thoroughly familiarized themselves with the necessary precautions.

Fumigation with Liquid Hydrocyanic Acid

Liquid hydrocyanic acid is used for general fumigation of entire buildings, mills, warehouses and factories. Fumigation of flour mills can be accomplished by direct application of the fumigant to every piece of milling machinery, through a special arrangement of tubing and dust-proof pressure nozzles known as the machinery piping system. Observations and studies over a long period of time indicated that 95% of all insect infestation in flour mills was confined to the machinery. This led to the obvious conclusion that for best results the hydrocyanide gas should be injected directly into the machinery. This project was started and completed by the American Cyanamid and Chemical Corp. in conjunction with the laboratories of the U. S. Department of Agriculture at Manhattan, Kansas, resulting in the perfection of the machinery piping system of fumigation with liquid hydrocyanic acid.

Chloropicrin

This is one of the older better-known fumigants. It is especially useful in buildings not sufficiently well constructed to confine other gases. Chloropicrin has earned a well-established reputation as an effective grain fumigant and it is well adapted to *spot treatment* of individual machines, elevator legs and conveyors. Chloropicrin is widely used for rodent control. Its tear gas vapors make it an excellent agent for this purpose. Very small quantities properly placed quickly drive rodents from warehouses and other places.

Entoleter Infestation Destroyer

In recent years, there has been increasing attention to methods of control of insect infestation in flour mills. Progress has been made by the discovery of a machine called the entoleter. This machine uses a centrifugal force principle which has been found highly effective in the destruction of insect life in bulk flour, prepared flour mixes, corn meal, spices, powdered soups and similar granular products. Flour is directed through a spout into a machine between 2 flat steel discs or plates that revolve on a central shaft at 3500 revolutions per minute. Small, round, hardened steel parts are closely spaced in two concentric rings between the two discs. The impact of the flour against the revolving discs and posts and against the housing of the machine is so great that all stages of insect life, including the egg, are killed. The treated flour is carried off through an outlet spout or is dropped directly into a bin. The entoleter machine ranges in capacity from 10 to 75 bbls. per hour. Where flour is to be stored for any length of time in storage bins, it is good insurance to sift it or to run it through a centrifugal machine before placing it in the bins. The following table shows the speeds at which various types of cereal products are being treated.

Flour, flour compounds, spices, powdered soups, and other finely ground products, 3500 r.p.m. (revolutions per minute)
Corn meal, farina, semolina, soy meal and grits, and similar granular products, 1750, 2500, 3500 r.p.m.
Dog food and other similar products, 2500, 3000, 3500 r.p.m.
Wheat (dry) 1750 r.p.m.
Wheat (tempered) for insect destruction only 1750 r.p.m.
Wheat (tempered) where entoleter is used as a pre-break as well as insect destroyer, 3000, 3500 r.p.m.

There is a very slight milling effect on the flour due to the action of the centrifugal entoleter. However, baking quality is not affected. Products such as farina, corn meal and semolina are not broken up too severely at 3500 revolutions per minute. The coarser the product is, the lower the speed at which all insect life can be destroyed.

Infestation Sources

The more important of these sources are insects entering with the grain stream, or from infested clear or low grade flour used in blending operations, or from returned flour in second-hand bags, or from infested milling machinery.

Operation of the Entoleter

This equipment is not intended to be used to recondition a product already heavily infested and unfit for human consumption. Naturally, all food products should be manufactured with the greatest care to minimize insect infestation, but despite all precautions and the best available methods, it is a well-known fact that there is a constant hazard of a small amount of infestation, for the most part in the form of eggs and tiny larvae present in the freshly milled product. The primary function of the entoleter is to destroy these eggs and tiny larvae and thereby eliminate any further development of infestation from within the product.

The amount of this infestation under proper conditions should be so small at the time of manufacture that it is of little consequence whether it is removed or not after being destroyed; but if left alive, it can easily develop to a point where the product is seriously contaminated before reaching the consumer.

The entoleter does not remove or add anything to products treated. As a matter of fact, no process of destroying insects has yet been successfully devised that separates insects from the product. For example, fumigation and heat sterilization, either electronic or steam, do not remove the insects.

Further, any process which attempts to remove the insects without destroying them is not entirely dependable. The tiny larvae get through and often reinfest the product. The commonly used rebolter is an excellent means of dressing finished products by removing inanimate foreign matter, but it is by no means sure protection against all infestation even where very fine silks can be used. Moreover, coarser products such as corn meal, farina, grits and the like cannot be bolted through fine enough silks to remove all forms of infestation, and yet these are basic food products requiring the same high quality standards as flour.

Aside from continuous protection to the finished product, the entoleter can be of real value in controlling the degree of infestation in food plants if used for the treatment of the various raw materials. In flour mills it can be used on wheat and blending stocks; in miscellaneous food plants on all the ingredients which are subject to infestation.

Treating the Finished Product

Application of the entoleter on incoming materials as they are received at the mill does not eliminate the need for further treatment before packing or storage. This is due to the persistent presence of insects, even in the cleanest mill, in wood floors, conveyors, boots, elevators and other parts of the structure and equipment, where cracks and hidden surfaces permit insect life to enjoy relative immunity.

Since the most thorough housekeeping methods cannot completely eliminate or control these breeding places, infestation develops within the mill even when it is controlled at the major sources. This leaves the product exposed to reinfestation at many points along the production line. For absolute protection of the finished flour, it is important therefore that it be treated by entoleter as near as possible to the packing equipment, preferably directly over the packer bins.

Processing of Wheat Germ

Wheat germ is a mill by-product particularly susceptible to infestation. The use of the entoleter on wheat germ after it has been flattened out by passage through rolls is out of the question since it is too severely broken up. On the other hand, the entoleter has been found to be quite satisfactory when applied to the stream containing the whole wheat germ as a substitute for the roll operation. In other words, the stream containing a mixture of whole germ, bran and middlings can be passed through the entoleter with the result that the whole germ itself is not disturbed except that it is thoroughly cleaned, and the accompanying infestation is destroyed, whereas the middlings and bran particles are reduced sufficiently so that when the mixture is passed through a sifter, with the proper screens or sieves, the germ can be tailed over almost in a pure state.

Entoleter as a "Pre-Break" on Tempered Wheat

The possibility of using the standard entoleter operating on tempered wheat at 3500 r.p.m. as a pre-break, as well as an insect destroyer, was first tried at Kansas State College years ago. Today it is in commercial use for this purpose with results that are almost incredible; namely the production of a higher percentage of patent flour of lower ash and better baking quality.

In this application, the entoleter treatment is followed by a good aspiration which removes the dirt dislodged by it as well as the insects destroyed. The aspirated stock is then classified according to size and distributed to the various rolls.

The Present Machine Compared to the First Electrical Entoleter

Briefly, the present mechanical entoleter as compared to the original electrical unit, which utilized a high voltage corona discharge to destroy insect infestation, is far simpler to install and operate, more efficient, requires only a fraction of the space and the initial cost. Maintenance cost and operating cost are all very much lower.

The Entoleter Compared to the Electronic Heat Process

The previously mentioned advantages of the present mechanical entoleter over the first electrical device apply as well to the recently developed electronic method of insect destruction. The latter employs high frequency waves for destruction of insects by heating the product to about 140° F or to a temperature at which insect life cannot survive.

Aside from the fact that heat in any form is often objectionable to food manufacturers not only because of moisture loss but also because of its effect on their products, the electrical energy required to heat all of a steadily moving stream of product from room temperature to 140° F is so great that this method is obviously expensive. For example, this method requires 1.5 kilowatt hours per 100 pounds to satisfactorily destroy insects in cereal products, whereas the entoleter requires from 0.015 to 0.075 kilowatt hour per 100 pounds depending on the size of the machine, type of product, or speed required to accomplish the insect destruction. The energy required by the entoleter apparently is only 1 to 5% of that required by the electronic heat method.

FOOD HANDLING

Illness due to food may be caused directly by the presence of certain microscopic forms of plant and animal life, by chemicals and by inherently poisonous plants themselves, and indirectly by adulteration of the food with various foreign substances, dirt and filth. Each of these agents will be considered separately.

Foods which have been allowed to stand for some time after cooking, particularly if they are not cooled and kept refrigerated, are likely to cause food infections. The introduction of harmful bacteria which cause epidemics of disease, as well as those which cause food spoilage and waste occur in practically the same manner.

Microbiology of Food

Micro-organisms are microscopic living plants or animals. They are of utmost importance to the food handler because they may infect foods and thus spread disease. Those that cause illness are called pathogenic. Some types do not cause disease at all. These are called non-pathogenic. They are undesirable because they may cause food to spoil, as for example, the souring or curdling of milk. Still others are highly beneficial, as for example, those inhabiting our intestinal tract and assisting digestion.

Bacteria

The most important forms of life which are concerned in food poisoning are called bacteria. These are so small that they cannot be seen with the naked eye. In size they measure about 1/25,000 inch. Each bacterium is a single small living organism.

Sometimes these bacteria exist singly as small round or oval-shaped bodies called cocci. When they exist as a pair, they are called diplococci. Sometimes they exist in long chains, each organism attached end to end to others. These are called streptococci. In clusters, they are described as staphylococci. Rod-shaped forms go under the name of bacilli. These exist in single cells or in chains. Spiral forms are called spirilla.

There is another significant agent often associated with bacterial life called a virus. These are probably minute forms of life, too small to be seen under the most powerful optical microscope. However, they can be seen with the recently invented electron microscope. Sometimes these forms are known as filterable viruses. They are minute organic forms intermediate between living cells and organic compounds but have the property of increasing in amount only in the presence of living cells. They can pass right through the pores of porcelain tubes.

Bacteria are found distributed all through nature. They are in the soil in enormous numbers. They float about in the air, and enter our mouths and lungs when we breathe. They are found in water, and on utensils of every kind. Of course, they are in our food. They are not found to any appreciable extent at considerable depths in the earth, extremely high altitudes, or in healthy normal tissues of plants and animals.

Bacteria are alive. They need food and water, just as any plant does. They give off waste products. Certain kinds that need air are called aerobes. Some grow only in the absence of oxygen and are called anaerobes. Some can grow under either condition. These are the facultative aerobes and anaerobes. Most types prefer a neutral food, one that is neither acidic nor basic, as, for example, meat sauce, eggs and milk. Some will grow on more acid products. Like most forms of life, bacteria need a comfortable temperature. At freezing temperatures, they are rather sluggish. Between 70 and about 100° F (a temperature of summer weather up to body temperature), they multiply rapidly. If the environmental conditions are satisfactory, they may double every half-hour. This means that in 24 hours, one organism might have 17,000,000 descendants. The ordinary types cannot grow very well when the temperature rises much above 100°.

When conditions of growth are not favorable, certain of the rod-shaped bacteria (bacilli) change their form. They become resistant to the harmful effects of their environment and thereby can survive under very difficult conditions, such as cold, heat or dryness. When they are in this condition or form, they are called spores, existing even on good food without any action. They will exist in this form when the food is too dry, or too cold, or too hot. As soon as a spore secures a medium with the proper moisture, warmth, and the environment just right for its growth, the spore-forms develop into what is called a vegetative condition. It is in this condition that microbic activity is at its height.

Bacteria have no legs, fins or wings. A few types do have one or more thread-like appendages (flagella), which help to propel them. However, most of them can only move from one place to another as the vehicle on which they exist moves them. Water can carry them, air can blow them, flies, animals and people can transport them.

They can be killed by heating them to temperatures above scalding. Not only is temperature important but the length of time the micro-organism is exposed to it is also significant. Therefore, there is no such thing as a so-called death point. Always, both temperature and time must be considered. A low temperature for a long time may be equivalent to a high temperature for a short time. But both factors should always be considered. If micro-organisms are exposed to sudden cold or if they are dried very quickly, a large percentage of them will die. Usually, death of all of them will not be accomplished, and accordingly we cannot expect a dried or refrigerated product to be sterile. The more resistant bacteria will survive.

When a food is heated to a temperature of about 140 to 185° F and held there for 30 minutes, all pathogenic or harmful bacteria are killed, although some harmless ones may survive. This process, called pasteurization, is usually applied to milk and fruit juices.

When subjected to heat treatment great enough to kill every single micro-organism, the product is said to be sterile. Some foods and food equipment are more easily sterilized than others. For example, an acid product like tomato juice or apple cider can be actually sterilized by what is only a pasteurization treatment for milk. On the other hand, milk, meat and peas must be heated in autoclaves at temperatures of 240° F for 20 minutes or longer, depending on the size of the package, to kill all the vegetative and spore-forms.

Micro-organisms can also be killed by various chemicals. The amounts to be used vary with the particular chemical and the character of the food itself. The concentration of the chemical may be expressed in percentage, as for example, 1/10 of 1% sodium benzoate in tomato catsup. It is often expressed as parts of the chemical to one million parts of the food, usually abbreviated to read "p.p.m." This designation is usually used for chemicals that are present in very small amounts, such as chlorine in drinking water.

Sometimes, a chemical is used in such small amounts that its concentration is not great enough to kill micro-organisms, except perhaps the least resistant forms, but it makes the food unfavorable for microbic growth. When chemicals are used this way, they are called preservatives, for example, sodium benzoate as above. Acid in pickles and sauerkraut makes the food an unfavorable medium for bacterial growth.

Sometimes it is impossible or even unnecessary to kill all the organisms that might be present. In such cases the treatment that destroys the vast majority of the microbic population is called disinfection. At the present time, this treatment is coming to be known as sanitization. Disinfection may be brought about by fumigating with some gas, e.g., sulfur dioxide, or by washing with hot alkaline, or acid, cleaning mixtures of chemicals, or by exposing the product to sunlight or ultra-violet light.

Yeasts and Molds

Besides bacteria, there are other forms of plant life which are similar to and are usually studied together with bacteria. These are the yeasts and molds, tiny plants with no green coloring matter. In considering food-handling, both yeasts and molds respond in general to about the same conditions that are conducive to bacterial infection. They usually grow on more acid products, like fruits, and are easily destroyed by heat.

Yeast cells can grow only in those foods that contain sugar. The sugar is changed into carbon dioxide and alcohol. This process, called fermentation, is the chief method for producing alcohol. On the other hand, yeast cells which are commonly found in the air, very often attack sweet foods, e.g., orange juice, condensed milk, plums, etc., and spoil them by changing their sugar into alcohol and carbon dioxide.

Under favorable conditions, yeast cells reproduce very rapidly by a process known as budding. A single cell gives rise to one or more small buds which in turn grow to maturity.

Conditions necessary for the growth and reproduction of yeast cells are as follows:

a Food containing sugar.
b Warmth. They are checked by cold and destroyed by boiling.
c Air. The air must be dissolved in water for the yeast to utilize it.
d Moisture.
e Yeast cells grow as well in the light as they do in the dark.

Molds are also tiny plants which are visible to the naked eye only when present in great numbers. They form filamentous, cottony, fiber-like mats of tangled threads, called mycelia, of varied color. We see blue mold sometimes on bread, and always on Roquefort cheese. These mold growths start from mold spores. The latter are small uni-cellular bodies which look somewhat like yeasts.

When foods are attacked by molds, they rapidly change in flavor, odor, and appearance, and eventually become completely decayed. Molds in themselves, however, and the foods attacked by them are not necessarily unwholesome. As is commonly known, some molds are used to flavor cheese. On the other hand, if the mold content in catsup becomes excessive, the catsup is considered unfit for human consumption because of its high content of decomposed tomatoes. The presence of mold in certain products may indicate contamination and poor processing.

Molds reproduce by forming tiny seeds called spores. One mold plant may produce thousands of spores. Each spore is about the size of a yeast cell.

Conditions necessary for the growth and reproduction of molds are as follows:

a Food. They may attack moist wooden containers, cloth sacks, paper bags, etc.
b Moisture. This is the most important condition.
c Warmth. They are checked by cold and destroyed by boiling.
d Air. Molds will attack and penetrate all types of foods. However, in the case of liquid foods, they appear only on the surface since that is in contact with the air.
e They grow best in dark places, because dark places are usually moist places.

Counting Bacteria

We have seen above that bacteria are very prevalent in nature, and that some are useful and some are harmful, and some have no effect. Sanitarians have learned from experience that a count of the number of bacteria present is a useful index of sanitation. For example, it has been found that a safe drinking water supply will have less than thirty bacteria per cubic centimeter (a volume measure about the size of a small thimble). If the number is greater than this, there is indication of pollution.

One of the most important bacterial aids to sanitation is the determination of a particular type of bacterium exemplified by Escherichia coli, and generally referred to as the coliform type of organism. This type of microbe inhabits the intestinal tract of human beings and is discharged in bowel evacuations. Typhoid fever organisms are likewise found in bowel discharges. So, if these coliform organisms are found in or on a food, they indicate that the food has possibly been contaminated with fecal matter. Of course, the fecal matter may not have come from a person sick with typhoid or another disease involving the intestinal tract, but inasmuch as it might have come from such a person, and since the idea of fecal contamination is repulsive anyway, the sanitarian is warranted in requiring that this type of organism be absent from food.

Bacteriologists have worked out an ingenious simple method of counting bacteria, which is most useful in determining the bacterial content of food. If a nutrient jelly is melted and then a known amount, usually 1 cubic centimeter, of water or milk that contains the bacteria is mixed with it, the jelly, when hardened will hold the bacteria suspended throughout the mass. As the bacteria multiply, their numbers increase to such an enormous extent that in about 2 days a small white, circular patch becomes visible on the jelly disc. This colony indicates that originally 1 bacterium was present there. By counting these colonies, we have a figure for the number of bacteria that were originally present. For example, a *count* of 50 colonies means that 1 cubic centimeter of the solution under examination had 50 bacteria in it.

These counts are usually made on jellies that contain an extract of seaweed, agar-agar. This jelly is liquid when warm. It is poured into a circular, flat dish, called a Petri dish, about 4 inches in diameter and about ½ inch deep. In this way, the colonies can easily be seen through the flat glass cover plate.

Sometimes, it is advantageous, especially in the examination of milk samples, to prepare the bacteria in a special mount so that each single organism can be seen distinctly. This enables us to tell what kinds of bacteria are present, and helps to trace the cause of the lack of sanitation.

Food Poisoning by Micro-organisms

Micro-organisms have quite definite nutritional requirements. They can grow only when they find themselves in an environment that furnishes the proper food and which, at the same time, does not inhibit their metabolism (digestion, assimilation and elimination). An organism may lie dormant on a food until the environment becomes favorable for its growth. Many human foods furnish just the proper assortment of substances to meet these microbic requirements.

When bacteria and all other micro-organisms are thriving and multiplying, they eat the food on which they exist. This causes a breakdown of the constituents of foodstuffs, and produces the end-products which are characteristic of the life processes of particular organisms. Some of these changes are desirable, such as the fermentations of sugars to make vinegar. Other effects are only obnoxious, as the souring of milk or the molding of bread. A few are harmful, as the growth of pathogenic bacteria in milk or meat products. Pathogenic micro-organisms are dangerous, not only because their growth on the food may produce massive doses of infection, but also because their metabolism may elaborate toxins. Toxins are poisons produced by the micro-organisms. Illness caused by toxic products formed by micro-organisms in the food itself is called food poisoning or intoxication. If the food served merely as a vehicle for conveying the microbes themselves into the body, the disease produced is called a food-borne infection.

A foodstuff begins to undergo changes in composition very soon after it is removed from its natural habitat. These changes are produced to a large extent by processes within the food itself, caused by certain bodies called enzymes. This process is known as autolysis. This effect is favored by exposure to oxygen, sunlight, moisture and wormth, and is further accelerated by certain contaminants such as iron and copper. These changes are the cause of more or less loss of quality.

Enzymatic effects may start with a loss of freshness and possibly of some color and flavor. Then comes a more serious destruction of some of the vitamins, an oxidation and splitting of the fats with the development of rancidity, and then an attack on the proteins to form various decomposition products such as ammonia and other noxious chemicals. These seldom develop to the point of causing illness because the food becomes offensive long before these end-products accumulate sufficiently to become harmful.

The old idea that the decomposition products of food were responsible for the ordinary gastro-intestinal upset is no longer held. It was believed that putrefaction caused the formation of toxic substances called *ptomaines.* We now know that ordinarily putrefaction does not necessarily cause food poisoning. For example, Limburger cheese is an excellent food.

The composition of a given food determines to a large extent its susceptibility to attack by disease-producing organisms. In general, foods usually consist of a relatively large amount of water, carbohydrates in the form of sugars, proteins, minerals, acids and vitamins. These constituents are scattered through all the common foods in varying proportions, giving to each its characteristic value as well as its health hazard. For example, milk spoils very rapidly because it is an excellent medium for the growth of disease germs; meat and bread spoil more slowly; sugar and flour do not spoil appreciably during their commercial life.

The very fact that some foods, like milk, eggs and meat broths are such excellent foods for human beings means that they contain the essential nutrients for micro-organic life as well.

By far the most frequent and important types of food poisoning are those caused by bacteria. Some of these were in the food before it was processed, and some are introduced afterward, even during the serving. Certain bacteria are so prevalent that, under ordinary conditions, food cannot be protected from them. When bacteria get into food, the latter is said to be infected. An infected food usually is one in which the organism can grow.

A food is infested when forms of life, more highly organized than bacteria, make it their abode. For example, flies infest food by depositing their eggs in it. The small worm, trichina, infests food by encysting in the muscle meats. The tapeworm may infect human beings who eat infested fish from fresh water, particularly perch, tullibees and pike undercooked.

When food becomes soiled or when dirt or harmful foreign matter of one kind or another gets into it, the food is said to be contaminated. Usually contamination refers to relatively dry foods, like bread or meat, whereas pollution is used with reference to fluids as water. Dirt in itself may be harmless but may be the carrier of disease germs.

There are a great many diseases, many of which occur where there is a breakdown of body functions. But in this case, we are concerned with diseases which may be transmitted by infested or infected foods. These diseases are spread by intestinal discharges; by discharges from the mouth, nose, and ears; from animal to man, by animal parasites; and by contact of man with man. These are called communicable diseases. The U. S. Public Health Service lists 62 of these.

Food and food utensils have directly and indirectly been the cause of many serious and often fatal diseases because they were the vehicles or instruments or agents for spreading infection.

Toxins

Food poisoning is also caused by bacteria which produce toxins in food. The most important of these are discussed below.

a. Botulism is caused by Clostridium botulinum, a spore-forming anaerobic organism. It grows best under conditions where most of the air is shut off. In the process of its growth, it produces a most powerful poison which remains in the food. In canned foods, a gas is usually produced and the cans are sometimes bulging at the ends from the inside pressure. The chief causative foods are canned vegetables and meats where the acidity is low. Mortality rate is about 70%.

Botulism, formerly known as *sausage poisoning,* has been largely associated with meat products. However, during the early stages of the canning industry, a great deal of trouble was experienced due to improper processing. The National Canner's Laboratory conducted a large amount of research and overcame this problem so that now most commercially canned foods can be used with confidence. However, foreign goods which are soldered by hand and processed under questionable supervision should be regarded with suspicion. Home canned foods have been responsible for botulism more often than commercial products and should be avoided by food establishments. Any canned or preserved foods which show evidence of spoilage should not be consumed by man or animal. If the food has a mushy, soft appearance, a foul odor, gas, or acid formation, it should be destroyed. Such food should not be tasted nor should any other food concerning which there is suspicion.

Control: One should never serve food taken from a can which bulged at the ends, nor preserved food which had a disagreeable and unnatural odor. Foods which have been kept over for future use should be thoroughly cooked before such use. Fortunately, heating to a boiling temperature destroys the toxin if it is present.

b. Staphylococci bacteria produce a toxin which causes food poisoning with usual symptoms occurring within 2 to 6 hours following the ingestion of the food involved. The foods most commonly involved are custard-filled pastries, cream-puffs, éclairs and napoleons. This toxin is more heat-resistant than others, though not as toxic, judging by the lower number of fatalities. Inasmuch as the organism is easily killed by heat, a sample may be bacteriologically sterile and yet toxic.

In nearly all cases of bacterial food poisoning, the food is of a kind which requires little or no cooking, or which, if it has been cooked, has been contaminated afterwards and then kept under conditions favoring the growth of bacteria. The foods most commonly involved are custards, meats, poultry, *creamed* foods, salads and foods with low acidity. The bacteria causing these food poisonings are most often of the Salmonella and Staphylococcus types. All of them, except the one causing botulism, are readily killed in a few minutes at boiling temperature.

Foods which have been allowed to stand for some time after cooking, particularly if they are not cooled and kept refrigerated, are likely to cause food infections. Foods which have been allowed to stand for some time should be recooked, especially canned and green vegetables. Food infection is the most important of the food poisonings which confront the food industry and the nation at large since it is the most likely to occur frequently and with mass infection.

The symptoms usually come on within from 1 to 24 hours after eating the food and consist of various combinations of digestive tract disturbances, such as vomiting, diarrhea, pain in the abdomen, headache, weakness and sometimes fever.

Infestation by Protozoa

Amoebic dysentery is caused by a tiny one-celled animal, Entamoeba histolytica. This organism exists in two stages in the intestinal tract; the active stage, or that phase in which it may cause active disease; and the cystic stage in which it is transmissible to man. The organisms die soon after leaving the body. The cyst is quickly killed by drying, but will live for a month or longer in clear water. Infection takes place only when some of the cysts are swallowed. Under certain conditions, they invade the intestinal wall and cause ulcers. Naturally, the symptoms vary with the extent of the ulceration. It can be communicated as long as the cysts are excreted in the stools.

Some of the sources and modes of infection are: infected food handlers, polluted water supplies, infected feces exposed to flies, or human fecal material used as fertilizer in gardens. Food handlers and flies are common sources of infection. Polluted water from wells, streams, springs, or cross-connections between sewers and pure water supplies in cities may be sources of infection. An infected food handler may infect a whole family, many persons in an institution, or occasional guests at a hotel or cafe. The long period of incubation and the irregularity of symptoms often make it almost impossible to locate the source of infection. About 7 to 10% of the people are infected in the temperate zones. The disease is more common in the tropics and is more prevalent in the summer than in the winter. The symptoms vary from mild digestive disturbances to severe fatal bloody diarrhea. The incubation period for this type of dysentery is 7 to 100 days; however, some infected persons never develop the symptoms.

Control: Personal cleanliness is most important, especially hand washing after visiting the toilet, since many persons may be carriers of the parasite without knowing it. Good water supply, adequate sewage disposal systems and inspection of piping in establishments are important to be sure there are no drain or leaks or connections by means of which the water could be contaminated. Prevent fly breeding and, if privies are used, make certain that they are of modern fly-proof construction located so as not to pollute well or other water.

Poisoning by Chemicals

In recent years, cases of food poisoning from the accidental contamination of food and drink with poisonous metals and other chemical substances have been much more thoroughly investigated in the light of the shortage of materials and the use of substitutes to supplant them.

Cadmium Poisoning

In the past few years a number of cases of food poisoning from the metal cadmium occurred. Several refrigerator renovating companies were plating the freezer compartments, known in the trade as evaporators, of old refrigerators with the metal cadmium. When the refrigerators were used long and were defective, some of the water vapor containing minute amounts of sulfur dioxide, the condensed refrigerant, dissolved some of the cadmium metal. This attacked metal fell into the trays containing water or other food containers in the refrigerator and so poisonous cadmium was incorporated into the ice cube or food. In practically all cases the victims were attacked within ½ hour with symptoms of nausea, severe cramps and vomiting.

In another outbreak of cadmium poisoning involving two adults and three children, a cadmium plated metal ice-tray of a renovated refrigerator was used to congeal some freshly prepared gelatin. In this case, the citric and tartaric acids in the original gelatin powder dissolved some of the cadmium. Food poisoning occurred when the contaminated gelatin was eaten.

Lead Poisoning

The swallowing of lead in sufficient quantities to give acute symptoms of nausea, vomiting and abdominal pain which may occur within as short a period as 10 minutes is rare. In the repair of food pipe lines, equipment and utensils, etc., the use of lead solder is prohibited, since acid foods or liquids will dissolve small amounts of lead which may result in its accumulation in the body. The soldering of food utensils and equipment should be done with 100% block tin.

Arsenic Poisoning

Accidental arsenic poisoning in man from the consumption of foods is very rare. When swallowed in appreciable quantities, arsenic gives rise, within 15 to 30 minutes, to symptoms of vomiting, diarrhea and severe abdominal pain. Arsenic poisoning has been of particular concern to fruit-growers and consumers of sprayed fruit.

The government, in order to minimize the consumption of the lead arsenate on fruits, has set the following tolerances for spray residue: lead 6.7 parts per million and arsenic 3.4 parts per million. Fruit containing more lead or arsenic than these tolerances must be rewashed before shipment to the consumer's market.

Antimony Poisoning

Practically all outbreaks of antimony poisoning that have been reported occurred from the consumption of foods cooked in enamelware, generally of the cheaper gray variety. Yet several years ago, New York City experienced an outbreak of antimony food poisoning from potato salad which was kept in an enamelled utensil. The vinegar present in the potato salad dissolved sufficient antimony to give symptoms of nausea and vomiting within 1 to 2 hours.

An analysis of enamelware manufactured by thirty-one different companies has shown that 65% of the enamelware tested contains antimony. These results indicate that enamelware should not be used for storing or cooking acid foods.

Fluoride Poisoning

Sodium fluoride has been the cause of more deaths from accidental poisoning than any other chemical agent. In practically all cases, the chemical was mistaken for such foods as corn starch, flour, whole milk powder, skim milk powder, baking powder and such things as bicarbonate of soda, Rochelle salts, etc.

Cyanide Poisoning

In 1930, Dr. H. Williams reported acute cases of cyanide poisoning from hotel silver polish. The use of cyanide in polishes is prohibited in New York City.

A simple test for cyanide is performed by moistening sodium picrate test paper in water and holding it above the suspected polish, but not in contact with it. In 5 to 10 minutes, the paper will turn first orange and then brick red, showing the presence of cyanide.

Zinc Poisoning

Cases of poisoning due to the accidental contamination of food with zinc are rather uncommon. Foods, especially if distinctly acid, as in the case of gelatin desserts, will dissolve zinc when in contact with it. The higher the acidity of the food is, the more zinc will be dissolved. The amount dissolved will also be greater if the zinc is in contact with the food over a long period of time. Under no circumstances should galvanized (zinc plated) utensils be used in the preparation or storage of acid food such as fruit juices, kraut, fruit syrups, etc.

Foreign Substances and Filth

As a rule, people are more or less fastidious concerning what they put into their mouths. Therefore, food should be produced, manufactured and serviced with care to conform to this fundamental public demand. However, when a person engages in the food business, he is apt to lose some of this feeling about food and treat it as any ordinary commodity. As a result, he may become careless and handle the food that he sells more carelessly than that which he himself eats.

This leads to such practices as the servicing of food with dirty hands, to preparing food without preventing hair from falling into it, to manufacturing when the plant swarms with flies, water and rats, to working decayed portions (just plain rotten food) in with the batch, and to permitting insect larvae, wings, leg fragments and excreta to become part and parcel of what he sells. Probably most filth accompanies the food as it is prepared for market by the original producer. If not originally packed in a clean manner, and freed from any contamination, this filth will accompany the product all the way to the ultimate consumer.

Adulteration

The consumer understands that a pure food is one that has not been changed from that which experience has taught him to expect as to its wholesomeness, safety, palatability and nutritive quality. Scientific and legal considerations distinguish between types of harmful and fraudulent mishandling of foods as follows:

Infection is the presence of undesirable, viable micro-organisms which have entered the food from some source during production or handling, and proliferate to spoil the food or make it dangerous to health, e.g., typhoid bacteria in oysters.

Contamination is the presence of any undesirable substance, organic or inorganic, viable or inert, which is foreign to natural food, for example, dirt in candy.

Adulteration may be any one of the following:

the addition of a substance which reduces or lowers the quality or strength, for example, water in milk, or the substitution of one product for another, for example coconut oil for butterfat; or

the abstraction of any valuable constituent, for example, skimming the fat from milk; or the addition of coloring matter or other substances to conceal inferiority, for example, yellow dye in plain noodles to make them appear to contain eggs; or

the addition of poisonous substances, for example, insecticide sprays on food; or

the presence of any portion that is filthy, putrid, or decomposed, for example, spoiled tomatoes in catsup.

Misbranding is the labeling of a product to be what it is not in identification, quality, or amount, for example, milk bread which contains no milk, or misrepresentation of weight on a label.

Sophistication is fraudulent treatment of a foodstuff to conceal its poor marketability, for example, cleaning and polishing of moldy fruit to simulate freshness.

Adulteration usually does not involve directly any appreciable health hazard. When adulteration is done deliberately, the intent is to cheapen the product. This may constitute too generous an addition of water to the product or maybe the substitution of a cheap ingredient to replace a more costly one. The offense is mainly an economic one. Practically, the only health aspect is the reduced food value.

CHAPTER XIV

COMPOSITION OF FOODS AND PREVENTION OF CAKING

COMPOSITION OF COMMON VEGETABLES
(Edible Portion)

Name	*Per Cent by Weight*							*Milligrams Per Kilogram* [1]	
	Protein	Carbohydrate	Fat	Crude Fiber	Ash	Calcium	Phosphorus	Iron	Copper
Artichoke (globe)	2.9	11.9	0.4	3.2	1.1	0.031	0.087	20.0	3.1
Asparagus	1.8	3.3	0.2	0.8	0.7	0.025	0.039	10.0	1.4
Beans (dry)	22.5	60.0	1.8	4.4	3.5	0.160	0.471	95.2	6.9
Beans (green)	2.3	7.4	0.3	1.9	0.8	0.046	0.052	9.3	1.0
Beans (baked)	6.9	19.6	2.5	1.4	2.1	0.052	0.155	32.0	2.3
Beets	1.6	9.7	0.1	0.9	1.1	0.029	0.039	23.6	1.9
Beet leaves	2.0	4.2	0.3	1.4	1.7	——	——	35.5	0.9
Cabbage	1.6	5.6	0.3	1.1	1.0	0.045	0.029	3.5	0.5
Cantaloupe	0.6	9.3	0.3	2.1	0.6	0.017	0.015	5.1	0.6
Carrots	1.1	9.3	0.4	1.1	1.0	0.056	0.046	6.4	0.8
Cauliflower	1.8	4.7	0.5	1.0	0.7	0.123	0.061	9.4	1.4
Celery	1.1	3.3	0.1	1.0	1.0	0.078	0.037	6.2	0.1
Chard	2.6	4.0	0.4	0.8	1.2	0.150	0.040	31.0	1.1
Corn (Canned)	2.8	19.0	1.2	0.8	0.9	0.006	0.103	4.7	1.0
Cucumber	0.8	3.1	0.2	0.7	0.5	0.016	0.033	3.3	0.6
Eggplant	1.2	5.1	0.3	0.8	0.5	0.011	0.034	5.0	1.0
Kale	3.9	6.0	0.6	1.2	1.7	——	——	—	—
Lettuce	1.2	2.9	0.3	0.7	0.9	0.043	0.042	7.0	0.4
Mushrooms	3.5	6.8	0.4	0.9	1.1	0.017	0.108	31.0	17.9
Onions	1.6	9.9	0.3	0.8	0.6	0.034	0.045	4.0	0.8
Parsnips	1.6	13.5	0.5	2.5	1.4	0.059	0.076	10.7	1.2
Peas (dry)	24.6	62.0	1.0	4.5	2.9	0.084	0.400	57.0	14.0
Peas (green)	7.0	16.9	0.5	1.7	1.0	0.028	0.127	21.0	2.4
Peppers (green)	1.1	4.6	0.1	1.4	0.7	0.006	0.026	4.0	1.0
Potatoes (white)	2.2	18.4	0.1	0.4	1.0	0.014	0.058	9.1	1.7

Name	Per Cent by Weight							Milligrams Per Kilogram[1]	
	Protein	Carbohydrate	Fat	Crude Fiber	Ash	Calcium	Phosphorus	Iron	Copper
Potatoes (sweet)	1.8	27.4	0.7	1.3	1.1	0.019	0.045	9.0	1.5
Pumpkin	1.0	5.2	0.1	1.2	0.6	0.023	0.059	9.3	0.3
Radishes	1.3	5.8	0.1	0.7	1.0	0.021	0.029	8.3	1.6
Rhubarb	0.6	3.6	0.7	1.1	0.7	0.044	0.031	7.6	0.5
Rutabagas	1.3	8.5	0.2	1.2	1.1	0.074	0.056	6.6	1.5
Sauerkraut	1.7	3.8	0.5	1.0	5.2	0.040	0.025	4.0	0.4
Spinach	2.1	3.2	0.3	0.9	2.1	0.067	0.068	38.5	1.2
Squash (Hubbard)	1.4	9.0	0.5	0.8	0.8	0.019	0.015	5.0	0.4
Tomatoes	0.9	3.9	0.4	0.6	0.5	0.011	0.026	6.0	0.7
Turnips	1.3	8.1	0.2	1.3	0.8	0.064	0.046	6.2	0.9
Turnip tops	2.9	4.2	0.4	1.2	1.8	—	—	—	—
Watermelon	0.4	6.7	0.2	1.0	0.3	0.011	0.003	2.3	0.7

[1] Parts per million

COMPOSITION OF NUTS
(Edible Portion)

Name	Per Cent by Weight							Milligrams Per Kilogram[1]	
	Protein	Carbohydrate	Fat	Crude Fiber	Ash	Calcium	Phosphorus	Iron	Copper
Almonds	21.0	17.3	54.9	2.0	2.0	0.239	0.465	40	12.1
Brazil nuts	17.0	7.0	66.8	3.9	3.9	—	—	40	13.9
Butternuts	27.9	3.5	61.2	—	—	—	—	68	12.0
Chestnuts (fresh)	6.2	42.1	5.4	1.8	1.3	0.034	0.093	41	6.0
Coconuts (dry)	6.3	31.5	57.4	3.4	1.3	0.059	0.155	20	6.9
Hazelnuts	—	—	—	3.0	2.1	0.287	0.354	41	—
Peanuts	25.8	24.4	38.6	2.5	2.0	0.071	0.399	20	9.6
Pecans	11.0	13.3	71.2	2.2	1.5	0.089	0.335	26	13.6
Pistachios	22.3	16.3	54.0	—	3.2	—	—	79	12.0
Walnuts (English)	18.4	13.0	64.4	1.4	1.7	0.060	0.360	21	10.0
Walnuts (black)	27.6	11.7	56.3	1.7	1.9	—	—	60	—

[1] Parts per million

COMPOSITION OF COMMON FRUITS
(Edible Portion)

Name	Per Cent by Weight							Milligrams Per Kilogram [1]	
	Protein	Carbohydrate	Fat	Crude Fiber	Ash	Calcium	Phosphorus	Iron	Copper
Apples	0.4	14.2	0.5	1.2	0.3	0.007	0.012	5.0	1.0
Apricots	1.1	13.4	0.3	3.1	0.5	0.014	0.025	25.0	1.8
Avocados	2.1	7.4	20.1	—	0.9	——	——	—	—
Bananas	1.3	22.0	1.3	1.0	0.8	0.009	0.031	17.6	2.1
Blackberries	1.3	10.9	1.0	2.5	0.5	0.017	0.034	10.0	1.6
Cherries	1.0	16.7	0.8	0.2	0.6	0.019	0.031	4.0	1.4
Cranberries	0.4	9.9	0.6	1.5	0.2	0.017	0.011	4.4	0.9
Dates (dried)	2.1	78.4	2.8	5.5	1.3	0.065	0.056	36.0	3.8
Figs (dried)	4.3	74.2	0.3	7.3	2.4	0.162	0.116	29.0	3.5
Grapefruit	0.6	12.2	0.1	—	—	0.021	0.020	2.7	0.3
Grapes	1.3	19.2	1.6	4.3	0.5	0.019	0.031	7.3	0.9
Lemons	1.0	8.5	0.7	1.1	0.5	0.036	0.022	6.0	0.4
Oranges	0.8	11.6	0.2	—	0.5	0.045	0.021	5.2	0.8
Peaches	0.7	9.4	0.1	3.6	0.4	0.016	0.024	3.5	0.9
Pears	0.6	14.1	0.5	2.7	0.4	0.015	0.026	3.5	1.0
Pineapple	0.4	9.7	0.3	0.4	0.3	0.018	0.028	3.7	0.7
Prunes (dried)	2.1	73.3	7.3	1.5	2.5	0.054	0.105	29.0	4.1
Raisins	2.6	76.1	3.3	1.7	3.4	0.064	0.132	29.0	2.0
Raspberries	1.7	12.6	1.0	2.9	0.6	0.049	0.052	8.8	1.4
Strawberries	1.0	7.4	0.6	1.4	0.6	0.041	0.028	6.8	0.2

[1] Parts per million

COMPOSITION OF CEREALS
(Edible Portion)

Name	Per Cent by Weight							Milligrams Per Kilogram [1]	
	Protein	Carbohydrate	Fat	Crude Fiber	Ash	Calcium	Phosphorus	Iron	Copper
Barley (pearled)	8.5	77.8	1.1	0.3	1.1	0.020	0.181	20	4.0
Bread (white)	9.2	53.1	1.3	0.5	1.2	0.027	0.093	9	3.4
Bread (whole wheat)	9.7	49.7	0.9	1.2	1.3	0.050	0.185	30	5.0
Bread (rye)	9.0	53.2	0.6	0.5	1.5	0.024	0.148	16	—
Corn meal	9.2	75.4	1.9	1.0	1.0	0.018	0.190	9	2.0

Name	Per Cent by Weight							Milligrams Per Kilogram[1]	
	Protein	Carbohydrate	Fat	Crude Fiber	Ash	Calcium	Phosphorus	Iron	Copper
Flour (white)	13.3	72.7	1.5	0.7	0.6	0.020	0.092	10	1.7
Flour (rye)	6.8	78.6	0.9	0.4	0.7	0.018	0.289	13	4.2
Flour (buckwheat)	6.4	77.9	1.2	—	—	0.010	0.176	12	—
Hominy	8.3	79.0	0.6	0.9	0.3	0.011	0.144	9	1.9
Macaroni	13.4	74.1	0.9	0.4	1.0	0.022	0.144	12	—
Oatmeal	16.1	67.5	7.2	0.9	1.9	0.069	0.392	38	5.0
Rice (white)	8.0	79.0	0.3	0.2	0.4	0.009	0.096	9	1.9
Soybean (meal)	40.0	7.0	21.0	—	—	—	—	—	—
Tapioca	0.4	88.0	0.1	0.1	0.1	0.023	0.090	16	—
Wheat (whole)	11.1	75.5	1.7	1.8	1.8	0.045	0.423	50	6.0
Wheat (bran)	14.6	46.5	3.8	11.3	5.0	0.120	1.215	78	11.7

[1] Parts per million

COMPOSITION OF MEATS AND POULTRY PRODUCTS (Edible Portion)

Name	Per Cent by Weight							Milligrams Per Kilogram[1]	
	Protein	Carbohydrate	Fat	Crude Fiber	Ash	Calcium	Phosphorus	Iron	Copper
Bacon	10.5	—	64.8	—	4.4	0.006	0.108	24	5.0
Beef (steak)	18.9	—	18.5	—	1.0	0.012	0.216	30	1.0
Chicken	21.5	—	2.5	—	1.1	0.012	0.232	32	3.5
Duck	18.0	—	19.0	—	—	0.010	0.200	23	5.0
Eggs	13.4	—	10.5	—	1.0	0.067	0.180	30	2.3
Egg white	12.3	—	0.2	—	0.6	0.015	0.014	1	—
Egg Yolk	15.7	—	33.3	—	1.1	0.137	0.524	86	1.2
Fish (cod)	11.1	—	0.2	—	1.2	0.012	0.120	6	5.5
Fish (halibut)	18.6	—	5.2	—	1.0	0.020	0.200	10	2.3
Fish (salmon)	22.0	—	12.8	—	1.4	0.025	0.250	12	2.0
Fish (tuna)	26.6	—	11.4	—	1.7	0.030	0.300	15	5.0
Gelatin	91.4	—	0.1	—	1.0	—	—	—	—
Ham	19.8	—	20.8	—	0.8	0.012	0.215	30	—
Liver (beef)	20.4	1.7	4.5	—	1.6	0.017	0.213	79	21.5
Lobster	16.4	0.4	1.8	—	—	0.018	0.188	9	7.0

Name	Pro-tein	Carbo-hydrate	Fat	Crude Fiber	Ash	Cal-cium	Phos-phorus	Iron	Cop-per
	Per Cent by Weight							*Milligrams Per Kilogram*[1]	
Margarine	1.2	—	83.0	—	3.0	—	—	—	—
Mincemeat	6.7	60.2	1.4	—	—	0.085	0.175	30	—
Mutton (leg)	19.8	—	12.4	—	1.2	0.010	0.270	30	4.0
Oysters	6.2	3.7	1.2	—	1.5	0.052	0.155	45	30.8
Pork (chops)	16.6	—	30.1	—	0.7	0.010	0.180	25	3.1
Shrimp	25.4	0.2	1.0	—	—	0.096	0.292	27	—
Turkey	21.1	—	22.9	—	1.0	0.030	0.420	45	1.8
Veal (cutlet)	20.3	—	7.7	—	1.1	0.013	0.228	30	2.5

[1] Parts per million

COMPOSITION OF DAIRY PRODUCTS
(Edible Portion)

Name	Pro-tein	Carbo-hydrate	Fat	Crude Fiber	Ash	Cal-cium	Phos-phorus	Iron	Cop-per
	Per Cent by Weight							*Milligrams Per Kilogram*[1]	
Butter	1.0	—	85.0	—	3.0	0.015	0.017	2.0	—
Buttermilk	3.0	4.8	0.5	—	0.7	0.105	0.097	2.5	—
Cheese (American)	29.0	0.3	36.0	—	3.4	0.930	0.683	13.8	1.8
Cheese (cottage)	20.9	4.3	1.0	—	4.2	0.995	0.374	—	—
Cheese (Swiss)	27.6	1.2	35.0	—	4.8	0.999	0.845	12.5	1.3
Cheese (cheddar)	27.7	4.1	36.8	—	4.0	0.990	0.682	12.5	—
Cream	2.5	4.5	18.5	—	0.5	0.090	0.080	2.2	—
Ice cream	4.0	20.0	14.0	—	0.9	0.150	0.120	—	—
Milk (whole)	3.3	5.0	4.0	—	0.7	0.120	0.093	2.4	0.2
Milk (skim)	3.4	5.1	0.3	—	0.8	0.122	0.096	2.5	0.2
Milk (evaporated)	8.7	10.2	8.2	—	1.5	0.250	0.200	5.3	1.2
Milk (condensed)	8.8	54.1	8.3	—	1.6	0.300	0.235	6.0	1.0
Milk (dried)	26.1	38.0	26.5	—	6.0	0.920	0.710	1.5	1.5

[1] Parts per million

PREVENTION OF CAKING OF POWDERED FOOD PRODUCTS

The ingredients most commonly used to retard caking are listed in the order of their effectiveness:

SUBSTANCES

Tri-calcium phosphate	Calcium sulfate (anhydrous)
Magnesium carbonate	Corn starch (dry)
Unslaked lime	Dextrose (anhydrous)
Sodium bicarbonate	Vegetable shortening (melting point, 92–98° F)
Di-calcium phosphate	

CHAPTER XV

PRESERVATION OF FOOD BY DEHYDRATION AND FREEZING

Principles and Methods of Dehydration of Vegetables

Prior to the year 1941, only a few vegetables were dehydrated, and of those few, the total production was small. During World War I, the need for dried products brought about a well established and important industry, but the production fell off considerably after that. Great hopes were expressed for the continuance of this business during peace time years, but they did not materialize. Perhaps the failure was due to the lack of technological knowledge in producing a sound palatable product. However, during World War II, dehydration of vegetables again became essential and this time expanded into a huge enterprise, with millions of pounds being produced. With due credit for the tremendous technological advancement in the science of dehydration, the consumer demand seems to have dropped off considerably since the end of World War II. We believe that the high cost of manufacture, resulting in high retail prices, and the steep competition with fresh vegetables may be responsible for the loss of consumer interest. However, there will always be some demand for this type of merchandise, but it will have to be manufactured with great care and with modern equipment, in order to bring about the quality and shelf-life expected by the consumer. This chapter will cover such information and knowledge sought by those desiring to enter into the business of dehydration of vegetables.

The expression *dehydration* usually refers specifically to a controlled process of drying vegetables and fruits with the aid of a forced circulation of heated air. One may well ask what the purpose of taking fresh vegetables, removing their natural water content and manufacturing them into a dehydrated product is. There are many reasons, some of which are the following:

In many parts of the country, there is a surplus of vegetables which can be taken up by dehydration that might otherwise be spoiled or wasted. Dehydrated vegetables are used in the manufacture of dried soup mixtures, etc. Dehydrated potatoes can be made into French fried and mashed potatoes. Dry cake and bakery mixes can be developed using dehydrated potato flour, carrot powder, sweet potato powder, powdered spinach, etc.

A person entering into this business must be assured of a reasonable supply of raw materials, preferably secured close to the dehydrating plant. There should be plenty of labor, fuel and electricity, and a good supply of pure water, as well as good sewerage accommodations. Onion and garlic drying plants should be located where the odors will not be a nuisance to the neighborhood. The building itself in which the dehydration takes place need not be expensive or elaborately built, but it should have cement floors and sufficient built-in drainage lines. Storage rooms should be dry, with plenty of ventilation and should be insect and rodent proof. Potatoes, sweet potatoes and onions may be kept on hand for at least a ten-day supply since these products do not spoil quickly. However, cabbage, carrots, turnips and beets should be dried right after harvesting. Serious loss in quality and nutritive value will take place if fresh products are mishandled. If unavoidable delays are experienced in processing sensitive vegetables, they should be refrigerated.

Preparation of Raw Material

Root crops as received from the field are sometimes contaminated by mud and other dirt. The cleaning operation can be initiated by means of a shaking screen device to loosen up and remove this accumulated dirt. While there are any number of different types of water washing equipment, maintained by different companies, the rotary type drum washer or shower sprays are recommended.

Root Peeling

Before the demand for dehydrated potatoes and carrots was as great as it is today, the only method of removing the peel was by the use of an abrasive machine which consisted essentially of a cylindrical abrasive-lined drum. The base or floor plate of this drum revolves at a regulated speed or number of revolutions per minute and is provided with water sprays. When the drum is in operation, the potatoes or roots are dumped into it, and the revolving floor pushes the vegetables against the sides and bottom of the abrasive walls. The flow of water flushes the removed skin through a waste drain opening at the bottom. A side door is used for ejecting the vegetables. The continuous type of abrasive peeler works on the same principle as the single charge unit just described. Here the potatoes pass through a winding course and over the rotating abrasive cylinders. The peeling and trimming loss of potatoes with the use of abrasive peeling machines may run higher than 20%, which is a considerable loss. The need was felt for a better method that would reduce the loss of peeling, and so the newer methods utilize boiling in saturated salt brine solutions and flame peelers, which subject the potatoes to a temperature of 2000° F for a period ranging from 15 to 30 seconds. The charred-out surface is then removed by rigorous scrubbing and washing, leaving the product clean and peeled.

Lye or caustic soda peeling is a method resulting in a smaller loss. The potato is submerged in a hot solution of lye for a short time then washed with acidulated water until all of the lye has been removed. There is also a process called retort peeling which is a steam heat process consisting of exposing the product to steam under pressure for a short period, in this way loosening the skins so that they can be removed by scrubbing. Potatoes are subjected to steam under pressure of 50 to 60 pounds per square inch for 30 to 45 seconds which corresponds to 300°F.

Vegetable Trimming

After the root crops have been peeled and cleaned, the residual skin, eyes, discolored areas, insect injury, etc., must be removed. The trimming loss will depend upon the efficiency of the peeling and trimming workers. The trimming of leafy vegetables involves hand removal of discolored leaves and injured parts. With cabbage, mechanical corers are used on the complete head of cabbage. The disadvantage of this method is that, if loose heads are being processed, the whole cabbage may fall apart, causing waste.

Cutting of Vegetables

After the trimming operations, the vegetables pass on to equipment for cutting them into desired forms and sizes. Potatoes may be made into strips or julienne type, slices, cubes, or other shapes desired by market demand. Slices usually are 1/8 to 1/4 inch thick; strips are 3/16 to 3/8 inch wide and not less than 3/4 inch in length; cubes are 3/16 to 3/8 inch on a side or 1/2-inch cubes. The cubes are the preferred form, and the slices are next. Leafy vegetables, with the exception of cabbage, usually do not require division into smaller pieces after trimming. Cabbage is shredded into pieces 1/8 to 1/4 inch in width. The machine used for this purpose is the same as that used by manufacturers of sauerkraut. The knives on the cutting machinery should be kept sharp in order to assure regular-shaped and well defined cut surfaces. The equipment necessary for making riced potatoes may be a rotary screw or plunge type. The action consists of mashing the product and forcing it through perforations not exceeding 3/16 inch in diameter.

In preparation of material for ricing, precooking can be carried out on the sliced, whole, halved or quartered product. The sliced product is probably the best for cooking since it lends itself to easy manufacture. Cooking in flowing steam in a steam cabinet or a retort, or passing the prepared material on a belt through a continuous steam blanch are methods also employed. Whatever process is used, thorough cooking is essential. The best results from ricing or putting through a perforated disc are obtained while the product is still hot. A cool product, upon ricing, may become pasty or gummy, and it may then be difficult to obtain a uniform spread on the drying trays.

Vegetable Blanching

The terms *blanching* and *scalding* as used in this chapter, refer to the practice of heating the prepared vegetable in boiling water, hot brine solutions, hot alkaline solution or in live steam for a short period prior to the principle process of further treatment. The object in blanching is to check development of undesirable colors, flavors or odors, as well as the loss of vitamins during dehydration and storage. It also results in a better finished product, as reconstituting it destroys bacteria and the enzymes which bring about the color changes.

Enzyme Activity

Inadequate blanching is believed to be at least part of the reason for the deterioration in quality that so often occurs in dehydrated vegetables. Since the enzymes, which are substances present in all living cells, accelerate chemical reactions, the importance of perfect blanching cannot be overestimated. Tests to determine the presence of enzymes must always be made with the purpose of destroying them.

Simple Blanching Test

Peroxidase Tests

There are two different tests that may be used to determine the presence of peroxidase enzymes. They have also been advocated to determine the extent of blanching of vegetables before dehydration.

One of these methods uses guaiacol, in the presence of which, the appearance of a red color shows insufficient blanching. The other test requires the use of benzidine. In this case, a blue color usually means that the heat treatment should be extended because the enzymes have not yet been inactivated.

Both of these tests may be carried out in white porcelain

evaporating dishes, in which the color changes can be readily noted. Always run a blank test on the vegetable.

The guaiacol test requires 2 grams of vegetable which has been ground with 10 cubic centimeters of distilled water. To this mixture add 1 cubic centimeter of a 1% alcoholic solution of guaiacol and 1 cubic centimeter of 0.30% hydrogen peroxide free from acetanilide. After shaking this combination, a red color will appear if the product was not blanched long enough.

The benzidine test is conducted by mixing 2 grams of ground vegetable with 10 cubic centimeters of distilled water and then adding 1 cubic centimeter of benzidine solution (0.5% benzidine in 50% ethyl alcohol) and 1 cubic centimeter of 3% hydrogen peroxide. If a blue color is produced it may be evidence that the product was insufficiently blanched. It is known, however, that metallic salts such as iron and copper have the ability to react under similar conditions to bring about a blue color. Consequently, a positive test is subject to question. The contact of vegetable tissues with metallic equipment, or the presence of such salts in the product, may cause errors in interpreting the results. In spite of the fact that these quality control tests are not foolproof, they serve as a useful means of checking on the blanching operations. Every dehydration plant should run them regularly, because a slight variation in time or temperature may result in improper blanching and consequently, poor products.

Catalase Tests

The simplest catalase test is to place a small quantity of vegetable fragments in a glass test tube and add enough 3% hydrogen peroxide solution to cover the material at least 1 inch. If no gas is given off within a few minutes, the catalase enzyme is considered to be satisfactorily checked. When a stream of little gas bubbles rises to the surface after the hydrogen peroxide is added, it is evidence that the heat treatment should be carried out either longer or at a higher temperature.

A more accurate catalase test, which enables a better comparison of the amount of gas produced, is carried out by grinding up in a mortar a 1-gram sample of the blanched material, together with an equal weight of clean sand and 0.6 gram of calcium carbonate. This mixture is diluted with 10 cubic centimeters of distilled water and the grinding then continued for two more minutes. Next, 4 cubic centimeters of this mixture are added to 8 cubic centimeters of hydrogen peroxide in a Smith fermentation tube. This tube is so constructed that the quantity of gas evolved in it can be measured.*

Sometimes a catalase test will be positive even if the products have been thoroughly blanched. For example, the catalase content in the eyes of some potatoes is highly resistant to heat.

Methods of Blanching

The three common methods of blanching vegetables are steam, hot water and series blanching. Steam blanching consists of exposing the prepared product to the temperature of live steam. In water blanching, the product is submerged in boiling water for a period of time. Series blanching is a modification of water blanching, in that the soluble solids leached from the products during the blanching operation are not allowed to accumulate above 4%, and this ratio is maintained by gradual introduction of fresh water and removal of the solids-saturated water. Different products require different blanching times all of which should be predetermined on a small batch. Thermometers should be placed on various parts of the blanching equipment, to aid in obtaining uniformity of heat distribution. The operator should always strive to maintain a temperature as close as 212° F as possible in his blanching processes. Of the various types of blanchers, the one most commonly used is the belt blancher. If this is to be continuous tray blanching, the blanching area must be increased in order to obtain capacity. Blanching in tanks of boiling water, although a simple operation, is generally acceptable only for small scale production, but does not lend itself to a high quality product.

**Note:*

For specific details of a more quantitative test see paper by R. R. Thompson, *Ind. Eng. Chem., Anal. Ed.* 14, 585, July 15, 1942.

For further detailed information concerning tests for this purpose see W. V. Cruess and M. A. Joslyn, *Proceedings of the Institute of Food Technologists,* 1942.

Vegetable Sulfuring

Sulfuring of white potatoes, cabbage, carrots, turnips, etc., is now considered a necessary part of the production of dehydrated vegetables. The use of solutions of sulfur dioxide has resulted in the retention of quality, higher finishing temperatures in dehydration and, consequently, shorter drying times. There are many ways to accomplish the sulfuring treatment, such as immersion of the vegetable in water to which the chemical has been added, or by spraying the trayed product, or by any well-known dipping technique. The idea is to get a small percentage of sulfur dioxide on the vegetable.

Principles of Scientific Drying

The dehydration of foods is a drying operation consisting of a controlled method of evaporation of nearly all of the water present in the fresh product. This evaporation, however, is a complex process, and an attempt will be made to explain it in the paragraphs that follow.

The air about us is one of our most important raw materials since it is utilized materially in the functions of drying. On the one hand, air currents carry the heat in from its source, such as a blower or a stove, to the place where the food is being dried, and on the other hand, it carries off the moisture coming from the food.

The water content of the cells of vegetables and fruits has salts, sugars and other substances dissolved in it. Water evaporates fast, but when salts, sugars, etc., are in solution, the process becomes a slower one. That is why it takes almost as much time to drive off the last traces of moisture as to drive off the bulk of the water.

The heat necessary to bring about the change of water to vapor is called *latent heat of vaporization,* and when the vapor condenses to a liquid, it gives up this latent heat of vaporization.

Use of Air in Dehydration

The vaporization of water in the vegetable depends upon the flow of warm air and the subsequent removal of the vapor with the same air. The heated air, which has been forced in by means of a blower, gives up part of its heat to vaporize the liquid, and is thereby cooled. Vaporization of water in this forced-draft drying equipment is essentially known as an adiabatic process. The amount of heated air required to evaporate a known amount of water can be accurately calculated. Methods of calculating this can be found in texts dealing with the subject of heat transfer.

Recirculation of Air

Although it is advantageous to recirculate the used warm air in fruit dehydration, it is not practical in the drying of vegetables. The disadvantage lies in increased drying time, which reduces output, and inasmuch as vegetables must be reduced to about 4 or 5% moisture content, the use of recirculated air is practically impossible. Besides, there is not enough saving of fuel to advocate the use of recirculated air for vegetable drying.

Relative Humidity

To determine relative humidity, place a wet cloth wick around the bulb of a thermometer. The bulb of a second thermometer and that of the one with the wet cloth, when exposed in a stream of dry air, will register different readings. The wet bulb thermometer will read many degrees lower than the dry thermometer. This is because the evaporation of moisture from the wick takes heat from it and from the thermometer in proportion to the rate of evaporation. This difference in registered temperature increases the lower the relative humidity of the air. For example, if the dry bulb thermometer reads 150° F, and the wet bulb thermometer reads 100° F, this difference of 50° F indicates 18% relative humidity. This information is obtainable from charts and tables on relative humidity. Vegetable dehydrators should be equipped with a wet and dry bulb thermometer located at the air entry and exit ends of the drying tunnel.

Air Velocity and Drying Rate

The velocity of the air varies with the volume of air per minute, and the higher the velocity, the faster will be the rate of drying. By air velocity is meant the rate of motion in linear feet per minute. For example, an air velocity of 500 feet per minute means that the air is traveling at the rate of 500 feet per minute. Velocity of air is a very important consideration for successful dehydration of vegetables. If the velocity is low, the rate of drying will be very slow and the output poor. Hot air, containing the correct amount of moisture, which is kept flowing over the food at the proper rate of speed will do a good job in removing water from the food. Each vegetable or food has a critical temperature for drying, resulting in the best quality. If this temperature is not watched carefully, discoloration, off-flavors and deterioration result. It is well to remember that at the outset of drying, the temperature may be considerably increased, whereas towards the end of the drying period, the temperature must be carefully watched.

When properly blanched pieces of vegetable are exposed to the warm air currents, they act, at first, like a sponge full of water. The moisture closest to the surface evaporates very quickly, and if the air velocity is reasonably high, e.g., 1000 feet per minute, the temperature inside the vegetable remains pretty close to the wet bulb temperature of the air. It doesn't take long before the surface of the vegetable becomes much drier than the inner part, and when this happens, the moisture must come up from the bottom until the whole piece is dried. As the drying proceeds, the cell walls collapse, and this is responsible for the shrinkage that takes place during the drying. The rate of drying depends upon the thickness of the vegetable.

Case-Hardened Vegetables

When the expression, *case hardening* is used, it means that the surface has formed a hard resistant coating which prevents moisture from coming up to the surface to be evaporated. In dehydrating cut vegetables, there is little chance of this occurring. If it did happen, the pieces would be recognized by a wet center, and longer drying would be required than would be normal for that vegetable. With fruits, however, case hardening is apt to occur, especially if the dehydration is started at high temperatures.

The air velocity across the trays determines the speed of drying. Air velocity of 500 feet per minute results in a longer drying period than that of 700 feet per minute. However, a very high velocity will not result in a corresponding high speed of drying, since, after a certain point, the speed of drying is not proportional to the increased amount of air. Air velocity has two effects. One is on the drying rate, and the other is on the speed of removal of the moist air. Generally, the greater the volume of air circulated, the less it will fall in temperature for a given amount of evaporation, and faster drying will result.

Thickness of Layer on Trays

The pieces are piled on one another, and those on top dry first. Those on the bottom will dry very slowly, if the air flows only across the top. In some types of dehydration, the air flows through the layer of pieces, getting good circulation and drying. In general, the drying rate is very much affected by layer thickness, and ways and means of baffling the air streams in the tunnel must be worked out to bring about faster drying.

Scorching Damage

Certain vegetables such as sweet potatoes, sweet corn and spinach, will stand much higher drying temperatures without scorching than onions or cabbage. Potatoes, carrots and turnips are sensitive to high-temperature drying. Scorching may also be the result of overdrying even though the temperature is correct. The damage is recognized by the development of a yellow or brown color and a burnt flavor or both. The general rule for maintaining color and flavor quality is to keep the temperature of the material low, but the air velocity as high as possible. Delay in drying for several hours in a warm atmosphere, high in humidity, will bring about poor quality. The product may become sour and moldy. Therefore, drying should start promptly and proceed rapidly. Poor quality may result where drying is conducted in a counterflow tunnel dehydrator where the tunnel is so long or the flow of air is so slow that the air temperature drops to only 5 to 10° F above the wet bulb temperature at the cool end. In good drying, the wet bulb temperature should be at least 20 to 40° F lower at the cool end.

Dehydrator Types

Conveyor Belt

On this type, the wet material is placed on a traveling conveyor belt, and hot air is forced through the layer of material and through the conveyor belt itself. The flow of air may be changed around in the different sections of the dryer, and the thickness of material may be adjusted at will. Conveyor belt driers must be built for thorough circulation of air, otherwise they will not be efficient. Uniform loading is important. If the direction of air flow is reversed in successive sections of the conveyor units, good drying is obtained.

Tunnel and Truck Dehydration

Counterflow Arrangement

Many types of tunnel dehydrators are in use, but most of them can be regarded as modifications of the counterflow, parallel flow and compartment styles.

In the counterflow type, the air flows through the tunnel in the direction opposite to the travel of the moist trayed material. The air entry, where drying is good, is called the hot end. The cool end naturally is the wet end. The wet end is the point or the door through which the material or trayed trucks are fed.

The counterflow type of dehydrator is the most common one on the Pacific coast for drying prunes, raisins and vegetables. Since the rate of evaporation from the nearly dry product is very low, the temperature drop of the air near the dry end of a counterflow dehydrator must be relatively slight. The wet end of the dehydrator does not lose moisture rapidly, since the air is cool and moist. The drying at the cool end is slow compared to that at the hot end. The baffling, or uniform distribution of the warm air coming in a counterflow tunnel, must be expertly arranged; otherwise, the drying time will be many hours longer than is good for the product.

Parallel Flow Arrangement

In this air flow arrangement, the direction is the reverse of that in the counterflow tunnel. Here the hot end is the wet end, and the cool end is the dry end. At the end of the tunnel, the product will lose moisture readily, since the air here is at its highest temperature. On the other hand, the drying is slow at the cool end, since the air passing over is low in temperature and high in humidity. Under these conditions, it becomes almost impossible to dry the vegetable to less than 10% moisture content. If the rate of loading the wet material on the trays were decreased, it may be possible to get good drying at the wet end, provided there is good baffling and a high air velocity.

Counterflow Parallel Arrangement

Counterflow provides the highest possible rate of drying at the dry end, whereas the parallel provides best evaporation at the wet end of the tunnel, and combinations of the two now find a wide use for the dehydration of vegetables. A simple design to accomplish this is the construction of a relatively short, parallel tunnel in which a large percentage of the total evaporation takes place. The product which may still contain as much as 50% moisture is then discharged into the counterflow tunnel constructed to be somewhat longer. Here the air flow and heat input may be substantially less, since comparatively little evaporation will be required. In some designs, the exhaust air from the counterflow tunnel is used to make up a part of the fresh air required by the parallel flow tunnel. In this two-tunnel operation, there will be additional handling of trucks and therefore some increase in the labor cost.

Compartment Arrangement

A less common arrangement of a continuous truck and tray dehydrator is a combination compartment and tunnel. Here the air flows through the trucks and is transverse to the longer axis of the tunnel. This arrangement makes it possible to circulate the same air through each truck several times and to reheat the air at each stage. For example, if several cabinet dehydrators are placed end to end with doors between units omitted, a truck load of wet material may start at one end of the group and progress through one compartment after another, emerging completely dried at the other end. If a blower is used to supply fresh air to the dry end compartment, and the exhausted air from each compartment is used to supply the fresh air requirement of the next one, there is a general countercurrent flow of air through the whole group. Since each compartment can be provided with a separate heater, this makes the regulation of air and temperatures at different stages of drying possible. The trucks used in compartments must fit very tightly along the tunnel walls to avoid short circuiting of air through useless channels.

Condensation of Humid Air

Any type of dehydrator can be modified so that water vapor is condensed from the exhaust air, which is then returned and used over again as fresh air. The dehumidification of the exhaust air is supposed to result in a better quality product, but its cost has discouraged wide application.

Combined Blanching and Pre-Drying

Some experimental work has been done on brief pre-drying at very high wet bulb temperatures, the object being to produce a longer blanching action with a substantial removal of the easily evaporated water. For example, exposure of cut vegetables to a current of air at 250° F and a wet bulb temperature of 200° F for 6 to 10 minutes may evaporate as much as half of the total moisture, and blanch at the same time. The process has possibilities, but it has not yet reached commercial development.

Sources of Heat for Dehydrators

Two general systems are commonly used to heat dehydrators: the direct-combustion heater and the indirect heater. With the former, the fuel is burned inside the dehydrator, and the gaseous products of combustion are circulated along with the drying air. In the indirect system, auxiliary heating surfaces for transferring heat to the drying air are used. The latter completely isolate the products of combustion from the drying air and so complete combustion is not essential, as far as the product is concerned. Fuel-oil burners should be installed in accordance with the recommendations of the National Board of Fire Underwriters.

Direct-Combustion Heaters

Direct heaters are limited to liquid or gaseous fuels. The rate of combustion of these fuels can be readily controlled over a wide range and so temperature can be controlled. Obviously, complete burning is essential with the direct heater, especially with oil, because imperfect combustion might result in smoking, which would seriously affect quality of product. Since there are not stack or transmission losses, optimum efficiency can be obtained, and a minimum of equipment is required for satisfactory performance. Most of the dehydrators on the west coast use this system.

In its simplest form, the heater consists of a *premix burner* located openly in one of the air passages of the dehydrator. The intensity of flame and hence the dehydrator temperature can be controlled by a manually operated globe valve placed in the fuel supply line. This simple system, however, has many limitations. Temperature regulation is tedious and combustion efficiency is likely to be poor. Close temperature regulation, however, can be accomplished by substituting a modulating control valve for the manual control valve.

A basic rule applicable with any fuel, is that combustion should occur in a relatively high temperature zone with a minimum of chilling from outside. If an open gas flame is exposed to an air stream of relatively low temperature, the partially burned gases in the outer surface of the flame may be chilled below the ignition point and swept away, with a resulting decrease in the efficiency. Fuel oil, especially the heavier grades, cannot be satisfactorily burned in an open flame, because the chilling effect of the surrounding air currents will cool the burning oil particles below their ignition point, and smoking will result.

A satisfactory way of burning either gas or oil is through the use of a combustion tube. This device consists essentially of a refractory-lined chamber of ample proportions to permit generation of the maximum heat required while the flame is confined to the combustion chamber. In the design of the chamber and burner, care must be taken to prevent direct impingement of flame on refractory lining, as this might result in either damage to the refractory or, with oil, the formation of coke-cones, with consequent smoking.

Because the products of combustion are to be mixed with the drying air, any amount of air necessary for complete combustion in direct heaters can be used without impairing the efficiency of the system. This factor is especially important with oil, because oil requires a considerable amount of excess air to insure complete combustion and prevent smoking. As a rule, heavy oil can be completely burned with 250% excess air. More than this amount of air may chill the flame and produce deleterious effects. The design of the secondary air opening varies with the burner used and should be provided for in accordance with the manufacturer's recommendations.

Indirect Heating System

Indirect heating systems can be divided into two general types. First, and simplest from the standpoint of equipment, are those with heat-exchange surfaces receiving heat directly from flue gases or products of combustion. The second, a somewhat more complicated type, employs a medium such as a liquid or vapor to transfer heat from its source to the heat-exchange surfaces.

The first type, as one might expect, is inefficient and difficult to control. It can, however, utilize a variety of fuels. Temperature regulation is difficult even when gas or oil is used. However, complete combustion is not essential for successful performance of the indirectly heated dehydrator. Some smoking can be tolerated, because the products of combustion do not come in direct contact with the drying air.

If coal or wood is used with this system, a fair degree of control can be obtained by either wasting some of the products of combustion to the atmosphere before they come in contact with the heat-exchange surfaces, or by diluting and cooling the flue gases with outside air prior to their flowing over these surfaces. Obviously these methods of control are wasteful. Although this type of indirect heating has its applications, it is of little value in commercial use.

On the other hand, the second type of indirect system offers practically ideal control characteristics. The heat-transfer medium is usually steam, but may be hot water or oil. This method of heating offers a clean, even source of heat with any type of fuel, either liquid, solid or gaseous. The boiler or accumulator is a source of heat that can be used instantaneously to cope with variations in demand. Regulation of temperature within the dehydrator can be accomplished by a relatively simple method which is applicable over a complete range of dehydrator sizes. On the other hand, the equipment is considerably in excess of that required by other methods, and the over-all efficiency is considerably below unity because of boiler and transmission losses and will in all probability be less than 60%. Initial cost, upkeep and overhead are correspondingly high.

Heat exchange surfaces are normally copper-finned coils. Relatively large amounts of heat can be transferred to air in motion by a compact heat-exchange surface of this type.

If a coil is to be built for use in a dehydrator, it is suggested that care be taken in the design to provide ample surface for adequate heat transfer, because the performance of the dehydrator is at stake. Rapid and complete elimination of the condensate and air is also necessary to promote coil efficiency.

Some large steam-heated dehydrators have been successfully

designed to use steam for driving the blowers and to utilize the exhaust steam in the heating system. The blowers are driven by bleeder-type turbines, operating at a back-pressure of from 10 to 50 pounds per square inch. Such an arrangement may reduce power costs very considerably. Each turbine may be designed and operated so that the steam flow through it will normally be less than the minimum steam demand of the corresponding heaters, so that additional high-pressure steam will always have to be bled into the heating system. Automatic controls will then adjust the auxiliary steam supply to satisfy the heater demand without affecting the operation of the blower.

Mechanical Movement of Air in Dehydrators

Fans generally used for the mass movement of air in dehydrators can be divided into two classes: propeller, axial-flow or disk fans and centrifugal or rotary impeller fans. Disk fans are generally limited to delivery of air against relatively low static pressures, although some special designs are capable of developing a pressure of 10 inches of water or higher. Centrifugal fans, on the other hand, can be used against either high or low static pressure. Installations in which low static pressure disk fans are suitable include cabinet driers, low-performance tunnel driers and belt-conveyor driers. Centrifugal fans can also be employed on high-performance tunnel driers, bin driers and in general on installations where the static pressures are substantially higher. A centrifugal fan consists essentially of an impeller or fan wheel and a casting or housing. The design of these parts can be varied to suit peculiarities of the load. The casing has either of two functions or both; one is to transform velocity pressure into static pressure and the other is to collect and conduct the air to the point of discharge, or fan outlet.

Centrifugal or Rotary Impeller Fans

Centrifugal impellers can be roughly divided into three classes: those with forward-curved blades, those with straight radial blades and those with backward-curved blades, sloping away from the direction of rotation. Impellers formed in this way are termed slow-speed, moderate-speed and high-speed types, respectively. Their ranges may be overlapping; therefore, each of these types may be used for the same work. Fan characteristics in each group may, however, be varied by changing the depths of the blades. Fans with deep blades will develop relatively high static pressures, whereas fans having short or shallow blades will handle large volumes at low pressures. This latter type is sometimes designated as a volume fan.

Although the forward-curved-blade centrifugal fan is widely used on tunnels originally designed as fruit dehydrators, they are not suited to that purpose. The probable reason for their original adoption lies in the fact that the relatively low shaft speed required for a given discharge was applicable for use with steam or gasoline engines.

The forward-curved fan has many characteristics that are unsuited to dehydrator service. First, the fan power increases when the static pressure of the fan is reduced; therefore, the fan absorbs maximum power at free discharge. However, at the normal operating static pressure, which is usually a condition near the point of maximum mechanical efficiency, the fan power may be less than one-half of free discharge power. Second, the fan has an unstable capacity or volume characteristic when operating at static pressures approaching the shut-off condition. Third, the slow speed of rotation necessitates an appreciable speed reduction if a standard electric motor is used to drive the fan. Systems having variable static pressure loading, as found in tunnel dehydrators and bin driers, are best served with fans having backward-curved blades. This type of fan, which is sometimes known as a limit-load or non-overloading fan, has operational characteristics which, at any given speed and specific volume, prevent it from absorbing more than a limited amount of power. This is of special importance when the fan is used in dehydrating systems where the static pressure in the tunnel drier may be suddenly decreased as a result of opening of doors, or, in the case of a bin drier, where the static load can be changed by removal of the product from the bin.

Propeller Axial-Flow, or Disk Fans

Unless very special designs are used, the disk-type propeller fan is generally suitable for use only where the static pressure is less than 1 inch of water pressure. There are two general classes of disk-type propellers: those with thin, steel curved blades, and those with air-foil sections similar to those used on airplane propellers. The former type is available commercially, and the latter can be readily built of wood. Regardless of the type used, the entrance and exit cones should be streamlined and the propeller shrouds should be built with minimum clearances in order to secure the highest efficiencies.

High-speed disk fans are likely to be very noisy, and this factor may restrict their use. When properly applied, they are efficient and constitute a relatively cheap means of obtaining a high volume of air movement with a minimum of equipment.

Although simple in appearance and operation, propellers are intricate in design. An improperly designed propeller may be extremely inefficient, wasteful and sometimes very dangerous. However, when properly designed, they can be readily constructed and are efficient and safe. Those who contemplate their use should make use of publications that contain adequate design information.

Bin Type Finishing Driers

The bin-finisher provides an efficient and inexpensive means of finally drying vegetables to a low moisture content. It is compact and has a high heat value. If it is to be used as a final drier, better use can be made of one of the preceding dehydrator regulators of whatever type it represents. If a bin-drier is added to the system, and if sufficient heating facilities are available, then the plant capacity is potentially increased. With the use of a bin-drier, the moisture of the produce in the dehydrator can be reduced to approximately 10 to 15%. There are two basic types of bin-driers, the batch and the continuous process. The continuous drier usually consists of a single unit, whereas the batch process drier is a multibin-drier.

The drying air is introduced through louvers at or near the bottom of the bin. The bin must be of air-tight construction and have an exhaust port located at some position above the active bin level. A gate is provided in the bottom of the bin to permit gravity flow removal of the products. To facilitate removal and to prevent arching or clogging, shaking devices are sometimes provided on the outside of the bins. Such a shaking device is called a pulsating vibrator.

When the bin is in operation, the partially dried material is allowed to enter in a steady stream, at such a rate that it keeps the bin completely filled. A draft of heated dry air is forced through the louver inlets at the bottom, and the product dries progressively upward; therefore, the dried product at the bottom is removed at the same rate as the bin is loaded from the top.

The batch-process bin-finisher is somewhat similar in design. A gate is located at the bottom to permit dumping of the entire bin. Three or more bins are usually required, as will be explained later. In operation, the entire bin is filled with the partially dried product. The material dries progressively upward, and, as soon as the material near the top is dried to the required moisture content, the entire bin is dumped, and a new cycle is started.

Operation of the Bin-Driers

During normal operation, the product, as loaded into the bin-drier, contains not over 15% moisture. This partially dried product is of sufficient stiffness so that it will not crush, and will form a solid mat. Therefore, care should be taken not to tamp the product during loading operations. The charging depth may range between 2 and 6 feet. Under this condition of loading, air can readily be forced through the material by standard-design blowers.

As a product loses its moisture, its rate of drying becomes slower and slower. The latent heat necessary for evaporation will be required at a proportionally slower rate. Since the air is the source of heat supply and the transfer medium, the required quantity and velocity of air become correspondingly smaller. The air bathes the product and carries away the diffused moisture as it is released.

The depth of bed will determine to a large degree the batch retention time. Usually, bins are operated at the upper limit of the usable temperature range. If excessive bed depths are used, the dried product on the bottom may be injured by prolonged exposure to maximum temperature. Therefore, to reduce the batch-retention time, relatively high velocities should be used.

An undesirable condition that may arise is illustrated by the following example. Let us assume that a deep-fed bin-drier is being used to finish potatoes. Assume further that the potatoes have been scraped from drying trays of the preceding stage in a room having a temperature of 60° F. It is probable then that, when the bin is filled with these potatoes, the average bin temperature will be in the neighborhood of 60° F. Assume that air at a relatively low velocity and temperature is circulated through this bed. The air will pick up moisture and gradually drop in temperature until, at a certain point in the bed, it can no longer dry the product. From this point, the cool product will gradually reduce the air dry bulb temperature until, at a point within the bed, the air becomes saturated. Further penetration of the air, while in contact with the cold bed, will reduce the air temperature below the dew point; in this way moisture will condense upon the product. If this condition arises, the wet product may be injured. The condition can be remedied by increasing either the temperature or the velocity or both, by decreasing the bed depth, or by providing auxiliary internal bin heating devices.

An operator can detect faulty bin operation by taking samples from the uppermost surface of the product periodically throughout the drying cycle and analyzing them for moisture content. The time intervals should be fairly close because the region of condensation, which is likely to be fairly shallow in depth, might otherwise not be detected.

Regardless of the velocity and temperature of air employed, a bin-drier will ultimately carry the product through the full depth of the bed to a condition approximating equilibrium with the entering air. This, of course, is based on the assumption that heat loss due to radiation, convection and conduction from the outer surfaces of the bin are relatively small.

Batch-drying is intermittent in operation; therefore several bins are usually provided to permit continuous drying operations. At least three bins are usually employed and, while one is being loaded, the others may be used in drying or unloading. Any number of units can be provided to suit plant capacity.

Air Desiccators

Bin-drier operation will vary considerably between humid and arid regions. If high relative humidities are prevalent in the plant locality, it will most likely be essential to employ some means of drying the air before it is introduced into the bin-finisher. A unit for this purpose is called an air desiccator or air dehydrator. The former is the preferable designation because the other may lead to confusion.

One of the most satisfactory means of drying air is by the use of a chemical adsorption or absorption drier. Units of this type are capable of delivering air at fairly high temperature and low absolute humidity. This method obviates the necessity of reheating the dehumidified air, as would be required by a dehumidifying unit using refrigeration. This factor may be of considerable importance in the selection of an air-desiccating method.

When an air desiccator is used with a bin-finisher, it may become desirable to recirculate all or part of the drying air. In multibin operation, the air expelled from each bin will have a different absolute humidity, because the products in the bins are at various stages of dryness. In practice, however, the air from each bin is discharged into a common return duct. If the absolute humidity of the mixed air in the return duct is less than that of the outside air, then it should be recirculated; if not it should be entirely wasted to atmosphere. On *borderline* installations this condition may vary from day to day; consequently, repeated daily checks should be made.

Sources of Heat

The air supply to bin-driers is preferably heated by steam. Direct-fired units are not recommended, because the product may acquire a disagreeable taste from the flue gases. There is also a fire hazard with the open flame, which may cause dust explosions. The amount of heat required by a bin-drier is relatively small, and, the heat wasted by not recirculating it is negligible. The problem of recirculation should be decided on the basis of absolute humidity only.

Spray-Driers

Spray-drying is based upon the general principle that moisture is removed rapidly from a finely divided substance suspended in an atmosphere capable of absorbing the water vapor and of supplying the heat necessary for its evaporation. Under these conditions, it may be assumed that rapid evaporation maintains the particle temperature at or near that shown by a wet bulb thermometer until drying is substantially complete, after which the powdered product is removed from the heated zone so quickly that thermally induced damage is avoided, even when high dry bulb temperatures are employed.

This necessary situation is attained in practice through the use of a large drying chamber, usually cylindrical in form with a cone bottom, but sometimes in the shape of a box. The cylindrical form ranges in commercial practice up to about 20 feet in greatest diameter and 40 feet in over-all height. Collectors that accumulate the powder are attached to, or built into, the drying chambers. Fans, heaters and baffles induce a rapid whirling flow of hot air or flue gas through the drying chamber, collector and exhaust duct. The product to be dehydrated is introduced into the hot-air stream near its inlet in the form of a fine mist produced by a spray nozzle. Moisture is removed in a few seconds, and the resulting fine powder is deposited in the dust-collection system, from which it is removed, cooled and packaged, usually in a continuous operation.

The spray-drying process is particularly adapted to the production of dried milk and eggs. Certain vegetable purees are readily dehydrated for use in soups, baby foods and similar products. Tomatoes and most of the fruits have not so far been successfully spray-dried on a commercial scale, presumably because of their high content of low-melting hygroscopic sugars, which impart to the dried product a tendency to stick to the equipment rather than to flow smoothly through it.

Vacuum-Driers

Water can be evaporated from moist substances at relatively low temperatures in a vacuum, and substances that are extremely sensitive to heat, such as serums, are accordingly dried under these conditions. Equipment for vacuum processing must be of heavy construction to withstand the pressure of the atmosphere, and all closures must be fitted with extreme precision to avoid leaks.

Most types of dehydration equipment can be built to operate under vacuum. Rotary-drum driers and rotary-kiln driers have been so used for some time. A type of cabinet dehydrator adapted for vacuum operation is perhaps used more extensively for food products than others. It consists of a cast-iron shell in which shelves internally heated by steam, hot water, hot oil or electricity are placed. Trays loaded with material to be dehydrated are placed on the shelves. Heat transfer is slow and inefficient in the absence of a vehicle such as moving air, and time for complete dehydration is relatively long. A recent development in this field employs a specially designed fan to circulate the attenuated water vapor present in the dehydrator over a heater and then over the charge. In this way, evaporation is greatly accelerated and the equipment, though operating batchwise, may prove to have a substantial capacity in terms of dried product. Another recent development is a continuous tunnel operating under vacuum. Neither of these devices has as yet been extensively used in commercial dehydration.

The principal handicap to practical use of vacuum-driers for dehydration of fruits and vegetables, namely, the difficulty of transmitting heat to the product rapidly enough to evaporate a large quantity of water effectively, can be overcome to a great extent if the drier is used only for the finishing step in a multistage dehydration process. Preliminary steps may remove 95% or more of the moisture originally present in the material; from that point, the vacuum-drier is able to operate very effectively. A combination process of this nature is being used commercially for the final dehydration of certain fruits, which are first dried in hot air to the usual moisture content for dried fruits, about 20 to 25%.

Dehydration of fruits or vegetables to final moisture contents of only 1 to 2% can be accomplished in one of two ways; by the use of highly desiccated air as the drying medium, or by the use of a vacuum-drier. The choice between these two methods may be determined by availability of the equipment or by the sensitivity of the nearly dry product to oxidation by warm air.

Rotary-Drum Driers

Drum-drying provides means for rapid dehydration of solutions, slurries, purees and the like, spread in a thin layer on a heated surface from which the dried product is continuously removed. The equipment consists of an internally heated revolving drum, upon the outer surface of which the substance being processed is spread at one point, while the dried product is removed by means of a blade at another. The temperature and rate of revolution of the drum are so chosen that drying is completed in the time required for the material to travel from the point of spreading to the blade that removes the product.

The operation can be carried out in air or in a vacuum. In the latter case, heat-sensitive products are better protected, since lower operating temperatures can be maintained. The material being dehydrated can be spread on the surface of the drum by spraying it, by means of a *doctor* blade, or by use of two drums placed parallel with a small clear space between them. In the latter case, both drums can be used as driers, and even spreading on the large drum can be attained.

Many products are removed from the drum in the form of continuous sheets, which are conveyed to a mill operating in a suitably conditioned atmosphere where the dry product is reduced to a powder and then packaged.

The drum-drying process is widely used for the production of dried milk and, to a lesser extent, for certain fruit and vegetable purees. The form of the product is confined to thin flakes and powders.

Rotary Driers

Rotary driers are rotating drums inclined at a slight angle from the horizontal. The wet substance is fed in at the high end and is conveyed with constant tumbling toward the low end, from which the dried product is continuously removed. Drying is facilitated by a current of air, either counterflow or parallel flow, induced by suitably situated fans, and heat may be simultaneously supplied on the outside of the drier. Construction is of either the single-or double-shell type. The latter affords greater protection of the product from damage by burning. Double-shell driers are sometimes provided with louvers in the inner shell through which air is admitted in direct contact with the material.

Rotary driers are efficient and extremely useful in drying many substances, but they have found little application in the field of fruit and vegetable dehydration. The constant rolling or tumbling of the wet product destroys the characteristic form of the pieces (slices, strips or cubes) and may result in production of a large proportion of powder. Rotary driers have been used successfully for the drying of precooked ground meat.

Changes in Material as Drying Progresses

The fundamental factor at the base of all combination systems is the enormous change in properties of the material as it dries. A freshly blanched piece of vegetable is soft and easily crushed. It is indeed a mass of water held together by a tenuous cell structure. Its vapor pressure is nearly equal to that of pure water. In contrast, the same piece of vegetable after drying to 5% moisture is greatly shrunken in volume and distorted in shape, is hard and may be very brittle, and its vapor pressure may not be more than 10 or 15% of the vapor pressure of water. These properties all change as the drying progresses, but not all at the same rate.

Hence the conditions which would be *ideal* for drying change as the piece dries. When the piece is very wet, the water will evaporate very rapidly if a great deal of air is supplied, and that air may be very hot without burning the piece. At the same time, the piece will be too tender to stand more than a minimum amount of handling. Essentially, similar conditions prevail while at least the first half to three-quarters of the weight is being lost through evaporation. As the moisture is further reduced, the piece, which will have already undergone most of the shrinkage in volume that it will experience, becomes tough or leathery. The rate of evaporation may not be more than a half or a third of what it was at first, even though drying conditions are kept the same; increasing the air velocity past the piece has no perceptible effect on the rate; the temperature of the piece rises almost to the dry-bulb temperature of the air. *Ideal* conditions now demand a lower temperature than at first, and the use of a high air velocity would be a needless waste of power. However, the piece will now withstand handling, tumbling or deep piling without damage. Conditions substantially similar to these prevail down to a moisture content of about 10 to 15%. Below that point, and more and more clearly as the moisture content falls further, conditions change again. The piece becomes hard and brittle. The rate of further evaporation is determined almost entirely by the thickness of the piece and the dryness of the air. The rate will perhaps become less than 1% of what it was at the beginning of the drying. *Ideal* conditions now call for a slow current of dry air, not too warm, and very little mechanical disturbance of the product.

The usefulness of different kinds of dehydrators at different stages of drying will depend not only on these physical factors but also on economic ones. It is an economic waste to use a large and expensive piece of equipment to perform work that a small cheap one will do as well. Consideration of the labor or complications involved in rehandling the product may favor one kind of transition over others.

Examples of Multistage Systems

Out of the very great number of possible combinations, the following may be mentioned. Not all of these are now in commercial use.

1 Three-stage conveyor-belt drier followed by a drying bin. The bin itself may be continuous in operation, in which case it is essentially a fourth stage of the conveyor, arranged for very deep piling.

2 Three-stage conveyor-belt drier followed by a continuous vacuum-finishing drier.

3 Counterflow tunnel followed by a finishing bin. This combination has been used in commercial operations for some years.

4 Cabinet drier followed by a finishing bin. Use of the bin releases the cabinet from the part of the run on which it is normally most inefficient.

5 Rotary drier followed by vacuum-drier. The rotary drier may be used for most of the drying of products which are tough enough to stand the tumbling action, but it tends to reduce a nearly dry product to powder.

6 Parallel-flow predrying tunnel, counterflow secondary tunnel and finishing bin. This is a combination which is being widely adopted for large commercial operations.

7 Parallel-flow predrying tunnel followed by two-stage conveyor-belt drier. The tunnel takes the part of the cycle for which the conveyor drier is least well adapted. The second stage of the conveyor drier can be designed for deep piling, as in a finishing bin.

8 Center-exhaust tunnel followed by finishing bin. The combination is equivalent to that of No. 6 above, except in the manner of handling the product between the first and second stages.

Temperature Reduction to Maintain Quality in Dehydrated Products

The quality of dehydrated vegetables can readily be injured before shipment by failure to reduce the temperature of packaged materials and to keep them cool. Holding dehydrated cabbage at 120° F for a week, for example, results in almost complete spoilage. Measures must be taken to insure that such products will be cooled to 90° F or less within 24 hours after drying has been completed. While they are on the inspection belt, freshly dehydrated vegetables cool to a considerable extent. Measurements of the temperature of vegetables should be made after they have cooled on the belt. Such measurements can be made by thrusting a thermometer into the center of a filled carton or can and allowing 10 minutes before the reading is made. Measurements on rate of cooling of potato strips in cans have demonstrated that the center of a full five-gallon can, when filled at 105° F and placed in air at 85° F, will cool to 90° F in 7 hours. Cartons and cans should be kept apart or unstacked, as far as possible, until they are cooled to 90° F or lower. A period of several hours may be needed. Temperature of the air is, of course, a factor. Canners commonly make use of cooling rooms. When dehydrated vegetables are shipped from a cool area through a hot one, close packing will be helpful, provided the cases, when shipped, are close to the air temperature at the point of shipment.

The heat-diffusing and heat-conducting properties of various dehydrated foods have been measured recently at the Western Regional Research Laboratory. The measurements on strips of dehydrated white potatoes, which are representative, approach the corresponding properties of sawdust, which is used commercially as a heat insulator. Cooling rates are directly proportional to the diffusivities. When the containers are cans, the average diffusity is increased to about 3 times that of dehydrated vegetables in foil-protected cartons. This is because the steel sheets are highly conductive and will aid in the removal of heat from the stack. No such increase in the rate of cooling occurs in unstacked single cans, since the steel is on the outside and at once comes to room temperature.

The peaks of storage rooms should be ventilated in summer, and cases should not be stored next to the roof. In a ventilated warehouse, a pronounced increase of temperature from the ground level upward has been observed. The outside air was

85° F; at the floor level, it was also 85° F; at a point at 8 stacks high, it was 93° F; and 6 feet above it was 98° F.

Refreshing and Cooking Tests on Dehydrated Vegetables

Dehydrated vegetables should be soaked in water for some time before they are cooked, otherwise they are apt to remain tough and shriveled. This procedure is known as refreshing or reconstituting the product. Only enough water to bring about plumping should be used or enough to approximate the amount of water originally present before drying. For potatoes and peas, this may be about 3 fluid ounces of water to each ounce by weight of dried product, and for spinach and cabbage perhaps 6 to 8 fluid ounces per ounce of dry weight. Powdered vegetables usually cook well without preliminary soaking. Vegetables, prepared by slicing them very thin or shredding just before dehydration, may not even have to be soaked first, since during drying they are practically precooked.

Cooking Reconstituted Vegetables

The refreshed product is allowed to drain off excess water, salted to taste and then cooked slowly until tender. Dehydration definitely removes some of the delicate flavors of the fresh products, and it will be found desirable to add suitable seasoning, such as strips of ham, bacon or smoked beef, a clove or two of garlic, or a little basil and sliced onion to bring out flavor. Tomato sauce is a pleasing addition to some products. Powdered pumpkin may be cooked with milk and sugar thickened with egg before filling the pie crust. White potatoes, after refreshing, may be fried, scalloped in the same manner as the fresh, or may be cooked until soft and then mashed with milk and butter.

Laboratory Refreshing Test

For rapid assaying, use a fifty-gram sample. For peas, potatoes, corn and sweet potatoes 200 cubic centimeters cold water are added. For other vegetables, use 300 to 500 cubic centimeters depending upon the drying ratio. Use about 300 cubic centimeters for root vegetables, 400 to 500 cubic centimeters for string beans, tomatoes, spinach and cabbage. If they float, keep them submerged with the aid of a weight and wire screen. The sample should be left for at least 12 hours, at room temperature, then drained for 3 to 4 minutes on a screen and weighed again. A little experience with different vegetables will give one an idea of the refreshing values.

The samples from the refreshing tests may be salted to taste, and each vegetable cooked for a specified time. About 30 minutes may be required for potatoes, peas and corn, and 60 minutes for most other products. Once the length of time has been determined for each variety, comparative tests should be made for flavor, texture, color, odor and taste. It is important to determine the drained weight.

Packaging Equipment and Methods

After final inspection and culling, dehydrated vegetables and fruits are packaged. If they are to be powdered, a hammermill will be needed, also a dehumidifier, if the atmospheric humidity in the room is high. Vibrators are often used in packing to produce a higher bulk density. A simple type consists of a low platform, set diamond-shape, with the pivot placed diagonally. Strips along the two back edges form a stop for the cans. Power is supplied by a one-third-horsepower motor operating at a few hundred revolutions per minute. An eccentric is mounted on a low-speed shaft, and this is connected to the back corner of the platform by a rod. In another type, a spring mounted platform is vibrated by a cam in a motor shaft. The vibrator can be equipped with a weighing scale. Weighing scales should have fifty-pound range, attachment for indicating the amount over or under and a tare weight.

Gassing is required for cabbage or carrots that are to be sold under government contract. Flavor, color and vitamins are retained through the elimination of air and its replacement by nitrogen or carbon dioxide. Vacuum packing is not used because five-gallon cans require a very solid fill if they are to be sealed under vacuum. A maximum limit of 2% of oxygen is recommended for sealed, gas-filled containers, and the analysis should be made at least 12 hours after the cans are filled and sealed. Air can be displaced to a point at which the oxygen content is below 2% by the use of (1) the cylinder-and-meter method, (2) the vacuum chamber method, or (3) the carbon-dioxide-snow or *dry ice* method.

Cylinder-and-Meter Method

The most common method of removal of air has been by the use of gas run from a cylinder through a reducing valve, a rubber tube and a metal purge tube thrust to the bottom of the can. The amount of gas is controlled by the pressure setting of the reducing valve, the duration of gas flow and the occasional testing of the gas at the top of the can with a burning match. If the match goes out, the can is considered to be sufficiently low in oxygen content. While this method commonly results in analysis showing below 2% oxygen, the time is not accurately measured, and the cans have been found to contain as much as 8% oxygen.

Tests have shown that the introduction of an iron-case dry-gas meter, of stock design, between the cylinder and the purge tube improves the method. The stock meter used in testing had one dial that measured 1 cubic foot per revolution. In the displacement of air by this method, the lid of the can is placed over the opening; the can is gassed, and, after removal of the tube, the lid is attached to the can by seaming to form an hermetic seal. Soldering is an effective method of closing cans but is a slower method than seaming. The following results have been obtained in tests of the cylinder-and-meter method of gassing:

Cubic Feet of Gas	Time in Seconds	Per Cent Oxygen Content of Can
2	60	0.8–1.0
1	30	1.2–1.6
1	10	1.8–2.4

These results were obtained with dehydrated cabbage and carrots at typical loads per can.

Vacuum-Chamber Method

Another process of removing air from five-gallon cans and replacing it with carbon dioxide or nitrogen is the vacuum-chamber method. This method requires equipment similar to that required in vacuum canning. One type of equipment has a flat plate with a space on it for one or more cans. Cans are placed on the plate and are covered by a heavy metal chamber or bell provided with a counterpoise. For the evacuation of the air in a chamber, a vacuum pump is provided. The effectiveness of the equipment depends largely on the closeness of the joint between the chamber and the plate, which governs the leakage of air inward, and on the degree of vacuum to which the pump can exhaust the air in the chamber and the cans.

For a single-unit chamber, the pump should have a capacity displacement of 100 cubic feet per minute, and should produce a vacuum of 29.5 inches of mercury. The highest degree of vacuum can be attained by the use of any of a number of rotary pumps. The next type of pump, in order of attainment of vacuum, is the two-stage steam ejector. At least 75 pounds per square inch of steam pressure is necessary for efficiency. The first cost is lower than those of rotary or reciprocating pumps, but the demand on steam between evacuations diminishes the utility of this device. Single-stage reciprocating pumps will not produce the required vacuum, but double-stage reciprocating pumps will. Some double-acting reciprocating pumps can be easily and inexpensively converted to double-stage pumps.

Modern gassing chambers of the bell or horizontal type hold the lid away from the can during the operation by means of an externally operated chuck or by an electromagnet. The procedure is carried out in the following way. The can is filled with a weighed quantity of dehydrated vegetables, and a lid is set in place. The chamber is closed. A vacuum of at least 29.5 inches is drawn on the chamber. This will take 20 to 40 seconds. The valve to the vacuum pump is then closed, and a valve to the cylinder of gas is opened, releasing the vacuum. A pressure of 1 to 2 pounds per square inch is built up in the bell, which will prevent the entrance of air when the chamber is opened. The chamber is opened, and the can is removed and hermetically sealed immediately. Cans should be handled by the corner edges, since any bellows action of the sides caused by handling will draw air into the can before it is sealed.

In practice, the time for evacuation is 20 to 40 seconds, for breaking the vacuum, about 20 seconds, and for a cycle, including opening and closing of the chamber and sealing a pair of cans, about 90 seconds. In tests, oxygen content of the atmosphere in cans of carrots and cabbage sealed by means of a double-stage reciprocating pump, with a 29.5 inch vacuum, and one chamber, was found to be 0.5 to 1.6%. Three to four cans may be evacuated, gassed and sealed per minute in modern-type machines.

Gassing by Means of Dry Ice

The solid carbon-dioxide-snow or dry ice method has been developed at the Western Regional Research Laboratory as a means of gassing carrots or cabbage with a minimum amount of equipment. The method has not yet been tried out commercially. Unlike water ice, dry ice changes directly into a gas when it is heated. It is shipped in fifty-pound blocks and packed in quadruple corrugated cartons. Losses in shipment and storage are approximately 6 to 10% per day, respectively. Suitable storage bins can be made with hardwood frames, sheet metal or plywood walls for the lining, six-inch layers of cork or other insulation on the bottom and sides, and a four-inch-thick kapok, duck-covered pad as the cover.

One-fourth pound of ground dry ice is placed in the bottom of a five-gallon can and distributed uniformly. Then the vegetable is weighed into the can, which is placed in a water trough equipped with a conveyor, and the lid is placed on the opening in the can. At once, an evolution of carbon dioxide takes place in the air, and later the gas escapes around the lid. Two cubic feet of gas are formed in the sublimation of the dry ice. After 6 to 12 minutes in water, all the solid carbon dioxide will have become gas, and the can is capped after a short period of standing in the air.

The atmosphere in five-gallon cans of cabbage or carrots

has repeatedly been brought to 0.8 to 1.0% oxygen by this method. Care must be taken to avoid the sealing of cans in which dry ice, not in the gaseous state, is present; dry ice in a sealed can will cause cans to bulge later. The ordinary precautions for handling dry ice should be observed.

Sealing

Dehydrated vegetables and fruits that are packaged in tin cans are hermetically sealed, even when packed in air. Formerly the stud hole was covered with a lid, and a tight joint was made by soldering, but recently, because of the shortage of tin and the lower percentage of tin in solder, the trend has been to use square cans with mechanically formed seams. These seams are of two types, double seam and single seam. With both, a sealing compound is used on the surfaces subject to closure, in order to produce an hermetic seal. An hermetic seal may be tentatively defined as one that is gasproof, moistureproof and bacteria-tight, both initially and throughout a common period of handling and storage. Shipments of five-gallon cans of dehydrated products are frequently made across mountain passes 7000 feet high. In a trip of this sort a sealed can filled at sea level would be subjected to 3½ pounds per square inch of internal pressure. Tests show that when a leaky can is packed with carbon dioxide at sea level, shipped over a 5 thousand-foot pass, and returned to sea level, the oxygen content may be expected to increase 4%. Sealing machines for five-gallon square cans have been developed by several companies and can be leased. Machines will seal at rates between six and eighteen cans per minute.

Packing in Cartons

The equipment and methods recommended for cartons vary somewhat from one supplier to the other. An inner carton of banding chipboard must be assembled and taped at the bottom, the carton is then filled, and the top is taped. A lead-foil bag is expanded into shape on a mandrel, and the filled carton is slid into the bag on an inclined chute. The package is placed in an erect position and passed on a roller conveyor through a heat sealer, where the envelope is deflated by suction or some other method. The *ears* of the envelope are folded over, a strip of folding U-shaped chipboard is placed over the bottom and the sides. The unit is thrust into the outer carton, which is then sealed, and the seams at the end are covered with cloth tape.

Compression to High Density for Packaging

In addition to the saving in shipping space that results when foods are dehydrated, further saving can be achieved by compression. When a ship is loaded with cargo at the rate of 1 ton per 40 cubic feet, the holds are filled and at the same time the ship carries a full load. The objective in the compression of dehydrated foods should therefore be 55 to 60 pounds per cubic foot as the minimum density. This calculated objective takes into account the fact that the finished package is a little less dense than unpackaged cakes. Investigations have shown that such densities can be attained without loss of quality in the reconstituted product.

Even higher densities are to be preferred if an oxygen-sensitive product is compressed and packaged in hermetically sealed containers. Dehydrated cabbage, carrots and tomato-juice cocktails are examples. The denser the pack is, the lower will be the ratio of oxygen in the can to the dry fruit or vegetable.

A further advantage of compression is the diminished tonnage of steel and tin required when cans are used, since doubling the density halves the metal required.

In addition to increased density, rehydration to original size and shape is an objective, with a low percentage of fines or small particles. In tests, fines that pass through a four-mesh screen after rehydration have been kept under 5%. Weighed samples of blocks were rehydrated and screen analyses made of the cooked product. The low percentage of fines prevents a mushy texture and maintains palatability equal to that of unpressed, rehydrated products. The table shows the conditions under which fruits and vegetables have been compressed with 5% or less of fines. The time required for reconstitution is not increased over that required for unpressed foods.

COMPRESSION OF DEHYDRATED FRUITS AND VEGETABLES

Fruit or Vegetable	*Per Cent Moisture*	*Temperature of Pressing in ° F*	*Pressure for Blocking in (Pounds per Square Inch)*	*Densities (Pounds per Cubic Foot)*[1] Initial	After Compression	Compression Ratio (To 1)	*Approximate Content of Packages (Pounds per Five Gallons)*	*Per Cent Reduction in Bulb*[2]
1	2	3	4	5	6	7	8	9
Apple nuggets	2.0	170	450	12.5	64	5.1	12.0	71
Do	2.0	75	850	12.5	53	4.2	12.0	64
Apricot halves	25.0	75	150	42.0	79	1.9	28.0	44
Beet cubes	4.1	160	650	25.0	62	2.5	17.0	57
Beet slices	5.7	120	650	12.5	64	5.1	8.0	80
Beet strips	4.2	120	650	15.0	57	3.8	10.0	72
Carrot cubes	5.2	160	650	19.0	62	3.3	17.5	50
Carrot slices	4.8	160	650	6.0	56	9.3	8.0	75
Carrot strips	3.7	160	650	10.5	47	4.5	10.0	62
Onion flakes	3.0	140	650	6.0	61	10.1	12.0	60
Prunes (whole)	28.0	75	150	48.0	78	1.6	36.0	27
Rutabaga slices[3]	6.9	140	850	11.0	59	5.4	12.5	68
Tomatoes (spray-dried juice)[4]	4.0	78	1500	30.0	60	2.0	27.0	31

[1] Densities before compression were measured on shaken but unpressed material (column 5). Densities of commercial shipments are frequently higher.

[2] The percent reduction in bulk equals the difference between the weights of a five-gallon can of compressed vegetables and the corresponding weight of the uncompressed vegetable divided by the former weight. In computing the net weights of packages of the compressed foods, allowance is made for 15% of unused space in cans and for 5% in cartons.

[3] The tentative moisture content of rutabagas is 5%, at which blocking will produce a lower density than that found.

[4] Spray-dried tomato-juice cocktail is packed 4 pounds per No. 10 (3 quarts) can.

High densities in the pressed block require higher pressure when the dried fruits and vegetables are low in moisture content. Furthermore, with less moisture content, any given pressure will produce a higher density and a more cohesive block if the dried product is hot instead of cold. With onions, beets, rutabagas and carrots, it has been found that blocks tightly wrapped in cellophane, with the wrapping sealed, need not be kept in a press to cool. During the cooling, the densities under such conditions decrease from 61 to 59 pounds per cubic foot for onions and from 62 to 58 pounds for carrots.

Sun-dried apricots, cut and pitted, can be compressed by hand at room temperature to a bulk density of 42 pounds per cubic foot. A pressure of 150 pounds per square inch at room temperature has resulted in a density of 79 pounds per cubic foot. Higher pressures and an elevated temperature are required for high densities when fruit products are dehydrated to as low as 2 or 3% moisture content.

Presses and Processes

Hydraulic presses have been used in Germany for the compression of mixed vegetables, herbs, carrots, cabbage and dried sauerkraut. Continuous tile presses are reported to be in use, in which pressure is imparted by two cams acting at different stations in the machine. Knuckle-joint presses are used in the United States to compress hops, and screw presses are in use for compressing dehydrated-egg powder. Some free-flowing powdered soups have been smoothly handled in industrial tableting machines on an experimental basis. The latter are not well suited for use in forming blocks of dehydrated vegetables because: the vegetables are not free-flowing, and therefore, automatic and uniform charging of the molds cannot be accomplished in typical machines; and such tableting machines ordinarily operate at compression ratios of 2 to 1, or less. The compression ratio is the ratio of the volumes of the filled mold before and after compression. It can be computed by dividing the density after compression by the density before compression.

The capacity of a press is usually stated in terms of the total safe working pressure between the platens. Molds may have one cavity or multiple cavities. Dried fruits and vegetables have been formed into blocks 6 by 3 by 1 inches in size. Blocks 2¼ inches with 1 edge rounded have been proposed, since they lend themselves well to packing in standard five-gallon cans at the rate of sixteen blocks per can. Such a block weighs 2 pounds and can be reconstituted into fifty servings of carrots. Military recipes are in terms of one hundred servings for companies, and probably the most suitable blocks should contain weights corresponding to this number of servings. Blocks can also be formed with indented scorings like those on milk-chocolate slabs to facilitate breaking the blocks into smaller units of known weight.

To reduce friction, the surfaces of molds must be very smooth, a condition which can be produced by a surface grinder. It is frequently the practice to follow this by nickel or chromium plating and polishing. The surface of new dies must be lubricated with salt-free, moisture-free, edible oils or fats, such as commercial hardened shortenings or special edible lubricants. Continuous use of lubricants may be required to reduce the friction as the fruit or vegetable moves by the plunger past the sides of the mold. Friction resulting from unpolished, inadequately lubricated steel surfaces diverts part of the pressure from the foodstuff to the sides of the mold; in this way the effective pressure is decreased at the bottom of the mold, and poor cohesion and laminated, leafy brick may result.

Sometimes very fine dried materials or even powder are compressed, for example, soup mixtures and dehydrated tomato-juice cocktail. Most dry powders have been processed satisfactorily on industrial tablet-compressing machines. Dried tomato juice of 4% moisture content has been readily compressed hydraulically by 1500 pounds per square inch at room temperature to a density of 60 pounds per cubic foot. The cakes were reconstituted to a 6.5% juice simply by beating with an egg beater and then adding boiling water and heating for 1 minute. The dry compressed cakes crush easily.

Dehydrated vegetables of diced, sliced, stripped or flake form yield high-density blocks that readily rehydrate. The maximum pressure has been maintained for a period, or *dwell,* of 1 minute. Blocking has been effected ordinarily at temperatures from 140 to 160° F. The purpose of an elevated temperature is to produce a pliable state of the vegetable for compression so that it will reach a high density without breaking into small pieces to any considerable extent. Firm blocks of somewhat lower densities have also been produced at ram speeds of 1 to 2 inches per second followed by a *dwell* of 15 seconds.

Fine material passing a four-mesh screen, caused by compression under the conditions shown in the tabulation, did not exceed 5% of the rehydrated vegetables for any form of beets, carrots, onions or rutabagas. In general, the cohesion of the block was only fair at lower pressures. Higher pressures were not required for good cohesion nor were they necessary for adequate compression.

A slight expansion of blocks ½ inch thick occurs in that dimension when they are removed from the hot molds. This expansion is slight because these pieces cool rapidly. An estimate of the time required for cooling thick blocks of carrots in air is afforded by the following data:

A block 2½ by 1⅜ by 6¾ inches of compressed, diced carrots cooled at the center from 140 to 90° F in 3 hours, in air at 80° F. The block was cellophane-wrapped, confined in a holding press, exposing the narrow sides and ends only. Removed at 90° F, it held its shape.

Two-pound blocks of diced carrots were formed in a mold 6 11/32 by 4⅜ inches of suitable depth. The conditions were 160° F, 1200 pounds per square inch, and a thirty-second period at that pressure. The blocks were removed, wrapped in paper and cellophane, and sealed. After compression, the dimensions over the wrapping were 6 17/32 by 4½ by 2 3/16 inches. The expansion in a direction perpendicular to the movement of the ram was 5/32 inch on a 6½ inch length, with allowance for thicknesses of paper and cellophane. The immediate expansion parallel to the movement of the ram was ⅜ to ½ inch. An additional expansion resulted before it was stopped by the wrap and by air cooling. Holding presses prevent the latter expansion.

Actual practice will be a compromise between the goal of most desirable density, on the one hand, and, on the other, the obtainable equipment and its economic adaptation to use.

Blocks should be wrapped and sealed at once in kraft paper or cellophane. If the packing case is large enough to permit slight swelling of the blocks, they may be packed immediately, but if very close adherence to the dimensions of the mold is required, the blocks must be cooled under light pressure, to a temperature of about 70° F.

If compression is carried out with the wrapper inserted in the mold, complete cohesion of the block is not required. With this latter method of compression and wrapping, the wrapper must maintain its integrity as a moisture-vapor resistant sheet without tears, pinholes or other breaks.

Handling of Specific Vegetable and Fruit Crops

Vegetables

In the following pages information is presented on the dehydration of specific vegetables. The process and, to some extent, the equipment required in dehydration vary with the product, and in addition the products vary in requirements for rehydration. The information that follows is more detailed for those crops that are dried in larger quantities than for others. For convenience, information bearing on certain steps has been assembled in the following six tables.

The first table shows the type of piece, blancher loadings and time of blanching for 18 vegetables. Since it is important that the plant operator know the moisture content of the material to be used, the approximate ranges in moisture content of the vegetables are shown in the second table, together with averages. From these percentages the ratios, in weight, of water in the vegetables to *bone-dry* matter have been calculated and are shown also. Bone-dry matter must not be confused with the finished product, which contains a low percentage of moisture, as shown by the maximum percentages permitted under government procurement specifications. The ratio of water to dry matter in the raw product is useful to the operator, because it shows him how much water is contained in the raw commodity and makes readily calculable the weight of water that must be removed.

BLANCHER LOADING AND BLANCHING TIME FOR SPECIFIC VEGETABLES

Vegetable	*Type of Piece*	*Approximate Blancher Loading Per Square Foot in Pounds*	*Suggested Blanching Time in Live Steam (210–212° F) in Minutes*
Beans (lima)	Whole, shelled	1.5	[1] 5
Beans (snap)	Cut	2.0	5–10
Beets	Whole		[2,5] 25–40
Beets	Cubes, slices, strips	3.0–4.0	[2] 4–10
Cabbage	Shreds	1.0–1.25	[3] 2–4
Carrots	Cubes, slices strips	3.0–4.0	[2,5] 2–4
Celery	Slices	1.0–1.25	[3] 1–2
Chard (Swiss)	Trimmed leaf	0.8	[3] 2
Corn (sweet)	Whole kernels on cob		4–8
Kale	Trimmed leaf	0.8	[3] 2
Mustard greens	Do	0.8	[3] 2
Onions	Shreds, slices	1.25	[4] 1–1.5
Parsnips	Cubes, slices, strips	3.0–4.0	[2] 6–10
Peas (green)	Whole, shelled	1.5	5
Potatoes	Cubes, slices, strips	4.0	[2] 2–10
Potatoes (riced)	Quartered		20–25
Do	Sliced	4.0	6–10
Rutabagas	Cubes, slices, strips	3.0–4.0	[2,5] 5–10
Spinach	Trimmed leaf	0.8	[3] 2
Sweet potatoes	Cubes, slices, strips	3.0–4.0	[2,6] 6–10
Tomatoes	Slices	1	2–3

[1] Large-seeded lima beans may require 1 to 2 minutes longer for proper blanching.

[2] Government specifications require blanching until the peroxidase system is inactivated.

[3] Government specifications require blanching until the catalase system is inactivated.

[4] Government specifications do not require blanching of onions.

[5] As the peroxidase test is interpreted at present, it is usually impossible to destroy the peroxidase system in rutabagas and beets in less than 30 minutes, but if blanched as suggested above, the dried products will probably keep well in storage.

[6] Sweet potatoes should not be permitted to come in contact with iron during blanching.

MOISTURE CONTENTS OF FRESH AND DEHYDRATED VEGETABLES

Vegetable	*Approximate Moisture Content of Fresh Vegetable* [1] Range Per Cent	Average Per Cent	*Ratio of Moisture Content to Bone-Dry Matter* [1] Range	Average	*Per Cent Maximum Moisture Content of Dehydrated Vegetable*
Beans (lima)	58.9–71.8	66.5	1.4– 2.5	2.0	5
Beans (snap)	78.8–94.0	88.9	3.7–15.6	8.0	5
Beets	82.3–94.1	87.6	4.6–15.9	7.1	[2] 5
Cabbage	88.4–94.8	92.4	7.6–18.2	12.1	[2] 4
Carrots	83.1–91.1	88.2	4.9–10.2	7.5	[2] 5
Celery	89.9–95.2	93.7	8.9–19.8	14.9	[2] 4
Chard (Swiss)	89.9–92.9	91.0	8.9–13.1	10.1 [3](10.8)	4
Corn (sweet)	61.3–86.1	73.9	1.6– 6.2	2.8	5
Kale	81.4–91.2	86.6	4.4– 9.3	6.5	4
Mustard greens	86.7–95.7	92.2	6.5–22.2	11.8 [3](12.2)	4
Onions	70.2–95.2	87.5	2.5–19.8	7.0 [3](8.8)	[2] 4
Parsnips	72.6–89.2	78.6	2.6– 8.3	3.7	[2] 7
Peas (green)	56.7–84.1	74.3	1.3– 5.3	2.9	5
Potatoes [4]	66.0–85.2	77.8	1.9– 5.8	3.5 [3](4.0)	[2,4] 7
Do [5]	———	———	———	———	[2,5] 8
Rutabagas	86.1–91.8	89.1	6.2–11.2	8.2	[2] 5
Spinach	89.0–95.0	92.7	8.1–19.0	12.7	4
Sweet potatoes	58.5–82.7	68.5	1.4– 4.8	2.2 [3](3.0)	[2] 7
Tomatoes	90.6–96.7	94.1	9.6–29.3	15.9	5

[1] Calculated from data of Chatfield and Adams.
[2] According to Government procurement specifications.
[3] Data in parentheses obtained at the Western Regional Research Laboratory, Albany, Calif.
[4] Cut.
[5] Riced.

The drying ratio and its converse, the drying yield, are shown for 18 vegetables in the following table. These ratios and yields have been calculated from the changes in the moisture content during the drying step. The drying ratio is the ratio of the weight of the material entering the dehydrator to the weight as it leaves the dehydrator commercially dry. The drying yield is the converse of the drying ratio and is expressed in percentage. These ratios are useful to designers of dehydrators, to operators, and also to prospective operators who wish to compare yields of product from various types of raw materials, since it can usually be assumed that the moisture content of the blanched, prepared material is approximately the same as that of the raw material.

DRYING RATIOS AND DRYING YIELDS FOR SPECIFIC VEGETABLES [1]

Vegetable	*Drying Ratio* [2] Range	Average		*Drying Yield* [3] Range Per Cent	Average Per Cent	
Beans (lima)	2.3– 3.3	2.8		29.6–43.2	35.2	
Beans (snap)	4.4–15.8	8.5		6.3–22.3	11.6	
Beets	5.3–16.1	7.6		6.2–18.6	13.0	
Cabbage	8.2–18.4	12.6		5.4–12.0	7.9	
Carrots	5.6–10.6	8.0		9.3–17.7	12.4	
Celery	9.5–20.0	15.2		5.0– 9.8	6.5	
Chard (Swiss)	9.5–13.5	10.6	[4](11.3)	7.3–10.5	9.3	[4](8.8)
Corn (sweet)	2.4– 6.8	3.6		14.6–40.7	27.4	
Kale	5.1– 9.7	7.1		10.2–19.6	13.9	
Mustard greens	7.2–22.3	12.3	[4](12.6)	4.4–13.8	8.1	[4](7.9)
Onions	3.2–20.0	7.6	[4](9.4)	5.0–31.0	13.0	[4](10.6)
Parsnips	3.4– 8.7	4.4		11.3–28.8	22.5	
Peas (green)	2.1– 5.9	3.6		16.7–45.5	27.0	
Potatoes	2.7– 6.3	4.2	[4](4.7)	15.7–36.0	23.6	[4](21.3)
Rutabagas	6.8–11.5	8.7		8.6–14.6	11.4	
Spinach	8.7–19.2	13.1		5.2–11.4	7.6	
Sweet potatoes	2.2– 5.3	2.9	[4](3.7)	18.6–44.6	33.8	[4](26.7)
Tomatoes	10.1–28.7	16.1		3.4– 9.4	6.2	

[1] Calculated from data in preceding table.

[2] Drying ratio = weight of prepared, blanched material entering drier divided by weight of dried material leaving drier.

[3] Drying yield = weight of dried material leaving drier divided by weight of prepared, blanched material entering drier x 100.

[4] Data in parentheses obtained at the Western Regional Research Laboratory, Albany, Calif.

The operator is more directly interested in the overall shrinkage ratio, i.e., the weight of unprepared raw product required to yield one weight unit of finished product which meets specifications. This may also be expressed as the reversed ratio, usually as a percentage, and is then known as the overall yield. The overall shrinkage ratio is always substantially higher than the drying ratio, and the overall yield lower than the drying yield, because all weight losses incurred at various steps of the process, such as culling, washing, peeling, trimming and inspecting, must be discounted. Estimates of such losses are presented in the discussion of certain vegetables but it must be remembered that these losses vary widely.

Suggested tray loadings are shown in the following table for various systems of air flow. These loadings cannot be regarded as specific recommendations; instead they are to be regarded as suitable for trial. Experience with various materials will indicate necessary modifications. Loading for cross circulation and through circulation of air are listed separately.

DEHYDRATOR-TRAY LOADINGS SUGGESTED FOR SPECIFIC VEGETABLES AND DIFFERENT SYSTEMS OF AIR FLOW

Vegetable	*Type of Piece*	*Cross Circulation of Air Per Square Foot in Pounds*	*Through Circulation of Air Per Square Foot in Pounds*
Beans (lima)	Whole	1:5	3–4
Beans (snap)	Cut	1.5	6–12
Beets	Cubes	1.3–1.8	6–12
	Slices	1.3–1.5	2–3
	Str'ps	1.5	6–12
Cabbage	Shreds	.75–1.3	6–8
Carrots	Cubes	1.3–1.8	6–12
	Sl'ces	1.0–1.3	2–3
	Strips	1.2–1.5	6–12
Celery	Slices	1.0–1.5	6–12
Chard (Swiss)	Trimmed leaf	.75–1.3	3–6
Corn (sweet)	Whole kernel	1.5–1.8	3–4
Kale	Same as for Swiss chard	——	——
Mustard greens	..do	——	——
Onions	Slices, shreds	1.0–1.3	6–8
Parsnips	Same as for beets	——	——
Peas (green)	Same as for lima beans	——	——
Potatoes	Cubes	1.3–1.8	6–12
	Slices	1.0–1.3	2–3
	Strips	1.2–1.5	6–12
	Riced	1.0	——
Rutabagas	Same as for carrots	——	——
Spinach	Same as for Swiss chard	——	——
Sweet potatoes	Cubes	1.3–1.8	6–12
	Slices	1.0–1.3	2–3
	Strips	1.5	6–12
	Riced	1.0	2–3
Tomatoes	Slices	1.3–1.8	2–3

When a bin drier is used for finishing the dehydration of vegetables, the most suitable depth of loading will vary from 2 to 6 feet and must be determined by trial. Loading depths of not over 3 feet are suggested for cabbage, sweet corn and tomatoes. The air entering the bin finishing drier should have a dry bulb temperature of 120 to 130° F and a relative humidity of 10% or less.

Variations between varieties and within a single variety due to maturity, cultural conditions or storage conditions make it necessary to determine safe operating temperatures by trial. The general principle to be followed is that the finishing temperature shall be carried as high as possible without damage to the product. To serve as a guide the temperature conditions for different systems of dehydration are shown in the following table for various commonly used types of drying equipment. Types of equipment and their operation are discussed in previous sections of this manual. For some of the vegetables, information is included on storage of the dried product after it has been packaged.

The following table contains the weights of certain dehydrated vegetables that can be packaged in standard containers.

WEIGHTS OF SPECIFIC DEHYDRATED VEGETABLES REQUIRED TO FILL A CASE CONSISTING OF TWO FIVE-GALLON CANS [1]

Vegetable	*Form*	*Net Weight Per Case of 2 Five-Gallon Cans in Pounds*
Beets	Slices	20
Do	Cubes	34
Cabbage	Shreds	10–14
Carrots	Cubes	35
Do	Slices	25–26
Celery	Ground	40–50
Onions	Slices	18–26
Do	Powder	40–50
Potatoes	Cubes	30–31
Do	Slices	18
Do	Strips	20–30
Do	Riced	34
Rutabagas	Cubes	30
Do	Slices	25
Sweet potatoes	do	24
Do	Riced	35

[1] Five-gallon cartons are cased individually, and the weight per carton is half the weight per case of two cans. Standard loadings have not been established.

In preliminary experiments, tests on the dried material aid in selecting the best operating conditions. The object of rehydration tests is to find the best conditions for preparing the product, and the object of quality tests is to place an estimate on the acceptability or desirability of the reconstituted product. In ex-

TEMPERATURE CONDITIONS SUGGESTED IN THE DEHYDRATION OF SPECIFIC VEGETABLES IN THREE TYPES OF DRIERS

[Cool End and Wet Bulb Temperatures Should Not Exceed the Temperatures Indicated]

	Truck and Tunnel Driers				*Conveyor-Type Drier Through Circulation*						*Cabinet Drier*			
	Parallel-Flow Predrying Tunnel also Primary Section of Center-Exhaust Tunnel			Counterflow Tunnel Used With or Without Predrier also Secondary Section of Center-Exhaust Tunnel	First Section		Second Section		Finishing End		Starting Temperature		Finishing Temperature	
Vegetable	Hot End ° F	Cool End ° F	Wet Bulb [1] ° F	Hot End [2] ° F	Dry Bulb ° F	Wet Bulb ° F	Dry Bulb ° F	Wet Bulb ° F	Dry Bulb ° F	Wet Bulb ° F	Dry Bulb ° F	Wet Bulb ° F	Dry Bulb ° F	Wet Bulb ° F
Beans (lima)	180	160	110	150	180	110	160	100	150	90	180	110	150	90
Beans (snap)	180	160	110	150	180	110	160	100	150	90	180	110	150	90
Beets	200	160	120	150	200	120	160	100	150	90	200	120	150	90
Cabbage	180	150	110	135	180	110	150	95	135–145	85–90	165	100	135	80
Carrots	200	170	120	150–160	200	120	170	105	155–160	95–100	200	120	155	90
Celery	180	160	110	125–135	180	110	160	100	125–135	90	165	100	125	80
Chard (Swiss)	200	180	120	170	200	120	180	110	170	100	200	120	170	100
Corn (sweet)	180	165	110	165	180	110	170	105	165	95	180	110	165	95
Kale	200	180	120	170	200	120	180	110	170	100	200	120	170	100
Mustard greens	200	180	120	170	200	120	180	110	170	100	200	120	170	100
Onions	165	145	100	[3] 130–140	165	100	145	95	130–140	85–90	165	100	130	80
Parsnips	200	170	120	155–160	200	120	170	105	155–160	95–100	200	120	160	95
Peas (green)	180	160	110	150	180	110	160	100	150	90	180	110	150	90
Potatoes	200–220	160	120	145–155	180–200	120	160	100	150	90	200	120	150	90–95
Rutabagas	200	170	120	150–155	200	120	170	105	150–155	95–100	200	120	150	90
Spinach	200	180	120	170	200	120	180	110	170	100	200	120	170	100
Sweet potatoes	200–220	170	120	160	180–200	120	170	105	160	95	200	120	160	95
Tomatoes	200	165	120	150	200	120	165	105	150	90	200	120	150	90

[1] The wet bulb depression at the cool end should be at least 30° F.

[2] For wet bulb temperatures see column 10 of figures. The wet bulb depression at the cool end should be at least 25° F.

[3] Varies with variety.

perimental work in which types, varieties, extent of blanching, and any of the other numerous factors that vary, or can be varied, are being tested, the rehydration and quality tests must supply the final measure of success or failure. In the discussion of specific vegetables which follows, the time of either soaking or boiling, or of both soaking and boiling, and the approximate drained weight of rehydrated product that have yielded the most desirable final product are frequently presented.

Lima Beans

Most of the commercial production of lima beans is located along the Atlantic seaboard. They require a growing season of about 4 months and fairly high temperature, and are planted later than snap beans in the northern states. In 1942 the average yield per acre in states that produce green lima beans was about ½ ton on the shelled and a ton on the unshelled basis. The dwarf or bush type is most commonly used in commercial canning and freezing, and for dehydration this type is probably better than the pole type because it can be harvested mechanically. In the north, Henderson, Fordhook, and Burpee are the common bush lima beans. Siebert, King of the Garden, and Large Green Seeded are the popular pole varieties. In the south, Henderson bush lima and the Sieva or Carolina pole lima are commonly grown.

Ordinarily, green lima beans are vined by machine in the harvest fields. After they are vined they are mixed with cracked or crushed ice and packed in lug boxes. The safest method of handling is to process them as soon as possible. Delay should not exceed 8 hours. The shelled beans should not be held in water; this treatment may cause undesirable odors and flavors in the product.

The operations required in the preparation of lima beans are cleaning, sorting out of over-mature white beans, washing and blanching. Green lima beans are blanched in live steam for at least 5 minutes. The large-seeded types, such as Fordhook, may require 1 or 2 minutes longer. The approximate blancher-loading and blanching time are shown in the preceding tables. If there is any delay between blanching and dehydration, the product should be cooled in water at 65 to 70° F. The delay period between blanching and dehydration should not exceed 1 hour. Information on shelled lima beans and on dehydrated limas can be found in the tables.

Dehydrated green lima beans present no unusual problem so far as packaging and storage of finished product are concerned. In rehydrating green lima beans it is recommended that 8 parts of water by weight be added to 1 part of the dried material and boiled gently for 30 minutes. At the end of 30 minutes the beans should be drained through an eight-mesh strainer and weighed. The weight of drained beans should be approximately 3 to 4 times the weight of the dried vegetable. The odor, flavor and color should be those of freshly cooked lima beans. The texture should be soft and tender.

Snap Beans

Snap beans are an important vegetable crop and rank fourth among those that are canned commercially. Nearly half of the 462,000 ton commercial crop produced in 1942 was processed. The leading states in production for processing are New York, Oregon, Maryland, Arkansas, Wisconsin and Michigan. Florida leads in production for the fresh-vegetable market. An early fall crop is harvested in Florida and harvesting continues with the early- and late-spring crops in the southern states. Summer crops are harvested in the states farther north. In California the early crops are harvested during the months of April, May, June and July, and the late crops during the months of August and September. The average yield in Oregon, where the entire commercial production was used for processing, was 5.3 tons per acre in 1942, which was higher than that of any other state in that year. The average yield in Pennsylvania the same year was 2.85 tons per acre for snap beans grown for the fresh-vegetable market and 2.10 tons per acre for snap beans grown for processing.

Only about one-seventh of the commercially grown snap beans are produced on pole-type vines; the remainder are produced on the dwarf, or bush, vines. Almost 90% of the total are green beans; the remainder are yellow or wax beans. Recommended varieties of green bush beans for dehydration are Asgrow Stringless Green Pod and Dwarf Horticultural; Bountiful and Stringless Black Valentine are acceptable but inferior to the two just named. A recommended pole variety is Lazy Wife. Stringless Kidney Wax is a recommended wax bean of the bush type. Varieties of snap beans are numerous, and it is probable that other high-yielding, excellent-quality varieties are suitable for dehydration.

Harvested beans should be handled in well-ventilated packages, such as hampers, with ample circulation of air between them. There should be as little delay as possible between harvest and processing. Beans should be free from surface moisture if they are to be held more than 1 day before processing; they cannot be iced because ice might cause water-soaked areas that resemble freezing injury.

Quality is important. Snap beans should be picked when the seeds make up 10 to 20% of the weight of the entire pod. They should be graded closely for stage of maturity, in order to insure a uniform dehydrated product. After they are graded and sorted, the beans are snipped mechanically and then washed thoroughly. Snap beans should be blanched immediately after they are washed. To stop the blanching action, it is advisable to spray the blanched beans with cold water and thus reduce the temperature to 65—70° F. This is especially important in case of delay following blanching, which should never exceed 1 hour.

Information on snap beans can be found in the preceding tables in this section.

Dehydrated snap beans present no unusual problems so far as packaging and storage of the finished product are concerned. In rehydrating snap beans it is recommended that 8 parts by weight of water be added to 1 part of dried product, which should soak for 4 hours. After the period of soaking, the beans should be boiled slowly for 30 minutes and then drained through an eight-mesh strainer and weighed. The weight of drained beans should be approximately 2 to 3 times the weight of the dried vegetable. The product should be tender, green in color and similar in flavor to cut fresh beans.

Beets

Beets are a widely grown, hardy, cool-season crop. High temperature during growth adversely affects their color and texture. Since beets of deep uniform color are most suitable for dehydration, the fall crop is probably more suitable than the crop grown in the spring. The fall crop is generally used for canning, and beets for this purpose are raised in the cool districts of New York, Wisconsin, Oregon and Michigan. About two-thirds of the national crop is processed. Beets for the early fresh-vegetable market are grown extensively in Texas, and for later market in New Jersey and Pennsylvania. The varieties that are solid, or nearly solid, dark-red or reddish-purple color, such as Detroit Dark Red, Morse Detroit and Ohio Canner, are considered best for both canning and dehydration. When such beets are dried, the color deepens to a purplish-black and the lighter zones become nearly indistinguishable. Varieties with pronounced light and dark zones in the roots are not considered desirable.

Beets should be harvested before they begin to develop woodiness. They should be topped before they are stored and should be sound and free from injury. They should be stored in an atmosphere of high humidity, 95 to 98% and at a temperature of 32° F. Late beets of high quality will keep from 1 to 3 months under these conditions. Prior to storage the beets should be washed, and storage time should be as short as possible. Ventilated containers, such as crates, are preferable for storage.

Grading to various sizes will effect considerable saving in waste during the peeling operation. Peeling losses will run as high as 30% with field-run beets; grading will reduce the loss to as low as 13%.

Beets can be blanched either whole or after they have been cut into slices, cubes, or strips. Blanching time will vary with size of beets or pieces and with characteristics of the particular lot, and must be continued, according to present government specifications, until the peroxidase system is inactivated. After whole beets have been blanched they are peeled, trimmed and then cut. There should be no delay between blanching and dehydration, but if delay is unavoidable the material should be cooled by a water spray.

Information on government specifications for beets can be found in the preceding tables in this section.

In rehydrating diced beets, it is recommended that 15 parts of water by weight be added to 1 part of beets and boiled gently for 30 minutes. Longer boiling may cause deterioration in flavor and color. At the end of this period the diced beets are drained by means of an eight-mesh strainer and weighed. The weight of drained beets should be approximately 5 to 8 times the original weight of the vegetable.

Dehydrated beet slices are treated in the same manner as diced beets; the drained weight will be 4 to 5 times the original weight. The rehydrated product should be soft and tender in texture and mildly sweet. The color should be a uniform deep red and the odor should be that of cooked fresh beets.

Cabbage

Cabbage is widely grown, with heavy concentrations in New York, Texas, Wisconsin, Florida, California and Pennsylvania, in the order named. Of the 1.4 million tons of cabbage produced in 1942, these six states grew approximately four-fifths. The northern states, which produce the late crop, supply nearly all of the cabbage used in sauerkraut. Approximately 15% of the yearly cabbage crop is used for this purpose. The average yield in states that produce late cabbage was 10 tons per acre in 1942, as compared with 7.6 tons for all cabbage-growing states. The harvest in the late states generally starts about June and continues until November. The winter and spring crops in the southern states are harvested from November to May.

The Savoy variety, a green-leaved, loose-headed type, has an attractive appearance and good flavor when dried, and a higher vitamin content than the solid-head, white types, but is not grown extensively in this country because of low yields. Midseason varieties are the late strains of Copenhagen Market and its wilt-resistant strain, Marion Market; All-Seasons and its wilt-resistant strain, Wisconsin All-Seasons; and Glory of

Enkhuizen. All of these are grown for the mid-season market and especially for kraut making. At present it is not definitely known how these varieties compare for drying purposes. Danish Ballhead and Wisconsin Hollander are the most important varieties grown for the late market and for storage, but not for kraut. Varieties of the Danish, domestic and pointed-head types are satisfactory for drying also, but may turn yellow upon dehydration unless special care is taken during blanching and drying. In general, they require less blanching and lower drying temperature than the Savoy varieties.

Cabbage is usually stored at low temperature range, maintained by ventilation with cold outside air. The storage house must be insulated to retard passage of heat through walls and roof, and it must be well ventilated to carry away the large amount of heat and moisture given off by the cabbage. Many of such storage houses can be found in New York, Pennsylvania, Michigan and Wisconsin. The cabbage is stored in bins, usually made with slatted floors and solid partitions, and often several tiers high, one bin above the other. The best method of storage is on slat shelves with the heads one or two layers high, but the expense of this method may be prohibitive. Cold storage is too costly for this crop. The ideal storage conditions are 32° F with 90 to 95% humidity. Late-crop cabbage will keep 3 to 4 months if properly stored. The longest-keeping varieties belong to the Danish Ballhead class. Early cabbage does not keep more than 3 to 6 weeks at 32° F.

Cabbage is usually handled in bulk or crates, the former method being most common. Since the outer leaves are relatively high in color and nutritive value, unnecessary stripping should be avoided. Trimming consists of removal of the outer bruised and discolored leaves, and waste will range approximately from 15 to 37%. After trimming, the core is removed with rotary cutters especially designed for this purpose or by subdividing the head with a knife in such a way as to cut out the core.

The cabbage is next washed by strong sprays of water in order to clean out the dirt and grit. By means of a rotary cutter, it is coarsely shredded into pieces not less than ⅛ inch nor more than ¼ inch wide. Shreds that are too fine will collapse during blanching and if dried in this condition, the material will stick to the loading surface and require a longer drying time.

Prior to blanching, cabbage shreds should be washed with sprays of clean water. Since cut leaves lose ascorbic acid content rapidly, it is necessary that the material be blanched immediately after shredding. If this cannot be done, the cut leaves must be held in a 1% salt solution. Under no condition should the cut cabbage be held for more than 1 hour between cutting and blanching.

The product must be blanched in live steam until the temperature throughout is not less than 190° F, or until the catalase system is destroyed. This may require 2 to 4 minutes or more. An undesirable color will develop in the finished product if it is overblanched. If a delay occurs between blanching and dehydration, it will be necessary to cool the product to 65 or 70° F by means of cold-water sprays; otherwise the color of the finished product may be injured. The blanched product should not be held longer than 1 hour before being dried.

Information on government specifications for cabbage can be found in the preceding tables of this section.

Hot packing of dehydrated cabbage tends to preserve the size of flakes by lessening breakage, because the dried shreds are more pliable when hot. The packaged product requires protection from both moisture and air. The vacuum-chamber-and-pump method of removing air is recommended, and it is preferable that the vacuum chamber be equipped with a sealing head; reduction below 2% of oxygen would make this type of equipment necessary. The reasons for a maximum of 2% of oxygen are urgent. The purpose of air displacement is to save the flavor, color and ascorbic acid by the elimination of oxygen. Analysis should be made at least 12 hours after the cans are filled and sealed.

In commercial practice, removal of air by displacement with carbon dioxide from a cylinder without a meter has produced irregular results, ranging from 1.5 to 8% of oxygen in the gas left in the filled can. Records obtained with an inexpensive iron-case gas meter have shown that 1 cubic foot of carbon dioxide will reduce the oxygen content of gas in the filled can to 1.8 to 2.4% in 10 seconds or 1.2 to 1.6% in 30 seconds; 2 cubic feet will reduce it to 0.8 to 1.0% in 60 seconds.

In general, dehydrated cabbage is one of the less stable of the dehydrated vegetables. In storage tests, Savoy cabbage that was blanched in steam, dried to 3.7% moisture content, and stored in carbon dioxide or nitrogen at 90° F became inedible within 16 weeks. At temperatures of 70 to 80° F, four varieties blanched and dried to below 3% moisture content retained both palatability and ascorbic acid without material loss. Tests with varied moisture content have shown that palatability and ascorbic acid are retained better at low than at higher moisture content.

In rehydrating dried cabbage it is best to subject it to a moderately long soaking, followed by a relatively short boiling or cooking period. When the material is to be used raw as salad or cole slaw, slightly longer soaking may be required. In rehydrating samples for testing, add 1 part by weight of dried cabbage to a consistent weight of water (10 to 15 parts). Soak for 1 hour; then simmer for 10 to 20 minutes in the same water. Drain through an eight-mesh strainer and weigh. The drained weight will be approximately 8 to 12 times the original weight of vegetable. Longer soaking will give a plumper product and slightly greater weight. Longer boiling may cause deterioration in flavor and color.

In preparing directions for use in homes and institutions, the producer should determine the amount of water and the soaking and boiling times necessary to yield a moist cooked product without excess water. A large part of the ascorbic acid will be lost if excess water is discarded. Cabbage is one of the best sources of ascorbic acid in the diet. While much is lost in the dehydration process, tests have shown that freshly dehydrated material contains 300 to 400 milligrams per 100 grams of dry cabbage.

Recently considerable interest has developed in a sulfite treatment in the dehydration of cabbage. Evidence has shown that this treatment is desirable for several reasons.

Carrots

Since carrots require a long growing season, their production is confined to areas with long periods of mild weather. Like potatoes and sweet potatoes, carrots can be stored and are available in the fresh state the entire year; hence they are not canned in large quantities. In 1942 the California production of carrots was almost 40% of the total national commercial crop of 506,000 tons. Texas, New York, and Arizona rank next in production. In most carrot-growing states, the crop is harvested from late spring to late fall. In California, carrots are grown during the entire year. In Arizona, Louisiana and Texas they are grown from late fall to early summer.

In carrots intended for dehydration, a deep uniform color, with as little difference as possible in character and depth of color between cortex and core, is a major requirement. Good to high quality and ability to stand up well under storage are also necessary. Red Core Chantenay combines desirable field and storage characteristics with deep orange color throughout the root, and is the variety chiefly used by canners and dehydrators in the east. Nantes also has rather uniformly deep orange flesh and high quality. Three varieties having fairly deep orange flesh with somewhat ligher yellow cores are Imperator, Danvers, and Long Orange. Imperator is grown chiefly in the west and southwest for shipment, and has been reported as satisfactory for drying in California.

Stage of maturity is highly important. Depth of color increases with age, but toughness and woodiness also appear and become pronounced in old material, which is hard and woody when cooked. The stage at which carrots are ordinarily harvested for storage represents the upper limit of maturity for dehydration.

Carrots are stored in fairly large quantities during the winter in those sections where the storage temperature can be held sufficiently low. They are sometimes held in cold storage, although prices do not usually justify this kind of treatment. Very light freezing may cause practically no injury, but carrots should be protected from severe freezing and are best stored at a temperature of 32° F. They are subject to wilting if the relative humidity is not 95% or higher; for this reason they are more easily kept in well-ventilated cellar or bank storage than above ground.

Before being placed in storage, carrots should be topped and all misshapen or injured roots sorted out. They should be kept in slat crates or ventilated barrels, and provision should be made for air circulation between the containers. Under good conditions they should keep 4 to 5 months. Washing is essential. Usually the roots are covered with mud, which increases loss due to mold and decay.

Grading to size reduces peeling waste. After grading, peeling can be done by passing the carrots through a lye, an abrasive or a flame peeler. Waste may run as high as 24% for abrasive peeling. Trimming is required to remove green ends and spots along the sides. Aggregate losses from peeling and trimming will vary between 20 and 30%. Dehydrated carrots are either sliced, cubed or stripped. Information on government specifications for carrots can be found in the preceding tables.

Dehydrated carrots require the same protection as cabbage; that is, they must be packaged in cans in which an inert gas has been substituted for air. The method is described under cabbage.

Dried carrots are intermediate in stability among dried vegetables.

When stored at 90° F in contact with carbon dioxide or nitrogen, samples of blanched dehydrated carrots containing 6 to 7% moisture were edible but of poor quality after 32 weeks. After 64 weeks, they had deteriorated further but were still edible. The carotene content of dried carrots is adversely affected by oxygen in storage but is not affected by variation in moisture content.

In rehydrating dried carrots it is recommended that 8 parts by weight of water be added to 1 part of dried vegetable and boiled gently for 10 minutes if diced or stripped, and for 20 minutes if sliced. After boiling, the carrots are drained carefully through an eight-mesh strainer and weighed. The rehydrated weight should be approximately 4 to 5 times the original weight for sliced, 5 to 6 times for diced, and 6 to 7 times for stripped carrots. Longer soaking will give a plumper product and slightly greater drained weight, but longer boiling may cause deterioration in taste, flavor and color.

Reconstituted carrots should be mildly sweet, free from sour or bitter taste, mild in fresh carrot flavor and soft and tender in texture. The color should be a uniform, brilliant, yellowish-red. The odor should be that of cooked fresh carrots, free from the so-called violet scent.

Celery

Celery is a cool-season crop, adapted to winter culture in the lower south and to growing either as an early-spring or late-fall crop in the central states. Farther north, in certain favorable locations, it can be grown throughout the summer. The localized distribution of this crop is due mostly to its soil requirements. Celery is grown on peat or muck, except in the irrigated areas in the west. The labor requirements for growing this crop are comparatively high. In California and central Florida, celery is produced in important quantities in the fall, winter and spring. Midseason and late-summer crops are grown mainly on the muck soils of Michigan, New York, New Jersey and Pennsylvania, with smaller acreages on the irrigated lands of the Pacific northwest. California, Florida, Michigan and New York grew approximately 85% of the 1942 commercial celery crop in the United States. The late-spring celery in California yielded about 25 tons to the acre, as compared with an average of 15 tons for other celery crops of the same state during that year. The average yield for all states was 14 tons per acre.

Golden Self-Blanching and Golden Plume are well adapted to early production. For storage, Easy Bleaching Green, Utah, and Giant Pascal are best. Other excellent varieties are Crispheart, Emperor, and Winter Queen. Much of the late celery grown in the northern states, notably in Michigan and New York, is placed in storage where it can be preserved for a period of from 2 to 4 months. While in storage, celery must be kept as cool as possible without freezing. Additional requirements are adequate air circulation and high relative humidity (90 to 95%).

Steps in the preparation of celery for dehydration are fairly simple. The discolored parts and those affected by plant diseases are first trimmed out by hand; then the stalks are given a thorough washing. For soup mixtures, celery is finely shredded, both leaves and stalk, and spread directly on the drying trays. For other purposes, the leaves, which dry more rapidly, are trimmed from the thick, fleshy stalks by hand and dried separately, either whole or shredded. The stalks are cut by a rotary slicer into transverse slices about ½ to ¾ inch thick.

The method of blanching celery for dehydration is the same as for cabbage, with the exception that celery will require about 2 minutes. As is true of cabbage, present government purchase specifications require inactivation of the catalase system. Information on government specifications for celery can be found in the preceding tables.

Dehydrated celery is commonly ground to make celery salt and is also used as flakes in some soup mixes. The commercial preparations include celery leaves, stalks and roots dried separately. Dried celery can be ground to a powder in a hammer mill, and if it is to be ground a shaker screen will be needed. Because celery is comparatively low in sugar content, the powder is not nearly so hygroscopic as are onion and garlic powders. For this reason, celery salt is not likely to cake during grinding or use. Like other seasonings, dehydrated celery is packed in cans or cartons that protect it against moisture absorption and loss of aroma. It is not neccessary to pack it under inert gas. For wholesale purposes in domestic trade, cartons lined with two sheets of waxed paper are suggested. Ground celery is packed at the rate of 20 to 25 pounds to a five-gallon unit.

Rehydration tests are not feasible for leaf powder but may be desirable for the stalks. Plant operators who wish to make tests should regard pungency as a highly valuable characteristic, and texture as secondary in importance. Minimum amounts of water should be used, and the material should be soaked rather than boiled.

Chard

Swiss chard is a type of beet that has been developed for its tops rather than its roots and resembles beets in cultivation and distribution. Chard should be handled like other greens. The whole leaves should be cooled, preferably with crushed ice, and stored at low temperature (34 to 35° F) and high relative humidity (90 to 95%). A very short storage or holding period is advisable.

In the packing plant the material should be sorted and washed. All damaged and discolored leaves should be removed and the stems cut off. Trimming and sorting losses are usually about 50%. Blanching should continue until the catalase system is inactivated. Uniformity of loading in the blancher is important. If delay occurs following blanching, the product should be cooled immediately to approximately 65 to 70° F by cold-water spray or immersion. As is true of other greens, the leaves are preferably dried whole. Information on chard can be found in preceding tables.

Dried greens should be carefully packaged in order to minimize breakage of the brittle product. In rehydrating it is recommended that 1 part by weight of dried chard be added to 6 parts of water, soaked for 2 hours, and boiled gently for 5 minutes. At the end of the period of boiling the product should be drained by means of an eight-mesh strainer and weighed. The weight of rehydrated product should be approximately 3 to 4 times the weight of the original dried product. In quality, the rehydrated chard should resemble cooked fresh chard. It should have a pleasant flavor and should be free from unattractive odor.

Sweet Corn

The principal regions where sweet corn is produced commercially are the Great Lakes states, the Middle Atlantic states, and the New England states. It is one of the major canning crops; of the 1.4 million tons produced commercially in 1942, more than 90% was grown for processing.

Eighty per cent or more of the sweet corn used for canning

and freezing is of the hybrid varieties, most of it yellow. In total volume the white hybrids are small, but in certain districts, notably Maryland and Ohio, they are important. Of the open-pollinated varieties, a few have good quality and are adapted to widely separated regions. Stage of maturity is important, the best stage being that which an experienced canner would select for canning in whole-grain style.

Sweet corn loses sweetness rapidly after harvest, but if it is cooled quickly the rapidity of the change can be reduced. Submerging the corn in tanks of ice water or passing it through a vegetable cooler flooded with ice water is an excellent cooling treatment. Corn should not be handled in bulk because of its tendency to heat, but should be put in baskets, small bags or crates, and stocked so as to permit air circulation. The best conditions for holding the corn while awaiting processing are a temperature of 31 to 32° F and a relative humidity of 85 to 90%. Deterioration takes place so rapidly that sweet corn should not be held longer than is absolutely necessary, and certainly not more than a few hours.

Machines husk from 1½ to 2 tons of corn per hour. After the ears have been husked, they are trimmed and then washed with sprays of clean water. For best results from the standpoint of preservation of flavor and prevention of excessive loss of water-soluble nutritive factors, sweet corn should be blanched on the ear in live steam until the peroxidase system has been destroyed, as shown by a test applied to a water extract of the blanched corn after the skin and cob tissues have been filtered out. After the ears have been blanched, they should be cooled to 65 or 70° F by immersion in clean, cold, running water. The kernels are then cut from the cob. Continuous cutters are used, and, as the corn comes from the cutter, it can be spread immediately on the trays. The cut corn should not be washed before drying, as this will result in loss of sugar and other soluble constituents. Chaff and cob tissue can be removed by a fanning mill after the product is dried.

Information on dried sweet corn can be found in the preceding tables.

Dried sweet corn presents no unusual problem so far as

packaging and storage of the finished product are concerned. In dehydrating sweet corn, it is recommended that 4 parts of water be added to 1 part by weight of the dried product. Allow the material to soak for 2 hours, and then boil gently for 10 minutes. The drained weight should be approximately 2 to 3 times the weight of the dried vegetable. The product should be mild in corn flavor and soft and tender in texture. The color should be uniformly white or yellow, depending upon the original color. The odor should be that of freshly cooked corn.

Among the best varieties and hybrids for drying are: Aristogold Bantam, Evergreen, Bantam Evergreen, Golden Cross Bantam, Narrow Grain Evergreen, Country Gentleman and Silver Cross Evergreen. Among other varieties and hybrids giving a good product are: Golden Bantam, Golden Hybrid 2439, Ioana, Marcross, Spancross, Top Cross Whipple Yellow, Tristate Hybrid, Country Gentleman 19 x 24 and Narrow Grain Hybrid 26 x 15.

Kale

Kale is a cool-weather green vegetable grown as a fall crop, chiefly along the middle Atlantic coast. It is dried in much the same manner as other greens are dried and presents no special problems. (See discussion on Swiss chard and spinach.) Losses that result from trimming and sorting average about 55 to 60% of the original weight. Information on blanching and dehydration are included in the preceding tables. Methods of rehydration and testing for quality are similar to those used with other greens.

Mustard Greens

The basal leaves of the mustard plant are used for greens and make an acceptable dried product. The plant requires but a short time to reach the proper stage for use. Preparation for dehydration is similar to that for other greens. (See discussion on chard, kale and spinach.) The weight of material lost through trimming and sorting is less than for other greens and is usually about 30 to 40%. Information on blanching and dehydration is contained in the preceding tables. Dried mustard greens can be rehydrated and tested for quality by the method used for other greens.

Onions

The leading onion-producing area is in New York, where the 1942 commercial production was 163,000 tons. The national average yield per acre in 1942 was 7 tons, and the yield was several tons higher in areas of heavy commercial production. During the early stages of growth the temperature should be fairly cool and the moisture supply ample. In the early-producing states, the onion harvest begins in April and ends in June. The intermediate states harvest this crop during the summer months, and the late states—eastern, central and western—from August to October. The marketing period for late onions lasts 6 or more months after the final harvest, and extends to the late spring of the following year.

In New York the most popular varieties of onions are Early Yellow Globe and Ebenezer; in Colorado, Mountain Danvers; in Texas, Yellow Bermuda. The most common fault found with dehydrated onions is their lack of pungency. The Ebenezer, Barnett, Southport Yellow Globe, White Portugal, Red Creole, White Creole and Yellow Danvers Flat are very strong onions and make excellent dried products. The White Portugal makes an excellent dried product, but the yield of onions per acre is low. Sweet Spanish, the various Bermudas and similar types of onions are too mild for dehydration. Some of the more pungent varieties, for example, Australian Brown (Oregon Brown or Buckskin), may yield a bitter-flavored dried product under certain drying conditions, but under other conditions they make good products. Australian Brown has a high percentage of dry matter, is an excellent keeper, and could be used to extend the season of operation, but it is difficult to peel. There is considerable variation in percentage of dry matter among varieties of onions; in this respect the mild varieties are low, and the pungent varieties high.

In the northern states, onions are held in common storage during the winter months. Part of the crop in this section is cold-stored, because after March there is likelihood of sprouting. The Ebenezer, Australian Brown, Creole and various Globe varieties are the best keepers. The most important requirement in storage is controlled relative humidity, ranging from 70 to 75%. Higher humidities may cause root growth and decay in the form of neck rot. A temperature of 32° F is sufficiently low to keep onions dormant and reasonably free from decay, provided they are in sound condition and well cured when stored. Good ventilation should be provided. If the onions are packed in sacks, they should be set off the floor 2 to 4 inches with space provided around each sack. Sacks are usually stacked in pairs laid crosswise, five or six stacks high, and sometimes placed on a framework of shelves to provide ample ventilation. In some districts onions are stored in slatted crates. Good storage stock can be kept 6 to 8 months.

Onions are usually received in sacks containing up to 100 pounds, and to prevent overheating are stacked until used. Handling is much more simple for this product than for most vegetables, because blanching is not required. The onions are thoroughly washed and cleaned to remove soil and foreign material; then the outer discolored layers are removed, after which the root base and top are cut off. They should be cut mechanically into slices or shreds ranging from ⅛ to ¼ inch thick. In pilot-plant operations, peeling and trimming losses have varied from 6 to 15%.

Government purchase specifications for dehydrated onions do not require blanching of the raw material. There is evidence that storage life of the dried product is prolonged by blanching, but if onions are blanched they may fail to meet existing specifications, since the appearance of the blanched product is less attractive than that of the unblanched. Prolonged blanching may cause a marked loss of pungency. The cut material should not be held longer than 1 hour before blanching or dehydration. Information on government specifications for dried onions is given in the preceding tables.

Government specifications limit the moisture content of onion flakes and slices to 4% and, in practice, flakes are packed at 3.5% moisture content. Onion powder should have an even lower moisture content, in order to prevent lumping. Dried onions are fragile; this fact should be kept in mind during handling in order to keep the fines low. Onion powder and flakes will absorb moisture in air having a relative humidity about 25% and in the room where onions are ground the atmosphere must be air-conditioned to 30% or less for the best results. For domestic shipments cartons having double waxed-paper liners, or cartons made of laminated material, including a layer of lead foil, are also used.

Dehydrated onions are more sensitive to heat than are potatoes, which occupy a middle position among dried vegetables in regard to stability. Experience has shown that too high temperature or too long exposure to a lower temperature causes a brownish color.

Dried onions in both the powdered and the rehydrated flake or slice form are used largely as seasoning agents. Rehydrated flakes or slices can be used in salads or as seasoning in soups or other vegetable mixtures and in other recipes that require raw onions, or they can be stewed and seasoned as a vegetable dish. Soaking at room temperature or in a refrigerator in a minimum amount of water (enough to insure wetting of all pieces) is the most suitable method of rehydration, and a minimum of 2 hours is required. Unblanched onion pieces will take up enough water to make them turgid and crisp in that length of time. There is considerable transfer of pungent flavor to the water during soaking, and, in the producer's directions for use, the suggestion should be made that unabsorbed water be used in soup stock or meat gravy.

Rate and completeness of rehydration can be determined by soaking the product in 7 to 8 times its weight of water, draining, and weighing the rehydrated sample. After soaking 2 hours, properly dehydrated onion will have reached 5 to 6 times its original weight. Rehydrated onions should be crisp and succulent and should have a mildly sweet or sweet-bitter taste and a rich pungent odor. The color should be yellowish white to light yellow. Red onions should have yellowish-white centers with bright-red outer rings. Dehydrated onions should be cooked only after rehydration, and the time of boiling should be as short as possible. White varieties are recommended for the preparation of dishes such as buttered or creamed onions.

Parsnips

Parsnips are of minor importance as a commercial vegetable crop; the comparatively small acreage grown apparently reflects a lack of general popularity. The relatively poor quality of dehydrated parsnips, and the fact that parsnips will keep well in the ground if protected from freezing makes dehydration impractical in most cases.

In many parts of the south, parsnips are planted so that they will mature during the early summer. As they require warm weather and warm soil at planting time, they are usually planted late as compared with other vegetable crops, in other parts of the country. They may be left in the ground until used, and can be stored in cellars or pits at high relative humidity, for 2 to 4 months. When dehydrated, they are prepared and processed in much the same manner as carrots. They present no unusual problems so far as packaging is concerned.

In making the rehydrating test on dried parsnips, it is recommended that 9 parts of water by weight be added to 1 part of dehydrated slices, brought to a boil rapidly, and simmered for 5 minutes. Then the sample should be drained and weighed. Other samples should be simmered for longer periods. It is probable that the optimum quality will be attained in approximately 10 minutes, and that the drained weight will be 5 times the original weight of dried sample. The high-quality sample will be sweet, pungent, natural in flavor, and will have a light to moderate greenish-yellow color.

Green Peas

Green peas are important in both the canning and the freezing industries and are grown and processed in large quantities in the Great Lakes region, the Pacific northwest, New York, Utah and elsewhere. Wisconsin leads among the pea-growing states. Peas are planted early because they are essentially a cool-weather crop. Varieties differ in time required for maturity. Two types are commonly distinguished, canning and market garden.

Alaska is a popular canning pea. Other prominent canning varieties are Green Admiral, Perfection and Surprise. Thomas Laxton, Stratagem and Tall Telephone, which are commonly used for freezing, and Tall Alderman are market-garden varieties. In general, the market-garden varieties are best suited for dehydration. An important matter is the stage of maturity; the peas should be mature, but still green and tender.

Peas are commonly vined mechanically in the field and packed in lug boxes, in which they are conveyed to the plant. They should be cooled as quickly as possible if they are to be held for more than a few hours; the addition of crushed ice to the lug boxes is an effective method. The safest method is to process vined peas as soon after harvest as possible. They should not be held in water. If it is necessary to hold unshelled peas before processing, keep them in air having a temperature of 32° F.

Information on green peas is given in the preceding tables. Dried peas present no unusual problem in packaging.

In rehydrating dried peas, add 1 part by weight of dried material to 6 parts of cold water and boil slowly for 10 minutes. Then drain and weigh. The weight of rehydrated peas should be about 6 or 7 times the original weight. The rehydrated peas should be mildly sweet, and their flavor and odor should resemble those of the cooked fresh product. They should be soft and tender in texture and the color should be uniformly bright yellowish-green.

Potatoes

Of all the vegetables that are dehydrated, potatoes are the most important. The demand for dehydrated potatoes exceeds that for any other vegetable; in fact, the total present demand for potatoes is almost equal to demands for all other dehydrated vegetables combined.

The national commercial production of potatoes in 1942 amounted to 11 million tons. Potatoes are grown in all states. Maine, with a production of 1.3 million tons, or 12% of the national total, outranked all other states. Idaho with 918,000 tons and New York with 822,000 tons were next in importance.

Each of five other states—California, Minnesota, North Dakota, Pennsylvania and Colorado—produced over 500,000 tons. The entire production of these eight states was more than half the total national crop.

Potatoes are grown during all seasons in one part of the country or another. In some of the Southern states, the harvest starts early in the spring, while in southern Florida and in the lower Rio Grande Valley potatoes are harvested toward the end of the year. Early potatoes are a relatively unimportant part of the total production. The season of availability for the intermediate and late crops extends from the beginning of harvest in late summer to at least the late spring of the following year. Yields per acre are influenced to a large extent by climate and the use of irrigation. They vary widely from year to year and among the states. The 1930–39 average yield per acre for all states was 3.4 tons, compared with 3.9 tons in 1941.

For dehydration many operators prefer the type that becomes white and mealy with cooking. Experimental testing of potatoes grown in any given area, prior to the establishment of a new dehydration plant, is advisable as a means of determining (1) the operating conditions that make the best dried product, (2) the acceptability of this product for use, and (3) the yield of dried product obtainable from the raw material.

The variety of potato most widely grown in the United States is the Irish Cobbler. Because of its remarkable adaptability, it is grown to some extent nearly everywhere. The Triumph is second in importance and acreage, with large concentrations in the Middle West, and Katahdin is the third most important variety. The Green Mountain group is important in Maine and the neighboring states. Production of the Russet Burbank group of varieties is concentrated entirely in the Northwest. In California, the White Rose is the leading variety. Rural Russet and Rural New Yorker No. 2 are grown extensively in certain states in the region of the Great Lakes. The conditions under which these varieties are grown affect their suitability for dehydration.

A variety of potato, to be suitable for dehydrating, must be grown in a district and under conditions to which it is suited, and where it will give optimum yields. Russet Burbank, when grown in Idaho, for example, is excellent, but the same variety grown in Maryland, where it is not adapted, makes an inferior product. In the same way, potatoes satisfactory for dehydrating can be grown in a certain district only if varieties adapted to that district are used.

Limited tests have shown the following varieties to be among the best when grown in the states indicated:

Michigan: Rural Russet, best; Irish Cobbler, Green Mountain, Chippewa, Katahdin, and Sebago, very good.

New York: Green Mountain and Pioneer Rural best; Katahdin and Chippewa, very good.

Washington: Russet Burbank and Sebago, best.

Idaho: Russet Burbank, best; Sebago, Triumph, and Katahdin, very good.

Colorado: Triumph and Katahdin, best.

Maine: All the varieties named above produce excellent or very good stock for dehydrating.

Immaturity or other factors that result in high water content definitely impair quality for dehydration. Immature *new* potatoes are not suitable for this purpose.

Only mature tubers, free from disease and bruises, should be stored. A storage temperature of 40° F is low enough to keep mature potatoes dormant 3 to 5 months. At this temperature, however, they may become mildly sweet. If stored at 40° F or lower, the potatoes may show marked yellowing and browning at the center of the piece after being dehydrated. For short-time storage, a temperature of 50 to 60° F results in good texture, color and flavor in the cooked product. Potatoes stored at the lower temperatures should be held at 60 to 65° F for 3 to 4 weeks just prior to dehydration. Under these conditions, the sweet taste will be lost and a satisfactory dried product obtained. The relative humidity recommended for potato storage is 85 to 90%.

Potatoes are commonly stored in pits or large bins of 150 to 1000 bushels in the northern states, but in milder climates they should be stored in small units. Heat insulation and ample ventilation are needed to provide the best conditions for storage. Potatoes should always be stored in the dark because in the light they become green and unfit for food as a result of the development of solanine, a bitter poisonous substance. In most districts potatoes are handled and stored in bags. After delivery to the plant, measures must be taken to make sure that the crop will remain in good condition until used. An even, moderately cool temperature should be provided, in insulated rooms if necessary. The bags should be stacked so as to provide enough aeration to prevent heating. When the weather becomes cold, heat should be provided to prevent freezing. Since potato handling is almost an all-year job, maintenance of cool storage to reduce spoilage is important.

Thorough washing is the first step in the preparation of potatoes. This operation can be carried out by running them through a revolving corrugated-drum or squirrel-cage washer equipped with sprays. After washing, grading to size should be performed if abrasion or radiant-heat peeling is used, since grading speeds up the peeling operation and reduces waste. With lye or brine peeling, grading to size is less important.

Trimming is necessary after the peeling operation to remove the eyes and black spots or unsound or damaged portions that remain after the product has passed through the peeling machines. The total waste from peeling and trimming varies with the grade and size of the tubers, the peeling method used, and the care exercised in peeling and trimming. In commercial plants, this waste is rarely lower than 15%, and may rise to 30 or even 40% if the raw material is inferior, or the operation is conducted carelessly. With abrasive peeling the average over-all loss is about 23% and the range 18 to 27%. With lye, brine, or heat peeling, the losses may be considerably less, even as low as 12%.

The peeled and trimmed potatoes pass to mechanical cutters where they are cut into slices, cubes or strips (julienne style) in accordance with the form desired.

After being cut the potatoes proceed to a washer where the loose starch is removed. This operation is then followed by blanching. If delay between cutting and blanching is unavoidable, the cut material must be held under potable cold water or a clean 2% salt solution. This procedure will protect the cut product from discoloration as the result of enzymatic oxidation. The prepared material should not be held for more than 1 hour before blanching. If the potatoes are to be riced, the peeled and trimmed potatoes can be quartered or sliced and then blanched.

After being blanched, the material should be sprayed again with clean water in order to remove loose starch that might cause the pieces to stick together during dehydration. Delays between blanching and dehydration should be avoided. In any case, the material should not be held longer than 1 hour. Quartered potatoes that are to be riced are exposed to live steam until cooked, usually for 20 to 25 minutes. After this operation and while still hot, they are passed through a mechanical ricer directly to the drying trays.

Information on government specifications for dehydrated potatoes is given in the preceding tables.

Tin cans have been replaced to some extent by five-gallon square cartons as containers for dehydrated potatoes. Dehydrators will require between 250 and 380 five-gallon cans or cartons for every 10 tons of raw untrimmed potatoes that are to be dehydrated. Dehydrated potatoes are moderately sensitive to heat, being less susceptible than carrots and more so than sweet potatoes.

The color and texture of rehydrated potatoes are influenced by the variety, growing environment, maturity, predrying storage, predrying preparation procedures, time and method of blanching, drying conditions, and method of rehydration. Under-blanched dehydrated potatoes will rehydrate slowly and incompletely. If over-blanched, the pieces may disintegrate in part or entirely when rehydrated at boiling temperatures. The most unsatisfactory samples are those that partially disintegrate, forming a mush in the water while the centers of the pieces remain dry, hard and chewy. It is assumed that cubes, strips and slices are produced because the forms are desired for specific table preparations; the rehydration procedure should therefore preserve these forms. Samples of high quality will be rehydrated successfully by boiling without previous soaking, but since table preparations, such as fried, escalloped and hashed browned potatoes, are best when unboiled potatoes are used, the rate and completeness of rehydration at room temperature, as well as at the boiling point, should be determined.

Because of the great number of factors that affect quality, some of which can be overcome by changing the conditions of rehydration, it is recommended that each producer conduct a series of tests for rehydration and quality on each lot of potatoes. An acceptable product should be rehydrated to satisfactorily plump pieces without becoming mushy or watery. The rehydration ratio (the rehydrated weight divided by the dry weight) will vary with the size and shape of the piece, the time held in water and the temperature of the water. Rehydration ratios obtained with high-quality samples are given in the following table.

REHYDRATION RATIOS OBTAINED WITH HIGH-QUALITY SAMPLES OF DEHYDRATED POTATOES

Form	*Size in Inches*	*Boiled Without Soaking* 5 Minutes	10 Minutes	20 Minutes	*Boiled After Soaking Overnight* 5 Minutes	10 Minutes	20 Minutes	*Soaked Overnight, not Boiled*
Cubes	6/16	2–3	3–4	3–4	3–4	3–4	4–5	2–3
Slices	2/16	—	—	4–5	—	—	—	—
Strips	5/32	3–4	4–5	—	4–5	4–5	—	3–4

In rehydrating dried cut potatoes, add 8 parts of water to 1 part by weight of the product and boil separate samples gently for 5, 10 and 20 minutes. Drain carefully through an eight-mesh strainer for 2 minutes. The rehydrated weight should be approximately 2 to 4 times the original weight for diced or cubed, and 3 to 5 times for julienne strips. Soaking before boiling will give a plumper product and slightly greater drained weight in the larger pieces. Long boiling will cause deterioration in taste, flavor and color. The rehydrated product should be free from sour or bitter taste, mild in potato flavor, and soft and tender in texture. The color should be uniformly yellowish white to pale yellow. The pieces should be whole and nearly perfect in form.

The addition of 5 to 7 parts of half-and-half milk and water to 1 part by weight of dried riced potatoes should yield a product having the normal consistency of mashed potatoes. If four parts of liquid yield a material thinner than is normal, the dehydrated riced potatoes may be considered unsatisfactory. The material should be held in a double boiler for 30 minutes at room temperature, placed over boiling water for 10 minutes, or until heated throughout, and then whipped until light and fluffy. The whipped material should be reasonably free from lumps and should have a mealy texture. A product having a waxy or pasty consistency should be considered unacceptable.

Rutabagas

Rutabagas are grown chiefly in the northern states and require 4 months or more to reach maturity. American Purple Top, Bangholm, and Early Neckless are varieties commonly used for dehydration. Harvested roots can be stored satisfactorily for several months in caves or pits at 32 to 40° F and at high relative humidity (95 to 98%). Much of the commercial crop is waxed prior to storage, as a means of reducing wilting.

Preparation includes washing, peeling, trimming and cutting. Waste in preparation ranges approximately from 8 to 15%. Although present government purchase regulations specify inactivation of the peroxidase system by blanching, such inactivation requires almost half an hour in live steam, which is much longer than is necessary to obtain a dried product that will keep well in storage. Information on moisture content of fresh and dried rutabagas, on temperature and humidity conditions suggested for trial in dehydration and on packaging is presented in the preceding tables.

Dried rutabagas are packaged in much the same manner as dried potatoes, that is, in five-gallon cans or lead-foil-protected cartons. In Canada, dried rutabagas are packed in cans containing an inert-gas atmosphere. They are readily injured by high temperatures during storage.

In determining rehydration characteristics, add 8 parts of water to 1 part by weight of the dried product and boil for 30 minutes. The drained weight of rehydrated product should be approximately 5 to 6 times the weight of dried vegetable. It should have a flavor and odor similar to those of cooked fresh rutabagas, and the color should range from yellow to buff.

Spinach

Spinach is grown during the entire year, in one section or another of the country, as a late-fall, winter, spring or summer crop. The summer and winter culture of spinach is possible only where moderate heat or moderate cold prevails. The most extensive plantings of early or winter spinach for the fresh-vegetable market are in Texas and California. Pennsylvania and New Jersey grow the bulk of the summer crop. Approximately half of the commercial crop is processed, the major portion in California.

The varieties of spinach commonly grown for use as fresh vegetable are probably satisfactory for drying. Nobel, also known as Giant Thick Leaf, is very productive. The smooth leaves are easily cleaned, and it stands for a long time before going to seed. Long Standing Bloomsdale is the most widely grown Savoy-leaved variety. Virginia Savoy and Old Dominion are good varieties for fall planting, since they are hardy, but are not suited for spring planting.

Spinach is perishable and should be processed as soon as possible after harvest. Cooling quickly to 34 or 35° F in air at 90 or 95% relative humidity will help prevent spoilage. Crushed ice applied to the pack will hasten cooling and protect the quality over short periods of unavoidable delay.

Undue handling or bruising of the leaves of fresh spinach will result in large losses of ascorbic acid. The first step in the preparation is the cutting off of the roots. The loose material is then washed under cold water sprays as it is loaded on the conveyor belts. Next, all damaged and discolored leaves are removed and the stems cut off. Not more than 10% of the stem should remain on the spinach. Trimming losses for spinach range from 45 to 65%. After sorting and trimming, the leaves are thoroughly washed to remove the remaining sand and dirt.

Spinach should be blanched, as soon after preparation as

possible, in live steam until the catalase system is destroyed. Care must be taken to see that the leafy material is uniformly loaded in the blancher. If there is a delay between blanching and dehydration, the product should be cooled immediately to 65 or 70° F by means of water sprays or immersion in cold water. This is essential as a means of preventing discoloration, undesirable brownish color, in the finished product. The delay between blanching and dehydration should not be more than 1 hour.

Information on dehydration is presented in the preceding tables. The leaves are preferably dried whole. Various kinds of greens should not be dried together, nor should they be mixed after they are dried. Because of the limp condition of the blanched leaves, care should be taken to spread the material so that it will not pile up in bunches.

Government procurement specifications require dehydrated spinach to be packaged in hermetically sealed containers in which the air has been replaced by nitrogen or carbon dioxide gas. Spinach dried to 2% moisture content can be packaged in lead-foil-protected cartons if it is stored at safe temperature and for short periods, such as would commonly be possible in domestic use. For longer and more severe storage, such as would be necessary in shipment abroad, gas-packed sealed cans are advisable. In quality tests, unblanched spinach dried to 2% moisture content and stored at 90° F was found to retain its palatability after more than a year.

For the rehydration test the following procedure is recommended: To one part by weight of dried spinach add 6 parts of water. Soak for 2 hours and then boil gently for 5 minutes. The drained weight of the rehydrated product should be about 3 to 4 times the original weight. Longer soaking will do no harm, but longer boiling will cause loss of vitamins and flavor. The product should be mildly bitter, soft, tender, a uniform dark green, and free from foreign odor.

Sweet Potatoes

Sweet potatoes rank next to potatoes in importance as material for dehydration, and are an important crop in the south and in the Middle Atlantic states. Approximately 707,000 acres of sweet potatoes were harvested in 1942. Georgia, with 100,-000 acres, or 14% of the total, ranked first in acreage harvested, as well as in total production. The 1942 production of 1.8 million tons was only one-sixth as much as the potato production in that year. In the years 1930–39 the combined production of nine southern states, which outranked all other states, averaged 1.6 million tons or 80% of the total. The harvesting of this crop extends from as early as June to August in Georgia, and from as late as September to November in many of the other sweet potato-growing states. The average yield per acre for all states was 2.5 tons in 1942.

Both the moist *yam* and the so-called dry-flesh varieties are suitable for drying. The products, however, are different, and the two types should not be mixed. Numerous varieties are grown, but those that are of commercial importance are less than a dozen in number, and of these not more than four or five are of importance. Porto Rico and Nancy Hall are the chief varieties in the south. In New Jersey, Maryland, Missouri and Kansas, members of the Jersey group, Big-Stem Jersey, Maryland Golden, Yellow Jersey, Orange Little Stem and Vineland Bush, are of chief importance. These are not all equal in suitability for dehydration, and because of varietal differences, as well as differences in soil and climatic conditions, tests should be made prior to the establishment of a plant in a given area, to determine (1) the operating conditions that make the best dried product, (2) the acceptability of the product, and (3) the yield of dried product obtainable from the material.

Rather extensive trials have shown that Nancy Hall and Nancy Gold yield dried products of the highest quality. Porto Rico, Orange Little Stem, Maryland Golden, Big-Stem Jersey and Yellow Jersey all gave equally good dried products very close in quality to Nancy Hall and Nancy Gold.

Sweet potatoes require special handling to prevent loss during storage. Only well-matured stock should be used, and it should be cured for 10 to 14 days at temperatures of 80 to 85° F with a relative humidity of 85 to 90%. This may necessitate the use of artificial heat. After curing, the temperature is allowed to drop to about 55° F with a relative humidity of 75 to 80%. Short periods at temperatures of less than 50° F will do no harm, but exposure to low temperatures for long periods may cause certain types of decay, frequently characterized by the formation of dark spots which may show in the raw material or develop only during the drying process. Prolonged storage of sweet potatoes causes deterioration of the raw stock, with resulting poor quality of the finished product when dehydrated. Slatted crates, bushel baskets, or shallow bins are used for storage. The roots should be handled as little as possible while stored.

Sweet potatoes are first washed and then steamed for 10 minutes to facilitate peeling. After washing, grading to size should be performed if abrasion peeling is used, since grading results in considerably reduced waste. Abrasive peelers cause a heavy loss when used on either raw or steamed sweet potatoes; the waste in peeling and trimming will run from 15 to 30% for sound materials. Hand, brine or lye peeling can be used with considerable reduction in losses.

Sweet potatoes are cut into slices, cubes or strips. Government purchase specifications limit cubes to the three-eighths-inch dimension. Immediately after cutting, the pieces should be washed thoroughly by strong sprays of clean cold water in order to remove starch from the cut surfaces. If not blanched immediately, the cut material should be kept under running cold water or in a 2% salt solution. For material on which this treatment has not proved effective the use of a 1 to 2% citric acid solution has been recommended. Citric acid is not, however, readily available. In any case, the material should not be held for more than 1 hour before blanching.

If the sweet potatoes are to be riced or powdered, they are precooked before they are dried. To make the riced product, the sweet potatoes are given the same preliminary treatment as potatoes, except that instead of being cut in pieces, the whole tubers are steamed until thoroughly cooked. The cooked product is then passed through a ricing device having holes not over ⅛ inch in diameter. If the ricing is done while the potatoes are hot, there is less tendency for the strings to stick together.

The strings should fall from the ricer to the drying surface.

Sweet potatoes should be blanched as soon as possible after they are peeled and cut. Blanching in steam is recommended. The time must be adjusted to allow complete destruction of the peroxidase system, which requires 6 to 10 minutes. It is essential to avoid all contact with iron during blanching. Failure to observe this precaution will result in blackening of the product. The maximum delay allowable between blanching and dehydration is 1 hour.

Information on the moisture content of fresh sweet potatoes and on drying ratios and conditions suggested for dehydration and on packaging is contained in the preceding tables. Dried sweet potatoes are considerably less affected by heat and moisture than are dried potatoes. For export, five-gallon cans or lead-foil-protected cartons, without gas packing, are used.

The rate and effectiveness of rehydration of sweet potatoes can be determined by obtaining the drained weight of the rehydrated sample at various intervals of time and temperature in water. To obtain this drained weight add 6 parts of water to 1 part by weight of sliced sweet potatoes and boil gently for 30 minutes. Drain carefully through an eight-mesh strainer for 2 minutes and weigh. The rehydrated weight should be approximately 2 or 3 times the original weight of the dried vegetable. Rehydrated sweet potatoes should be sweet, free from bitter, strong or sour taste, mild in sweet potato flavor, and soft and tender in texture. Their color should be that of cooked fresh sweet potatoes of the same variety, free from dark spots and streaks.

Soaking before boiling will give a plumper product and slightly greater drained weight. Longer boiling may cause deterioration in taste, flavor and color. Products that have been stored for some time at 90° F may absorb water more slowly than freshly dried material. Soaking before boiling will compensate for this more satisfactorily than will increasing the time of boiling. There is a tendency toward disintegration of pieces when boiled 30 minutes, but a raw taste and flavor may result from shorter boiling.

Tomatoes

Tomatoes are widely grown and rank first among crops that are canned. They are highly nutritious and comparatively easy to produce; they yield well and are consumed in large quantities. Dehydration of tomato slices of fleshy varieties in the types of equipment used for other cut vegetables is possible, but the product is hygroscopic and does not keep well except in tightly closed cans. Tomato puree can be dried on drum driers and juice can be spray-dried to form a powder. There is considerable interest in the latter products but so far commercial production has not been developed. Progress has been made with the technical problems of tomato dehydration, but because of the high water content of tomatoes the method is not highly feasible from the economic standpoint. Moreover, tomatoes and tomato juice can be easily preserved by canning in glass containers, of which there is an ample supply.

Fruits

The information presented here on fruits is concerned chiefly with dehydration in tunnel and cabinet dehydrators, by means of currents of heated air. Sun drying, evaporation in kilns, spray drying, vacuum drying, and other methods are also, in greater or less degree, established commercial methods of producing dried fruits or dried-fruit products, and some of these methods are also applied to vegetables.

Most fruits should be harvested frequently and then dried as soon after harvest as possible. Apples, apricots, cherries, peaches, pears and nectarines should be picked from the trees and not allowed to fall to the ground. Prunes are harvested from the ground 2 or 3 times during each season. The trees are shaken for the second and third pickings. Figs are harvested from the ground and are usually one-third to two-thirds dried when harvested. Grapes are harvested from the vines by pulling or cutting off the bunches. All of the softer small fruits require special care in handling and will tolerate very little storage; such fruits are not dried commercially in large amounts. Apples, pears and cranberries can be stored over fairly long periods; pears must be dried as they ripen in storage. Citrus fruits have the advantage of good keeping-quality in storage; drying of citrus fruits, however, is still in the experimental stage.

Although high-quality fruits yield the best dried products, it is a common practice to dry packing-house and cannery culls. All fruits should be carefully washed. Apples and pears can be washed at the time of spray residue removal in a continuous machine. Other fruits may be washed under sprays or by immersion while moving on a belt. Growers with small quantities commonly wash fruit in boxes by immersion. Rotary washers should not be used for tender-fleshed fruit.

Blanching is seldom used in the preparation of fruits for drying, but is advisable for apricots, nectarines, peaches and pears if they are to be dehydrated rather than sun-dried. Prunes and some grapes are commonly dipped in hot lye solution as a means of making fine breaks in the skin; this effect, referred to as checking, facilitates drying. The treatment is not however a blanching operation, i.e., a thorough penetration by heat. For the most part, large pieces or whole fruits are dried, in contrast to the thin slices, cubes, or strips of vegetables. Fruits contain large quantities of juice, and when blanching is used it must be carefully controlled to avoid excessive loss of juice, called bleeding.

Blanching in a manner similar to that used with dehydrated vegetables has been suggested for certain cut fruits that are to be dehydrated, as a means of improving quality from the standpoint of cleanliness, freedom from insects, superior vitamin retention and better appearance. Sulfuring is, however, standard commercial pre-treatment for the lighter colored fruits, such as apples, apricots, golden-bleached raisins, peaches, pears, and nectarines. Prunes and figs are very rarely sulfured. Fruit is usually sulfured in a sulfuring house while on trays stacked on trucks. The trays must be nonmetallic. The trucks occupy most of the space but there must be sufficient open space to permit free circulation of the fumes. It is advisable to burn sulfur in a shallow iron pan placed at or slightly below the floor level between the entrance and the nearest truckload of fruit. The amount of sulfur burned and time of sufuring varies with the type and condition of the fruit. Exposure of sulfured fruit to sunlight for a few hours before it is put into the dehydrator is sometimes used as a means of improving color but this is not necessary if blanching is used.

Most of the commercial fruit dehydration industry is located along the Pacific coast. In areas where sun drying is feasible, particularly northern and central California, it is used as well as dehydration. Apricots, peaches, pears and raisins are for the most part sun-dried. Apples are dried in large quantities in New York and Washington, and cranberries are dried chiefly in New England.

Mechanical fruit driers have gone through a long period of development, and a large number of older types and designs are still in use. During the later years of World War I and in the period that followed, a number of new commercial makes of driers were introduced, chiefly of the tunnel type with forced draft. In California, where sun drying is common, the mechanical drier was originally looked upon as "rain damage insurance," and, although rains are infrequent there during the drying season, a series of rains, such as the destructive wet spell of 1918, could cause enormous losses. Size of operation and consequently size of mechanical installation varies, with capacities ranging from a few tons daily to 100 tons or more. The mechanical equipment also varies with the type of fruit; for example, a prune drier differs in certain respects from other types. The type most generally used is the counterflow tunnel with varying methods of heating, air flow and air propulsion. The fruit is conveyed on trays loaded on trucks.

The following table presents general information on fruit dehydration by mechanical methods. Because of the variation in equipment and the fact that fruits are not dried to the low moisture contents of vegetables, except in the form of nuggets and powders, the data vary considerably. As is true of vegetables, the temperature should be carried as high as possible without injury to the product. Experience and the application of principles discussed in vegetable dehydration are necessary in practical operation.

PREPARATION AND DEHYDRATION OF FRUITS[1]

Fruit	*Type of Piece*	*Tray Loading Per Square Foot Pounds*	Pretreatment[2] Method	Pretreatment[2] Time in Hours	*Maximum Finishing Temperature[3] in °F*	*Per Cent Moisture Content of Product*	*Per Cent Yield, Unprepared Basis*
Apples[4]	Slices	1.5	Sulfur	½–2	155	15–20	10–15
Apricots[5]	Halves	2.0	..do	1–2	155	15–20	15–20
Berries[6]	Whole	1.0–1.2	...	...	150	10–15	15–20
Cherries (pitted)[7]	..do	1.0	...	...	165	24–28	15–25
Cranberries[8]	Chopped	.8–1.0	...	...	150	5–10	10–15
Figs (Kadota)[9]	Whole	3.0	Sulfur	2–3	160	15–20	25–30
Grapes[10]	..do	3.5	...	...	160	15–20	20–30
Peaches	Halves	2.5	Sulfur	2–3	155	15–20	15–20
Pears[11]	..do	2.5	..do	3–5	150	10–15	10–20
Prunes[12]	Whole	3.0	Lye-solution dip	...	165	15–20	30–40
Walnuts	..do	...	...	...	([13])	....	....

[1] The data presented here were obtained from several sources and represent approximate commercial tunnel-drying practice. Government purchase specifications have been issued for a number of dried fruits and dried-fruit products.

[2] If feasible, sulfured fruit is often exposed to sunlight for several hours just before it is dehydrated.

[3] The starting temperature varies with the equipment and the fruit, and can be determined by trial.

[4] Apples are extensively evaporated in kiln driers. They are commonly sulfured by immersion for 15 to 20 seconds in 0.25% sodium bisulfite solution.

[5] A somewhat longer sulfuring time is required for sun drying.

[6] Strawberries, raspberries, loganberries, blackberries and others.

[7] Cherries should be dipped in sodium carbonate solution when dried whole. Royal Anne and sour cherries should be sulfured. Unpitted cherries will yield a higher percentage of dried product than that shown in the last column.

[8] Considerable amounts of drum-dried cranberry puree are produced commercially.

[9] Figs other than Kadotas are not sulfured.

[10] Light-colored grapes used to produce bleached raisins are dipped in hot lye solution to check the skins and then sulfured about 3 hours.

[11] Longer sulfuring (24 to 48 hours) is used for sun drying.

[12] Hot- or cold-water dip may be used when fruit is to be dehydrated. Lye dip is always necessary when fruit is to be sun-dried.

[13] Not over 110.

Dried fruits are usually stored in bins holding several tons, if they must be held for some time before delivery to the packer. Packers buy the dried fruits in bulk and grade, process, store and package them for distribution. Bulk dried fruits are usually handled in boxes. For domestic consumption, dried fruits are usually packaged in paper cartons or in wood or fiber boxes. Fumigation is usually necessary, to destroy insect infestation. Most dried fruits are pressed to some extent when they are packaged.

Apples

Apples constitute about one-fifth of all the fruit grown in the United States. Production is heavily concentrated in Washington; New York ranks second, and a number of other states produce substantial quantities.

Fresh apples can be stored for several months at 35 to 38° F. The length of storage period varies with the variety. Spray residue must be removed from apples before they are used; removal can be accomplished by means of a hydrochloric acid wash. Further preparation includes peeling, coring, trimming, slicing and sulfuring.

The best, although not the standard, method of grading dried apples is based on quality of sauce, pie, or other product. By this method, varieties of inherently good cooking-quality rank high, and such varieties are best for dehydration because all dried apples are cooked. In commercial practice, the whitest dried apples are rated highest regardless of variety. Whiteness, however, is to a considerable extent controlled by sulfuring. Late-maturing, firm-fleshed varieties are generally suitable for drying. Apples are sulfited by dipping for 15 to 20 seconds in a 0.25% solution of sodium bisulfite after slicing or by exposure to the fumes of burning sulfur for about 20 minutes after peeling but before slicing.

Apples are commonly dried in kilns, or evaporators. The fresh slices are spread about 18 inches deep on a slatted floor; hot air passes through the fruit, which is turned once or twice with a shovel, and the finished fruit (after about 20 hours of drying) is shoveled into bins. Apples can also be dried by mechanical, forced-air dehydration. This method is used extensively in Canada, where the center-exhaust system is advocated. Apple crisps, or nuggets, are produced by coarsely grinding evaporated apples and drying the small pieces to 3% moisture content or lower (government purchase specification) by the vacuum process. This product has much better keeping-quality, especially at higher temperatures, than ordinary dried apples.

Apricots

Commercial production of apricots is restricted to the Pacific coast, with 90% in California. During the years 1936–40 about 65% of the California crop was dried, 25% was canned, and 10% sold in the fresh state. Increasing amounts are being preserved by freezing. The production in the other states is sold for the most part on the fresh-fruit market. Almost half of the dried apricot pack is exported. The Blenheim and Royal varieties, which are indistinguishable, are preferred for drying because they bear well and the dried product is excellent in quality. Other varieties commonly dried are Tilton and Moorpark.

Fresh apricots cannot be stored more than 1 or 2 weeks, and for best dessert qualities they must be harvested when fully ripe. They are never peeled before being dried. Preparation includes washing, halving and pitting. Trays are loaded with a single layer of halves, laid with the cups up. The fruit is then sulfured and dried in the sun. If it is to be dehydrated, it should be blanched about 3 minutes prior to sulfuring. Apricot powder is made by grinding dried apricots and drying the ground material under vacuum to a low moisture content (2½% for government purchase).

Berries (Including Cranberries)

Strawberries are produced commercially in many states, with large commercial production in Arkansas, Louisiana and Oregon. Most of the commercial production of cranberries is located in Massachusetts, Wisconsin and New Jersey; Washington and Oregon produce the remainder, about 10% of the national total. Nearly all the commercial canning and freezing of berries other than cranberries takes place in the Pacific Coast states. Berries can be dehydrated but, with the exception of cranberries, few are dehydrated commercially. Cranberries are stored in refrigerated warehouses, but usually not longer than 3 months. In the preparation of puree for drum drying, cranberries are cooked in a small amount of water until the skins crack and are then put through a pulper to remove skins and seeds. The puree is dried on a steam-heated stainless-steel drum and comes off the drum in a continuous sheet which is ready to be pulverized.

Cherries

Sour cherries are produced mostly in the Great Lakes region and the eastern states, and most of the sweet cherries are grown in the Pacific Coast states. Michigan and California lead in production. Common varieties of sweet cherries are Bing, Royal Anne, and Tartarian. The leading sour cherry is Montmorency. Both sweet and sour cherries can be dehydrated but are not dried in large amounts. Fresh cherries can be stored for only a short period. In preparation for dehydration cherries are stemmed and then thoroughly washed. Pitting is done by a machine. Cherries can be dried without pitting; thus loss of juice is avoided. Steam blanching has been suggested but gives the fruit an undesirable texture. Royal Anne and sour cherries should be sulfured before they are dried.

Figs

Figs are produced for the most part in California, but to some extent in Texas. Four varieties are dried: Calmyrna, Adriatic, Kadota and Black Mission. The first three are light-colored and the latter black. Kadota is grown chiefly for canning and is not commonly dried. The dried Calmyrna is considered superior to the dried product of the other varieties. Figs are both sun-dried and dehydrated. They dry to a considerable extent while hanging on the tree. Most figs are dried without any pretreatment. Kadota may be sulfured. If they are to be dehydrated, exposure to the sun for 1 to 2 days is essential to remove the green color.

Grapes

Grapes rank third in production among fruits, i.e., below apples and oranges. California leads in production; in fact, 90% of the total national crop is produced there. New York and Michigan rank next, and a large number of states produce minor quantities on a commercial scale. Varieties can be classified, according to use, as raisin varieties (about three-fifths of the total), wine or juice varieties and table varieties. Raisin grapes should have a sugar content of about 23%. Nearly the whole national production of raisins is dried in California, where they are largely sun-dried without pretreatment, usually in the fields where they grow. Sulfur-bleached raisins are prepared by dipping seedless grapes in alkali to check the skin, sulfuring, exposing to the sun for 3 to 4 hours and then drying in the shade. A similar preparation procedure is used for golden-bleached raisins, but they are dehydrated artificially rather than sun-dried.

Peaches

In the production of peaches, California ranks first among the states and Georgia second. In California most of the clingstone peaches are canned, whereas the larger part of the freestone crop is dried. Of the total peach crop in that state, about one-third is freestone, and the most important varieties are Elberta, Lovell, Muir, and J. H. Hale. Most of the Elberta crop is sold in the fresh-fruit market; dried peaches are for the most part Muir and Lovell. In general, eastern-grown peaches do not make attractive dried products, but the freestone varieties can be successfully dehydrated.

Although mechanical dehydration is possible, nearly all of the commercial production of dried peaches is by sun drying in California. In normal times a substantial portion is exported. Preparation for drying is similar to that for apricots. Trays are loaded with a layer of halves, with cup sides up.

Pears

Pears are a minor fruit, in terms of tonnage. California, Washington and Oregon produce almost three-fourths of the national crop. The Bartlett is the leading variety and is the only one that is extensively dried. Few pears are dried commercially outside California, and most of the production there is by sun drying. Although the Bartlett pear is the only variety dried commercially, other varieties can be dehydrated artificially. The Kieffer, which is grown in eastern states, can also be dehydrated but has two objectionable features: difficulty of ripening and a gritty texture. Pears are first washed to remove spray residue. After ripening, they are prepared by cutting in half and removing the calyx. The halves are spread on trays, sulfured 24 to 48 hours, and then exposed to the sun for 2 to 3 days; drying is completed in the shade in tray stacks. If pears are to be dehydrated, it is advisable to steam blanch for 20 to 30 minutes prior to sulfuring.

Prunes and Plums

Prunes are important among dried fruits, most of the crop being dried. Sun drying is feasible and about half of the California crop is processed by that method. In states north of California the climate does not permit sun drying, and dehydration is the only method used. In normal times, dried prunes are exported in substantial quantities.

The French prune is most commonly dried in California; in Oregon and Washington, the Italian is the leading variety, and others are dried to a lesser extent. Plums other than prunes can be dried only with difficulty, but could be dried if the need arose. Prunes have the special advantage of drying well with the stone intact and without fermentation. The preparation treatment consists of dipping in a hot lye or sodium carbonate solution in order to check the skins. This procedure is always used with prunes to be dried in the sun but is not essential when dehydration is used. In fact, when the latter method is used a cold-water wash is more satisfactory. Prune powder is made by a process similar to that used to make apricot powder. The dried prunes are first pitted; then they are ground, and dried to a low moisture content in vacuum.

Preservation of Food by Freezing

Vegetable foods are being frozen for immediate and future consumption in ever increasing quantities. The purchaser of such foods finds that they have the advantage of better flavor, which seems to be lacking in the every-day run of fresh vegetables. The art of freezing has made tremendous progress with the aid of modern equipment. The preliminary process of preparation for freezing is very rapid and sanitary. When foods are frozen at temperatures from minus 5° to minus 20° F, the process is referred to as *sharp freezing.* The heat is withdrawn from the vegetables quickly by means of cold air blasts brought into the freezing rooms in many ways.

Slow freezing is far from satisfactory for vegetables. Food held in cold storage within freezing temperatures is definitely subject to spoilage brought about by the slow, but sure, process of growing bacteria and enzymes. Quick freezing stops these changes although in some instances, after thawing out, there may be a softening effect on the texture of the product. Quick freezing does retard the growth of bacteria, yeasts and molds, and it slowly reduces the number of bacteria originally present in the food. Enzymes also are retarded during the freezing period, but are not completely destroyed or inactivated. This effect is important inasmuch as vegetables in which the enzymes were not destroyed previous to freezing would result in an inferior frozen product.

Method of Freezing Vegetables

Mature garden-fresh vegetables are trimmed, washed and blanched for 1 to 5 minutes in hot water, then immediately chilled in cold water. At 212° F the enzymes catalase and oxidase are inactivated in 1 minute, and peroxidase in 3 to 6 minutes. After the draining off of excess water, the vegetable is placed in a sharp freezer room at –15 to –25° F for 8 to 12 hours, until completely frozen. It can then be transferred and stored at 0° F or even lower. The following vegetables are suitable for freezing:

Asparagus

Preparation: Cut tips 4 to 4½ inch lengths. For cut, use one-inch lengths. Use within 2 hours after cutting or it becomes woody. To hold for 24 to 48 hours, chill in ice water and refrigerator just above freezing.

Blanching: Steam for 4 to 7 minutes or boil in water for 2 to 3 minutes. Chill at once by plunging into cold or ice water.

Pack, freeze and store at 0° F.

Beans (Green)

Preparation: Prepare as for canning.

Blanching: Steam for 4 to 6 minutes or boil in water for 2 to 3 minutes. Chill in cold water. Drain.

Pack, freeze and store at 0° F.

Beans (Wax)

Preparation: Prepare as for canning. These beans do not retain flavor and color especially well during freezing and storage.

Blanching: Same as for green beans.

Beans (Lima)

Preparation: Harvest beans before maturity when proportion of white beans is high. Shell and blanch immediately as they lose their flavor, sweetness and vitamin C content rapidly.

Blanching: Steam for 3 to 4 minutes or boil in water for 1½ to 2 minutes. Chill, drain, freeze and store at 0° F.

Beets

Preparation: Wash, trim and dice. Freeze very young beets whole.

Blanching: Steam diced beets for 4 to 6 minutes, and whole beets for 6 to 8 minutes.

Chill, drain, freeze and store at 0° F.

Broccoli

Preparation: Harvest before buds open to avoid yellow flowers. Trim and use only tender portions of flower stems.

Blanching: Steam for 6 to 8 minutes or boil in water for 3 to 4 minutes. Chill, drain, freeze and store at 0° F.

Beet Greens

Preparation: Process young immature plants with partially developed roots before they wilt.

Blanching: Steam for 2 to 3 minutes or dip in boiling water for 1 to 2 minutes. Chill, drain, freeze and store at 0° F.

Brussels Sprouts

Preparation: Firm, compact, medium-sized and dark green-colored sprouts should be used.

Blanching: Blanch in same way as for broccoli.

Carrots

Preparation: Use only young tender carrots. Scrub and dice or leave whole.

Blanching: For diced carrots steam for 4 to 6 minutes or boil in water for 2 to 3 minutes. For whole carrots steam for 6 to 8 minutes or boil in water for 3 to 4 minutes. Chill, drain, freeze and store at 0° F.

Cauliflower

Preparation: Use solid heads. Trim and break into small pieces.

Blanching: Steam 4 to 6 minutes. Place in boiling water for 2 to 3 minutes. Chill, drain, freeze and store at 0° F.

Sweet Corn

Preparation: Harvest corn while in milk and before drought or starch stage. Husk, silk and trim.

Blanching: Blanch on cob for both cut and corn on cob. Steam for 16 to 20 minutes and place in boiling water for 8 to 10 minutes depending on size of ears. For whole kernel corn and cream corn, cut from ear as for canning. Chill, drain, freeze and store at 0° F.

Eggplant

Preparation: Wash, slice ¼ to ½ inch thick.

Blanching: Steam for 6 to 8 minutes or place in boiling water for 3 to 4 minutes. Drain, chill, freeze and store at 0° F.

Peas

Preparation: Shell, and sort out immature, over-mature, split peas and dirt.

Blanching: Steam 2 to 3 minutes or dip in boiling water for 1 to 1½ minutes. Cool, drain, freeze and store at 0° F.

Peppers (Sweet)

Preparation: Select thick-walled green or red fruit. Wash, halve, seed, and slice or dice.

Blanching: These may be frozen without blanching or may be blanched for 2 minutes.

Spinach

Preparation: Cut before flowers appear. Quickly prepare for freezing. Wash thoroughly and remove yellow leaves.

Blanching: Steam for 4 to 5 minutes or place in boiling water for 2 to 2½ minutes. Cool, drain, freeze and store at 0° F.

Other Greens (Mustard and Turnip Greens and Kale)

Greens must not stand until wilted before processing.

Preparation: As for spinach.

Blanching: As for spinach.

Squash (Winter)

Preparation: Fully ripe. Wash, cut in strips 1 to 2 inches wide and peel. Cut in ½ inch slices.

Blanching: Steam for 6 to 10 minutes. Place in boiling water for 3 to 5 minutes. Cool, drain, freeze and store at 0° F.

Turnips

Preparation: Select young tender roots. Slice or dice as for carrots.

Blanching: Steam for 6 to 10 minutes. Place in boiling water for 3 to 5 minutes. Chill, drain, freeze and store at 0° F.

Method of Freezing Fruits

Cranberries, blueberries, raspberries, loganberries, youngberries and plums may be frozen in original cases without the use of sugar.

Strawberries, cherries, currants and raspberries may be

frozen with the addition of sugar to the fruit in a ratio of 1 to 5, 1 to 4, 1 to 3 and 1 to 2. There is sufficient juice in the fruit to form syrup in a large enough quantity to cover them. This method is more satisfactory than packing in syrup since it retains the form of the whole fruit. Strawberries, peaches and raspberries however, may be crushed and may be treated with added sugar. This method helps to exclude the oxygen and retains better flavor. Peaches, plums, apricots and apples may be sliced, treated with a small percentage of sulfur dioxide, which keeps the fruit slightly bleached, and then packed in sugar syrup. These fruits do not have sufficient juice to permit the addition of dry sugar alone. The freezing of fruits results in chemical changes and, unless processed just right, the fruit, after thawing out, will look very poor. Some fruits have to be properly prepared for freezing to be edible after they are thawed out. Some of the fruits suitable for freezing and their method of preparation before freezing are as follows:

Apples

Preparation: Select firm, well matured fruit. Peel and slice. Slices may be put in a 3% salt solution to prevent darkening, or treated with sulfur dioxide before freezing, or may be put directly into a 40% syrup, or dry-sugared.

Apricots

Preparation: Select firm, well matured fruit. Peel and cut into halves. Can, sulfur, dip in boiling water, steam or put at once into a 40 or 50% sugar solution to prevent darkening.

Packing: Cover with a 40 or 50% sugar solution or cover with dry sugar, 1 to 3 or 1 to 4.

Blackberries

Preparation: Use firm ripe berries only. Over-ripe berries or those picked after rain, or in hot weather are not satisfactory. Wash, drain well and place berries, whole, in the container.

Packing: Pack in 40% syrup or in dry sugar, 1 to 3 or 1 to 4.

Dewberries, (Lucretia, Olympic) Youngberries, Boysenberries, Nectarberries and Loganberries

Preparation: Same as for blackberries.
Packing: Same as for blackberries.

Raspberries (Red)

Preparation: Use mature but not over-ripe fruit. Do not pick after a rain or in hot weather as they turn dark. Rapid handling is essential.

Packing: Cover with a 40% syrup.

Raspberries (Black)

Preparation: Same as for red raspberries.
Packing: For pies, use 1 part sugar to 3 or 4 parts berries.

Raspberries (Purple)

Preparation: Same as for red raspberries.
Packing: Cover with a 40 or 50% syrup or pack in dry sugar, 1 to 4.

Strawberries

Preparation: Use firm and ripe berries only. Discard green and over-ripe berries. Hull, wash and drain quickly.

Packing: Pack whole or cut in one-eighth-inch slices or crushed. Cover with 1 part sugar to 4 parts berries, or with a 40 to 45% syrup. Set whole berries in refrigerator over-night before freezing.

Blueberries

Preparation: Green and over-ripe berries are unsatisfactory. Remove all leaves and stems.

Packing: Cover with a 40 or 50% syrup. For pies, freeze without sugar. Dry sugar is not recommended.

Cantaloupes and Melons

Preparation: Select finest melons with firm ripe flesh. Cut into squares or balls for cocktails and salads.

Blanching: Do not blanch.

Cover with a 45% cold syrup and freeze. Store at 0° F.

Cherries (Red)

Preparation: Pick when tree-ripe. Immature fruits turn pale and over-ripe ones turn dark, after freezing. Wash, remove pits and pack in containers quickly.

Packing: For pies, cover with 1 part sugar to 4 parts pitted cherries; or cover with 60% cold syrup.

Cherries (Sub-Acid)

Preparation: Same as for sour cherries.

Packing: Same as for sour cherries.

Cherries (Sweet)

Preparation: Harvest fruit when it is ripe and firm. Immature fruits shrivel, fade and are tough after freezing. Wash, stem, sort and pack. They may be pitted.

Packing: Cover with 50% cold syrup, or 1 part sugar to 6 to 8 parts pitted cherries.

Cranberries

Preparation: Wash, sort and free of debris. Drain and pack into containers.

Packing: They keep very well without sugar. Freeze whole.

Figs

Preparation: Harvest at proper stage of maturity. Wash and sort out soft, green and over-ripe fruit.

Packing: Cover with a 40% cold syrup and freeze.

Nectarines

Preparation: Prepare the same way as for peaches.

Packing: Cover fruit with a 40% syrup and freeze.

Peaches

Preparation: Harvest at firm ripe stage, usually a few days later than for market, but while still firm. Avoid bruising. Prepare rapidly to prevent browning. Lye peel (1 part lye to 10 parts water) at 140° F or by dipping in boiling water. Chill in cold water and remove skin. Neutralize in 2% citric acid or a solution of lemon juice, if lye peeled. Cut in half and remove pits. Slice or crush and quickly cover with sugar to prevent darkening.

Packing: Cover quickly with 40 to 50% cold syrup. Syrup must cover fruit completely. Freeze at once.

Watermelon

Preparation: Select only the best melons with firm ripe fruit. Cut flesh into balls or squares.

Packing: Pack loosely in containers with waxed paper between each layer of pieces. Cover with a 40 to 45% cold syrup.

Storaging of Frozen Foods

After the fruits or vegetables are put through the quick freezing method, they are usually removed from the freezing room and stored in another cold room where the temperature is approximately 0° F. Unless frozen foods are packaged properly, they have a tendency to lose moisture and become dry during the storage period. This is very likely to happen if any currents of air are allowed into the room.

Cooking Frozen Foods

Frozen foods upon thawing out are highly perishable. Once they are taken from cold storage, they should be used as soon as possible. The tissue cells have been somewhat softened and disrupted by blanching and freezing, and therefore are subject to spoilage and deterioration in flavor and quality very rapidly.

Vegetables after thawing out may become mushy, lose flavor and darken within 24 hours. Fruits act like vegetables but keep much longer, especially if frozen in sugar or in sugar syrup. Once frozen foods are thawed, they should not be refrozen. They lose more flavor and quality and are not nearly as good when used.

Cooking of Frozen Vegetables

It takes about half the time to cook frozen vegetables as it does to cook fresh ones. Avoid over-cooking the frozen vegetables since they lose flavor and color. The cooking time is shorter because the blanching and freezing process has already softened the tissues. Vegetables may be allowed to thaw partially or may be cooked slowly until they are thawed out.

APPENDIX

WEIGHTS AND MEASURES, CHEMICAL GLOSSARY, BIBLIOGRAPHY, ABBREVIATIONS

Weights and Measures

Common Kitchen Measures

Equivalents of Capacity
(All measures level full)

3	teaspoons	= 1 tablespoon
½	fluid ounce	
16	tablespoons	= 1 cup
2	gills	
½	liquid pint	
8	fluid ounces	
1	liquid pint	= 2 cups
16	fluid ounces	

Liquid Measure

4	fluid ounces	= 1 gill
4	gills	= 1 pint
2	pints	= 1 quart
4	quarts	= 1 gallon
	(231 cubic inches)	
31½	gallons	= 1 barrel
2	barrels	= 1 hogshead

Cubic Measure

1728	cubic inches	= 1 cubic foot
27	cubic feet	= 1 cubic yard
144	cubic inches	= 1 board foot
128	cubic feet	= 1 cord*

* The use in trade of the so-called "face cord," which is less than the above, is illegal.

Dry Measure

(For fruits, vegetables, and other dry commodities)

2	pints	= 1 quart
8	quarts	= 1 peck
4	pecks	= 1 bushel
	(2150.42 cubic inches)	
105	quarts	= 1 barrel
	(7056 cubic inches)	

Linear Measure

12	inches	= 1 foot
3	feet	= 1 yard
5½	yards	= 1 rod
40	rods	= 1 furlong
8	furlongs	= 1 mile
1760	yards	
5280	feet	

Square Measure

144	square inches	= 1 square foot
9	square feet	= 1 square yard
30¼	square yards	= 1 square rod
160	square rods	= 1 acre
640	acres	= 1 square mile
36	square miles	= 1 township

Avoirdupois Weight

$27^{11}/_{32}$	grains	=	1 dram
16	drams	=	1 ounce
16	ounces	=	1 pound
4	quarters	=	1 hundredweight
20	hundredweights	=	1 ton
	25 pounds	=	1 short quarter
	28 pounds	=	1 long quarter
	100 pounds	=	1 short hundredweight
	112 pounds	=	1 long hundredweight
	2000 pounds	=	1 short ton
	2240 pounds	=	1 long ton

Common Rules of Measurement

Rectangle Area = length × width

Solid with rectangular sides Volume = length × width × height

Circle { Circumference = 3.1416 × diameter; Area = 3.1416 × radius × radius

Cylinder { Area (exclusive of that of ends) = 3.1416 × diameter × height; Volume = 3.1416 × radius × radius

International Metric System

In the international metric system the fundamental unit is the meter—the unit of length. From this the units of capacity (liter) and of weight (gram) are derived. All other units are the decimal subdivisions or multiples of these. These three units are simply related; e.g., for all practical purposes 1 cubic decimeter equals 1 liter and 1 liter of water weighs 1 kilogram. The metric tables are formed by combining the words *meter, gram,* and *liter* with the six numerical prefixes, as in the following tables:

One meter = 39.37 inches; 1 liter = 1.0567 liquid quarts; 1 gram = 0.035 avoirdupois ounce.

Units of Length

Millimeter	=	0.001	meter
Centimeter	=	0.01	meter
Decimeter	=	0.1	meter
METER	=	1	meter
Decameter	=	10	meters
Hectometer	=	100	meters
Kilometer	=	1000	meters

Units of Capacity

Milliliter	=	0.001	liter
Centiliter	=	0.01	liter
Deciliter	=	0.1	liter
LITER	=	1	liter
Decaliter	=	10	liters
Hectoliter	=	100	liters
Kiloliter	=	1000	liters

Units of Weight (or Mass)

Milligram	=	0.001	gram
Centigram	=	0.01	gram
Decigram	=	0.1	gram
GRAM	=	1	gram
Decagram	=	10	grams
Hectogram	=	100	grams
Kilogram	=	1000	grams

Units of Area

The table of areas is formed by squaring the length measures, as in our common system. A plot of land 10 meters square is called an *are* (meaning *area*). The side of 1 are is about 33 feet. The hectare is 100 meters square, and, as its name indicates, is 100 ares, or about 2½ acres.

Approximate Weights of Some Common Dry Commodities

	Pounds per Bushel
Apples	48
Beans	60
Beets	60
Carrots	50
Cranberries	32
Cucumbers	48
Onions	57
Parsnips	50
Peaches	48
Peanuts	22
Pears	58
Peas (Dried)	60
Peas (Green, unshelled)	56
Potatoes (White)	60
Potatoes (Sweet)	54
Tomatoes	56
Turnips	55

Ice: 1 cubic foot	=	57.2	lb
30 cubic inches	=	1	lb
Sugar, granulated 1 cup	=	½	lb
Butter 1 cup	=	½	lb
Lard 1 cup	=	½	lb
Flour 1 cup	=	¼	lb
Rice 1 cup	=	½	lb
Cornmeal 1 cup	=	5	oz

Raisins (stemmed) 1 cup	=	6	oz
Currants (cleaned) 1 cup	=	6	oz
Bread crumbs (stale) 1 cup	=	2	oz
Chopped meat (packed) 1 cup	=	½	lb

These weights are approximate only and should not be used to determine whether correct measure is given or received.

Approximate Weights of Some Common Materials

	Pounds per Cubic Foot
Coal	
Bituminous (piled loose)	44 to 54
Anthracite (piled loose)	50 to 57
Coke (piled loose)	23 to 32
Charcoal of pine and oak	15 to 30
Earth (common loam)	
Dry (loose)	72 to 80
Moist (moderately packed)	90 to 100
Soft mud (packed)	110 to 120
Masonry	
Brickwork	100 to 140
Roughly scabbled mortar rubble	140 to 150
Mortar (hardened)	103
Sand	90 to 117
Gravel	90 to 117
Trap (quarried, in loose piles)	107

	Pounds per 1000 Board Feet	*Pounds per Cubic Foot*
Woods		
Fir (Balsam)	2170	26
Hemlock	2330 to 2580	28 to 31
Maple	3250 to 3920	39 to 47
Oak	3080 to 4670	37 to 56
Pine (American White)	1830 to 2580	22 to 31
Pine (Yellow)	1920 to 3080	23 to 37
Poplar	1830 to 2580	22 to 31

The wood is supposed to be seasoned and of average dryness.

Useful Conversion Factors of U. S. Measures

To Convert From	*To*	*Multiply By*
Barrels (dry)	Bushels	3.281
	Pecks	13.12
	Pints	210
	Quarts	105.0
(liquid)	Gallons	31.5
	Pints (liquid)	252.0
	Quarts (liquid)	126.0
B.T.U.	Calories (gram)	252
Bushels	Barrels	0.3048
	Liters	35.24
	Pecks	4
	Pints	64
	Quarts	32
Calories (gram)	B.T.U.	0.00397
Cubic Centimeters	Drams (fluid)	0.2705
	Minims	16.23
	Ounces (fluid)	0.03381
	Pints (dry)	0.001816
	(liquid)	0.002113
Grains/Pound	Parts/Million	142.9
	Milligram/Kilogram	142.9
	Per cent	0.01429
Grams	Drams (Avoir.)	0.5644
	Grains	15.43
	Ounces (Avoir.)	0.03527
	Pounds (Avoir.)	0.002205
Grams/Liter	Grains/Gallon	58.42
	Parts/Million	1000
	Per cent (Weight/Volume)	0.1
Liters	Bushels	0.02838
	Drams (fluid)	270.5
	Gallons	0.2642
	Gills	8.454
	Ounces (fluid)	33.81
	Pecks	0.1135
	Pints (dry)	1.816
	(liquid)	2.113
	Quarts (dry)	0.9081
	(liquid)	1.057
Minims (fluid)	Cubic Centimeters	0.06161
	Drams (fluid)	1/60
	Milliliters	0.06161
	Ounces (fluid)	1/480
Drams (Avoir.)	Grains	27.34
	Grams	1.772
	Ounces (Avoir.)	0.0625 or 1/16
	Pounds (Avoir.)	1/256

To Convert From	*To*	*Multiply By*
Drams (fluid)	Milliliters	3.697
	Minims (fluid)	60
	Ounces (fluid)	0.125 or ⅛
	Pints (liquid)	0.007813
Gallons	Barrel (liquid)	0.03175
	Liters	3.785
	Ounces (fluid)	128
	Pints (liquid)	8
	Quarts (liquid)	4
Gills	Pints (liquid)	0.25 or ¼
	Quarts (liquid)	0.125 or ⅛
Grains	Drams (Avoir.)	0.03657
	Grams	0.06480
	Ounces (Avoir.)	0.002286
	Pounds (Avoir.)	1/7000
Grains/Gallon	Parts/Million	17.12
	Milligram/Kilogram	17.12
	Per cent	0.00171
Ounces (Avoir.)	Drams (Avoir.)	16
	Grains	437.5
	Grams	28.35
	Pounds (Avoir.)	0.0625 or 1/16
(fluid)	Cubic Centimeter	29.57
	Drams (fluid)	8
	Gallons	1/128
	Gills	0.25 or ¼
	Liters	0.02957
	Milliliters	29.57
	Minims	480
	Pints (liquid)	0.0625 or 1/16
Parts/Million	Per cent	0.0001
	Milligram/Kilogram	1.000
	Grains/Gallon	0.0584
	Grains/Pound	0.00700
Pecks	Bushels	0.25 or ¼
	Liters	8.810
	Pints	16
	Quarts (dry)	8
Percentages	Grains/Gallon	584.2
	Grains/Pound	70
	Grams/Liter	10
	Milligrams/Kilograms	10,000
	Parts/Million	10,000
Pints (dry)	Bushels	0.01563 or 1/64
	Liters	0.5506
	Pecks	0.0625 or 1/16
	Quarts (dry)	0.5
(liquid)	Cubic Centimeters	473.2
	Drams (fluid)	128
	Gallons	0.125 or ⅛

To Convert From	*To*	*Multiply By*
	Gills	4
	Liters	0.4732
	Minims	7680
	Ounces (fluid)	16
	Quarts (fluid)	0.5 or ½
Pounds (Avoir.)	Drams (Avoir.)	256
	Grains	7000
	Grams	453.6
	Ounces	16
Quarts (dry)	Bushels	0.03125
	Cubic Centimeters	1101
	Liters	1.101
	Pecks	0.125 or ⅛
	Pints	2
Quarts (liquid)	Cubic Centimeters	946.4
	Drams (fluid)	256
	Gallons	0.25
	Liters	0.9463
	Ounces (fluid)	32
	Pints (liquid)	2
Tons, Long (Avoir.)	Kilograms	1016
	Pounds (Avoir.)	2240
Tons, Metric	Kilograms	1000
	Pounds (Avoir.)	2205
Tons, Short (Avoir.)	Kilograms	907.2
	Pounds (Avoir.)	2000

AVOIRDUPOIS

Liquid Measures

Gal	*Qt*	*Pt*	*Fluid Ounces*	*Fluid Drams*	*Minims*	*Cubic Centi-meters*	*Cubic Inches*	
1	4	8	128	1024	61440	3785.332	231	1 Gallon
	1	2	32	256	15360	946.333	57.75	1 Quart
		1	16	128	7680	473.167	28.875	1 Pint
			1	8	480	29.573	1.804	1 Ounce
				1	60	3.696	0.225	1 Drachm
					1	0.061	0.003	1 Minim
			0.033815			1.000	0.0491	1 Cubic centimeter
				0.27052		1.000		1 Cubic centimeter
					16.2311	1.000		1 Cubic centimeter

Dry Weights

Pounds (lb)	*Ounces* ℥	*Drachms* ʒ	*Grains (gr)*	*Grams (g)*
1	16	256	7000	453.592
	1	16	437.5	28.349
		1	27.34	1.77184
			1	0.064798
	0.035274			1.000
		0.56438		1.000
			15.43235	1.000

Apothecary's Weight

(Apothecary's Grains and Avoirdupois Grains Are the Same)

Grains *gr*	*Scruples* ℈	*Drams* ʒ	*Ounces* ℥	*Pound* *lb*	*Grams* *g*
20	1				1.29598
60	3	1			3.88794
480	24	8	1		31.10350
5760	288	96	12	1	
		0.25721			1
	0.77162				1
15.43235					1

Apothecary's Fluid Measure

Minims *min*	*Drams* *fl* ʒ	*Ounces* *fl* ℥	*Pints*	*Gallon*
60	1			
480	8	1		
7680	128	16	1	
		128	8	1

Metric System

Weight

Milligrams	Centigrams	Decigrams	Grams	Kilogram
1.0	0.1	0.01	0.001	
10	1	0.1	0.01	
100	10	1	0.1	
1000	100	10	1	
			1000	1

Liquid

1 cubic centimeter (cc)	=	1 gram by weight		
100 cubic centimeters	=	1 deciliter	=	100 grams by weight
1000 cubic centimeters	=	1 liter	=	1000 grams by weight
		100 liters	=	1 hectoliter

Metric Conversion Table

Centimeter	=	0.3937	inch
Meter	=	3.28	feet
Meter	=	1.09	yards
Kilometer	=	0.621	mile
Inch	=	2.54	centimeters
Foot	=	0.305	meter
Yard	=	0.914	meter
Mile	=	1.61	kilometers
Square centimeter	=	0.155	square inch
Square meter	=	10.764	square feet
Square meter	=	1.196	square yards
Square kilometer	=	0.386	square mile
Square inch	=	6.452	square centimeters
Square foot	=	0.0929	square meter
Square yard	=	0.836	square meter
Square mile	=	2.59	square kilometers
Cubic centimeter	=	0.061	cubic inch
Cubic meter	=	35.29	cubic feet
Cubic meter	=	1.308	cubic yards
Cubic inch	=	16.4	cubic centimeters
Cubic foot	=	0.028	cubic meter
Cubic yard	=	0.765	cubic meter
Liter	=	0.0353	cubic foot
Liter	=	0.2642	U.S. gallon
Liter	=	61.023	cubic inches
Liter	=	2.202	pounds of water at 62° F
Cubic foot	=	28.32	liters = 7.484 gallons
Gallon	=	3.785	liters
Cubic inch	=	0.0164	liter
Gram	=	15.432	grains
Gram	=	0.0353	ounce
Kilogram	=	2.205	pounds
Kilogram	=	0.0011	ton
Ton	=	2000	pounds
Grain	=	0.0648	gram
Ounce	=	28.35	grams
Pound	=	0.454	kilogram
Ton	=	907.03	kilograms
Ton	=	0.907	tonne

Conversion of Degrees Centigrade Into Fahrenheit and Vice Versa

Degrees Centigrade = 5/9 (F — 32)
Degrees Fahrenheit = 9/5 (C + 32)

N.B.: In case of negative temperatures (below zero) the words *add* and *subtract* are to be interchanged.

Commonly Used for Equipment and Structures in Food Plants

Data for this table from The Hauser-Stander Tank Co., Cincinnati, Ohio and Forest Products Laboratory, Forest Service, U. S. Department of Agriculture.

Kind of Wood	*Weight per cu ft at 12 Per Cent Moisture*	*Absorption of Food Ingredients * gm per strip 4 in by 1 in by ¼ in*							*Per Cent Permanent Expansion, Caused by Food Ingredients*						
		Water		Dilute Acetic Acid	Salt Brine		Cottonseed Oil		Water		Dilute Acetic Acid	Salt Brine		Cottonseed Oil	
		Cold	Hot	Cold	Cold	Hot	Cold	Hot	Cold	Hot	Cold	Cold	Hot	Cold	Hot
Cypress (Tidewater Red)	32 lb	3.8	5.4	4.3	1.6	5.7	2.1	4.4	1.0	2.0	3.5	2.5	1.0	0.0	—2.1
Douglas Fir	30 lb	5.1	7.7	3.4	2.6	6.5	1.2	4.4	0.5	0.5	2.5	2.2	1.1	0.0	—2.4
Yellow Pine (Long Leaf)	41 lb	4.9	5.8	2.9	1.9	5.8	1.6	2.2	2.5	3.0	2.8	1.5	1.0	0.0	—3.6
Redwood	30 lb	5.9	6.4	4.5	5.1	7.9	3.6	5.2	3.5	13.0	7.0	5.5	1.1	0.0	—0.9
Hard Maple	44 lb	8.2	7.9	6.1	5.8	8.3	1.1	4.4	1.5	0.5	4.4	3.5	1.1	0.0	—2.9
White Oak	48 lb	6.5	8.3	4.4	2.2	4.8	0.7	1.1	1.0	0.0	1.5	1.0	1.1	0.0	—1.7

Kind of Wood	*Weight per cu ft at 12 Per Cent Moisture*	*Physical Effects of Food Ingredients ***				*Color Imparted to Food Ingredients*					*Taste Imparted to Water*
		Water	Dilute Acetic Acid	Salt Brine 10 Per Cent	Cotton-seed Oil	Water Cold	Water Hot	Dilute Acetic Acid Cold	Salt Brine 10 Per Cent Cold	Cotton-seed Oil Hot	
Cypress (Tidewater Red)	32 lb	None	None	None	None	No	Yes††	No	Yes	No	No
Douglas Fir	30 lb	None	None	None	None	No	Yes††	No	No	No	Yes
Yellow Pine (Long Leaf)	41 lb	None	None	None	None	No	Yes††	No	No	No	Yes
Redwood	30 lb	None	SS†	SS†SB	SS†	Yes	Yes	Yes	Yes	Yes††	No
Hard Maple	44 lb	None	None	None	None	No	Yes††	No	Yes	No	No
White Oak	48 lb	None	None	None	None	Yes	Yes	Yes	Yes	Yes††	No

* Observed after 1 week with cold liquids and 8 hours with hot (boiling).
** Observations made after immersion for 1 month in the cold liquid.

†SS = slightly soft; SB = slightly brittle.
††Yields color after a long period.

Baumé and Specific Gravity Conversion Table

For Heavy Liquids
American Standard:

$$\text{Sp. gr.} = \frac{145}{145 - {}^{\circ}\text{Bé}}$$

at 60° F

° Bé	Sp. gr.	° Bé	Sp. gr.
0	1.0000		
1	1.0069	36	1.3303
2	1.0140	37	1.3426
3	1.0211	38	1.3551
4	1.0284	39	1.3679
5	1.0357	40	1.3810
6	1.0432	41	1.3942
7	1.0507	42	1.4078
8	1.0584	43	1.4216
9	1.0662	44	1.4356
10	1.0741	45	1.4500
11	1.0821	46	1.4646
12	1.0902	47	1.4796
13	1.0985	48	1.4948
14	1.1069	49	1.5104
15	1.1154	50	1.5263
16	1.1240	51	1.5426
17	1.1328	52	1.5591
18	1.1417	53	1.5761
19	1.1508	54	1.5934
20	1.1600	55	1.6111
21	1.1694	56	1.6292
22	1.1789	57	1.6477
23	1.1885	58	1.6667
24	1.1983	59	1.6860
25	1.2083	60	1.7059
26	1.2185	61	1.7262
27	1.2288	62	1.7470
28	1.2393	63	1.7683
29	1.2500	64	1.7901
30	1.2609	65	1.8125

For Light Liquids
American Standard:

$$\text{Sp. gr.} = \frac{140}{130 + {}^{\circ}\text{Bé}}$$

at 60° F

° Bé	Sp. gr.	° Bé	Sp. gr.
10	1.0000	51	0.7735
12	0.9859	52	0.7692
15	0.9655	53	0.7650
18	0.9459	54	0.7609
20	0.9333	55	0.7568
21	0.9272	56	0.7527
22	0.9211	57	0.7487
23	0.9150	58	0.7447
24	0.9091	59	0.7407
25	0.9032	60	0.7368
26	0.8974	61	0.7330
27	0.8917	62	0.7292
28	0.8861	63	0.7254
29	0.8805	64	0.7216
30	0.8750	65	0.7179
31	0.8696	66	0.7143
32	0.8642	67	0.7107
33	0.8589	68	0.7071
34	0.8537	69	0.7035
35	0.8485	70	0.7000
36	0.8434	71	0.6965
37	0.8383	72	0.6931
38	0.8333	73	0.6897
39	0.8284	74	0.6863
40	0.8235	75	0.6829
41	0.8187	76	0.6796
42	0.8140	77	0.6763
43	0.8092	78	0.6731
44	0.8046	79	0.6699
45	0.8000	80	0.6667

31	1.2719	66	1.8354	46	0.7955	81	0.6635
32	1.2832	67	1.8590	47	0.7910	82	0.6604
33	1.2946	68	1.8831	48	0.7865	83	0.6573
34	1.3063	69	1.9079	49	0.7821	84	0.6542
35	1.3182	70	1.9333	50	0.7778	85	0.6512
						86	0.6482
						87	0.6452
						88	0.6422
						89	0.6393
						90	0.6364

Used for sugar syrup and acids and liquids heavier than water.

Used for ammonia, gasoline and other liquids lighter than water.

CHEMICAL GLOSSARY

COMMON AND CHEMICAL NAMES OF VARIOUS MATERIALS

A

Acacia gum—gum arabic
Acetate of lime—calcium acetate
Acetic ether—ethyl acetate
Acetin—glyceryl monoacetate
Acetyl salicylic acid—aspirin
Acetylene tetrachloride—tetrachlorethane
Adeps lanae—lanolin
Alcohol—ethyl alcohol
Aldehyde—acetaldehyde
Alumina—aluminum oxide
Alum—potassium aluminum sulfate
Alundum—fused aluminum oxide
Ammonia, aqua—ammonium hydroxide
Aniline oil—aniline
Animal charcoal—bone black
Aqua fortis—nitric acid
Aqua regia—nitric and hydrochloric acid
Argols—crude cream of tartar
Arsenic, red—arsenic disulfide
Asphaltum—mineral pitch

B

Baking soda—sodium bicarbonate
Banana oil—amyl acetate
Barium white—barium sulfate
Baryta—barium oxide

Barytes—barium sulfate, natural
Basafor—barium sulfate, precipitated
Bauxite—aluminum oxide, hydrated
Benzene—benzol
Black boy gum—accroides gum
Black hypo—lead thiosulfate, impure
Blanc fixe—barium sulfate, precipitated
Bleaching powder—calcium hypochlorite
Blue lead, sublimed—basic lead sulfate
Blue stone } —copper sulfate
Blue vitriol }
Boiled oil—linseed oil, boiled
Bone black—animal charcoal
Boracic acid—boric acid
Borax—sodium tetraborate
Brazil wax—carnauba wax
Brimstone—sulfur
British gum—dextrin
Bromo "acid"—tetrabromfluorescein
Burnt sugar coloring—caramel color
Butanol—butyl alcohol
Butter color—annatto
Butter of antimony—antimony chloride
Butyric ether—ethyl butyrate

C

Cadmium yellow—cadmium sulfide
Calcium phosphate—calcium phosphate, monobasic
Calomel—mercurous chloride
Carborundum—silicon carbide
Capsicum—red pepper
Carbolic acid—phenol
Carragheen—Irish moss
Catechu—cutch
Caustic potash—potassium hydroxide
Caustic soda—sodium hydroxide
Ceresin wax—ozokerite and paraffin mixture
Chalk—calcium carbonate
China clay—aluminum silicate
China wood oil—tung oil

Chinese wax—insect wax
Chloride of lime—calcium hypochlorite
Cholestrin—cholesterol
Chrome green—lead or zinc chromate plus ferric ferrocyanide
Chrome yellow—lead chromate
Cinnabar—mercuric sulfide
Citronella oil—verbena oil
Cognac oil—oenanthic ether
Colloidal clay—bentonite
Collodion—nitrocellulose *solution*
Cologne spirits—ethyl alcohol, pure
Colophony—rosin, pine resin
Columbian spirits—methyl alcohol
Colza oil—rape seed oil
Copper aceto arsenite—Paris green
Copper arsenite—Scheele's green
Corn sugar—dextrose, cerelose
Corn syrup—glucose, mixture of dextrin and dextrose and maltose
Corrosive sublimate—mercuric chloride
Corundum—aluminum oxide
Cream of tartar—potassium bitartrate
Cresol—cresylic acid
Crude oil—petroleum, crude
Cyanamid—calcium cyanamide

D

Dead oil—creosote oil
Degras—wool grease
Dekalin—decahydronaphthalene
Dextrose—corn sugar, cerelose
Dope—pyroxylin *solution*
Dutch-liquid—ethylene chloride

E

Earth, infusorial—diatomaceous earth
Earth wax—ozokerite, mineral wax
Egg oil—egg yolk
Elaterite—mineral rubber
Epsom salts—magnesium sulfate

Emery—impure aluminum oxide
Ester gum—glycerol ester of rosin
Ether—ethyl ether
Ethyl nitrite—nitrous ether

F

Fir, balsam—Canada balsam
Fixed white—barium sulfate
Flaxseed—linseed
Flea seed—psyllium
Fluorspar—calcium fluoride
Fool's gold—iron pyrites
Formalin—formaldehyde, 40% solution
French chalk, talc—magnesium silicate
Fuchsine—magenta
Fusel oil—amyl alcohol
Fuller's earth—aluminum silicate, hydrous

G

Galena—lead sulfide
Gasoline—petroleum spirit
Glance pitch—manjak
Glass, water—sodium silicate
Glauber's salt—sodium sulfate ($10H_2O$)
Glycerin—glycerol
Glucose—corn syrup
Glycol—ethylene glycol
Graphite—plumbago
Green soap—soft soap
Green vitriol—ferrous sulfate
Grain alcohol—ethyl alcohol
Ground nut oil (Arachis oil)—peanut oil
Gum lac—shellac
Guncotton—nitrocellulose
Gypsum—calcium sulfate
Guignets' green—chromium oxide, hydrated

H

Heavy spar—barium sulfate
Hematite—iron oxide

Hexalin—cyclohexanol
Hexamine—hexamethylenetetramine
Hydrosulfite—sodium hydrosulfite
Hypo—sodium thiosulfate

I

Ichthyol—ammonium sulfo ichthyolate
Indene—para-cumarone
Indian gum—karaya, gum
Indian red—ferric oxide
Isinglass—pure fish gelatin
Italian red—iron oxide, red
Ivory black—bone black

K

Kauri gum—copal, gum
Kaolin—aluminum silicate
Kieselguhr—diatomaceous earth

L

Lanolin—purified wool grease
Lanum—lanolin
Lead chromate—chrome yellow
Lead sulfide, basic—white lead, sublimed
Lemon, salts of—potassium binoxalate
Lemon yellow—barium chromate
Licorice—glycyrrhiza
Ligroin, light—petroleum ether
Lime, dry—calcium oxide
Lime, slaked—calcium hydroxide
Limestone—calcium carbonate
Litharge—lead monoxide
Liver of sulfur—potassium sulfide
Lithopone—zinc sulfide and barium sulfate
Lunar caustic—silver nitrate

Lye—sodium hydroxide
Lysol—cresol-soap solution

M

Magnesia—magnesia oxide
Magnesite—magnesium carbonate, natural
Magnesium silicate—talc
Maize oil—corn oil
Malt sugar—maltose
Metol—methyl-para-aminophenol sulfate
Methanol—methyl alcohol
Microcosmic salt—sodium ammonium phosphate
Milk sugar—lactose
Mineral wax—ozokerite
Mineral pitch—asphalt
Minium—lead oxide, red
Mirbane oil—nitrobenzol
Muriatic acid—hydrochloric acid
Myrtle wax—bayberry wax

N

Naphtha (petroleum)—petroleum distillate
Naphtha, solvent—coal tar naphtha
Naples yellow—lead antimonate
Nickel salts, double—nickel ammonium sulfate
Nickel salts, single—nickel sulfate
Niter—potassium nitrate

O

Oil of bitter almond—benzaldehyde from bitter almond nut
Oil of mirbane—nitrobenzol
Oil of mustard—allyl isothiocyanate
Oil of wintergreen—methyl salicylate
Oleic acid—red oil
Olein—glyceryl trioleate, natural
Oleum—sulfuric acid, fuming
Orange mineral—lead oxide, orange red
Orpiment—arsenious sulfide, yellow

P

Paraffin oil—mineral oil, petrolatum liquid
Paris white—calcium carbonate
Paris blue—ferric ferrocyanide
Pearl ash—potassium carbonate
Petrol—gasoline
Petrolatum—petroleum jelly
Plaster of Paris—calcium sulfate
Plumbago—graphite
Prussian blue—ferric ferrocyanide
Prussiate of potash, red—potassium ferricyanide
Prussiate of potash, yellow—potassium ferrocyanide
Prussic acid—hydrocyanic acid
Pyramidon—amidopyrine
Pyrethrum—insect flowers, powdered
Pyroligneous acid—wood vinegar, acetic acid
Pyrolusite—manganese dioxide
Pyroxylin—nitrocellulose

Q

Quicklime—calcium oxide
Quicksilver—mercury
Quinol—hydroquinone

R

Red lead—lead tetroxide
Red oil—oleic acid
Red oxide—ferric oxide, red
Rochelle salt—potassium sodium tartrate
Rosin—pine resin, colophony
Rottenstone—tripoli
Rouge—ferric oxide

S

Sal ammoniac—ammonium chloride
Satin white—reaction product of hydrated lime and alum

Salt—sodium chloride
Saltpeter—potassium nitrate
Salts of vitriol—zinc sulfide
Scale wax—paraffin wax, low melting point
Silica—silicon dioxide
Slaked lime—calcium hydroxide
Sod oil—degras
Soda ash—sodium carbonate
Soda (washing)—sodium carbonate ($10H_2O$)
Sodium bisulfite—sodium acid sulfite
Sodium phosphate, dibasic—disodium phosphate
Soft soap—potash soap
Stearin—Glyceryl stearate
Sublimed lead—lead sulfate, basic
Sucrose—cane sugar, beet sugar
Sugar of lead—lead acetate
Sulfuric ether—ethyl ether

T

Talc—magnesium silicate
Tartar emetic—antimony potassium tartrate
Tetralin—tetrahydronaphthalene
Theobroma oil—cocoa butter
Titanium dioxide—titanium oxide
Toluol—toluene
Triacetin—glycerol triacetate
Tripoli—natural amorphous silica not diatomaceous
Tripolite—diatomaceous earth
Train oil—whale oil
Trinitrophenol—picric acid
Turkey red oil—castor oil, sulfonated

V

Venetian red—ferric oxide
Vermilion—red mercuric sulfide
Verdigris—copper acetate, basic
Vitriol—sulfuric acid

W

Water glass—sodium silicate
White arsenic—arsenic trioxide
White bole—China clay, kaolin
White lead—lead carbonate, basic
White metal—Babbitt metal
White wax—beeswax, bleached
Whiting—calcium carbonate
Witherite—barium carbonate, natural
Wintergreen oil, synthetic—methyl salicylate
Wood alcohol—methyl alcohol

X

Xylol—xylene

Y

Yacca gum—accroides gum

Z

Zinc white—zinc oxide
Zinc yellow—potassium zinc chromate

ABBREVIATIONS

avoir.	avoirdupois
b.p.	boling point
Bé	Baumé
C	Centigrade
cc	cubic centimeter
c.d.	current density
c.p.	chemically pure
cu in	cubic inch
cu ft	cubic foot
d.	density
dil.	dilute
dr	dram
F	Fahrenheit
fl dr	fluid dram
fl oz	fluid ounce
g	gram
gal	gallon, gallons
gr	grain, grains
hr.	hour
kg	kilogram
l	liter
lb	pound, pounds
m.p.	melting point
min.	minute
min	minims
oz	ounce, ounces
pt	pint, pints
Q. S.	A quantity sufficient to make
qt	quart, quarts
r.p.m.	revolutions per minute
sec.	second
sp. gr.	specific gravity
sq. dm.	square decimeter
tbsp.	tablespoonful
tsp.	teaspoonful
U.S.P.	U. S. Pharmacopeia
V.	voltage
wt.	weight

BIBLIOGRAPHY

W. V. Cruess, *Commercial Fruit and Vegetable Products,* New York, McGraw-Hill Book Co., Inc., 1938.

A. E. Leach, *Food Inspection and Analysis,* New York, John Wiley and Sons, 1923.

A. L. Winton and Kate Barber Winton, *Structure and Composition of Foods,* New York, John Wiley and Sons, 1935.

Julius E. Wihlfahrt and Robert W. Brooks, *A Treatise on Baking,* New York, Standard Brands Inc., 1935.

D. K. Tressler and C. F. Evers, *The Freezing Preservation of Foods,* New York, Avi Publishing Co., 1943.

Paul Richards, *Cakes for Bakers,* Chicago, Bakers Helper Co., 1945.

Saul Blumenthal and Dr. F. Fiene, *Handbook of Food Manufacture,* New York, Chemical Publishing Co., Inc., 1938.

Saul Blumenthal, *Food Manufacturing,* New York, Chemical Publishing Co., Inc., 1942.

H. Bennett, *Chemical Formulary,* New York, Chemical Publishing Co., Inc., 1945.

F. W. Tanner, *Bacteriology & Mycology of Foods,* New York, John Wiley & Sons Inc., 1919.

C. Campbell, *Canning, Preserving, Pickling,* Chicago, The Canner Publishing Co., 1937.

A. G. Woodman, *Food Analysis,* New York, McGraw-Hill Book Co., Inc., 1924.

W. A. Rooker, *Fruit Pectin,* New York, Avi Publishing Co., 1928.

Bulletins and Papers

A. W. Bitting, "Commercial Canning of Foods," *U. S. Dept. Agr. Tech. Bull.* **196** (1915).

A. K. Balls and U. S. Hale, "Peroxidase Darkening," *Ind. Eng. Chem.*, **27**, 335 (1935).

G. L. Baker and P. B. Meyers, "Pectin Jelly Making," *Del. Agr. and Exp. Sta. Bull.* **144** (1926).

C. R. Fellers, "Apple Juice Extraction," *Mass. Agr. Exp. Sta. Tech. Bull.* **15** (1928).

J. S. Caldwell, "Evaporation of Fruits," *U. S. Dept. Agr. Bull.* **1141** (1923).

G. B. Ridley, "Tunnel Driers," *J. Ind. Eng. Chem.*, **13**, 453 (May, 1921).

E. H. Wiegand, "Recirculation Driers," *Ore. Agr. Exp. Sta. Cir.* **40** (1923).

M. A. Joslyn and G. L. Marsh, "Role of Peroxidase in Deterioration of Frozen Fruits and Vegetables," *Science,* **78, 174** (1933).

"Vegetable and Fruit Dehydration," *U. S. Dept. Agr. Misc. Pub.* **No. 540.**

Food Journals

The Fruit Products Journal, Avi Publishing Co., Inc., New York

Food Industries, McGraw-Hill Publishing Co., New York

The Canner, Canner Publishing Co., Chicago

The Glass Packer, Ogden Publishing Co., New York

The Western Canner & Packer, Miller Freeman Publications, San Francisco

Canning Age, Ogden Publishing Co., New York, New York

INDEX

F

H

Q

R

S

www.ingramcontent.com/pod-product-compliance
Lightning Source LLC
LaVergne TN
LVHW100022220926
848483LV00001BA/98